Nose Dive

Also by Harold McGee

On Food and Cooking:
The Science and Lore of the Kitchen

The Curious Cook:
More Kitchen Science and Lore

Keys to Good Cooking:
A Guide to Making the Best of Foods and Recipes

Nose Dive

A Field Guide to the World's Smells

HAROLD McGEE

PENGUIN PRESS

NEW YORK

2020

PENGUIN PRESS
An imprint of Penguin Random House LLC
penguinrandomhouse.com

Illustrations of chemical structures by Alice Phung

LIBRARY OF CONGRESS CATALOGING-IN-PUBLICATION DATA

Names: McGee, Harold, author.
Title: Nose dive : a field guide to the world's smells / Harold McGee.
Description: New York : Penguin Press, 2020. | Includes bibliographical
references and index.
Identifiers: LCCN 2020003220 (print) | LCCN 2020003221 (ebook) |
ISBN 9781594203954 (hardcover) | ISBN 9781984881878 (ebook)
Subjects: LCSH: Odors. | Smell.
Classification: LCC QP458 .M42 2020 (print) | LCC QP458 (ebook) |
DDC 612.8/6—dc23
LC record available at https://lccn.loc.gov/2020003220
LC ebook record available at https://lccn.loc.gov/2020003221

Printed in the United States of America
10 9 8 7 6 5 4 3 2 1

Designed by Amanda Dewey

To all the chemists past and present
whose pursuit of flying molecules
made this book possible

Contents

....

Preface · MY FIRST GROUSE

....

No matter where or how you're reading these words, at this very moment there's a world swirling all around you, and into you, that teems with the makings of delight, disgust, understanding, and wonderment. It's an invisible nimbus of flying molecules: countless specks of matter launched into the air we breathe, whizzing at highway speeds, whose presence we perceive as smells. This book is about those specks and smells, and about making the most of our access to them.

Many fine books have been written on our sense of smell, on the pleasing aromas of food and drinks and perfumes, on the nature of disgust. Here I've put together something different: a guide to the wide world of smells, nice and not, and the airborne molecular specks that stimulate them. Since the specks are representative bits from throughout the material cosmos, I like to call this wide world the *osmocosm*, from *osme*, the ancient Greek for "smell" or "odor," with its inner resonance and hint of wizardly magic. The osmocosm contains multitudes—thousands of different molecules at least, possibly thousands of thousands. There's more to it than even the most sensitive among us can experience. And much or all of it is inaccessible to the many people whose sense of smell has somehow been impaired. But no matter how much of it we happen to notice, we're always immersed in the osmocosm. It's a fundamental feature of the world we inhabit. It's well worth exploring, even if only in imagination and thought.

The general term for any airborne molecule is *volatile*, a word that derives from

the Latin for "to fly" and was first applied centuries ago to birds and butterflies
and other winged creatures. It was one of these original volatiles, a flavorful
wild bird, that drew me into exploring the world of molecular volatiles. Let me
explain how it did, and how I hope you'll use this guide to become a smell explorer yourself.

My longtime beat is the science of cooking. Back in 2005, when experimental
cooking was the talk of the restaurant industry, I traveled to Spain and England to get a taste of its innovations. The leading avant-garde chefs, the Adriàs
and Rocas and Heston Blumenthal, aspired to give diners an unforgettable meal,
with long menus of novel dishes that were variously startling, amusing, baffling,
and sometimes delicious. It was a stimulating few days. But my most memorable
mouthful came near the end, during a very traditional British lunch with Fergus
Henderson and Trevor Gulliver at their London restaurant St. John.

It was early fall, so I ordered grouse, a game bird just then in season that I'd
never had the chance to taste. It came roasted whole, rare and plainly, on a slice of
toast, with a tuft of fresh watercress. I expected to enjoy it, but not that the first
bite would leave me speechless. It did. I was completely absorbed, first by intense
sensation—a meatiness that was almost too strong to be pleasant, and edged with
bitterness—and then by confused emotion. I was momentarily paralyzed, unable to
say a word to my tablemates. They looked at me with some concern, but then Fergus smiled, nodded, and said: "Ah, of course. Your first grouse."

I'd always been interested in understanding what makes foods delicious, but
that experience impressed me like no other with the power of flavor to trigger
strong feeling—and to persist. That grouse was still in my mouth many hours later
as I tried to concentrate on a performance of Shakespeare's *Tempest*.

Another moment several years later impressed me with the power of aroma
alone. I had managed to grow what looked like an oversize taste bud on the tip of
my tongue, maybe an eighth of an inch across: a good joke for a food writer! Eventually I saw a specialist who advised removing it. He gave me a local anesthetic,
snipped it out, then cauterized the wound with an electrical instrument that
burns and seals the blood vessels. There was a puff of smoke, and I smelled the

typical aroma of beef on a very hot grill, burned but also slightly decomposed. A surprise, but it made perfect sense: it was the aroma of grilled McGee! Another good joke. And as I had that lighthearted thought, I got light-headed, then leaden-limbed, and broke into a cold sweat. The physician quickly reclined the chair, and in a couple of minutes I was fine again, just embarrassed. I had thought that I was taking the experience in stride, even relishing the irony, but my body ambushed me. Another unforgettable moment and smell.

The usual cultural touchstone for connecting flavor with emotion is the morsel of madeleine cake that Marcel Proust's narrator dunks into a cup of linden-blossom tea in the first volume of his novel *À la recherche du temps perdu*, or *In Search of Lost Time*. That bite surprises the unnamed narrator with a shudder of "exquisite plea-sure" that he eventually traces back to tasting the same combination in idyllic childhood. My shudders weren't exactly pleasurable: they seemed more likely to be instinctive warnings. The grouse was so strong and funky that it might have been spoiled; cauterized tongue probably evoked the misery of my tonsillectomy twenty years before. But was that all they meant? I felt there was something else going on.

My ruminations eventually led me to a less celebrated passage of Proust that resonated much more. In the fourth volume, *Sodom and Gomorrah*, the narrator indulges in a favorite drink and is struck by the sensations it provokes:

> The orange squeezed into the water seemed to yield to me, as I drank, the secret life of its ripening growth, its beneficent action upon certain states of that human body which belongs to so different a kingdom, its powerlessness to make that body live, but on the other hand the process of irrigation by which it was able to benefit it, a hundred mysteries revealed by the fruit to my senses, but not at all to my intellect.

Once again a taste of food catches the narrator's attention and triggers a feeling of elusive significance. But this time it's not about his past life—it's about the food. The orange somehow evokes the mystery of its creation and its goodness for alien creatures like us. The narrator doesn't follow up this intimation the way he does the pleasure of the madeleine. But if he did, then his *recherche* would shift away from lost time and toward found fact, toward the natural histories and inner workings of fruit and animal.

Proust's orange encouraged me to see my taste of grouse as an invitation to consider *its* mysteries. It was a call to stop and think and learn, to ask, *Why* did that bird have such a strong and distinctive flavor?

So I did ask, and I learned. Unlike domestic ducks and pigeons, British grouse are true game birds, living in the wild on open heathlands, constantly scrounging for food and evading predators, often infected with gut parasites that make them more easily scented by foxes and dogs. They're driven from cover and shot on the wing, their carcasses matured—"hung"—for several days, guts included, to tenderize them and intensify their flavor. In 2007, I made a pilgrimage to western Scotland and shared an unforgettable weekend with St. John's game supplier Ben Weatherall and his family. I spent hours on Overfingland Heath watching the birds, marveling at their explosive takeoff when flushed from the low brush, and their dazzling speed and maneuvers to hug the rolling hillsides until they were out of sight. No wonder their flight muscles are so dark with flavor-making metabolic machinery! I chewed the astringent, bitter heather they live on, and smelled the heaviness in the cool storeroom where they matured.

Harsh wild feed, powerful and well-exercised flight muscles, damaged guts leaking feed and digestive juices in a carcass on the cusp of decomposition: these are elements that combine to give traditional grouse its distractingly strong flavor. Having grown up in the prepackaged, sanitized, deodorized late twentieth century, with that initial bite I was in a way tasting flesh for the first time, at some level recognizing in my mouth the transfixing funk, chemical and emotional, of animal life and struggle and death. Yes, maybe a health warning, but at the same time so very much more! I felt my hunger for understanding satisfied, the experience retroactively enriched.

And I got to wondering what meanings I might find hinted at in more ordinary eating experiences. Of course most foods just taste of themselves, as we expect them to from past experience. It's the unusual and incongruous that capture the attention. I'd often been struck by the ways that unrelated foods can seem to echo each other. Parmesan cheese can taste like pineapple. What connection could there possibly be between old cow's milk and ripe tropical fruit? Raw oysters can taste like cucumber. Sherry wine can taste like soy sauce, corn tortillas like honey—specifically chestnut honey. Even odder are foods that echo inedible things: the seaside in green tea, horse stables in wines, sweaty feet in some Swiss cheeses.

Thoughts of sea and stable and feet, and of grilled McGee, highlighted the fact that the flavor echoes I'd perceived are similarities specifically in *smells*. Our sense of smell is the bridge between our experience of foods and our experience of the larger world. It normally accompanies every breath we take through the nose. We sense smells in the world when we breathe in, and flavors in the mouth when we breathe out. And smell gives us detailed information about what's around us or about to be swallowed. Pinch the nose and we can taste sweetness and sourness on the tongue, but we can't tell a citrus soda from a cola; we can't tell that bread in the toaster is going from brown to carbon black.

It seemed to me that to understand the flavors of tea and wine and cheese would mean delving into the smells of oceans and animals and feet, to find out why *they* have the smells they do. This was a daunting prospect—but then increasingly exciting. In fact, why stop at smells that happen to be obviously echoed in foods? Why not savor things in the larger world the way we can savor food and drink, smelling them actively and curiously, learning about their volatiles and their backstories, and using that knowledge to experience them more fully?

I got hooked. It was a kick to sniff at everything I could think of, then connect those immediate, personal sensations with precise laboratory identifications of the flying molecules that trigger them, and through the molecules, with the larger scientific understanding of the world's workings. I often felt a sense of wonderment at those workings and at the collective achievement of humankind in figuring them out. Despite its longtime reputation as one of the lowest of human faculties, smell clearly has the power to engage us with the world around us, to reveal invisible, intangible details of that world, to stimulate intense feeling and thought: to nudge us into being as fully and humanly alive as we can be.

So I became an amateur smell explorer and immersed myself in the osmocosm. I went on a ten-year sniffing expedition through the world and through the scientific literature. I've written this book to share what I've learned: to point out and delve into smells that are out there to be noticed, and relate what those smells can tell us about how they came to be, about the otherwise insensible workings of the world. Not just food and drink and roses, but compost and sodden flowerpots, asphalt and laptops, old books and dog paws, the myriad mundane yet revelatory things that fill our lives. There's a rich world of sensations and significance out there, intangible and invisible and fleeting, but vivid and real.

N ow that I've explained the eccentric path by which I came to write this book, I should explain the eccentricities of the pages that follow: why they're filled with what look like mashups of ingredient labels and tasting notes, and why the first chapter begins with the unsmellable Big Bang.

Smell is such a powerful and revealing sense because it detects actual little pieces of things in the world. Those little pieces are volatile molecules, so little that they're able to break away from their source and fly invisibly through the air to reach our nose. To begin to understand a thing's smell, then, is to identify the many volatile molecules it emits. Its overall smell is a composite, created by the component smells, or "notes," of its most prominent volatile molecules. When different things seem to echo each other with shared component smells, it's a sign that those things have some volatile molecules in common. And the chemical identities of the molecules are keys to why they're there. They're tokens of the processes that created them.

So: much of this book is anchored in volatile chemistry. And chemistry of any kind is seldom an inviting subject for anyone but chemists! But I am not a chemist, and the chemistry in this book is not an end in itself. It's a means of gaining insights into your own personal experience in the physical world, a means of smelling more and having a sense for what the smells mean. In fact, many of these molecules are longtime friends that have been pleasing or annoying you all your life without your knowing they exist. We know and recognize and appreciate these significant bits of the world by smell, but they haven't been properly introduced, individually and by name. The names that chemists have given them can be confusing at first, but they do have their own logic. And when we meet up with smells and named molecules often enough, the names begin to stick. These days many beer lovers can tell you about the esters and volatile phenols in their favorite ales; cannabis connoisseurs know their terpenes; craft perfumers, their aldehydes.

Because each chapter describes dozens of different things, each emitting many volatile molecules and component smells, I've distilled the relevant information in the smell tables that you'll see throughout. They're designed to make it easy for you to control your chemical exposure. Most of the tables have three columns. The first column lists a handful of related items of interest: particular parts of the body, or flowers, or cheeses. The second column lists some of the component smells that contribute to each item's overall smell. These may look like tasting notes from ads and reviews, but they're not merely subjective impressions. They're the smells of specific molecules that have been objectively identified as significant volatiles in that item. Those molecules are listed in the third column.

If you're mainly interested in the component smells you might get a whiff of on your skin or in some Parmesan shavings, and you don't want to be distracted by the chemistry, then just stay to the left and center of the tables. Simply paying attention to these nuances of smell can be rewarding. In a 1948 poem, the Scot Hugh MacDiarmid poked fun at the chemical approach of modern "osmology" while praising plain attentiveness: "a flower's scent by its peculiarity sharpens / Appreciation of others."

But if you're curious about *why* a daisy smells so unlike a rose, why your skin sometimes has a metallic tang or Parmesan can seem both fruity and a little sickening, then look to the rightmost column to see what particular molecules are involved, and in the surrounding text to see where they come from. These details further sharpen appreciation with understanding.

So much for navigating the tables; now for navigating the book as a whole. I've written this guide both for casual browsing and for learning about the greater osmocosm. It's organized not by smells but by the familiar things in our world that emit them. So you'll find the human body in chapter 6, flowers in chapter 10, cheeses in chapter 19. You're welcome to head right for favorite or despised or newly encountered smellables. Or browse the pages and tables and see what catches your eye.

For readers who want to explore the world of smells more systematically and refresh their understanding of just what a molecule is, I've organized the chapters in the sequence that helped me get my own nonchemist bearings among volatile molecules, and I hope it will do the same for you. It emerged from pondering smell echoes: if oysters can smell like cucumbers, then which was the first to carry that particular molecule? And did something else carry it before either of them? I came to realize that, like everything in the physical world, the molecules of smell have histories that are part of the ongoing history of creation itself, the evolution of the cosmos as a whole. That evolution started billions of years ago in the mystery of the Big Bang, before there was a single molecule of any kind, and ever since has moved in the direction of greater molecular diversity and complexity.

When I looked into the early history of the cosmos, I was fascinated to find that some of the molecules that we smell every day existed long before there was any creature around to smell them, before there was even a planet Earth for creatures to live on. They're among the simplest molecules, just a handful of atoms, as easy to grasp as H_2O. Some of them are also the source of smells produced by most forms of life. And as life has diversified over the eons, so have the volatile molecules it emits.

The simple is easier to understand than the complex, and it's a stepping-stone to grasping complexity. So I've structured this book in five parts that introduce the molecules of smell a few at a time, roughly as they emerged. I invite the novice smell explorer to imagine yourself alongside the Chef of the cosmos, superhuman but with a human nose: sniff the stew of matter and energy as it cooks over the eons, notice how its smells develop, and get to know the increasingly complex—and pleasant!—molecules they arise from.

Part 1 starts with the sparse primordial volatile molecules of outer space, the sulfurousness of Earth and its early single-cell life, and the basic starter set of volatiles and smells shared by all living things. Part 2 documents how animal bodies, ours included, owe most of their smells to their mobility and the communities of microbes they carry along with them. Part 3 celebrates the creativity of the plant kingdom and its terrifically diverse volatiles and smells, fresh and woody and flowery and fruity. Part 4 describes the smells that emanate from the planet's waters and soils, and from life's remains when they're transformed into smoke and tar, fuels and plastics. And Part 5 concludes with the smells that humankind loves and pursues for their own sake, in fragrances and foods and drinks.

Welcome, then, to the osmocosm, the world aswirl right under our noses.

Introduction · A SENSE FOR
THE ESSENTIAL

> The smell of a person's body is the body itself which we breathe in through our nose
> and mouth, which we suddenly possess as though it were the body's most secret
> substance and, in short, its nature. The smell which is in me is the fusion of the
> other person's body with my body. But it is the other's body with the flesh removed,
> a vaporized body which has remained completely itself, but which has become a
> volatile spirit.
>
> · Jean-Paul Sartre, *Baudelaire*, 1947

Vaporized bodies, secret substances: these are smells? Well, something along those lines, yes! Smells may be everyday, ordinary sensations, but the more closely you look at them, the more extraordinary they become. Jean-Paul Sartre, like Proust a French student of the sensuous, captures their strange, ghostly corporeality in this passage about women and perfumes in Charles Baudelaire's poetry. When we smell another person's body, we literally bring a portion of that body into our own body, into tissues in our head, which then signal its presence to our mind. This is true whether we smell a lover or a stranger, a sewer or a rose. When we smell something, it's because particles of that thing—its vaporized, airborne, volatile molecules—enter us and momentarily become part of us.

That's a disturbing thought. No wonder we instinctively hold our breath when we smell something disgusting. But it's also an eye-opener, a nostril-dilator. It means that smell connects us directly and intimately with the substance of the world we live in. It means that, even though smell has been widely considered among the least valuable of the human senses, one that our pets are far more talented at exercising, it can bring more to our lives than we've given it credit for.

Before we plunge into the smells of the world, let's start at home, in our own heads, and get better acquainted with how smell works and what it has to offer us.

Molecular senses for a molecular world

When Sartre wrote about the smell of a woman as her "vaporized body" or her body's "volatile spirit," what he was really talking about was her volatile *molecules*. Molecules are invisibly small particles of matter, the diverse building blocks that make up things in the physical world and give them their substance and specific qualities. Taste and smell are molecular senses: they detect and report on the presence of particular molecules in the air around us and in our mouths. For all their greater prestige, our senses of vision and hearing aren't as directly in contact with the things of the world: they register only light waves or waves of air pressure whose movement has been influenced by their presence. The sense of touch does bring us in direct contact with physical objects and materials, but only in bulk form; it can't distinguish particular molecules the way smell and taste can. Smells and tastes are our most direct, intimate, and specific encounters with the molecules that make up the world.

Like everything else in the physical world, our bodies are also made of molecules, and our senses of taste and smell work by means of their own specialized molecules, the taste and smell receptors. Taste receptors reside primarily on taste buds on the tongue. They're on the lookout for a handful of particular molecules, or pieces thereof, that dissolve in the mouth's watery saliva from the foods that we put in our mouths, or other materials that we choose to chew on or suck or lick. We have around fifty or so different kinds of taste receptors, and they give rise to a handful of taste sensations—the familiar sweet, sour, salty, and bitter, and the less familiar umami, or savory. All of these indicate the likely suitability of foods and drinks for our nourishment.

Smell arises from two patches of sensitive skin placed out of sight in the front of our head, behind and a little below our eyes. Their total area is less than a tenth the area of the upper tongue surface, around a square inch. Its four hundred or so different kinds of odor receptors recognize molecules carried in the air we breathe in and out. Smell is on the lookout not for a handful of particular molecules, but for *whatever* molecules happen to be in the air that might be significant for our well-being, whether the aroma of ripe strawberries in a bowl or the smokiness of a forest fire miles away. It doesn't bother to notice the great majority of the air's molecules, the nitrogen and oxygen and carbon dioxide and water, because their presence isn't significant—they're always there. But it's very sensitive to the molecules that come and go, that give clues to what's going on around us. Since its few hundred receptors can work with one another in many different combinations, smell could theoretically distinguish among many millions of different molecules and molecule mixtures.

Smell is more versatile than taste. It's more open-ended, broader, more specific, and more sensitive. And it's much more informative, because things in the world are made up of many different kinds of molecules—far more than the dozens that taste can notice.

Smells arise from mixtures of flying molecules

As Sartre said of bodies become smells, the molecules that we breathe in and smell are *volatile*, a term that in chemistry means "tending to evaporate," to escape as a gaseous vapor from solid or liquid materials. Smell molecules must escape their source—a person's body, food, or drink, a tree, a fire—and fly through the air to reach the smell receptors up in the nasal cavity. Most of the molecules in the things around us are too big and heavy to fly, or they cling too strongly to other molecules, so what we're actually able to smell is a selection of the molecules in things, the ones that evaporate from the surface and escape. These volatiles are representatives of the bodies that emit them, but they leave the bodies behind.

And most things emit mixtures of volatiles. There isn't any such thing as a single apple molecule, a single potato molecule. Apples and potatoes are made up of many different kinds of molecules—water, starch, sugars, proteins, fats, minerals, acids, DNA, pigments, phytochemicals that repel insects, and on and on. An apple and a potato each gives off dozens of volatiles. Their distinctive smells come from their different mixtures.

Because even the simplest smells arise from a composite of volatiles, smells are often likened to a musical chord, a combination of several different notes that we hear as a single recognizable sound. Another analogy, closer to home, is to something cooked. You combine tomatoes, olive oil, garlic, and basil, and those flavors come together to give the flavor of red sauce. You may or may not register the aroma of each ingredient, but each contributes to the distinctive flavor of the sauce. Well, each of those ingredients is in turn a composite of molecular ingredients that come together to give it *its* distinctive flavor: of tomato, olive oil, garlic, basil. It's these molecular ingredients that we'll explore in this book.

We can't actually see these flocks of mingled volatiles, but they're easy enough to imagine and connect to our everyday experience. I live in a hilly neighborhood of San Francisco, and often see the air and its currents become visible as fog spills over Twin Peaks and flows down toward the bay. That got me to thinking: If single odor molecules were visible in the same way that the trillion-molecule fog droplets are, and were somehow color-coded to reflect their tremendous diversity, then from my window I would see rainbowed odor plumes constantly forming and dissipating, puffs and swirls and masses moving, disappearing, reappearing, blending, from the jasmine vine and lemon tree and fir and eucalyptus in neighboring yards, from the rooftop shingles, open windows, sidewalks, dogs and their owners, cars and buses, cyclists struggling up the hill. . . . And when I actually smell the nearby flowers and trees, or the smoke from a chimney, it's because traces of those molecule plumes have wafted directly to the air around me, where I can pull them into my nose.

When I take a break from sensory deprivation at my desk and go for a run, I see and hear and feel a lot, and I smell a lot too. The smells are more intermittent than the smoothly shifting visual scene and the noises and pavement-pounding and wind, but there are always many different ones, appearing and disappearing in the space of a few breaths as I move into and out of the plumes of mixed volatiles.

Some emanate from sources I can see as I pass by them. A Thai restaurant. A bakery. The fresh asphalt on a repaved street. Both dank and freshly sawn wood in an old house torn open for remodeling. A mix of rubber and engine oil from an auto repair shop. Trash cans. A fetid street drain. A freshly mowed front lawn. A shopping cart jammed with grimy bedding.

Other smells I recognize even though their sources are invisible. Marijuana smoke. A shorting electrical transformer. Steer manure in someone's garden. Flowers, heady, heavy. The exhaust from a clothes dryer. Vapors from kitchens and backyard grills: burnt toast, fried fish, fried onions, red sauce, coals just ignited with lighter fluid, grilled chicken, beef. As I run upwind on a dry sidewalk, the smell of rain on pavement, harbinger of a wet finish line.

Each of us has experiences like this every day—passing encounters with plumes of volatile molecules.

Smells and flavors are in our head

Even if we're ready to think about smells as mixtures of molecules, and smell as the most specific and discriminating of our molecular senses, it's not at all easy to smell smells as mixtures. In everyday experience we smell *things*, objects and materials in the world that have simple, instantly recognizable smells: manure and flower smells, beef and chicken smells, qualities that, as Sartre said, seem to be the individual spirit of the body they evaporate from.

This impression results from the fact that the encounter of taste and smell receptors with molecules is just the first step in our perception of a smell or taste. Though we say casually that foods "have" flavors and flowers "have" smells, in fact what they have are volatile molecules. Sensations and perceptions, smells and tastes and flavors, are products of our brain. The brain doesn't simply register the direct reports of the receptors, but actively creates smells and tastes and flavors by filling out those reports with many other kinds of available information, and especially information from its database of past experiences.

When the smell and taste receptors register their target molecules from something in the mouth or the air, they send electrical pulses, each a tiny fraction of a second long, to particular receiving areas in the brain. Nerve cells in these areas then collect and organize these signals and send their own signals to various other areas, and those areas in turn communicate with each other. Eventually—in a crowded split second—the brain processes the many streams of signals and integrates them into a sensation, which neuroscientists term the odor or flavor "image" or "object" that we can consciously perceive. And part of that sensation is an association with the thing that triggered it.

So we don't normally experience the aroma of coffee as the mixture of the many different volatile molecules that create it. We experience it as . . . coffee.

Why does the brain handle the receptors' reports the way it does, bringing in other reports from the eyes and ears and memory banks, and presenting our conscious minds with an executive summary? Because it evolved as the organ that coordinates all of our biological functions to help us survive in a complex, ever-changing world. For all its remarkable powers, the human brain can't keep tabs on everything that's going on at every moment. So it has to simplify and focus. The senses constitute a system for constantly collecting data about the immediate sur-roundings, paying special attention to changes (hence smell's lack of interest in nitrogen and oxygen and water), quickly aggregating, editing, and comparing them with a database of past experiences, and coming to a quick decision about how to act. Taste and smell above all enable animals to recognize nourishing food and ingest it, to recognize toxic or spoiled food and avoid or expel it, to sense potential danger from nearby predators or fires and escape it, to distinguish kin from strang-ers, the healthy from the sick. Their purpose was not to dissect the smell of coffee into its volatile ingredients, or ponder the nuances of an orange or a grouse.

But neither was the original purpose of hearing to develop spoken language or music! People who love coffee and perfumes and many other aromatic materials do in fact dissect and ponder. It doesn't come naturally, but it's doable and rewarding.

Catching the brain at work; noticing mixtures and echoes

Smells are triggered by mixtures of volatile molecules, then shaped and presented as simplified conscious perceptions by the actively editing, synthesizing brain. We get hints of all this in moments when there's something unusual about a smell, some kind of discrepancy or discordance or surprise, when the brain has to work harder at coming up with a suitable odor image.

One day a couple of years ago I got home from a long run in the fresh and var-ied air, walked into the kitchen, and soon noticed that something wasn't quite right. At first the air seemed simply stale. As I sniffed, it seemed even less pleas-ant, as if the kitchen had bad breath. I thought that something in the larder might be spoiling—one of the fresh tomatoes, or an onion, both of which can rot disgust-ingly. I checked: the produce was fine. The smell bothered me more and more. The sink drain? No. I wondered whether someone had forgotten to flush the toilet around the corner. No. Maybe a mouse had gotten into the wall and died there—something I'd dealt with decades before. No easy way to check that.

Finally, scanning the kitchen and sniffing, I saw the source of the smell. In plain sight on the table, on a plate, under a glass dome: a soft-ripened cheese from Vermont, bound in a thin hoop of sprucewood. I'd bought it the day before, unwrapped it and set it out to recover from its hibernation in the store's cheese case, and managed to forget it. I put my nose next to the plate and sniffed: sure enough, that was the smell. Stronger, and now it smelled like cheese—or an aspect of cheese. I lifted the dome, took a deep sniff, and smelled that strong stinky smell, but others as well, including ammonia, which for some reason hadn't filled the room the way the stink did. The immediate mystery was solved.

I'd been experiencing the active and fallible brain at work. Having forgotten that there was a cheese on the table, mine did its best to make sense of that abnormal smell filling the room, pulling up possible scenarios from my past encounters with smells like it.

Because I had molecules and the brain on my mind, I wanted to see how the cheese itself tasted after this unusual introduction. I cut a hole in the top and tasted an oozy spoonful. Even though the source was now right in my mouth, the stinky note seemed much reduced, and milky, meaty, piney, and fruity aromas took over the center stage, along with salty and tart tastes and the creamy texture. The smell of the cheese in my mouth was very different from its smell in the kitchen air.

A pretty mundane experience, but a lot of food for thought! Smells are triggered by mixed volatiles, and the brain does its best to make sense of all the inputs available to it. There were many different sides to that cheese, and my brain wasn't comfortable with one of them. It must share some volatile molecules with morning breath, rotting vegetables, stagnant drains, excrement, and dead animals. Not so nice. On the other hand, it also seemed to share volatiles with meats and ripe fruits. Why was it only the not-so-nice smell that I noticed before I saw the cheese? When I put a spoonful right into my mouth, why did that note get weaker instead of stronger? And how does cheese making manage to coax from bland milk stinky and ammonia smells on one hand, and meaty and fruity smells on the other?

This brief kitchen mystery was a rare experience, but I'm intrigued all the time by foods whose smells suggest or echo or rhyme with very different things in the world—because my brain notices some feature that they seem to have in common. Cheese and dead animals, cheese and ripe fruits: sometimes, startlingly specifically, pineapple! Coffees and wines that smell like a horse stable. Fresh green tea that smells like chicken, and an hour later, dregs that smell like the seashore. Blue borage flowers from the garden that taste like oysters. A black salt mined in the Himalayas that smells like cooked eggs. Curious!

You don't have to know anything about the molecular nature of smell to notice these resemblances. But if you do know a bit, then you can begin to investigate what they might mean. Molecules are made, not born; they're evidence of the processes that formed them. This is obvious in the case of the fried-fish and tomato-sauce smells that I catch on my runs, made by cooks combining ingredients and applying heat. It's less obvious but just as true for the smell of coffee and horse stable, mountain rock and egg. What are the volatile molecules they share, and how did they come to be made in such different things? Smell echoes are clues to the invisible dynamism of the world.

Identifying the world's volatile molecules

Happily for the curious sniffer, chemists have been hard at work cataloging the molecules that emanate from foods and stables and rocks. Beginning in the late 1940s, they've developed and refined machines that do to smells what a prism does to light: separate what seems to be a single simple sensation into its component subsensations. Neutral white light is a mixture of all the different colors of the spectrum, and a prism makes that spectrum visible by spreading out the mixture into its different wavelengths. Machines called gas chromatographs do the same thing to volatile molecules.

The chemists' gas chromatograph starts with a sample of volatile molecules collected from a food or object or place, and separates the kinds of molecules from each other by how volatile they are—by how much energy they need to escape solids and liquids and become a gas. The volatility of a substance corresponds roughly to its boiling point: the higher its volatility, the less energy it needs to escape, and the lower its boiling point. Alcohol is more volatile than water, and boils at a much lower temperature, 173°F instead of 212°F (78°C instead of 100°C). So if you slowly heat a mixture of alcohol and water, when it approaches 173°F, the vapor coming off its surface will contain more alcohol than water. By the time it gets to 212°F, most of the alcohol will already have boiled off, and the vapor will be mainly water. This is how distillers start with beer that's just 5 percent alcohol and make whiskey that's

40 percent: using an apparatus called a still, they heat the beer and collect the alcohol-rich vapors that come off at temperatures well below 212°F.

A gas chromatograph is something like a still, but one that's designed to work on mixtures of molecules with many different volatilities. You inject the sample at the inlet—the sample may be tiny, a fraction of a gram—and eventually the volatile molecules exit one by one at the outlet. In between, the sample is carried in flowing hydrogen or helium gas into a long coiled tube lined with a complex absorbent material. The tube sits in an oven whose temperature slowly rises. Volatile molecules in the sample initially stick to the tube lining, and then escape it and pass to the end of the column at different times that depend on their volatility. As pulses of different volatiles exit the outlet, they can then be passed into another instrument, a mass spectrometer, that analyzes their chemical makeup, allowing the chemist to correlate "retention times" in the column with specific molecules.

The chromatograph's volatile pulses can also be passed into a tube that leads to the nose of a very patient human detector, who sniffs and names the kind of smell that she detects at a given retention time. This combination of machine and sensory analysis is called gas chromatography-olfactometry, or GC-O. (The preferred scientific term for the sense of smell and the act of smelling is *olfaction*.) It allows chemists and sensory scientists to analyze a sample smell—of a flower or a steak or the air near an industrial hog farm—and come up with a list of the specific volatile molecules in the sample, *and* what each of those molecules smells like. Then we know two things: what molecules are involved in these smells, and what set of single-molecule smells are somehow aggregated in our brains to form the smell of the flower or steak or stink.

GC-O is a brilliant invention. Hundreds of scientific papers have been published with lists of volatile molecules and their associated smells. It's this steadily growing scientific census of the world's volatile molecules that I'll be drawing on as we explore the smellable world.

Talking about smells: which came first, lemongrass or ant?

Most GC-O smelling is done by people who have been trained to respond quickly to the isolated volatiles that come through the smelling port, completely stripped of any real-life context, and describe their smell in a second or two, before the next one comes along. I've tried my nose at it a few times, and for the untrained it's a high-anxiety ride that reminded me of Lucille Ball not keeping up on the candy assembly line. Over and over again I would recognize a smell as being familiar, but—like the cheese in my kitchen—not be able to identify it or come up with a precise description. During the half-hour run of a sample of fried ground beef, I hit the detection button around eighty times, and was confident of my description for maybe ten or twenty. And what a mix! Individual peaks on the chromatograph variously smelled like cooked vegetables, crayons, styrene plastic, nail polish remover, toast, sulfur, raw green leaves, soap, maple syrup, bread, sweat, manure, nuts, and—most immediately obvious to me—strawberries. Strawberries!

GC-O is a very cool tool, but it confronts us with the biggest challenge of delving into smells and their meanings. We can only come up with a description for molecule mixtures that we recognize from experience, or for individual molecules that we recognize as being a prominent part of familiar mixtures. Where we can recognize a real elephant for the first time if we've seen a photo or sketch beforehand, we can't recognize and evaluate a flavor or smell unless we've actually experienced it or something like it before. So we necessarily describe them by reference to what we've already tasted and smelled, as I did for my fried-beef run.

Particular volatile molecules are regularly described as smelling grassy, or floral, or fruity, or meaty, or fecal, because they contribute to the characteristic and commonly experienced smells of grass, flowers, fruits, meats, or excrement—or somehow trigger the same brain circuits that these materials do. And because many particular volatile molecules are found as part of the mix in a variety of different materials, different GC-O sniffers may give different descriptions to the same volatile molecule, and one sniffer may give several different names to a single molecule. There are several volatiles that can smell both catty and fruity because they're found in both litter boxes and mangoes. (Yes, crazy! We'll get to that.) Others get described as soapy but also green-leafy, fresh-cilantro-like: because they're major volatiles in both soap and cilantro.

Our reliance on experience means that what we smell and how we talk about it both depend on the happenstance of our individual lives. Several years ago I had the chance to hear the Brazilian chef Alex Atala give a talk about little-known ingredients from the Amazon. As part of his presentation, he passed out samples of Amazonian ants. Like many people in the audience I expected an "interesting" flavor, not a delicious one. But we were happily surprised to find that they taste like a combination of lemongrass and ginger: originally Asian flavors that have spread to cosmopolitan parts of the West. But, said Atala, you need to realize: to the people of Amazonas, the ants are delicious because they taste like ants. As he later wrote about a woman who had made him an ant broth, "When I made Dona Brazi try things that do not exist in the Amazon, such as lemongrass and ginger, she laughed and said that they tasted just like ants."

How we register and name and think about smells depends on where we've happened to encounter them first. That's a tremendous limitation on our thinking—and on our potential for enjoyment! Not just because there's no way to experience everything in the world, but also because experience itself is limited. While many people love cilantro in the Asian and Mexican cuisines that embrace it, many others find it to be disgusting, probably because they first smelled cilantro's most prominent volatiles in soaps and haven't been able to get past its identification with something that doesn't belong in the mouth.

Once we realize this subjectivity and relativity in the experience of smell, we can then take it into account and make an effort to focus on the objective aspects of the experience. The GC-O says there are strawberry volatiles in fried beef, and my nose tells me that Atala's ants produce some of the same molecules as ginger and lemongrass. So I can make a conscious effort to notice fruity aroma in my hamburger; and I can try to figure out why it could possibly be that fruits and meats, ants and plants, seem to be emitting the same volatile molecules from such different bodies, belonging to entirely different kingdoms of living things.

The expansiveness of secondhand perception

It used to be common knowledge that human beings have a lousy sense of smell. Dogs perform amazing feats of tracking people through the woods after just a sniff of discarded clothing. Genetic studies demonstrate that we have fewer than half the number of smell receptors that our pets do. And as we've seen, whatever information our impoverished receptor team is able to muster is so massaged by the brain that our conscious minds seldom get it straight from our own mouth, or nose. What molecules are actually out there? There's no way to tell for sure. So no arguing about taste, or smell. Whatever.

In 2004, the eminent Yale neurobiologist Gordon M. Shepherd published a paper titled "The Human Sense of Smell: Are We Better Than We Think?" Shepherd argued that receptor counts are no indicator of what we can do with a sense. We have fewer receptors devoted to hearing than many other animals, but we're the species that developed speech and music. Instead, our real strength is in what our brains can do with the senses. Tastes and smells are just isolated signals until the brain turns those signals into integrated perceptions, and Shepherd points out that no other animal devotes more brainpower to smell and taste than humans do. We may not be very good at sniffing out an odor trail in the woods, but we do make fine discriminations between degrees of charring in seeds of the coffee tree, the qualities of fermented grapes from different patches of land, and expensive dabs of plant and animal fluids that we apply to our skin. In areas where people have become sufficiently interested in smells—in perfumery and wine and food connoisseurship—they're impressively sensitive to nuances.

Shepherd's paper and a later book, *Neurogastronomy*, suggested that we can do more with the sense of smell—more fully realize its inherent potential for intimacy and specificity—because we *think*. We aren't born knowing how to use our senses, how to see or hear or smell. We learn to use the capacities of our body from infancy on, mostly unconsciously. As creatures that think, we can choose to extend that learning consciously. The same meddling in the brain that compromises the accuracy and objectivity of our direct sensory experience can also enlarge that experience.

Of course people have been intrigued by smell and have speculated about it for millennia. Above all they have tried to find some kind of order in the tremendous variety of smells, some categories or general qualities by which they could be organized. Early philosophers in Greece and China and India came up with just a handful of categories, starting with "pleasant" and "unpleasant." From the seventeenth century on, the categories multiplied, as scientists, physicians, and then perfumers and food and drink makers got involved, each group focusing on qualities that were relevant to their professions. Recent decades have brought renewed attempts by chemists and perfumers to find the "true" basic categories, and in just the last few years attempts by data jockeys who load all the smell systems they can find into computers to tease out whatever hidden patterns might be there. In the world of food and wine, great effort has gone into the devising of "flavor wheels" that graphically break up composite flavors into their constituents, to guide enthusiasts in savoring the nuances. The first modern flavor wheels were devoted to whiskies and then wines, but you can now find them for among other things beer, cheese, coffee, tea, olive oil, chocolate, maple syrup, oysters, and tap water. And aroma wheels for fragrances.

The pioneering sensory psychologist James J. Gibson called this sharing of information "perception at second hand," "the process by which a human individual is *made* aware of things" with the help of other people, rather than becoming aware through one's own direct perception. It allows us to get past the limitations of our particular lives and take advantage of the accumulated experience and understanding of generations of others.

Acquiring a new nose and world

It can be a bewildering and nose-numbing prospect to learn about smell through flavor wheels and molecular rosters. That is unless you're a sensory scientist or chemist or some sort of food or scent professional or a devoted amateur, in which case they can be utterly, endlessly absorbing. And even transforming. The French sociologist Bruno Latour has made the case that by devoting yourself to learning a few of the components in a complex mixture, you develop not just your mind, but also your body, and the world that you experience. It isn't just an intellectual exercise.

In a 2004 essay called "How to Talk about the Body," Latour analyzed what happens when an aspiring perfumer trains to become better at recognizing and working with basic fragrances. Veteran perfumers put together odor training kits with dozens of single fragrance notes, selecting and arranging the notes so that the novices can learn to recognize progressively finer differences among them. This may sound like a simple process of rote memorization in the service of professional training. But because it involves the work of an otherwise naive sense, one that hasn't had the occasion to experience these smells and their differences before, Latour argues that this process of deliberate, guided smelling produces a newly discriminating sense, *and* thereby a newly accessible part of the real world.

> Starting with a dumb nose unable to differentiate much more than "sweet" and "fetid" odours, one ends up rather quickly becoming a "nose" (*un nez*), that is, someone able to discriminate more and more subtle differences and able to tell them apart from one another, even when they are masked by or mixed with others. It is not by accident that the person is called "a nose" as if, through practice, she had *acquired* an organ that defined her ability to detect chemical and other differences. Through the training sessions, she learned to have a nose that allowed her to inhabit a richly differentiated odoriferous world. Thus body parts are progressively acquired at the same time as "world counter-parts" are being registered in a new way.
>
> Acquiring a body is thus a progressive enterprise that produces at once a sensory medium and a sensitive world.

Acquiring a new nose may sound like a painful surgical procedure, but it's happening even as you read this introduction—I hope painlessly! Of course it's a metaphor for developing relevant parts of the brain. Sure enough, a number of studies have found significant differences in the brain structures and activities of trained perfumers and wine experts. A physical nose is just the visible stiff portal to a hidden, dynamic sensory system whose operating rules and memory banks and databases and connections are being updated with every breath, to keep up with the invisible roiling osmocosm it's reporting on. The more experience and information that it has to work with, the more of that world it can notice and bring to our attention.

Now that we have a general sense for what we smell and how, and how we talk about and learn about smells, let's see what some secondhand sniffing around can do for us.

Part 1

······································

SIMPLEST SMELLS

Chapter 1 • AMONG THE STARS

....

> The intellect is empty if the body has never knocked about, if the nose has never
> quivered along the spice route. Both must change and become flexible, forget their
> opinions and expand the spectrum of their tastes as far as the stars.
> > • Michel Serres, *The Five Senses*, 1985

Yes, the stars!

The sensory spread that's laid on for us every day of our lives went onto the fire around fourteen billion years ago and has been simmering around the stars ever since. Our universe is a stew of matter and energy, and some of the molecules that we smell and taste today bubbled up in it very early on, long before the simplest form of life.

It may sound crazy to sniff and slurp through airless interstellar space, but generations of astronomers have opened the heavens for us to imagine just that. So: You're standing somewhere under the open sky, on a clear night, away from city lights. After you let your eyes adapt to the darkness, you can make out hazy patches here and there, perhaps under Orion's belt in the winter, or the band of the Milky Way in Sagittarius in summer. Zoom your mind's eye in on those indistinct patches, and borrow from the telescopic images you've seen of nebulas in deep space: dramatic swaths and swirls of light set in star-studded blackness, sometimes backlighting darker swirls. These are immense clouds of stardust, diffuse matter that has been driven out of stars as they burned, burned out, collapsed, and exploded. The bright clouds glow with energy; the dark ones coldly absorb it.

Now release a super-volatile emanation of yourself. You're a space-time traveler, an assistant to the Chef of the cosmos, disembodied except for chemical senses sensitive enough to sample—and robust enough to withstand—its primordial flavors. Fly light-years into the stew, plunge into those dusty clouds, and open up.

You taste mineral saltiness, and bitterness, and sharp acids, and even sweetness. You feel and smell the irritating pungency of ammonia cleaner, and the stink that it dispels. You catch the heady smells of solvents, of alcohols, of campstove fuel. Vinegar. Eggs. A hint of fruit!

By earthly standards that doesn't sound like an especially delicious composition. But it's intriguing. What are those familiar molecules doing out there? And why just those? To start so way out and way back helps stretch both our understanding and our sense of wonder. It shows that the smells and tastes to come, the various earthly creatures that produce their own, and the perfumers and cooks who modify and multiply them are all participants in the original, ongoing project of the cosmos: the unfolding of matter's possibilities.

This chapter is about the initial stages of that unfolding, the fires of the stars and their flavorful ashes.

Recipe for the universe: mix matter and energy, and cook

How did volatile molecules that we smell every day come to exist both here and in outer space? It's quite a story, one that emerges from the collective observations and thinking of thousands of scientists from many countries over many decades. It involves the birth of the cosmos as a whole and the origins and evolution of life on Earth. And at the heart of this nondenominational, transcultural creation story is a cosmic version of cooking.

Consider making caramel on your stovetop. You start with a single ingredient, white crystals of table sugar, which taste simply sweet and have no aroma. Put the sugar in a pot, apply heat energy, and stir. After a few minutes, you've turned the solid crystals into a colorless liquid. Still no aroma. Keep heating, and that liquid turns pale yellow—and begins to smell. It gets light brown, then progressively darker and stronger-smelling. In the end you've made a dark syrup that's sweet but also sour and bitter, and richly aromatic. From one substance you've made many: from simplicity, complexity.

A similar process cooked up the entire universe as we know it. The original recipe from the Chef of the cosmos goes something like this. Mix a dozen kinds of elementary particles together with four fundamental forces, and set aside. After a few hundred million years, the particles have combined to form atoms, a hundred different kinds. After another long stretch, many of those atoms have combined to form molecules—and the mix begins to smell. Some of the molecules combine to form particles of dust, and the dust clumps up to form planets. At least one planet, our own, produces increasingly complex molecules, then collectives of molecules that somehow come alive—and these generate a vast bouquet of new volatiles for the Chef to savor, caramel included. So: from a handful of elementary particles the Chef has made countless kinds of molecules, with countless qualities.

This primordial cooking underlies all of our experience, mundane and miraculous. To understand why volatile molecules exist at all for us to smell, and why they exist where they do, let's start in the pristine cosmic kitchen as the Chef gets things going. No smells yet, but just wait.

Cooking up stars

However the known universe came into being, most astrophysicists agree that it did so around fourteen billion years ago in an explosive flash at an unimaginably high temperature. From the moment of this "Big Bang" the universe expanded outward. As it expanded it cooled down, and the kinds of matter and energy that we know on Earth began to appear. In the first fraction of a second emerged packets of electromagnetic energy called photons, which we know as light and heat and radio waves. Along with photons appeared three kinds of raw matter, the subatomic particles that combine to make atoms: protons and neutrons that form the central nucleus of the atom, and electrons that orbit around the nucleus. It's the different numbers of subatomic particles in atoms that give us the hundred-odd different elements with their different qualities: hydrogen, carbon, oxygen, and so on. One solitary proton forms the simple nucleus of atoms of hydrogen, so hydrogen was the first element to be born, followed by nuclei of helium and a bit of lithium.

After only a matter of minutes, the continuing expansion of the universe cooled and slowed everything down to the point that the protons and neutrons no longer had enough energy to fuse together to make heavier atomic nuclei. The evolution of matter paused, for some hundreds of millions of years.

But during that long hiatus, one of the universe's fundamental forces worked inexorably to reenergize matter. Gravity is a force that acts between any two bodies of matter, tiny or huge, and pulls them toward each other. In the newborn three-element universe, neighboring atoms gradually felt each other's gravitational pull. They gathered into clusters, clusters into more crowded clusters, all the while moving faster and faster, bouncing off each other with more and more force, releasing more and more heat energy as they did.

As the universe as a whole continued to expand and cool off, gravity created hot pockets of densely crowded atoms, some of them so dense and hot that they began to emit enough energy to glow. This was the first generation of stars.

Cooking up chemical elements in stars

The material richness of our world is a reflection of its chemical complexity, its countless combinations of the hundred-odd chemical elements. The first stars had just three elements to work with. They generated nearly all the rest by becoming fantastic self-adjusting, self-destroying, billion-degree ovens.

Imagine a member of that first generation of stars. As gravity causes its matter to crowd together and collide with ever-increasing force, its temperature and energy increase. At a few million degrees, the conditions are right for two hydrogen nuclei to fuse into a single helium nucleus. This reaction releases energy—which jolts the nuclei into moving fast enough to resist the gravitational force. Fusion and gravity balance each other, and the star can burn with a steady flame like this for billions of years, using hydrogen nuclei as fuel and producing helium nuclei as the residue. When it has consumed most of its hydrogen fuel, the fusion reaction slows down, gravity begins to dominate again, the largely helium core of the star begins to contract, the temperature rises—until the helium nuclei can become the new fuel, fuse to form yet larger nuclei, and again balance gravity so that steady burning can continue. Now we have oxygen and carbon: two of the primary chemical players in the saga of life and the osmocosm.

Then the cycle of contraction, temperature rise, and new fusion repeats again and again, at ever-escalating temperatures. The star takes on an onionlike structure, with portions of the newly formed elements surviving in the outer, cooler layers. After twenty-five new elements have been cooked up, gravity finally prevails over fusion and forces the star into a final burst of creativity—and generosity. It contracts the star's core to such an extreme density and temperature that the core explodes and becomes what's called a *supernova*. The energies released are so extreme that they trigger the formation of ninety-odd more elements. And the explosion blasts these and the first twenty-six into interstellar space, the space between stars.

The supernova thus serves up its creations to the calmer cosmos at large. And it's here that the elements can manifest their individual qualities, explore their affinities for each other, join up, and initiate the next stage in the unfolding of matter's possibilities—the stage in which the first molecules of smell emerge.

Cooking up molecules between stars

Molecules are the stuff of our world, the substance of nearly everything that we see and touch, taste and smell. They're simply combinations of elements, two or more atoms that have joined with each other in a specific arrangement. Given a hundred elements to work with, the number of possible arrangements is—well, astronomical. It's with the birth of molecules that the cosmos attained entirely new levels of complexity.

Molecules are products of the electromagnetic force, the attraction between particles of opposite electrical charge. The nucleus of an atom carries a positive electrical charge thanks to its protons. Electrons that orbit the nucleus carry a negative electrical charge, and it's the attractive force between positive protons and negative electrons that keeps the electrons in orbit. Molecules are the structures that result when the nuclei of different atoms *share* orbiting electrons with each other. That electron sharing is the bond that holds them together in a stable structure. Some molecules consist of just two or three atoms—for example, carbon monoxide, CO, and water, H_2O—while DNA molecules include many thousands. Most volatile molecules have between a few and a few dozen.

The electromagnetic force isn't strong enough to withstand the energies at work in a star. It readily forms molecules at the moderate temperatures that we experience in everyday life, between fire and ice, where atoms are moving slowly enough that they can bump into each other and bond without immediately getting knocked apart again. In the near-motionless cold of deep space, and with atoms few and far between, it can take many years for those atoms to encounter and react with each other. More favorable interstellar regions are "giant molecular clouds," the smudgy chiaroscuro patches familiar from telescopic images of the constellations Orion and Sagittarius. These are remnants of supernovas and aged stars that gravity has slowly pulled together, along with new stars that are beginning to burn nearby. They harbor regions denser with atoms, and temperatures around those of our kitchens and ovens. As their name indicates, these clouds are where astrochemists have had the best luck at detecting cosmic molecules.

Molecules in open space exist because their atoms happened to bump into each other and stick. The most abundant atoms in space include hydrogen (abbreviated H), oxygen (O), and carbon (C), whose particular electron-sharing tendencies naturally lead to the formation of small molecules like oxygen gas (O_2), water (H_2O), and carbon monoxide and dioxide (CO and CO_2). Carbon atoms also readily bond with each other to form long chainlike molecules, as well as six-cornered ring molecules. The chains and rings readily nestle together with others of their kind, and can aggregate to form ever larger masses: cosmic soot. The dark swirls in the molecular clouds are a mixture of carbon soot and similar aggregates of primordial minerals. These various particles make up what's called interstellar dust.

The individual grains of interstellar dust are microscopically tiny, but their influence on the development of the cosmos is huge. They provide a solid surface to which free-floating atoms and small molecules can stick. They thus act as gathering places that encourage chemical activity, new reactions, larger molecules. On them the material world became increasingly diverse, complex, capable of further development. And to the nose of the cosmic Chef, it became aromatic—billions of years before our own sun began to shine.

In 2020, the roster of known interstellar molecules numbered more than two hundred. Here I'll note only the few dozen molecules that we also sense in everyday life, along with the everyday materials that they dominate and so "smell like" to us.

Detecting the smells of space

At last, smells! But how can we possibly know—not imagine, but know—what molecules are so far out there in space?

By the telltale traces they leave in the energy that the cosmos constantly rains down on our planet. Astrochemists are connoisseurs of electromagnetic radiation, and in particular the visible light, infrared light, and radio waves that originate in stars and galaxies and reach us across the vastness of space. In one year these forms of radiation travel staggering distances, so when we see stars and galaxies, we are seeing deep into both space and time, into the past history of the cosmos.

Astrochemists collect the faintest light and radio emissions with telescopes that are far more efficient and sensitive than our eyes or radios. They then pass them through the electronic equivalent of a prism that separates them into their component colors, or frequencies. The pattern of frequencies—the spectrum—is a kind of fingerprint that makes it possible to identify the kind of matter that emitted it, and the kind of matter that might have absorbed some of it on its long journey to Earth.

Stars give off mainly high-energy electromagnetic radiation—as we know from our sun's blinding visible light and the UV that can burn us. The space between the stars is cold, so most atoms and molecules there don't have enough energy to radiate. Instead, they tend to absorb radiation from the stars. When they do, this produces a dark absorption line in the spectrum coming toward us from the stars behind them. The cold matter may then re-radiate some of the energy it absorbed in a lower-energy form, often in the infrared and radio parts of the electromagnetic spectrum.

Because the radiating and absorbing properties of matter can be studied in the laboratory, scientists can compare the spectra in starlight with lab spectra, and identify the materials in the stars and in space around them. Such is the power of this approach that 150 years ago, French and English astronomers discovered the existence of a previously unknown element a hundred million miles away in the sun, decades before it was found on Earth. That element was helium—named from the Greek for "sun," *helios*—which is abundant in stars but scarce on our planet.

The smallest smellables:
sulfurous, ammoniacal, ozonic

Let's start with the simplest cosmic molecules for which we have smell receptors, those made of just three or four atoms. (Two-atom sodium chloride is salty, hydrogen chloride sour, but no two-atom molecules are aromatic.) We don't have receptors for water, H_2O, carbon dioxide, CO_2, or nitrous oxide, N_2O, though all are important in the air we breathe. But two other simple volatiles have very familiar smells and will turn up often in our explorations of the osmocosm.

Eggy, sulfurous hydrogen sulfide, H_2S, combines the elements hydrogen and sulfur in a molecule whose smell we can detect in very small traces, possibly because higher concentrations can be irritating and even fatal. We typically identify the smell as "eggy" because it's the characteristic note of freshly cooked eggs, or when it's stinkily strong, "rotten-eggy" because it escapes from decomposing organic matter of all kinds. But Earth's volcanoes and hot springs have been emitting this molecule from long before the earliest organisms or eggs. Better to call its primeval smell "sulfurous" or "sulfidic." Hydrogen sulfide is the simplest example of a general rule in our osmocosm: the presence of a sulfur atom in a volatile molecule lends a distinctive quality to its smell, one that can be unpleasant when dominant, but appealing when blended with others. Sulfur volatiles are what give garlic and onions and cabbage their strong identities, but they also contribute to the aromatic appeal of roasted meats and coffee, and "exotic" notes in some fruits and wines.

One other three-atom sulfur volatile detected in space is **sulfur dioxide**, SO_2. It isn't as common in everyday life as hydrogen sulfide, and is more irritating than aromatic, but it's unmistakably sulfurous.

SOME SMELLABLE THREE- AND FOUR-ATOM INTERSTELLAR MOLECULES

Smells	Molecules
boiled eggs, sulfurous	hydrogen sulfide, H_2S
irritating, sulfurous	sulfur dioxide, SO_2
ammonia	ammonia, NH_3
fresh, pungent	ozone, O_3

Ammonia, NH_3, with an atom of the element nitrogen at its center, was one of the first molecules detected in interstellar space. It's also found in the atmospheres of the gas giant planets, Jupiter, Saturn, Uranus, and Neptune— and in household cleaning products, overripe cheeses and salamis, underripe animal manures, and urine. Its smell is that of simple unscented household cleaner, which is about 30 percent ammonia. Smelling salts are also made from ammonia because it's irritating and triggers strong physical reflexes; prolonged exposure can be fatal. Just as volatile molecules with a sulfur atom tend to share a common sulfurous quality, many nitrogen volatiles hint more or less strongly at the sharpness of ammonia or urine. Most of them include *amine* in their chemical names: keep an eye out for the various *methylamines*.

Pungent, fresh ozone, O_3, was named by a German chemist from the Greek root for "smell." It's a very reactive molecule, so we seldom smell it directly except after nearby lightning strikes or power-line arcing or prolonged use of high-voltage laser printers, whose electrical energy can force three oxygen atoms together instead of the usual two. The smell may come from its strong oxidizing effects on other molecules in the air or even in the nose.

Carbon chains and rings: the backbone of life and its smells

The majority of primordial molecules larger than four atoms contain carbon, the fourth most abundant element in the cosmos after hydrogen, helium, and oxygen. Carbon is the backbone of life on Earth, and that's because it's the most versatile of elements, the one through which the creativity of matter, its potential for new forms, is most readily expressed. Its versatility is already fully evident in interstellar space, where for billions of years it has been generating a preview of what life on Earth would come to smell like.

Thanks to its particular complement of electrons, each carbon atom can form as many as four bonds with other atoms. Carbon atoms readily form long open chains and complex closed-chain ring structures, because each atom can bond to two others—thus becoming part of a network—and still have two bonds left over for other elements. Even when carbon atoms bond only to each other, they can do so in many different arrangements. This is why pure solid carbon can be either amorphous or highly organized—particles of black soot, or soft, slippery pencil-lead graphite, or brilliantly clear, hard diamonds.

Carbon atoms are found in most of the millions of naturally occurring molecules that scientists have cataloged. Chief among these are the molecules of life. The physical structures and chemical machinery of all living things on our planet are made of molecules that are largely carbon. Fossil fuels come from living microbes and plants that were buried hundreds of millions of years ago, so coal and petroleum and natural gas, most plastics, and many other industrial chemicals, including solvents and lubricants, are mainly carbon.

It's also thanks to carbon that there are so many smells in the air for us to enjoy. The bonds between carbon atoms in carbon chains are electrically symmetrical, and common side bonds with hydrogen nearly so. This means that most carbon-chain molecules tend to have weak electrical asymmetries, and therefore weak attractions to water and other molecules that are asymmetrical and have electrically positive and negative ends. When carbon chains and water are mixed, the strong electrical attraction of water molecules for each other squeezes the carbon chains out into separate clusters. A familiar example is the cloud of separate oil droplets in an oil-and-vinegar salad dressing, or the layer of oil at the top that the droplets slowly aggregate into. Oil molecules are long, heavy carbon chains that stay put atop the watery vinegar, but shorter and lighter chains are mobile enough to escape into the air, where we can inhale and sense them. It's thus the electrical dissimilarity of carbon chains to water that helps make short carbon chains in natural materials *volatile*, or prone to go airborne and become smellable.

As we'll see in chapter 3, Earth's teeming carbon-chain volatiles fall into a handful of broad families. Their first few members emerged long, long ago, from the basic properties of carbon and four other elements. Here's an introduction to these pre-Earth pioneers and some surprisingly earthy smells—and to an easy shorthand for visualizing invisible carbon structures. Take a look at the table drawings on the facing page. Rather than label every atom in a molecule, chemists often just outline the zigzag backbone or ring formed by the carbon-carbon bonds, with double bonds getting double lines. The line tips and corners indicate carbon atoms, and hydrogens are mostly omitted. I'll include sketches of selected volatiles throughout, to give you a clearer idea of their family relationships, and how just an atom or two can change what a molecule smells like.

Carbon-hydrogen molecules: fuels and solvents

Hydrocarbons are molecules made up only of carbon and hydrogen atoms. The carbon atoms bond to each other to form either variably long straight chains, or rings, usually of six members, and then fill in their remaining side bonds with hydrogen. They're the simplest family of carbon chains and so probably among the earliest molecules to form in interstellar space. In human history, they're largely post-industrial. We're familiar with hydrocarbons because they burn well and mix well with oils and greases: they're good fuels and good solvents.

Methane, which we know as natural gas, consists of one carbon and four hydrogen atoms: CH_4. Methane itself is odorless, but the version with a free bond, $-CH_3$, is important as a component of many odorous molecules; this one-carbon group is indicated with the prefix **methyl-** (often written as a separate word) in chemical names. Along with the high-temperature torch fuel acetylene, C_2H_2, methane is flammable but odorless: a potentially dangerous combination, so manufacturers add traces of stinky sulfur volatiles to both to make them detectable.

Faintly sweet-smelling ethylene is a two-carbon hydrocarbon detected in space; its formula is C_2H_4. It was burned in nineteenth-century gas lamps and is an important hormone in plants, where it stimulates both fruit ripening and the deterioration of stored produce.

SOME SMELLABLE CARBON-HYDROGEN INTERSTELLAR MOLECULES

	Smells	Molecules and shorthand structures
=	faintly sweet	**ethylene**, C_2H_4
⬡	sweet, gasoline	**benzene**, C_6H_6
	mothballs, lighter fuel	naphthalene, $C_{10}H_8$

Sickly-sweet-smelling, solvent-like benzene is a six-carbon ring molecule. On Earth it's produced from fossil fuels and used for many industrial purposes. We seldom smell it now because it's known to be carcinogenic and its use is regulated.

Mothball- and lighter-fluid-like naphthalene is a double-ring ten-carbon molecule. In addition to its use to kill clothes moths and as a fuel in cigarette lighters and campstoves, wine lovers know a modified version of naphthalene as the prized "kerosene" note of well-aged Rieslings! Naphthalene is the smallest "polycyclic aromatic hydrocarbon," or multi-ring hydrocarbon. PAHs with four or more rings are nonvolatile components of the soot created by incomplete burning of materials like wood, coal, and tobacco. They're toxic, as naphthalene can be (its use in mothballs is prohibited in some countries).

Carbon-sulfur and carbon-nitrogen molecules: sulfurous and fishy

Simple carbon-chain molecules that include sulfur or nitrogen are distinctive: they bear at least some resemblance to hydrogen sulfide and ammonia.

Rotting-cabbage methanethiol (pronounced "methane thigh-all") is the one-carbon, one-sulfur molecule CH_3SH, which can result when hydrogen sulfide, H_2S, reacts with methane, CH_4. It's hard to imagine anything more organic than a decomposing vegetable, and yet that characteristic molecule is present among the stars. Like hydrogen sulfide, methanethiol is a volatile we'll come across frequently, on land and sea and in the air around us, even emerging from us, as a common by-product of life itself. We're even more sensitive to it than to hydrogen sulfide, and like hydrogen sulfide it's toxic. **Ethanethiol**, the two-carbon relative of methanethiol, is a bit less aggressively sulfurous, and on Earth it contributes to the smells of some raw fruits and cooked vegetables. In other chemical names, *thio* indicates the presence of the sulfur-hydrogen (-SH) couple, and often a sulfurous or otherwise unusual smell.

SOME SMELLABLE CARBON-SULFUR AND CARBON-NITROGEN INTERSTELLAR MOLECULES

Smells	Molecules
rotting cabbage, sulfurous	methanethiol, CH_3SH
cooked cabbage, onion, sulfurous	ethanethiol, C_2H_5SH
fishy, ammonia	methylamine, CH_3NH_2

Fishy-smelling methylamine is a combination of methane and ammonia, with the carbon and nitrogen atoms bonded to each other. It's the simplest of the *amines*, a group of nitrogen-containing molecules that are characteristic of animal metabolism and animal smells. Though not all amines are derived directly from ammonia, their name is. The word *ammonia* itself comes ultimately from the name of the Egyptian god Amun; a Roman temple once associated with Amun, located in modern-day Libya, was near a rich mineral deposit of nitrogen-containing salts.

Carbon-oxygen molecules: the families

So far the smells of interstellar space are mostly unpleasant. They're either "chemical" because the molecules there are the same as the materials that we use on Earth to burn or clean or fumigate, or they're disgusting because they're prominent here in decomposing plant or animal remains. All of these primordial volatiles have something in common: an absence of oxygen atoms. Bring oxygen, the universe's third most abundant element, into the structures of carbon chains, and the smell register begins to shift.

As we'll see, oxygen was critical to the development of life on Earth, and rare is the molecule in plants and animals that doesn't include atoms of oxygen along with carbon and hydrogen. The molecules that contain all three are relatively rare in space, but they do exist, and point the way to chemical themes that living things will explore in great and often pleasing variety.

Carbon chains can take on oxygen atoms in several different ways, and each way forms the basis for a family of molecules wherein the different family members are simply carbon chains of different lengths. And as we'll see, different chain lengths can mean very different smells.

Oxygen prefers to form two bonds with other atoms. So if one oxygen atom forms two bonds to the same end carbon atom on a chain, the result is a member of the **aldehyde** family; if an oxygen does the same to any carbon atom other than the end carbon, the result is a **ketone**. If one oxygen forms only one bond to the end carbon, and uses its second bond to bond to a hydrogen atom, the result is an **alcohol**. If two oxygens attach to the end carbon, one with a double bond and the other with a single bond plus hydrogen, the result is a **fatty acid**—so called because this family provides building blocks for fat and oil molecules. And if a fatty acid and an alcohol react with each other to form a single combined molecule, with one oxygen linking the two carbon chains, the result is an **ester**.

Each of these families makes major contributions to the smells of our world, and each goes back many billions of years.

Carbon-oxygen molecules: irritants, solvents, vinegar . . . fruits!

Here are the primordial founders of the carbon-oxygen clans of volatile molecules. First the interstellar aldehydes:

Chemical, irritating formaldehyde is the one-carbon aldehyde, a preservative used in biology labs and embalming and manufacturing; it's a known carcinogen.

Fresh, green-apple-like acetaldehyde is the two-carbon aldehyde, found in many fermented foods, including yogurt and aged wines.

Earthy, cocoa-like, nutty, winey propanal is a three-carbon aldehyde, and is familiar to us because it's found in a number of fermented foods. (In chemical names, the aldehyde suffix is often abbreviated to -*al*, so *propanaldehyde* and *propanal* are names for the same molecule.)

Choking, acrid propenal is also a three-carbon aldehyde, but its first two carbons share two bonds instead of one. It's produced when we overheat oil on the stovetop, and it's toxic. Propenal is also called *acrolein*.

The smallest ketone, and the molecule that gave the ketone family its name, **solventy acetone** is a three-carbon chain and is commonly used in nail polish remover. It's also detectable on our own breath when we haven't eaten for a few hours; our body produces it when it's low on carbohydrate fuel and has to start burning fat for energy.

SOME SMELLABLE INTERSTELLAR ALDEHYDES AND A KETONE

Smells	Molecules
biology lab preservative	formaldehyde, aka methanal, CH_2O
fresh, green apple	acetaldehyde, aka ethanal, CH_3CHO
earthy, cocoa, winey	propanal, CH_3CH_3CHO
irritating	propenal, aka acrolein, CH_2CHCHO
solvent	acetone, CH_3COCH_3

Now the interstellar alcohols, two in number:

Vodka-like, solvent-like methanol and ethanol are the one- and two-carbon alcohols, both of them intoxicating, both toxic. Unflavored vodka and rubbing alcohol give us the purest experience of their smell. Methanol is known as methyl or wood alcohol and is extremely toxic; traces are found in the products of alcoholic fermentation: wines, beers, distilled drinks. *Ethanol* and *ethyl alcohol* are the chemical names for what we commonly call alcohol. After water, it's the primary component of all wines, beers, and distilled spirits.

The interstellar fatty acids, also two to date, are sour to the taste like other acids but also volatile:

Sharp, slightly vinegary formic acid is the one-carbon volatile acid, a chemical weapon found in ants and other insects but turned against them by the anteater, which relies on it to help digest them.

Sharp, vinegary acetic acid is the two-carbon volatile acid, and very familiar: it's the defining molecule in vinegar, produced from molecules of ethanol by particular bacteria that can grow in beers and wines.

SOME SMELLABLE INTERSTELLAR ALCOHOLS, ACIDS, AND ESTERS

Smells	Molecules
rubbing alcohol	methanol, CH_3OH
vodka	ethanol, CH_3CH_2OH
sharp, vinegar	formic acid, COOH
vinegar	acetic acid, CH_3COOH
fruity	methyl formate, CH_3OCHO
fruity, winey, rum	ethyl formate, CH_3CH_2OCHO
solvent, fruity	methyl acetate, CH_3OCOCH_3

Finally, and most remarkable of all, the interstellar esters, alcohol-acid fusions, three so far:

Solvent-like but fruity methyl formate, two carbons, and **ethyl formate** and **methyl acetate**, three carbons, have similar smells despite their different component alcohols (methyl, ethyl) and acids (formic, acetic). **Methyl acetate** is found in some nail polish removers and plastic glues. The solvent quality of these molecules shades into a kind of general fruitiness that's characteristic of alcohols like wine, brandy, and rum, and hints at the unequaled delights of fruits themselves. The ester family is the volatile specialty of ripe fruits and yeast fermentations.

New cosmic flavors in asteroids

So far we've been sniffing at the primordial molecules that assembled from atoms floating in open space or collected on dust grains. But there are others out there that are hidden from the astrochemist's view. Once gravity has pulled molecules together into dust grains, it pulls the grains together into increasingly larger bodies whose interiors can foster an even greater diversity and complexity of reactions and combinations—in part because they're shielded from molecule-breaking radiation. We have an idea of what's in them because we can actually get our hands on debris from Earth's local neighborhood.

The solar system includes our sun and materials that were too far away or moving too obliquely to be pulled into it. These escapee gases and dust crowded together in orbit around the sun, colliding and aggregating and eventually forming the planets. The leftovers of this process include *meteoroids*, rocky objects that range from the size of dust grains to a few feet or meters across; *asteroids*, up to a few hundred miles or a thousand kilometers across; and *comets*, asteroid-size "dirty snowballs" with a large proportion of ice and other frozen gases. And collisions continue. Past encounters with asteroids have caused major extinctions of life on Earth, and thousands of small meteoroids rain down on us every day, delivering many tons of cosmic matter. Though most are pulverized or incinerated by the passage into our atmosphere, some pieces of them—meteorites—fall intact to the ground, where astrochemists can analyze their interiors.

Many science museums have meteorite fragments on display. It can be a thrilling experience to touch them and imagine their slow growth as the solar system formed, the molecular explorations they fostered within as they grew, and their incandescent entry into our world. The most remarkable are *carbonaceous chondrites*, fragments of asteroid relics of the very early solar system, a mix of clay-like minerals, little spheres of silicon-oxide glass, and carbon-based molecules. Among the first and most prominent to be scrutinized was the Murchison meteorite, which fell on September 28, 1969, in Australia's Victoria Province. More than a hundred tons of fragments were recovered and sent to laboratories at NASA for analysis. These stony bits of solar debris turned out to contain many carbon chains and rings that Earth's living things use as building blocks to make DNA and RNA, life's blueprints, as well as proteins, life's chemical workhorses.

Astounding: it might be that some ingredients for life on Earth could have come from outer space! A little less astounding, but still remarkable: these building blocks were accompanied by many additional volatile molecules not detected in space whose smells now figure prominently in our everyday lives. Among them are basic carbon chains and rings "decorated" with additional atoms or carbon-hydrogen methyl groups that can strongly affect what they smell like. The decorations are often indicated in the molecule names, sometimes separated by a space—for example, "methyl butyric acid"—and sometimes run together—as "methylbutyric acid." Chemists have rules about these matters; my rule in this book will be to make the names as clear as possible for nonchemists.

Cheesy short-chain fatty acids are extensions of the one-carbon formic and two-carbon acetic acids detected in interstellar space. Propionic, **butyric**, and hexanoic acids are three, four, and six carbons long, and are common in aged cheeses and other fermented foods.

Sweaty, cheesy branched fatty acids like **methylbutyric acid** are variations on the short-chain acids that have an extra one-carbon methyl branch extending from the side of the chain. Like the straight-chain acids, they're reminiscent of cheese but also can smell like human sweat: that's where we're most likely to encounter them.

Almond-extract benzaldehyde is a six-carbon benzene ring with a one-carbon formaldehyde-like decoration on one corner. It has a pleasant smell familiar from the handy kitchen extract and from cherries.

Antiseptic phenol is that same benzene ring, but with an oxygen-hydrogen group on one corner, and a "chemical" smell; it's a frequent ingredient in cleaners and disinfectants.

 Tarry, barnyardy cresols are phenol molecules with a methyl group on one of the other ring carbons; they're common in fossil fuels and animal wastes.

Honey-like phenylacetic acid is a benzene ring with an acetic acid decoration on one corner; it's delightfully sweet and flowery.

Fishy or ammonia-like amines and pyridine are carbon-hydrogen-nitrogen molecules that are especially numerous in meteorites. There are a dozen or more amines, condensed from ammonia and various carbon chains, while **pyridine** is formed from ammonia and a six-carbon ring.

SOME SMELLABLE MOLECULES FOUND IN METEORITES

Smells	Molecules
sour, cheesy, vomit	propionic, butyric, hexanoic acids
sweaty, cheesy	methylpropionic, methylbutyric acids
almond extract	benzaldehyde, aka benzenemethanal
adhesive bandage, antiseptic	phenol, aka hydroxybenzene
barnyard, tar	cresols, aka methylphenols, aka methylhydroxybenzenes
sweet, honey, floral	phenylacetic acid, aka benzeneacetic acid
ammonia, fishy	dimethylamine, pyridine

Of course, all of these molecules are present in trace, probably unsmellable amounts. Nothing much of them would have been extracted from the golf-ball-size meteorite that an astronomer-winemaker in Chile placed in the barrel for his Viña Tremonte 2010 Cabernet Sauvignon "Meteorito." If we could detect them, however, we would get quite an aromatic extension to the volatiles detected by telescope, from the chemical to the unpleasantly animalic to the floral.

Primordial smells, familiar and strange

It's been a long supersensory trek through these first stages of creation, through physical extremes that are completely foreign to our own experience. We've met some of the very first molecules as they emerged in outer space billions of years ago. Yet the molecules aren't foreign to us at all. True, they're not an especially appealing bunch. Many are austere and harsh, qualities that seem a fitting reflection of their original birthplace. Ammonia, hydrogen sulfide, methanethiol, and propenal are suffocating and toxic. Our bodies reflexively reject them in order to survive. They offer not pleasure but a reference point. In eggs and cabbage and cheese, the sulfur volatiles are a kind of spice, a primordial pepper, interesting as trace accents. But then there are more appealing molecules. The solvents and fuels are toxic yet intoxicating, seductive. Familiar and pleasant are vinegar, hints of cheese, fruit, honey—their volatiles forming in a place where no plant or animal or microbe has ever been.

The fact that all these volatiles formed in the clouds of interstellar space means that they're elementary, universal molecules, molecules that can come together wherever the conditions are right. As we'll see, on Earth they're often the stripped-down residues of life's more complex workings. If we'd been assisting the Chef of the cosmos from the Big Bang on, so that the smells of our home were among the last we experienced rather than the first, then cooked eggs and rotting greens and vinegar and alcohols would remind us of dust clouds, atoms first meeting atoms long enough to start a relationship, their simple newborn offspring pointing the way toward the great molecular diversity of our world.

The smells of Earth will always be our reference points. All this imaginary sniffing around the cosmos is very much perception at second hand. But occasionally, in the right circumstances, it can bring some resonance, some added dimension to familiar sensations. It might be likeliest at a picnic or backyard cookout that lasts past sunset, or on a camping trip at evening, when those faraway infernos emerge from the twilight to remind us. Lighter fluid or stove fuel, scorched oil, a vinegar dressing, a deviled egg, a just-unwrapped cheese, a sip of wine or rum: all offer distant echoes of the early cosmos, sensible traces of the inherent, relentless creativity with which matter explores its own possibilities—the creativity of which we ourselves are both a product and an agent.

Chapter 2 · PLANET EARTH, EARLY LIFE, STINKING SULFUR

Now listen, and I'll explain the nature of
Avernian regions and the lakes nearby.
First, these are called "Avernian," that is, "birdless,"
Because they're fatal to all kinds of birds.
For whenever the birds fly out over those places
They forget their feathered oars, let their sails go slack,
And gently fall like raindrops to the earth
If the place is ground beneath, or into the water
If an Avernian lake awaits below.
There's one near Cumae, where sulfur-stinking mountains
Smoke, and bubble up with hot springs everywhere. . . .
And don't you see, besides, that the earth itself
Produces sulfur and the stink of asphalt?
And when those who search for veins of gold and silver
Probe with their iron picks the earth's deep secrets,
What odors does Scaptensula Mine exhale!
What evils breathe out of those lodes of gold! . . .
All of these tides of death the earth steams forth,
Breathing them out beneath the open sky.

> • Lucretius, *On the Nature of Things*, about 50 BCE

Earth, rocks, gunflint, sulfur, hydrogen: terrifying, primary, molar,
simple, primeval—I was going to say atomic—mineral odors. Here
lies our horror of chemistry, the reason our ancestors burned
alchemists and sorcerers at the stake, terrified by the common
ground shared by knowledge and death.

> • Michel Serres, *The Five Senses*

The Earth's physical depths. A legendary afterworld where the dead exist posthumously. The smells of sulfur. This cluster of associations is an ancient one in Western culture, possibly going back to the early Egyptian dynasties. And sulfur is in the mix for good reasons. That element crystallizes in lemon-yellow deposits around the mouths of volcanoes, those openings from which the inner Earth expels it. When it's heated, solid sulfur first melts into a yellow liquid, then turns red, the color of fire and blood. Apply a spark and the liquid ignites and burns with a blue flame—hence sulfur's alias *brimstone*, originally *burn-stone*. And as it burns, it emits choking, strong-smelling fumes. Today this bizarre behavior provides an amusing off-label use for little pellets of garden sulfur: pile a few and apply a match. Thousands of years ago, alongside the mysteries of death and the hidden violent world below, it helped inspire visions of underworlds and afterworlds where the unrighteous dead are punished in lakes of fire and brimstone.

Lucretius was a Roman poet who undertook to show how everything in the world and our experience can be explained as manifestations of a purely physical reality. In the popular belief of his time, the sulfurous bird-killing air of the region just west of Naples was a sign that the gates to Hades, the land of the dead, were nearby. Lucretius debunked this, pointing out that ordinary mines and hot springs exhale stinking and toxic fumes. Today we know that Avernus and Cumae sit among the sunken remains of a massive volcano, the miles-wide caldera of the Campi Flegrei, or "burning fields," and that volcanoes and hot springs all over the planet release the same gases.

Yet as Michel Serres says, even if we no longer subscribe to superstitions, there's still something uncanny and disturbing about smells from the mineral world. "Minerality" in modern winespeak may denote the smells and pleasant associations of moist stones, but *minerals* originally named materials brought up from mines. We detect the simple sulfur molecules that emanate from belowground and register even small amounts as stinking because they are indeed dangerous and potentially fatal to all air-breathing animals, birds and us alike, best moved away from. And they can be an unnerving reminder that, though we think of our planet as a largely hospitable oasis, a Mother Earth or Gaia whose meadows and forests and oceans teem with life, in fact life inhabits a very thin veneer on an austere conglomeration of asteroids.

But there's more to sulfur dioxide and hydrogen sulfide than deathliness. We met these simple three-atom molecules in interstellar space, where the first of them likely formed before there were asteroids or even grains of dust. And it appears that they may well have been key ingredients in the emergence of life on the young unveneered Earth, and so contributed to the eventual making of us. The first living things—the first aggregations of molecules capable of maintaining and multiplying themselves—had to make do with what little that harsh world had to offer. A number of molecules that can be deathly to us now meant life to some of them. And the same molecules still betoken the life of countless living things thriving all around us and inside us—invisibly, but smellably. We get whiffs of them today not only near hot springs and volcanoes, but also in swamps, in sewage, even at breakfast.

Hades on Earth: pot and water for the primordial soup

Our planet came to be after almost ten billion years of molecule building in space on dust grains and in asteroids. So was the newborn Earth as flavorful as the Murchison meteorite? Probably not, and for a simple reason: it was a molecule-breaking place. Scientists borrow from Greek mythology and name the first few hundred million years in the Earth's history the Hadean Eon: hellishly hot, hostile to chemical complexity.

Earth is thought to have been formed primarily by massive asteroid collisions, events that would release a tremendous amount of heat energy. When it was about a hundred million years old, it was hit by another body about the size of Mars, which vaporized enough of that object and Earth to form our moon. With the heat from collisions and additional energy released by the decay of radioactive elements, the newborn Earth would have run a temperature of several thousand degrees, and been a molten mix of simple elements and the smallest, most robust molecules, mainly water, carbon monoxide, carbon dioxide, and sulfur dioxide.

This period of heavy bombardment tailed off around 3.8 billion years ago with the decline of the local asteroid population in our sun's neighborhood. As the Earth cooled, its various elements and molecules separated into different regions and niches that persist to this day. The early atmosphere, the layer of gases that were heavy enough to be held by Earth's gravity, was very different from what it is now. There was none of the free oxygen that our life depends on. It was mainly carbon dioxide and monoxide, nitrogen—and water vapor.

That water would be crucial to the Earth's flavoring-up. There was plenty on the hot young planet, much of it probably delivered by incoming meteorites and comets. But for some time, the Earth's surface was a cross between a steam bath and a pressure cooker, and the surface water was either all vapor or liquid hotter than its usual boiling point. As cooling continued, the steam eventually condensed, and in a deluge that may have lasted for many years, it formed the Earth's oceans.

This was an epochal event in matter's exploration of the possible. Liquid water is unmatched in its ability to mix with and dissolve other molecules, and thus allow them to move and find and react with one another. And liquid water is a billion times more crowded with molecules than interstellar space. So in the relatively temperate Archaean ("beginning") Eon, the Earth's waters offered a newly encouraging medium for chemical evolution. The oceans dissolved minerals from the planet's solid crust, gases from the atmosphere, and gases venting from the hot interior. They may have taken in more advanced molecules from the waning rain of meteorites.

Thus Earth's waters came to host multitudes of different atoms and molecules, crowded together, free to move and collide and react. They were the starting stock for the primordial soup to come.

Whiffs of early Earth: sulfurous volcanoes and hot springs

This scientific account of the formation of the early Earth may lack the simplicity of the creation story in Genesis, but it's awe-inspiring in its own way, in the unfolding of purely physical forces and events that we recognize from everyday life, but on a spectacularly massive and powerful scale. Unlike the biblical Creator, who seems to have withdrawn from his creation, these forces continue to manifest themselves to this day. That's why, more than four billion years after the planet's birth, it's still possible to get a whiff of the largely unnoticed 99 percent of our planet and its hellish past. Just head for the vicinity of an active volcano, or, more comfortably, to a natural hot spring.

Volcanoes are the most spectacular manifestations of the energies pent up in the mineral planet. They get their name from Vulcan, the Roman god of fire and the forge. Volcanoes are breaches in the Earth's outer crust that release materials from the hot, pressurized mantle to escape in eruptions of hot gas, ash, and sometimes molten rock. The explosive force is created by trapped gases expanding as the molten rock rises. The gases are mainly superheated water—steam—with some carbon dioxide and sulfur dioxide, and traces of others. Volcanoes have played a major role in shaping Earth's surface from its earliest stages, and they've affected the evolution of living things as well. Of the several mass extinctions that have punctuated that evolution, some appear to have been caused by volcanic eruptions that may have lasted for millions of years, filling the atmosphere with sunlight-blocking, planet-cooling dust particles and droplets of sulfuric acid.

Today there are about fifty significant volcanic eruptions every year. Several times that number constantly emit fumes and give us a whiff of what the planet smelled like before life transformed it. I've been close to the top of Mount Etna in Sicily, ten thousand feet (three thousand meters) above sea level, and have gotten within sniffing distance of Hawaii's Kilauea. The volcanic smell is unmistakable, heavy and suffocating, acrid and sulfurous, thanks to sulfur dioxide, the choking component of volcano breath, and hydrogen sulfide, sulfidic and equally toxic. Sulfur dioxide emissions from Kilauea are so common and irritating that the state department of health has a standard scale for reporting their intensity. Tiny fractions of a gram in a cubic yard or meter of air can be toxic, and Kilauea can release as much as ten tons of sulfur dioxide every hour. In September of 2014, people on the west coast of Norway noticed the smell of sulfur in the air, and it turned out to have come from the fresh eruption of Iceland's Bárðarbunga volcano, eight hundred miles (1,300 kilometers) to the west.

There are so many volcanoes and geothermal areas in Iceland that the word *geyser* comes from there, as does a less familiar word for what the Norwegians smelled: *hveralykt*, hot-spring smell, which is usually dominated by sulfidic hydrogen sulfide. Underground temperatures that generate hot springs are not as extreme as volcanic temperatures, and water favors the conversion of some sulfur dioxide into hydrogen sulfide. The planet's many hot springs flow with water that has come into contact with hot areas of the crust or even molten magma, and carries various gases and minerals from that contact.

There are a number of hot and sulfur springs in California. My favorite is the Bumpass Hell area near Mount Lassen. It memorializes the improbably named miner Kendall Vanhook Bumpass, who found it but lost a leg in 1865 when he stepped through a thin crust into the near-boiling mud underneath. You can smell it on the trail long before you get there, and hear what sounds like the hissing roar of a steam-powered engine or factory. From its edge you look down onto a barren, mostly treeless swale, bleached white by the constant rain of sulfuric acid that forms when sulfur dioxide from the noisy vent dissolves into the air's moisture, and replaces the variety of minerals in the exposed rocks with sulfates. Bumpass's companion, the editor of the *Red Bluff Independent* newspaper, reported that "all the wonders of Hell were suddenly before us." This is a California version of the ancient Avernus—not deadly to birds or the tourists who wander over the boardwalks, but sulfurous enough to give you the flavor of the caustic early Earth.

Whiffs of early life:
ocean-bottom hot springs

If the early Earth was so toxic, so inhospitable to chemical complexity and to life, then how could life get going in the first place? It's an endlessly fascinating question, with lots of theories and as yet no clear answer. But important clues have come from the last few decades of research into environmental microbiology, the study of what microbes inhabit which niches in the world. It turns out that while Avernian volcanoes and hot springs and mines may be inhospitable to *us*, some microbes thrive around and in them and echo their smells, evidence that their ancestors probably did eons ago.

In the northwest corner of Wyoming, Yellowstone National Park occupies the vast expanse of a supervolcano that last erupted 3,300 years ago. It's pockmarked with hot springs, many of them brightly and variously colored, green and orange and pink and red and blue. In the mid-1960s, an Indiana University microbiologist, Thomas D. Brock, discovered that some of them harbored living bacteria at temperatures approaching the boiling point—far hotter than had been thought possible for any living things. The Yellowstone springs have since become famous among biologists for their vivid displays of life in hellish conditions, the colors all the visual signatures of various microbes thriving at the simmer, sometimes tolerating smelly hydrogen sulfide, sometimes generating it themselves.

In 1974 the biologist Robert D. MacElroy gave these and similarly hardy microbes the general name *extremophile*, meaning "lover of extreme conditions," with *extreme* meaning at or beyond the limits that we and most familiar living things can tolerate. Scientists then went on the hunt for extremophiles and found microbes that can tolerate extreme temperatures, acidities and alkalinities, dryness, radiation, and pressures, from the vacuum of outer space to the crushing weight at the bottom of the Mariana Trench in the Pacific Ocean.

With the development of the deep-water submersible vehicle *Alvin*, which later explored the wreckage of the *Titanic* and the Nazi battleship *Bismarck*, in 1977 an expedition first reached and took samples at a hydrothermal vent—an ocean-bottom hot spring—a mile and a half below the surface of the eastern Pacific near the Galápagos Islands. As the explorer Robert Ballard reported months afterward, they were astonished to find that on an otherwise barren ocean bottom, the vent supported "a dense biological community," including clams a foot across. This oasis immediately led the scientists to ask a basic question:

> What were the organisms eating? They were living on solid rock in total darkness.
>
> An answer to this question began to emerge later when the water samples obtained from inside the vents by *Alvin* were opened for analysis. . . . As the chemists drew the first water sample, the smell of rotten eggs filled the lab. Portholes were quickly opened. The presence of hydrogen sulfide was the key.

That initial stink eventually led scientists to deduce that the sulfide-rich vent water fuels the growth of hardy, pressure-resistant microbes, and these initiate a robust food chain capable of supporting those impressive clams. Subsequent deep-sea expeditions have documented dozens of undersea vents, studied many of them in detail, and suggested the possibility that life got its start as single cells that evolved in similar physical and chemical extremes on the early Earth.

Aside from the sulfurous springs of Yellowstone and Bumpass Hell, what does all this have to do with the smells of our everyday world? The biochemical systems that early life developed to make it on the early Earth survive today, and not just in extremophiles. Microbes living all around us, even inside us, scent our world with a set of molecules defined in what may have been life's boiling beginnings.

The key to molecule building: energy

Wherever and however life arose, there's one fundamental system on which all the others would have depended: a system for providing energy. It's pretty obvious that it would take energy to build large, complex, orderly structures from small, simple, disorderly collections of building blocks. On the early Earth, the main source of carbon for making carbon chains was carbon dioxide, CO_2. To start a chain by joining two molecules of CO_2, any chain-making system has to break the bonds between each carbon and at least one of its two oxygens, and then cause a new bond to form between the two carbons. Chemical bonds consist of electrons that two atoms share with each other, so breaking and forming bonds means pulling electrons away from some atoms and pushing them onto others. All this work of pulling and pushing takes energy.

What energy source did the first life draw on? The same one that all life still draws on: the natural flow of electrons.

Recall that atoms of the different elements contain different numbers of the subatomic particles, protons and neutrons in the nucleus, and electrons orbiting around the nucleus. Only a few of each atom's electrons are available for sharing with other atoms. And the elements differ in how strong a pull their nuclei exert on those bonding electrons. Many metals are happy to give up their bonding electrons altogether. That's why copper and iron are good at conducting electricity and heat: their electrons are free to move wherever they're pushed or pulled. At the other extreme, oxygen not only hangs on tightly to its own bonding electrons, it pulls hard on the bonding electrons of other elements. Metals are natural electron donors, and oxygen is a natural electron acceptor.

That is the potentially powerful energy source that early life managed to tap and control: the natural tendency of electrons to move from electron-donor elements to electron-acceptor elements. It's the same source that we tap in the batteries that power our flashlights and cell phones.

Living things don't contain solid batteries that deliver constant electron flow, but they do organize the local environment of their molecules to encourage electron flow from donor to acceptor atoms, and they coordinate that electron flow with the forming of carbon-chain bonds. Like our modern power plants and their smokestack emissions, this energy generation within microscopic cells generates its own chemical by-products, the altered electron donors and acceptors. And some of these we can smell.

The energetic versatility of sulfur

To assemble their chemical power plants, early living cells had to work with readily available elements in their vicinity, organize them to encourage electron flow from donors to receptors, and control that flow. Iron is an abundant and generous electron donor and probably catalyzed much early carbon chemistry. Control of that chemistry requires the involvement of intermediary elements, so that the electrons don't simply make one quick jump and get locked in the acceptor molecule. And abundant sulfur made an excellent intermediary element. It readily accepts a couple of electrons from hydrogen or iron to form hydrogen or ferrous sulfides, but it can also give up as many as six to oxygen in sulfur dioxide and compounds called sulfates. So it can shift back and forth between being an acceptor and a donor. In the primordial waters it was probably present in various forms that included sulfur dioxide gas exhaled from volcanoes, the sulfuric acid that forms when that gas reacts with water, both solid and dissolved metal-sulfate and metal-sulfide salts, particles of solid elemental sulfur, and gaseous and dissolved hydrogen sulfide. Whenever one of these molecules is transformed into another, electrons flow, and in the right chemical environment, energy can be tapped.

Sulfur was just one of several elements to play an important role in life's early energy mining on the young Earth, but it was prominent enough to mark some of the oldest rocks thought to carry traces of biological metabolism. In the Pilbara region of northwestern Australia are former ocean sediments dated to 3.5 billion years ago, rich in iron and sulfates from volcanic activity, and containing grains of fool's gold, or pyrite, iron disulfide, FeS_2.

It's currently thought that the first living cells probably evolved chemical systems to extract energy from whatever donor-receptor-intermediary elements were available in their local environments. And they probably evolved together in consortia, or cooperative networks, in which particular cell types would consume some resources but also generate molecules that could be resources for other types. Networks of different cells could thus flourish in conditions where individual types would starve. One of the first to be identified was the *sulfuretum*, a network of bacteria in which some use sulfides as electron donors, thereby generating sulfur or sulfates, while others use sulfur or sulfates as electron acceptors, thereby *re*generating sulfides. The different groups effectively recharge each other's batteries while generating their own energy. Sometimes these microbial consortia are visible to the eye: back in Yellowstone National Park, Octopus Spring is home to layered mats, each layer exploiting different molecules, nourished and protected by the layers above and below. A mille-feuille of microbes!

So in addition to being a primordial product of interstellar space and our planet's geology, hydrogen sulfide is both a component and a by-product of extremophile life, and probably has been since life arose. But of course we don't have to go to a hot spring to catch that sulfidic smell. In the next great leap of evolution, life came up with a far more effective power plant, remade the planet, and redefined *extreme*. The extremophile legacy and hydrogen sulfide are very much with us in everyday experience. We'll come across them often, in nature and our foods and ourselves.

Life learns to thrive on sunlight and water

The first forms of life likely used iron and sulfur minerals to generate a flow of electrons like that in a low-voltage battery. These early cells probably developed slowly and patchily, because the energy available to them was modest and they depended for it on local supplies of the right minerals. Life overcame these limitations by adopting two resources that were abundant and almost ubiquitous on the Earth's surface, and that remain the ultimate sources of energy and substance for most living things on Earth: sunlight and water.

Light is pure electromagnetic energy, and the sun's light is a by-product of its element-creating nuclear fusion reactions, tremendous quantities of photons poured onto the Earth's surface. Today, the sign of life tapping this energy source is visible all around us in the original solar panels: plant leaves and grass blades green with the pigment molecule chlorophyll. Chlorophyll looks green because it absorbs the part of the spectrum that looks red, thereby capturing a portion of the sunlight's energy, which it transfers to electrons in some of its atoms. Primitive versions of chlorophyll came to be organized into *photosystems* directing these energized electrons into a flow that could power the cell's biochemical machinery. In the process called *photosynthesis*, they're funneled directly into a system for synthesizing long carbon chains from simple carbon dioxide in the air and water. Photosynthesis was a huge advance for unleashing the creativity of living cells. With it they could build carbon chains, grow, and multiply many times faster than most cells before them.

After sunlight, the second abundant energy resource that early life learned to tap was the oxygen locked up in water, H_2O. Oxygen is unmatched among the elements in its pull on electrons and power to move them, but early life had no direct access to it. At some point a line of microbes probably represented by modern-day cyanobacteria (*cyano-* being Greek for "blue-green," the color of their pigments) managed to rejigger a photosystem to pry it out of H_2O. This system used light-energized electrons to break the bonds holding hydrogen and oxygen atoms together, then passed electrons and hydrogen atoms to a partner photosystem to generate more energy and build new carbon-hydrogen chains. It released the remaining oxygen atoms into the air as a waste product, unneeded for the subsequent steps in photosynthesis. The free oxygen was then available for use in other cell systems as a powerful electron acceptor. Above all it enabled cells to extract the maximum energy that goes into the building of carbon chains, by breaking them back down all the way to the original starting materials, carbon dioxide and water.

So these turbocharged microbes could build copious carbon chains in nearly any surface waters on the planet, wherever they had access to sunlight and water and carbon dioxide, with only a sprinkling of minerals needed for their photosystems. They could churn out carbon chains all day, store some of them as fuel reserves, and then efficiently break them apart—"burn" them—for energy with free oxygen to stay active during the sunless night.

Some scientists have likened the microbial invention of oxygen-generating—*oxygenic*—photosynthesis to the human invention of agriculture, when our ancestors went from being hunter-gatherers, scrounging whatever limited fare the surroundings had to offer, to farmers who grew and stored their own food supply, and used that supply to fuel the development of populous cities and civilizations. Oxygenic photosynthesis was so advantageous that it became the dominant mode of life both in the waters and on land. Those pioneering cyanobacteria gave rise to all modern sea and land plants, from microscopic algae to massive redwoods. And they transformed the planet's surface, its smells included.

The great oxygenation clears the air

If you and the Chef of the cosmos were keeping tabs from afar as oxygenic photosynthesis evolved, you would have seen the young Earth's appearance change from a hazy reddish brown to the familiar sparkling blue and white. If you leaned in closer for an occasional sniff, you'd have smelled its acrid stinkiness fade. These would have been the superperceptible signs of what geochemists have variously dubbed the Great Oxygenation Event (emphasizing the injection of oxygen into the planet's processes) or the Great Oxidation Event (stressing the chemical effects of that injection). When those microscopic cyanobacteria burgeoned and bubbled out massive quantities of such an electron-hungry, reactive element into the air, they altered the planet from the top of its atmosphere to deep below the rocky crust.

The first breaths of oxygen released by the cyanobacteria were taken up by metals and other elements in the oceans and seabeds and on land, which formed telltale rust-red layers of iron oxides that have been dated to as long as three and a half billion years ago. These spontaneous oxidation reactions doubled the number of different mineral compounds on the Earth to more than four thousand, and thanks to them surface rocks and ocean sediments are now dominated by oxide, sulfate (sulfur-oxygen), and carbonate (carbon-oxygen) compounds. Once the oceans and land were fully oxidized, unreacted oxygen gas began to accumulate in the oceans, and when it had saturated them to capacity, it poured into the atmosphere and became a significant presence there for the first time in the planet's history.

Oxygen's arrival cleared the air. It erased Earth's reddish smog-like haze by oxidizing methane and various methane by-products generated by volcanoes, decaying microbes, and the sun's ultraviolet light. It deodorized the fumes of the inner Earth and early life by reacting with hydrogen sulfide and ammonia to form sulfur dioxide and nitrogen oxides, which in turn react with moisture to form odorless acids and fall to the surface as acid rain. It thus made the air into today's mostly neutral medium for our sense of smell, through which volatile molecules fly to reach our olfactory receptors and our perception. And by forming ozone molecules, O_3, which absorb ultraviolet light and so block much of the sun's damaging radiation from reaching the Earth's surface, oxygen in the air made the land habitable for living things to smell and be smelled.

Aerobes and anaerobes, the dark side of hydrogen sulfide

What geologists call the Great Oxygenation Event is for biologists the Oxygen Crisis or Oxygen Catastrophe, because it caused one of the first mass extinctions of living things. The accumulation of oxygen triggered the planet's first ice age, preemptively oxidized minerals that many microbes had used as battery materials, and attacked the molecular systems of all life forms that had been evolving without the need to deal with its reactivity. In fact, oxygen can be toxic even for the living things that now depend on it to keep the inner fires burning, which include all plants and animals and many microbes: a fact that might be called the Oxygen Irony. Modern life forms inherit much of their biochemical machinery from early cells that evolved before the coming of oxygen, and oxygen and its by-products can damage that machinery. Hence the importance of the *anti*oxidant molecules in our foods that help limit that damage.

Microbes are tough and versatile forms of life, and so despite rising levels of the oxygen that was toxic to them, the non-photosynthetic types persisted. As the pioneering French chemist Louis Pasteur found when he first proved that microbes are alive, some of them absolutely require oxygen to grow, while others can grow only when oxygen is absent. He named these two kinds of microbes *aerobes* and *anaerobes*, from Greek roots meaning "air" and "life." Anaerobes thrive in geological underworlds and biological innerworlds, niches where oxygen is used up by aerobic microbes and not replaced: notably ocean sediments, stagnant fresh-waters and swamps, the digestive innards of oxygen-breathing animals, us in-cluded, and digestive wastes excreted from within. Because anaerobes can't take advantage of oxygen to extract the maximum energy from minerals or the detritus of other cells, their power-generating systems give off not carbon dioxide and wa-ter, but such volatiles as hydrogen sulfide, ammonia, and short carbon-chain rem-nants like cheesy butyric acid: echoes of the primeval smells of the mineral planet and cosmos.

It's their resident anaerobic microbes that give muck and swamps and sewers that common primeval stink. We're especially sensitive to hydrogen sulfide, whose smell the *Alvin* scientists and many others routinely describe as that of "rotten eggs." It's true that this erstwhile nourishing and dominant molecule is now one of the volatiles produced when microbes grow in damaged animal and plant tissues and break down their complex molecules into simple ones, so in our oxygenated world it's often a sign of death and decay.

Mirroring the effects of oxygen on anaerobes, hydrogen sulfide is a powerful poison for aerobic forms of life like ours. That realization began among cesspool and sewer cleaners of eighteenth-century Paris, where its smell was associated with suffocation and the blackening of metal coins in their pockets. Hydrogen sulfide interferes with our systems for generating chemical energy by binding to their critical iron atoms, just as cyanide and carbon monoxide do. There's evidence that some post-oxygenation mass extinctions were exacerbated by its massive re-lease from deep oxygen-poor ocean waters, or from brief burgeonings of sulfide-producing microbes after volcanic eruptions. We can detect hydrogen sulfide at very low concentrations, around one molecule in a billion air molecules. At just 10 parts per million, it irritates the eyes; much above that level it damages and chokes.

Spas, spa-cooked eggs, black salt: sulfidic goodness

Hydrogen sulfide and its smell also have their benign sides, especially when we can manage our exposure to them. In moderation, hydrogen sulfide is part of the appeal of the world's hot springs, which have widely been considered good for the health. Just a few miles from deathly Lake Avernus and the fabled gates to Hades are spas that go back to Roman times, their mineral-rich, odorous waters sought out as curatives. One theory was that these strong sulfurous smells—controlled by the goddess Mefite, whose name gave us the term *mephitic*, "noxious smelling"— drove illness from the body. That's unlikely, but scientists have determined that hydrogen sulfide is in fact a common minor by-product of the metabolism of most living things. Small amounts of it stimulate the germination and growth of plant seedlings and slow the deterioration of ripe fruits in storage. Traces produced in the human body can relax blood vessel walls, an effect that among other things contributes to successful erections. Nothing rotten about that.

In fact, "rotten eggs" is a misleading cliché for what hydrogen sulfide smells like. It's actually the smell of freshly cooked eggs. In 2013, chemists at Sejong University in Seoul monitored the volatile molecules released from eggs when they were hard-cooked and then held at room temperature for several days until they spoiled. Egg white proteins are rich in sulfur-containing amino acids, some of which happen to give off hydrogen sulfide when heat denatures the proteins. The hotter and longer the eggs are cooked, the more hydrogen sulfide is formed, and the stronger the smell. The hydrogen sulfide levels are high just after cooking, then drop drastically as the eggy aroma fades. Though hydrogen sulfide makes a comeback as microbes begin to colonize and rot the eggs, the leading volatile at this stage is methanethiol, another simple sulfur molecule that we encountered in space, and that here on Earth is usually described as smelling like rotting cabbage. Hydrogen sulfide and methanethiol often show up together in our lives, but it's methanethiol that more reliably signals rot.

A fine emblem of hydrogen sulfide's several identities is the black egg. Japan has hundreds of sulfurous hot springs, *onsen*, and many onsen establishments offer tender onsen eggs, gently cooked in spring waters that are usually around 180°F (80°C). The *kuro tamago*, or "black egg," of Owakudani, the "Great Boiling Valley," comes from hot springs a dozen miles southeast of volcanic Mount Fuji, which are rich in both iron and hydrogen sulfide. The two react on the carbonate eggshell to form black ferrous sulfide, FeS, which the porous shell retains, providing a dramatic visual contrast to the white egg inside. And long cooking causes the egg's own iron and sulfur molecules to react and deposit a film of dark FeS on the surface of the yolk. Sulfides on sulfides on sulfides!

There's a less common kitchen ingredient that's prized exactly because it spices foods up with a hydrogen-sulfide aroma. And it does so thanks to a mineral that reflects the chemistries of both deep-sea vents and energy-hungry microbes. *Kala namak*, Hindi for "black salt," is an edible salt mined in the Himalayan regions of India and Pakistan. It's used to flavor a variety of foods, notably the Indian snacks called chaat. Black salt is mainly sodium chloride, ordinary table salt, but it includes a variety of other minerals that were deposited along with it when the seas that once covered that region evaporated. Chunks of black salt are actually garnet-colored from the presence of a particular iron sulfide (Fe_3S_4) called greigite, which forms in both geological and biological processes, in hot deep-sea vents and also when certain bacteria use sulfate as an electron receptor and generate sulfides. When you dissolve black salt in water, it gives off a powerful eggy smell and throws a cloud of black particles as the greigite reacts with the water to form hydrogen sulfide and simple ferrous sulfide.

To think of hydrogen sulfide as the smell of eggs only begins to touch on the place it has in life and life's history. It's an edgy presence in our daily lives. It's as primitive a molecule as there is, one of the first formed in the cosmos. It was a dominant part of the characteristic harshness of the early Earth, and still is in places where the inner Earth breaks onto our surface today. It's a food and by-product of molecular cooperatives that first found a way to develop in that harshness, the early forms of life. It may have played a role in mass extinctions of later forms. It betokens death, but also life at its most rugged. Altogether, it's so much more than the accidental product of feathered animals that have been around for a geological blink of the eye! Rather than "eggy," the smell of an interstellar molecule and the early Earth and hot springs and swamps and black salt and cooked egg deserves a less contingent description, one that encompasses them all. *Sulfidic* it mostly is, from this page on.

Still it's brilliantly fitting that, given the role of hydrogen sulfide in helping early life hatch from the rigid mineral surface of the young Earth, we smell it most often when we crack open the mineral shell of a freshly cooked egg, and find the bland raw ingredients of new life firmed up and flavored for our own nourishment and pleasure.

Chapter 3 · LIFE'S STARTER SET

....

Carbon will play the game of complexity on a grand scale. It will become the
greatest hero of chemical and biological evolution.

· Hubert Reeves, *Atoms of Silence*, 1981

From the Big Bang onward, the basic substance of our cosmos has been exploring its potential for taking on new forms, generating new relationships and
organizations, reaching new levels of complexity. Atoms are heroic, as astro-
physicist Hubert Reeves calls them, in the sense that they have repeatedly overcome
or worked around the forces of the actual—inertia, stasis, entropy—to probe the
realm of the possible. And carbon has been the element in the vanguard of this exploration. Among all the elements created in the stars, its atoms are especially gregar-
ious, good at playing with others, responsive to opportunity.

It was carbon's persistent play on cosmic dust and rocks and planets that led it
to form diverse chains of atoms, that led some of those chains to organize into
groups, the groups into systems that could build new chains, the systems into the
self-contained, self-multiplying entities that we call living cells. A more than heroic achievement, one that inspires baffled awe. And it was followed by another,
almost as unimaginable: these fragile collectives of carbon chains managed to re-
model the Earth itself into a habitat far more suitable for their exploration of
matter's possibilities. In the process, Hero Carbon cleared the air of its first dominant smells, the malodorous sulfidic drone that dominated the previous chapter,

and began to replace it with the diverse volatile carbon molecules and smells that we know today.

This chapter is a quick introduction to life's simplest volatile molecules and the surprising range of sensations they give us. It's chemistry, but chemistry you can smell. Feel free to study or skim or skip it, or circle back from later chapters. You don't need it to begin exploring the world's smells. But a little chemistry will help you get your bearings in the less familiar world *of* smells, the osmocosm, and get to know some of its important landmarks.

Life's starter set of carbon-chain volatiles

Throughout our warp-speed tour from stars to planet to microbes, I've used the term "carbon chain" as shorthand for the complex molecules from which living cells construct themselves. Now that early cells have cleared the air, and cleared the way for carbon to jump to new levels of the complexity game, it's time to pause and get to know some of these chains. We'll begin with the molecules that all life forms, primitive and advanced, can release into the air during basic metabolism, the common work of keeping their biochemical machinery running. Think of them as the starter set of life's smells. We usually encounter them in groups, and when mixed together they create a kind of general volatile presence, the olfactory hum of life. Sometimes one or two will dominate with their own qualities and make a more particular impression. We'll find them contributing to many of the smells we enjoy— and many that we avoid! All of them are signs of life at work, and not all life works to the benefit of our own.

Though the carbon chains of living things can include hundreds and thousands of atoms, most carbon-chain volatiles are relatively small and simple. The basic metabolism of living cells tends to generate chains between two and four carbons long, and the commonest include only atoms of carbon, hydrogen, and oxygen. Chains longer than twelve or so carbons are too heavy to launch themselves easily into the air, and they also tend to nestle alongside each other and form even heavier clusters. They're not very volatile. So our starter set of life's volatiles is a bunch of lightweight, small, simple carbon chains. They fall into four main chemical families—the acids, alcohols, aldehydes, and hydrocarbons—and there are definite resemblances among family members. Of course carbon being the player it is, it makes plenty of other variations on these chains, with bends and branches and other kinds of atoms, and we'll get to them in due course. But the starter set gives us a pretty impressive range of smells on its own.

Why does basic metabolism happen to produce these particular volatiles? Because they are common fragments of a cell's carbon-chain machinery. All cells are constantly breaking down complex carbon chains into the simpler building blocks from which they're constructed, both to generate energy and to repurpose their chains for repairs or making new molecules. There are three main building blocks: sugars, which make carbohydrates; amino acids, which make proteins; and lipids, which make fats, oils, and related molecules that assemble into the waterproof membrane that encloses cells. Two of the three building blocks, the sugars and amino acids, are not volatile themselves. They're strongly attracted to water molecules and to each other, so they don't easily escape from cells into the air where we can smell them. But when they're broken apart for their energy, some of their *fragments* are volatile. Most sugars and amino acids are no more than six carbons long, so their fragments are smaller, typically four or two carbons. Lipids are less attracted to water molecules, so both they and their fragments do tend to be volatile. And lipid chains can be much longer. Fatty acids, which are the lipid building blocks for fats and oils and cell membranes, are often fourteen to twenty-four carbon atoms long. So the breakdown of fatty acids can produce a spectrum of chain lengths, from two carbons up to the twelve or so that approach the limit of being volatile.

With this background briefing on the nature of flying carbon chains, we're ready to get specific. Here's the starter set of life's smells.

Fragrant alcohols, sharp acids

The easiest way to get to know the basic carbon-chain volatiles and their families is through their most familiar representatives and smells. So let's start with two such smells: alcohol and vinegar. We know them most clearly from unflavored vodka and distilled vinegar, which are fairly pure solutions of just one carbon chain each, uncomplicated by other volatiles. Unflavored vodka is about 40 percent alcohol and 60 percent water. It smells like . . . strong alcohol, of course. And alcohol is a two-carbon molecule. Now take distilled vinegar, which the label will identify as 5 percent acetic acid. Acetic acid smells like . . . vinegar, of course. Like alcohol, acetic acid is also a two-carbon molecule. But these two two-carbon molecules don't smell anything like each other!

The difference in smell comes from a difference in what each molecule has on one end of its two-carbon chain. On both molecules, one carbon shares bonds with three hydrogen atoms. The other carbon, though, shares bonds with oxygen atoms. The **alcohol** carbon shares one bond with an oxygen atom, while the **acetic acid** carbon shares one bond with one oxygen atom and two bonds with a second oxygen. Those two extra bonds with oxygen are what make the difference between the smell of alcohol and the smell of vinegar.

And they're what make the difference between the alcohol family and the acid family. The members of each family have varying numbers of carbon atoms in their chains, but they have in common a family-defining number of oxygen bonds. For example, rubbing alcohol, often labeled with its chemical name isopropyl alcohol, is a three-carbon chain with the one oxygen bond on its center carbon instead of the end. That one additional carbon atom in the chain and the center-carbon oxygen bond is a big enough difference to make it undrinkably toxic—but it still smells pretty similar to drinking alcohol. If a three-carbon chain has three oxygen bonds on its end carbon, then it's not acetic acid but propionic acid, one that smells sharp like vinegar, but different: like Swiss cheese, the kind with big holes.

Let's pause with these four examples—two alcohols and two acids—and note a couple of things. First, volatile molecules in a family share some qualities: the alcohols smell similar to each other, the acids smell similar to each other. Chemical families are also smell families, at least to some degree. And when we describe the smells of particular molecules, we do so by association with the things in our everyday life that we most readily recognize them from. We say that acetic acid smells like vinegar because we encounter that molecule in vinegar, and three-carbon propionic acid smells like Swiss cheese because that molecule is prominent in that specific cheese.

Because these volatile molecules are such common by-products of living cells, we often encounter them in more than one thing and so may describe their smell with more than one association. One striking example of such multiple associations is the four-carbon butyric acid that comes after three-carbon propionic. Like propionic acid, it's prominent in cheeses, but specifically cheeses that have been aged for a long time. It's also prominent in human vomit. So the four-carbon acid can remind us of two very different materials, one often delicious, the other always disgusting. Most multiple associations are not nearly as extreme as this!

Now let's meet some other, less familiar members of the alcohol and acid families. Here's a table listing a dozen in each family, along with the smells that flavor chemists attribute to the different carbon-chain lengths. Just take a minute to browse and get a sense for these two families. Because there are many different kinds of chemical acids that aren't volatile molecules—for example, hydrochloric and sulfuric acids—the carbon-chain acids are usually specified as the *fatty* acids. Those shorter than six carbons are the *short-chain* fatty acids, and those with six to twelve carbons are the *medium-chain* fatty acids. (Animal fats and cooking oils contain the *long-chain* acids, up to twenty carbons and beyond.)

SMELLS OF THE ALCOHOL AND ACID FAMILIES

Carbon atoms in chain	Alcohols (1 end oxygen bond)	Fatty acids (3 end oxygen bonds)
1	alcoholic	sharp, pungent, fruity
2	alcoholic	vinegar
3	alcoholic	sharp, Emmental cheese, vinegar
4	ethereal, winey, whiskey	cheesy, rancid, vomit
5	pungent, fermented, whiskey	sharp, cheesy, sweaty, rancid
6	green leaf, fruity, apple	cheesy, rancid
7	fresh, floral, lemon	waxy, cheesy, dirty, fruity
8	orange, mushroom, melon	fatty, rancid, cheesy
9	fresh, floral, orange	waxy, dirty, cheesy
10	waxy, floral, orange	rancid, sour, fatty
11	fresh, waxy, floral, soapy, clean laundry	waxy, creamy, fatty cheese, coconut
12	soapy, waxy, fatty, earthy	mild, fatty, coconut

Pretty amazing, isn't it? Such a range of different smells and associations! From unpleasant and dirty and harsh to the ethereal and refreshing and clean and delicious, flowery and citrusy and tropical. All from a small set of simple carbon chains. It's just a taste of the virtuosity of our Hero Carbon—and of our discriminating nose and brain—that it can evoke such a wide swath of life with single small molecules.

And do you see the family resemblances? The alcohols, especially the longer chains, tend to share a lifting, "ethereal" quality, which gets expressed in alcoholic drinks, flowers, and some fruits, especially citrus peels. The interesting oddball is the eight-carbon alcohol, which is characteristic of mushrooms, inhabitants of the soil—not something we normally think of as either lifting or ethereal. But if you smell mushrooms with those qualities in mind, you might notice the resemblance, and perhaps begin to think of mushrooms as flowers and fruits of the soil—they are indeed "fruiting bodies" that release the spores that will hatch into the next generation.

The acids, on the other hand, are mostly the opposite of ethereal. They get their family name from a root meaning "sharp, biting," and many of them have this quality. The acids longer than two carbons also share cheesy, sweaty animal smells, with rancidity a recurring theme as well. The acids evoke these very different materials because they're a component in all of them: so likewise cheese and spoiling meat and sweat can evoke one another through their shared acids. This helps explain why strong-smelling cheese is an acquired taste for most people. It may be a carefully crafted food, but its volatile acids can remind us of things that don't belong in our mouths. With three oxygen bonds, the acids are as fully oxidized as they can be without breaking the end carbon off the chain and forming carbon dioxide. They're thus of little energy value in themselves, and indicators that the valuable molecules have already been exploited, often by energy-strapped anaerobic microbes that don't break them all the way down to carbon dioxide.

Most of the volatile acids are essentially unpleasant, and the alcohols don't get particularly pleasant until they're at least six carbons long. This pattern may reflect the fact that the shorter chains are often the by-products of building-block breakdown by microbes, or by long exposure to oxygen in the air. Some of these small molecules are actually chemical weapons that microbes deploy to deter other microbes from competing for the same resources. This is the case for the alcohol produced by yeasts in wine making and brewing, for example, or the acetic acid that bacteria generate from alcohol when wine and beer spoil—or when we intentionally make them into vinegar. The medium-length carbon chains, by contrast, are more likely to be volatiles purposely generated by plants, animals, and microbes as signals: the smells of flowers and fruits that are meant to catch the attention of insects and animals, often to attract them.

Liquid-fuel hydrocarbons, wide-ranging aldehydes

Now we know something about the basic carbon chains tipped with one or three oxygen bonds. But those end carbons could also have two oxygen bonds, or none at all. Those two possibilities give us the other two major families of carbon-chain volatiles. The *hydrocarbons* consist of hydrogen and carbon only, no oxygen— hence their name. And the family with two oxygen bonds is the *aldehydes*, short for "dehydrogenated alcohols" (an alcohol minus one hydrogen atom). Here's a table showing the hydrocarbon and aldehyde chains. Again, take a minute to peruse the qualities and patterns, and see what a difference an oxygen bond can make.

SMELLS OF THE HYDROCARBON AND ALDEHYDE FAMILIES

C atoms in chain	Hydrocarbons (0 oxygen bonds)	Aldehydes (2 end oxygen bonds)
1	none	chemical, pungent
2	none	pungent, ethereal, fruity, green, fresh
3	none	ethereal, earthy, winey
4	lighter fluid, gasoline	pungent, cocoa, malty, musty
5	none	fermented, bready, fruity
6	gasoline	grassy, green apple
7	gasoline, stove fuel	fresh, fatty, green, herbal
8	gasoline	waxy, citrus peel, green
9	none	waxy, rose, citrus peel
10	none	sweet, waxy, citrus peel
11	none or gasoline	waxy, soapy, floral, citrus, fresh laundry
12	none or gasoline	soapy, waxy, citrus, green

The hydrocarbons are a family apart from the others: they have either no smell or a particular "chemical" smell, the smell of various flammable liquids that we use as fuels and solvents. The smell-free one-carbon version, methane, one of the important atmospheric molecules on the early Earth, is what we know as natural gas. The smell we associate with natural gas comes from stinky sulfur compounds that producers add to it to make it smellable and so less dangerous. The three-carbon version, propane, is a common stove and torch gas, and the four-carbon version, butane, is used in lighter fluids. **Octane**, eight carbons, is the reference hydrocarbon for gasoline (which is actually a mixture of several hydrocarbon chains of varying lengths). We think of hydrocarbons as having chemical smells because we encounter them in prominent amounts only in fuels and solvents. Nevertheless, traces of some hydrocarbons are present in many foods and in our own bodies, and contribute to life's olfactory hum.

The aldehydes are the most wide-ranging carbon-chain family in the diversity of things they evoke, including fresh green foliage and green fruits, and more sophisticated foodstuffs that are first fermented or germinated, then baked or roasted: bread and cocoa and barley malt. The shorter-chain aldehydes take us in these directions, while the longer chains like **octanal** consistently evoke waxes and unscented soaps. This may be because beeswax and petroleum-derived paraffin waxes, which are mainly long chains of dozens of carbon atoms, always contain traces of oxidized shorter fragments. So do soaps, which are made by breaking up fats into their component fatty acids.

Ants, butter, goats: carbon-chain namesakes

So far I've been introducing the simple carbon-chain volatile molecules by their structure—how many carbons long they are, how many oxygen bonds they have— and by the materials that they're most prominent in. When we get out into our world, beginning in the next chapter, and find some of these molecules, I'll want to refer to them specifically, by name. Unfortunately, one of the obstacles to feeling at home with molecules is their names. They can be hard to keep track of, because they're so numerous and unfamiliar, and because many individual molecules have more than one. But you don't have to worry about learning or remembering molecule names. The important ones will become familiar by repetition as they come up in later chapters—acetic will evoke vinegar smell, hexanol green grass and leaves. And I'll make a point of giving a molecule's primary smell along with its name. What really matters is that you feel at home with the existence of these molecules, their membership in families of similar carbon chains, and the fact that specific molecules stimulate specific sensations.

But molecule names are useful, and there is a kind of logic behind them. Or a couple of different logics. Many of the original names for these carbon-chain molecules were rooted in their smells, and that gives us a hook to remember them by. The later logic is based on the number of carbon atoms in a chain, an approach that can accommodate an infinite number of molecules and helps us distinguish among molecules that smell alike.

The original system for naming carbon-chain molecules dates from the eighteenth century, when the earliest experimental chemists, mainly in France, first recognized the existence of different kinds of acids in different natural materials. The chemists named the acids for the materials each was found in or most smelled like. Back then scholars were partial to Greek and Latin, so they took the names from these ancient languages. This is how the one-carbon acid came to be called *formic*, from a word for "ant"; the two-carbon acid *acetic*, from a word for "vinegar"; the four-carbon chain *butyric*, from a word for "butter"; and so on, as shown in the table. Today these roots just aren't obvious in English.

ORIGINAL NAMES OF THE CARBON-CHAIN ACIDS AND ALCOHOLS

C atoms in chain	Names
1	formic acid (*formica*, ant) methyl alcohol or methanol (*methys*, wine)
2	acetic acid (*acetum*, vinegar) ethyl alcohol or ethanol (*ether*, the upper air)
3	propionic acid (*pro-*, ahead; *pion*, fat)
4	butyric acid (*butyrum*, butter)
5	valeric acid (*Valeriana*, aromatic root)
6	caproic acid (*caper*, goat)
7	enanthic acid (*oenanthe*, wild grape)
8	caprylic acid (*caper*, goat)
9	pelargonic acid (*Pelargonium*, geranium)
10	capric acid (*caper*, goat)

There was nothing systematic about these names, and nothing in them to indicate what a molecule's actual chemical structure is. By the twentieth century, chemists had recognized the existence of the carbon-chain families, each with many possible members. So they developed a new, systematic nomenclature based on the number of carbon atoms in a given chain. They let stand the names for the shortest chains, which are also the commonest, and then applied Latin roots for numbers to chains with five carbons and more: *penta-* for five, *hexa-* for six, *octa-* for eight, *deca-* for ten, and so on. Because several other original names were already so ingrained, they're still used as synonyms. So six-carbon hexanoic acid is also called caproic acid, named for goats; eight-carbon octanoic acid is also called caprylic acid, again named for goats; ten-carbon decanoic acid is also called capric acid—goats again, thanks to their richly smelly fats! In this particular group of acids, it's actually easier to distinguish the *hexa-* and *octa-* and *deca-* prefixes than the slight variations on *capr-*.

One last table now to summarize the roster of simple carbon-chain volatiles, their names and families and associated smells. It's here for browsing and for reference, to help place molecules that happen to intrigue you in later pages when you find them contributing to your experience.

NAMES AND SMELLS OF THE STARTER-SET CARBON CHAINS

Number of C atoms in chain	Hydrocarbons $\cdots -\overset{\overset{\displaystyle H}{\mid}}{\underset{\underset{\displaystyle H}{\mid}}{C}}-H$	Alcohols $\cdots -\overset{\overset{\displaystyle H}{\mid}}{\underset{\underset{\displaystyle H}{\mid}}{C}}-OH$	Aldehydes $\cdots -\overset{\overset{\displaystyle H}{\mid}}{C}=O$	Fatty acids $\cdots -\overset{\overset{\displaystyle OH}{\mid}}{C}=O$
1	methane: odorless	methanol: alcoholic	formaldehyde: chemical, pungent	formic acid: sharp, pungent, fruity
2	ethane: odorless	ethanol: alcoholic	acetaldehyde: pungent, ethereal, fruity, green, fresh	acetic: vinegar

continued

Number of C atoms in chain	Hydrocarbons $\cdots -\overset{\displaystyle H}{\underset{\displaystyle H}{\overset{\mid}{\underset{\mid}{C}}}}-H$	Alcohols $\cdots -\overset{\displaystyle H}{\underset{\displaystyle H}{\overset{\mid}{\underset{\mid}{C}}}}-OH$	Aldehydes $\cdots -\overset{\displaystyle H}{\overset{\mid}{C}}=O$	Fatty acids $\cdots -\overset{\displaystyle OH}{\overset{\mid}{C}}=O$
3	propane: odorless	propanol: alcoholic	propanal: ethereal, earthy, winey	propionic, propanoic: sharp, Emmental cheese, vinegar
4	butane: gasoline	butanol: ethereal, winey, whiskey	butanal: pungent, cocoa, malty, musty	butyric, butanoic: cheesy, rancid, vomit
5	pentane: odorless	pentanol: pungent, fermented, whiskey	pentanal: fermented, bready, fruity	valeric, pentanoic: sharp, cheesy, sweaty, rancid
6	hexane: gasoline	hexanol: green leaf, fruity, apple	hexanal: grassy, green apple	hexanoic, caproic: cheesy, rancid
7	heptane: gasoline, stove fuel	heptanol: fresh, floral, lemon	heptanal: fresh, fatty, green, herbal	heptanoic: waxy, cheesy, dirty, fruity
8	octane: gasoline	octanol: orange, earthy, mushroom, melon	octanal: waxy, citrus peel, green	octanoic, caprylic: fatty, rancid, cheesy
9	nonane: odorless	nonanol: fresh, floral, orange	nonanal: waxy, rose, citrus peel	nonanoic, pelargonic: waxy, dirty, cheesy
10	decane: odorless	decanol: waxy, floral, orange	decanal: sweet, waxy, citrus peel	decanoic, capric: rancid, sour, fatty

Number of C atoms in chain	Hydrocarbons $\begin{array}{c} H \\ | \\ \cdots-C-H \\ | \\ H \end{array}$	Alcohols $\begin{array}{c} H \\ | \\ \cdots-C-OH \\ | \\ H \end{array}$	Aldehydes $\begin{array}{c} H \\ | \\ \cdots-C=O \end{array}$	Fatty acids $\begin{array}{c} OH \\ | \\ \cdots-C=O \end{array}$
11	undecane: odorless or gasoline	undecanol: fresh, waxy, floral, soapy, clean laundry	undecanal: waxy, soapy, floral, citrus, fresh laundry	undecanoic: waxy, creamy, fatty cheese, coconut
12	dodecane: odorless or gasoline	dodecanol: soapy, waxy, fatty, earthy	dodecanal: soapy, waxy, citrus, green	lauric, dodecanoic: mild, fatty, coconut

Beyond the starter set

At the beginning of this chapter we breezed through a billion or two years of Earth's history to marvel at Hero Carbon's invention of photosynthesis, its liberation of free oxygen, that oxygen's transformation of the planet's minerals, waters, and air, and the oxygen-fueled proliferation of countless life forms that filled the waters, overgrew the land, and give us so much to smell. There's another critical plot line in this saga, one that our acquaintance with the simple carbon-chain volatiles helps highlight: the line that leads to our ability to tell a five-carbon chain from a six-carbon chain, or an alcohol from an acid, and to associate those molecules with their sources. That too is a marvel!

Chemical perception probably began with early anaerobic microbes that were able to detect significant molecules around them, the first step in being able to move toward nutrients and away from toxins. Eventually most creatures developed systems for detecting significant molecules of all kinds, not just nutrients and toxins: molecules that signify the presence of food sources, or potential mates, or predators, or hospitable places to mature or reproduce. Once microbes and plants and animals had colonized the land, the air became a medium of exchange among them, just as the waters had been for their ancestors. Airborne molecules became significant parts of the chemical environment, a vital source of information about the materials and creatures that emitted them, and so our amphibian ancestors began to develop and bequeathed to us a sense of smell that is marvelously wide-ranging and discriminating.

Our tidy table of the short and medium-length carbon chains already covers quite a swath of smells, but carbon is so virtuosic that there are many other variations possible on even these simplest of volatile chains. And our sensory systems are specific enough that we can smell the differences. Take the ten-carbon aldehyde **decanal**, for example. The simple chain is found in waxes and citrus peels. But tweak two neighboring inner carbons on that chain of ten so that they share two bonds instead of one, and you get several possible chains that are kinked rather than straight. Some of these **decenals** are found in and smell like cilantro leaves. Make a second pair of double-bonded carbons on that chain and you get kinkier **deca*di*enals**, some suggestive of deep-fried foods and chicken meat. Not much like orange peel! And **deca*tri*enals** smell like citrus peels again—but green ones—as well as seaweedy and painty.

These double-bond variations are called *unsaturated* molecules, because the double carbon bonds lower the number of hydrogen atoms along the chain: it's no longer fully *saturated* with hydrogens. (The same terms are applied to the longer carbon chains in our foods; land animal fats are typically composed of straight saturated chains, vegetable and fish oils unsaturated chains.) We're able to register these structural details in volatiles and associate them with their very different sources: ripe and unripe fruits, green leaves, materials heated in high-temperature oil.

SOME TEN-CARBON ALDEHYDE VARIATIONS

Molecule	Smell
decanal	sweet, citrus peel, floral
decenal	fatty, cilantro
decadienal	deep-frying, cooked chicken
decatrienal	seaweedy, painty

Another example: there's a whole group of chains that are like aldehydes, except that the two oxygen bonds are on the next-to-end carbon instead of the end carbon. This is the family of ketones, not nearly as common in our world as the other families. But one family member is familiar: acetone, the two-carbon ketone. We encountered it in outer space, and it smells like a chemical solvent, because it is: it's available as such by the can. But it's also a common by-product of living metabolism. As we'll see, we can sometimes smell it on each other's breath—where it indicates that we haven't eaten for hours.

And a third example, malodorous but familiar: straight carbon chains can have branches, commonly single-carbon methyl groups attached to an inner chain carbon. Four-carbon butyric acid smells like vomit and old cheese, five-carbon *methyl*butyric acid smells sweaty (see page 59).

So while the basic carbon-chain volatiles evoke an amazing range of things in our world, natural and man-made, they are in fact just the start. We'll get acquainted with plenty more in the chapters to come. To keep them as approachable as possible, I'll omit the numbers and letters that chemists add to some names to be absolutely specific about their structures. There's only one version of each of the straight carbon chains, but less regular molecules can come in a number of different structures. At the risk of occasional imprecision, I'll stick with the base names: so (E,E)-(2,4)-decadienal will appear simply as decadienal. And I'll resort to space-saving shorthand in the tables.

Time now to get out into that world of our actual experience. How about the soil, the forest, the seashore? Herbs and spices? Animals that aren't goats? Coming right up.

Part 2

ANIMALS

*Dependence, Mobility,
Microbiomes*

Chapter 4 · ANIMAL BODIES

··

Why is it that no animal is pleasant to the smell . . . and when they decay they are
unpleasant to the smell, but many plants when they decay and wither become still
more pleasant to the smell?

> • *Problemata* ("Problems"), about 200 BCE

No one is ever rendered speechless amidst the aromas of foliage and flowers; the
distinct odors of flesh sometimes make us gasp, leaving us breathless in the duel of
mingled bodies. Sweat, shroud. Here is the frontier or catastrophe, the border which
opens up or closes off what we might call instinctive repugnance: deep, pungent,
dense, black aromas, underground, in graves.

We like vegetable detritus well enough; animal excrement repels us, but not
always, it can be heady; when it comes to game, we can appreciate the smell of meat
that is high. Yet we flee from the stench of death.

> • Michel Serres, *The Five Senses*

It's an ancient observation: plants and animals smell different, and plants smell
better, especially in death. Two thousand years ago, a Greek text long ascribed
to Aristotle asked why that should be. Michel Serres is the modern French
philosopher who provided the exhortation for chapter 1: both nose and mind
should be open to adventure as far as the stars. So why should we flee from what
lies at our feet? Serres's answer is straightforward. To one degree or another, the
smells of animals remind us of our own animal nature and existence as material
objects, and so they remind us of our inevitable end, our entropic dissipation into
the fragments of crude matter that we smell on our fellow creatures. From sweat
to shroud.

That makes sense, as does the more practical and life-serving fear that decaying animal flesh could well harbor deadly microbes or toxins. Repulsion is a prudent default response. But Serres also notes that the smells of animal decomposition can be intriguing. "High meat"—strong-smelling, slightly decayed—is exactly what rendered me speechless in 2005 and instigated this book. The eminent Harvard primatologist Richard Wrangham once told me of living in the African bush with local guides who, like him, could smell a rotting animal corpse a mile away, but unlike him were always eager to find it and carry it to camp and cook and eat it. As we'll see when we get to fermented foods in chapter 19, decomposing animal flesh may well have been an important resource in human evolution, and some cultures prize it today. Wrangham's guides were hungry for meat. Smell explorers hungry for experience and understanding can likewise learn to set aside instinctive repugnance, track down the unpleasant smells of animals dead and alive, and be rewarded.

Why begin our exploration of the smellable world with animal flesh, not foliage and flowers? For a couple of reasons. You and I are animals, so we're starting right at home. And the world of animal volatiles and smells is relatively limited, mainly the standard starter-set carbon hum of life's metabolic exhaust, a sulfidic aspect familiar from mineral earth and early life, and just a few additional distinctive molecules. We may think of animals as more advanced forms of life than plants: active, *animate*, as their name suggests, whereas plants seem inanimate, passive. But animals move because they're less autonomous than plants are. They can't feed and build themselves and spin out volatiles from just air and water and soil and sunlight, as plants astonishingly do. Their few distinctive smells are mostly by-products of their dependence on other forms of life, the need to pursue and hijack them. The exceptions to this rule are the molecules that some particular animals synthesize specifically to communicate with each other. We'll get to the bespoke smells of skunks and stinkbugs in the next chapter.

Animals are indeed some of carbon's most advanced moves in the cosmic game of complexity. But when it comes to smells, they're more of a throwback. No less intriguing for that: after all, in a couple of chapters, we'll be sniffing at ourselves! Here we'll start with the generic smells of our fellow mammals, including those of our pets, which can be at least occasionally pleasant.

Animal life, motor molecules, potent residues

Once Hero Carbon had come up with photosynthesis around three billion years ago, it took a while to parlay the growing stockpiles of energy and carbon chains into the first animals. The original single-cell microbes could reproduce simply and quickly, readily adapting to new conditions. Individual cells could also cluster into groups. About a billion years ago, some that did so discovered the possibilities of cooperation: that groups could be more efficient at gathering nourishment, protecting themselves, escaping from the crowd and colonizing new niches, and exploring new ways for matter to be.

The result of these early experiments in group living was the evolution of three major kingdoms of multicellular life forms, each with a largely distinctive lifestyle. Members of the plant kingdom generally get their energy from the sun and their carbon from carbon dioxide in the air. Members of the fungus kingdom, which includes mushrooms and molds and yeasts, generally grow in direct contact with other living things, either while they're still alive or once they've died, and extract energy and carbon from them. Members of the animal kingdom also feed on other living things or their remains, but they actively move from one place to another to find them and take them into their own bodies to be dismantled. The first microscopic animals appeared in the seas, and over the eons diversified into many different forms, including fish, insects, amphibians, reptiles, mammals, and birds, in approximately that order.

Because their life is characterized by movement—searching, grasping, ingesting—animal bodies are especially well endowed with the molecules that perform these tasks: namely proteins and the molecules that help make and service proteins. It's the abundant remnants of proteins and their sidekicks that make animal smells stand out from the volatile hum of general living metabolism.

Proteins are one of the three most abundant kinds of carbon chains in living things, and the only ones that routinely include atoms of nitrogen and sulfur. The two other main chain types, carbohydrates and lipids, serve primarily as energy storage and structural materials. Proteins are above all the molecules of change and action. The class of proteins called *enzymes* are found in all living things. They're molecular machines: carbon chains that assemble other carbon chains or break them down, that extract energy from some molecules and use it to modify others. The systems that create animal movement are built from specialized motor and cable proteins. Every time a heart beats or lungs pull in air, countless motor molecules in the cardiac or diaphragm muscles scurry along cables, breaking bonds and forming new ones with each infinitesimal step. Weight for weight, animal bodies contain double or more the mass of protein molecules that stationary plants do.

Animal bodies are also unusually rich in the molecules that specify how to build proteins and that deliver energy to them. DNA and RNA, which encode the blueprints for proteins and direct their construction, are large molecules built up from carbon-ring building blocks called *purines*, which also contain nitrogen atoms. So too is the specialized energy carrier molecule ATP (adenosine triphosphate), which does the job of delivering the energy stored in carbohydrates and fats directly to the protein molecules. Thanks to their protein systems and the constant need for upkeep and energy, animal tissues contain five to twenty times more purines than plants do, weight for weight.

Though proteins and the molecules that service them are especially abundant in their bodies, animals don't stockpile these molecules the way they do energy-storing fats. They usually deal with a surplus by scavenging some energy from them, partly breaking down the simple carbon-chain portions, and then excreting the nitrogen- and sulfur-containing leftovers. Microbes in and on animal bodies do much the same. These remnants of proteins and purines, especially those that contain sulfur or nitrogen atoms, are among the volatiles that we're most sensitive to, and usually repelled by. So the power-generating systems of animals become the source of powerful and disgusting smells.

Animal volatiles:
sulfidic, cheesy, sweaty, tarry

Protein breakdown begins with the dismantling of these very large molecules into their building blocks, the amino acids. There are about twenty different amino acids. All have a nitrogen-carrying amine group, NH_2 (hence the term *amino*), along with a short-chain fatty acid portion (hence amino *acid*) and some kind of "side chain." When animal cells or microbes break amino acids apart to extract some of their bond energy, they produce fragments of the short-chain acids and the side chains. The amine nitrogen is usually split off in the form of ammonia, NH_3.

About half of the amino acids have simple straight carbon side chains that include or produce members of life's volatile starter set, among them sharp, rancid acetic and butyric acids: not pleasant but not specifically animalic. Similarly, a couple of amino acids include a sulfur atom, and these can give rise to sulfidic hydrogen sulfide, rotten-cabbage methanethiol, and/or methyl sulfides, all smelly but fairly common in nature.

SOME SULFUROUS SMELLS OF PROTEIN BREAKDOWN

Smells	Molecules
sulfidic	hydrogen sulfide
rotten vegetable	methanethiol
cooked vegetable	methyl sulfides

More specific volatile signatures of protein breakdown come from several amino acids with side chains that either branch or form hexagonal rings. Branched side chains have an additional carbon atom—a methyl group—projecting from one of the inner main-chain carbons. These can be broken off to form branched-chain fatty acids, three or four carbons long plus the one-carbon branch. And the side-ring fragments can also end up with a branch. Their smell qualities reflect the usually protein-rich materials in which we encounter them most often.

Sour, rancid, cheesy methylpropionic acid smells much like plain propionic acid.

Cheesy, dirty, fruity, sweaty methylbutyric acids come in two versions, depending on which chain carbon has the branch; it's the one with the alias **isovaleric acid** that notoriously smells much like **sweaty feet**.

Antiseptic, "chemical" phenol is a six-carbon ring with an alcohol-like oxygen-hydrogen group at one corner. It's also found in smoke and petroleum, easily manufactured, and so most familiar from its use in disinfectant cleaners, packaged adhesive bandages, and marking pens.

Barnyardy, tarry, "chemical" cresols are phenol rings with an added methyl branch on one or another corner. They're prominent in animal manures, and also smoke and petroleum products, especially asphalt and tar.

**SOME CHEESY, SWEATY, BARNYARD SMELLS
FROM PROTEIN BREAKDOWN**

Smells	Molecules
cheesy	methylpropionoic acid (branched carbon chain)
sweaty feet	methylbutyric acids (branched carbon chains)
antiseptic, bandage, medicinal, tar	phenol (carbon ring)
tar, horse stable	cresols (carbon ring)

These various amino-acid fragments and by-products make for quite a bouquet, not especially pleasant, and not especially repellent either. But none of them includes a single atom of nitrogen, the defining *amino* element. Bring it into the mix and you add another dimension altogether, one conjured by some of the molecular names. *Putrescine. Cadaverine.* Here we go, deep into animality.

Animal volatiles: ammoniacal, urinous, fishy, putrid

It turns out that nitrogen is a tricky element for the animal body to manage. When it's not paired up with itself to form N_2, the very stable molecule that constitutes most of the air we breathe, individual nitrogen atoms are very electron-hungry and reactive. The ammonia, NH_3, produced when any amino acid is broken apart, is toxic to the workings of animal cells in anything but trace amounts, so cells combine it with carbon dioxide to form an unreactive molecule called *urea*, which can be safely stored and eventually exported. Purine molecules left over from DNA and RNA and ATP breakdown are converted into *uric acid*, also storable, also exported.

Both urea and uric acid are soluble in water, not volatile, and don't smell. But living cells can break them apart in turn for energy and building materials, and in the process generate nitrogenous by-products that are volatile and stinky. The simplest of these is **ammonia** itself, the primal molecule that's found in outer space and volcanic eruptions, with a familiar **acrid kitchen-cleanser smell**. We have that association because concentrated ammonia is good at dissolving grease and disinfecting. Then there's a group of related molecules called *amines*, whose name comes from *ammonia*, and refers to ammonia-like molecules in which one or more of the hydrogen atoms in NH$_3$ is replaced by carbon.

Urinous, fishy methylamine, ethylamine, dimethylamine, and trimethylamine are common amine variations found in and on and excreted from animals. The qualities we associate with them arise from the fact that in everyday life we get the biggest hit of amines from the breakdown of urea and uric acid in urine, and from saltwater fish. The fish happen to accumulate a nonvolatile amine, trimethylamine oxide, to counteract the high salt levels in their environment (see page 386). Once the fish die, bacteria scavenge the abundant oxide for energy, and in the process produce volatile trimethylamine.

Unpleasant as the amines are, they have strong competition for the most offensive molecules on the planet. While all of the amino acids have one amino nitrogen that can eventually produce ammonia and amines, a few of them pack more than one nitrogen atom on every molecule, and produce very different volatiles. Three of these were discovered and named in the 1870s and 1880s by one man, the German physician Ludwig Brieger, whose research specialty was the chemistry of decay and putrefaction, particularly the volatile and toxic substances that they produce. A remarkable legacy!

The amino acid tryptophan carries its additional nitrogen atom in a side chain made of two carbon rings fused together. Some microbes can free this ring portion to make a stand-alone molecule, one that had been identified before Brieger in the ancient and smelly fermentation of certain plants that produces the deep blue dye *indigo*.

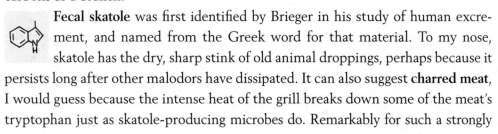

Mothball, "chemical" indole, named for indigo, is often described as having a fecal smell, but this is incorrect. That distinction belongs to a variation on indole that Brieger discovered, and that retains one of the chain carbons as a branch.

Fecal skatole was first identified by Brieger in his study of human excrement, and named from the Greek word for that material. To my nose, skatole has the dry, sharp stink of old animal droppings, perhaps because it persists long after other malodors have dissipated. It can also suggest **charred meat**, I would guess because the intense heat of the grill breaks down some of the meat's tryptophan just as skatole-producing microbes do. Remarkably for such a strongly "organic" smell, skatole has been found among the volatiles generated by purely chemical reactions in deep-sea hot springs.

When Dr. Brieger undertook a study of rotting meats, he discovered another couple of foul volatiles, one of them likely the most universally repulsive of them all. The amino acids lysine and arginine can be broken apart in ways that create straight carbon chains with a nitrogen atom at each end.

Mildly fishy putrescine, a four-carbon chain, turns out not to be particularly putrid. When exposed to the air, it's quickly oxidized to form a stronger-smelling ring compound, **pyrroline**, which is found in and smells of human **semen.** Pyrroline is also an important volatile in the smell of human skin and body residues after they've come into contact with dilute chlorine bleach (which oxidizes the amino acid proline to pyrroline), as well as the aroma of **boiled corn.**

Putrid, decaying-flesh cadaverine is the five-carbon chain appropriately named from the term for an animal corpse. It is indeed nauseating, strongly repellent to a range of animals.

SOME AMMONIA, FISHY, PUTRID SMELLS FROM PROTEIN AND PURINE BREAKDOWN

Smells	Molecules
ammonia	ammonia
old fish, urine	methylamine, ethylamine, trimethylamine
mothballs	indole (nitrogen-carbon ring)
dry feces, charred meat	skatole (nitrogen-carbon ring)
rotting flesh	cadaverine (nitrogen-carbon chain)
semen	putrescine (nitrogen-carbon chain), pyrroline (nitrogen-carbon ring)

While putrescine and cadaverine are prominent in decomposing animal tissues, traces of them are also found in the living animal body, in semen and in saliva, some produced as by-products of ordinary metabolism, some by microbial activity. We'll find them in and on ourselves. But if a single molecule can be credited with being the universal signal of animal death, it would be cadaverine.

Animals dismantled: the bouquet of death

Cadaverine may be its leading component, but the stench of death, the ultimate in animalic unpleasantness, is a composite of many volatiles generated by many agents. One simple factor in its repulsiveness is its intensity: turn the volume down to the barest hint and the same set of volatiles, in trace amounts, is the smell of the living animal body. When an animal dies, its own unmanaged enzymes begin to break down its tissues, microbes gain entry and feed and generate metabolic wastes, flies and beetles attracted by all these volatiles lay eggs that hatch into hungry maggots that produce their own wastes, and eventually solid flesh liquefies: a days-long process whose grossly visible progress you can see compressed to a minute or two in plenty of online videos. The stench encompasses the common starter set of general metabolism—alcoholic, sharp, rancid, cheesy, grassy, and fruity, along with all the animalic specialties from protein and purine breakdown, which dominate. The stench evolves, as forensic chemists have found by cataloging the volatiles emitted by animal cadavers over the course of hours, days, and weeks. This kind of grisly information can help investigators estimate how long crime and disaster victims have been dead.

There's good evidence that cadaverine and its fellow amines are the key ingredients in the stench of death, the special sauce that makes it as foul and repellent as it is to many animals (though not to carrion-eating insects or scavengers like vultures or sufficiently hungry humans). For one thing, evolution has given many animals, from fish to humans, a special set of receptors, distinct from the usual smell receptors, that have been dubbed the trace amine–associated receptors, or TAARs. They detect a variety of amines, including trimethylamine, cadaverine, and putrescine. In lab experiments, zebra fish swim away when a dead fish is placed in their tank, and do the same if the dead fish is replaced by a source of pure cadaverine. If a lab rat dies, its cage mates will ignore it for a few days before trying to bury it in the cage litter. But they'll bury it right away—and will do the same for a piece of wood—when either is sprinkled with a little putrescine and cadaverine. The TAAR that responds to cadaverine may be the primary death detector among vertebrate animals.

Encounters with the stench of death are relatively rare in the modern industrial world, so much so that it provides news headlines when it leads to the discovery of an unsuspected corpse. Raw meats are the flesh of dead animals, but they're handled in ways that minimize decay and generally have only a faint smell until they're cooked, a very dilute cocktail of volatiles mainly from fat breakdown. The big exception to this rule is fish, whose bodies decay much faster than those of land animals as a result of their life in the water, as we'll see in chapter 15. Trimethylamine is one of our major olfactory landmarks, familiar not just from fish and docks and beaches, but from our own bodies (see page 103).

It's easy for the smell explorer to get a good idea of death's stench. Put a scrap of raw meat into an empty jar, screw the lid on, and let it sit, unrefrigerated. After a few days, open the lid just a crack and take a cautious whiff. It won't smell quite the same as a whole cadaver, whose decomposition involves a much larger cast of creatures and makes for a much richer stew. But it should be plenty repulsive.

Anaerobic animal innards

Enough about the smells of death! However, the most prominent and familiar smells of active animal life are only slightly less unpleasant. Just conjure the smells of pet litter, zoos, stables, feedlots, hog farms. All of these flow from one of the defining features of animal life: its dependence on other life forms for nourishment.

The basic strategy of ingesting other life forms has given animals a very particular body topology. Animals have an inside, a gut, a chamber where their food is held while it's disassembled into carbon-chain building blocks. When the fertilized egg of a multicellular animal begins to develop into an embryo, one of the first things to happen is that the mass of cells folds to form an interior space, the future digestive system. That interior space soon becomes home to a multitude of other cells: single-cell microbes, which are everywhere and adaptable. Their numbers inside the gut come to dwarf the numbers of cells in the animal's own body.

It now seems likely that all multicellular life on Earth has had its outrider and inrider *microbiomes*—shifting communities of single-cell creatures that share resources with their host creature. Because they multiply and adapt and evolve much more rapidly, microbes help complex creatures cope with changing conditions, and they perform useful functions that their hosts might never develop. Creatures from termites to cows to humans depend on large gut communities to digest various food molecules, to make vitamins, and to protect against other mi-

crobes that cause disease. Some biologists suggest that the important unit in animal (and perhaps plant) evolution is not a particular species, but that species along
with its consortia of microbial partners, whose genes make critical contributions
to the fitness of the team.

The presence of an inner chamber of microbes has an important consequence for
animal metabolism and animal smells. Like all living things, animal cells build up
and break down complex carbon-chain molecules, and their basic aerobic metabolism mainly generates carbon dioxide and water. Along the way they also produce
traces of the various carbon-chain volatiles that we met at the end of the previous
chapter, the simple hydrocarbons and alcohols and aldehydes and ketones and acids
that contribute to the olfactory hum of life, its smokestack emissions.

But while animals themselves run their metabolism on oxygen, their interiors
are isolated from the aerobic world. They're oxygen-deprived, anaerobic, as the
Earth was before photosynthesis. And stuffed as animal innards are with the remains of other creatures ready for the breaking down, and with protein-rich secretions and sloughed-off cells from the animal's own gut lining, they're a paradise for
anaerobic microbes. As those microbes generate energy for themselves, they emit
typical by-products of anaerobic metabolism, molecules that we've encountered
before in space and on the early Earth—sharp and cheesy and rancid acids, sulfidic
hydrogen sulfide and methanethiol—along with the especially noxious products of
protein breakdown.

The animal body can absorb some of these molecules and put them to good
use, but some remain trapped in the mix of undigestible food materials and the
mass of trillions of microbial cells. To avoid having the residues of digestion and
metabolism and microbial life accumulate inside and interfere with its ongoing
function and movement, the animal body has systems for getting rid of them. The
jettisoned solid waste, animal excrement, is an intense source of anaerobic volatiles, and largely defines our experience of living animals and their smells.

Dispatches from within: excrements

When I was growing up in suburban Chicago in the 1950s, after Sunday afternoon
dinners that often centered on rare roast beef, my parents would take all four children on car rides into the countryside. We were grossed out by the smells of the

dairy farms and incredulous that, as my father was happy to insist, our delicious dinner could have come from those animals. We had a mantra for the moment we got the first whiff: "Red meat from a cow? Pee-YOO!"

I later learned that we were echoing an ancient exclamation at something rotten and stinking, the Indo-European root *pu*, from which stem "putrid" and "putrefy." Animal excrement is generally disgusting to us. But this is apparently a reaction that we learn, not an automatic biological reflex. Young children aren't repelled by excrement, and many mammals practice coprophagy, or excrement eating, some of our primate relatives included. In her 1983 book *Gorillas in the Mist*, Dian Fossey noted that gorillas of all ages have been observed eating dung, their own and others', fresh from the source: the animals "catch the dung lobe in one hand before it contacts the earth. They then bite into the lobe and while chewing smack their lips with apparent relish." Rabbits and some other plant-eating mammals get the vitamin B_{12} they need by routinely eating food twice, the second time after its residues have been enriched by their gut microbes. Studies of rabbits and mice have found that the presence of their excreted pellets in the cage tends to lessen aggressive behavior, lower heart rates, and offer "positive, comforting" effects, perhaps because they suggest familiarity and therefore safety.

Few primates of the human tribe ever smack their lips at the smells of animal excrement, and many find them overwhelmingly disgusting at close range in an enclosed space, as attested by a 2014 Associated Press news story, dateline Philadelphia: "A cross-country flight had to make an unscheduled landing when a service dog pooped twice in the aisle, sickening passengers with the odor." The scientists who actually choose to study animal excrement and its volatiles do so largely to figure out how to reduce the offensiveness of feedlots and hog farms. But there are also contexts in which these smells are less offensive, appealing in their own way, maybe even comforting when we can associate them with the open-air countryside and stables and small farms.

The smell of excrement from an animal's digestive system comes largely from the anaerobic microbes thriving in its oxygen-scant lower reaches. The microbes feast on food residues that haven't been digested and absorbed by the animal body, and also residues from the animal body itself, mainly the cells lining the digestive tract that are constantly being sloughed off and replaced, and the protein-rich mucus that lubricates it.

The physical mass of excrement can be up to half microbial cells, and the density of microbes in the lower digestive tract of animals is among the highest of any known niche on the planet. So there's a lot going on in there, and all that activity generates volatiles. Sulfurous hydrogen sulfide and methanethiol are almost always prominent because there's usually plenty of oxidized sulfur in the animal gut—all green plants carry sulfur-containing lipids, and intestinal mucus contains sulfated carbohydrates—and anaerobes use it as an electron donor to generate energy. Two protein by-products are especially emblematic of excrement: barnyardy cresol and fecal skatole. The generic excremental mix runs a close second in breath-stopping power to the smell of animal death.

Some animal excrements have a distinctive makeup and smell that can be traced to a particular diet or metabolism. **Horse excrement** is less offensive than many, and was even described as "sweet" by the eighteenth-century natural philosopher George Cheyne. The horse and its microbes digest its plant foods quickly and only partly, so much of its excrement is smell-less fiber. The volatiles are dominated by the carbon rings cresol and phenol, which we also encounter in asphalt and disinfectants, and which can therefore seem less specifically fecal. By contrast, cattle are endowed with several stomachs, including the microbe-packed rumen, and they have the habit of regurgitating the rumen contents for another chew to get the most out of their plant feed. The **excrement of beef and dairy cattle** is therefore rich in the full range of metabolic volatiles. Omnivorous **pigs** get some of their nourishment from high-protein animal materials, and they produce excrement especially rich in branched acids, sulfides, and carbon rings. For some reason the pig gut and its microbiome are notably prolific of fecal-smelling skatole, some of which is transported from the intestine and stored in fat tissues all over the body, where it can contribute to the special "pigginess" of pork.

Of course, there's a second major animal excretion. **Urine** is a fluid that animals excrete separately from the semisolid remains of digestion, though they often end up mixed with each other on the ground. It carries primarily the waste products of the animal's own metabolism, and in particular the nonvolatile nitrogen-containing urea and uric acid, along with small quantities of amines. Urine is practically microbe-free until it leaves the body, but once it does, microbes feed on its urea and uric acid and boost the emissions of ammonia and amines: hence the "urinous" character we ascribe to these volatiles.

Bird excrement has a distinctive and especially pungent smell because it combines the smells of digestive excrement and nitrogen disposal. Probably to help conserve water, most birds put their excess nitrogen exclusively into uric acid, which is much less soluble than urea and so can be excreted as a semisolid paste along with wastes from the digestive tract. Bird excrement is essentially dung and urine rolled into one. It's usually dominated by ammonia and amines, along with vinegary and cheesy acids.

SMELLS OF ANIMAL EXCREMENTS

Excrement of	Component smells	Molecules	Sources
most animals	cheese, vomit	butyric, propionic acids	microbes metabolizing various chains
	horse stable	cresol	
	sweaty feet	methylbutyric, methylpropionic acids	microbes metabolizing proteins
	sulfidic	hydrogen sulfide	
	rotten vegetables	methanethiol	
	ammonia	ammonia	microbes metabolizing proteins & urea/uric acid
pigs: added smells	dried excrement	skatole, indole	microbes metabolizing proteins, uric acid
	old fish	trimethylamine	
horses: main smell	horse stable	cresol, phenol, indole	animal metabolism; microbes metabolizing proteins
birds (guano): main smell	ammonia, old fish	ammonia, amines	microbes metabolizing uric acid

From manure to guano to CAFO

Immediately annoying as they are, the smells of animal excrement are also tokens of an otherwise largely insensible but existential crisis for much of life on Earth. They're a reminder of the inescapable reshuffling of matter and energy that keeps the great game of complexity going. The stench of the modern feedlot signals the fateful move by which Hero Carbon has managed to vault to new levels of invention, but at the price of devastating much of its achievement to date. That move was the arrangement of carbon chains into *Homo sapiens*, animals capable of mobilizing matter and energy on an unprecedented scale, and thereby damaging intricate ecosystems across the planet.

We have various terms for excrement, polite and impolite, one of them being "waste," from a root meaning "empty" or "desolate." In fact the strong smell of animal excrements signals its richness in diverse carbon and sulfur and nitrogen molecules, and so its value as nourishment for other living things. When early farmers discovered that value thousands of years ago, they made possible the long-term success of agriculture, and with it the development of civilization.

Our Stone Age ancestors must have been very familiar with the smells of excrements. Hunters would have used those smells as other carnivores do, to locate their prey. Dogs joined human communities something like thirty thousand years ago. Around ten thousand years ago, the first settled agricultural communities enlisted goats, sheep, cattle, and horses to take advantage of their milk and meat, hair and hides, and their plow-pulling muscle power. Archaeological remains indicate that their dung was used as a building material and a fuel for fire, uses that live on today in less industrialized parts of the world, along with fumigation to eliminate insect infestations and even recreational cow "chip" throwing. Excremental smells must have permeated the life of the early farmers.

At some point early farmers also observed that when excrement happened to be trodden into the soil, it improved the later growth and productivity of food crops—as we now know, because it replaces nutrients taken from the soil by the crops, and has beneficial effects on the soil's physical structure and biological diversity. Archaeologists have found that the organized application of excrement to crop fields goes back at least seven thousand years in Greece and central Europe.

Because this practice improves crop productivity slowly, and so represents a long-term investment of labor and resources in the soil, it may well have helped inspire the earliest ideas of land management, ownership, and property. The word *manure* itself has its origins in nothing at all to do with excrement: the roots are Latin words for "hand" and "work," and the compound of the two originally meant "to cultivate" or "to hold property."

So the smell of **manure**—excrement prepared for the soil and worked into it—is an ancient sign of the most fundamental forms of care and cultivation, labor and value, the practice of feeding the soil so that it will continue to feed us. It's milder, tamer, less aggressive than fresh excrement, generally made by mixing excrement with stable straw and feed residues and stockpiling it for some time to make it less concentrated and to eliminate disease microbes. The addition of low-nitrogen material, exposure to the air, and the metabolism of aerobic microbes all combine to reduce the levels of ammonia and amines and sulfides—though high-nitrogen **chicken manure** is noticeably more redolent of ammonia than cow and steer manure. These are the smells that we can still encounter in farm country, in plant nurseries and hardware stores, in our own backyards. Not exactly pleasant, but positive.

Highly unpleasant and negative are the raw, uncomposted, intense smells that emanate from **concentrated animal feeding operations**, or **CAFOs**, which confine and raise large numbers of animals—hundreds, thousands, hundreds of thousands—in a small area, and have come to dominate modern meat and dairy production over the last few decades. They accumulate huge quantities of excrement that can be smelled from afar. I live in central California and pass by the Harris cattle ranch on Interstate 5 near Coalinga whenever I drive between San Francisco and Los Angeles. Even with the car windows closed, I can smell it long before I see it. Tens of thousands of beef cattle are confined there, each animal generating some 65 pounds (30 kilograms) of urine and excrement a day. Today's formulated feeds usually supply more nitrogen than the animals would obtain from their natural diet of plants, so their excrement is especially rich in the most offensive volatiles, the branched acids, cresol, skatole, ammonia, and amines.

The obnoxious smell of industrial-scale animal production is a sign of the real dangers that concentrated excrement poses to workers and animals alike. High indoor levels of hydrogen sulfide and ammonia can be acutely toxic to eyes and lungs.

Skatole is also found in cigarette smoke and is known to damage lung DNA. Methane is combustible—it's the main component of natural gas—and has caused explosions in CAFOs. Some workers have been fatally overcome by excremental fumes. Because CAFO operators frequently dispose of excrement as cheaply as possible, dumping it into open storage lagoons or spraying it directly onto fields, even their neighbors can suffer health effects from the volatiles, and nearby soils and waterways can become badly polluted.

It's exactly because CAFOs are offensive and harmful that the volatiles of animal excrement have been so well studied. Crazily but appropriately, chemists borrow the terminology of top, middle, and base notes from the perfume world (see page 477) to describe the smells of CAFOs. The top notes, very volatile and quickly dispersed, are ammonia and hydrogen sulfide. The more persistent middle notes include amines, thiols and sulfides, aldehydes and alcohols and ketones. The constantly present base notes are the short-chain straight and branched acids, cresol and other phenolics, and skatole. In a 2006 study of swine and beef cattle operations, barnyard cresol was identified as the primary offensive odor, and could be detected as far as ten miles (sixteen kilometers) downwind. It's probably the first long-distance hint I get of that I-5 Eau de Coalinga.

So the smells of CAFOs are smells of modern industrial agriculture, different from manure in both quality and significance. They're still organic, manifestations of the basic workings of living things, but they're the smells of a rupture in the system that in nature and traditional agriculture returned matter and energy from the soil to the soil. They're the smells of organic matter and energy isolated and withheld from the broad cycle of life on Earth.

Why don't CAFOs compost their excrement into manure? Because it doesn't pay. Chemists began to unravel which components of manure are essential plant nutrients in the nineteenth century; and then early in the twentieth, the Germans Fritz Haber and Carl Bosch figured out how to manufacture salts of ammonia in factories directly from nitrogen gas in the air. The key: using tremendous amounts of chemical energy stored in the remains of ancient plants—coal or petroleum or natural gas—to generate very high temperatures and pressures. So began the era of concentrated chemical fertilizers, which contributed to steep increases in agricultural productivity, and in turn to steep increases in the human population. Across the world today, the use of synthetic fertilizers outweighs manure by something like five to one, and they're so cheap that farmers apply them profligately: about half of their combined nutrients are lost from the fields they're applied to, and end up in the atmosphere or waters that drain from the fields.

The invention and triumph of synthetic fertilizers have been a very mixed blessing. They helped propel the rapid development of civilization and new technologies for manipulating matter, both means by which Hero Carbon has forged more complex forms and organizations at an unprecedented rate. But they also lie at the root of profound damage to the complex biological world that made human life possible in the first place. Overworking agricultural soils, eliminating more and more wild habitat to feed and house our billions, treating animals inhumanely, polluting soil and air and waters, altering the energy balance between Earth and sun and so causing global climate change: the stench of CAFOs signifies all this. It's a smell to be open to and pondered by anyone who wants to know how our world works, and how it doesn't.

There's one particularly odorous form of animal excrement that links the manure millennia with the synthetics centuries. *Guano* is a term now popularly applied to bird and bat excrement, but the word comes from the Quechua of ancient Peru, where it named the extraordinarily nitrogen- and phosphate-rich deposits of excrement from fish-eating seabirds—cormorants, pelicans, and boobies—on islands off the Peruvian coast. They're as concentrated as they are due to the birds' protein-rich animal diet and the dry climate, which dehydrates the excrement while minimizing the conversion of uric acid to ammonia and amines that otherwise evaporate. By roosting on the islands over the course of centuries, the birds built up deposits of guano as much as a hundred feet (thirty meters) deep, which were mined by Quechua peoples for some 1,500 years. An 1841 report for the English Royal Agricultural Society noted, "From the guano island near Arica such a stench proceeds that vessels are prevented by it from anchoring near the town."

Trade in guano boomed in the nineteenth century. Historians suggest that it helped stimulate the general adoption of soil nutrient concentrates in European and American agriculture and the decline in productive use of farm animal excrements, which became the polluting wastes that they now often are. Guano is still mined in a number of places in the world. I bought a bag once to smell the ammonia and fishy amines that may have begun as muscle molecules propelling silvery fish through the oceans years, maybe centuries, ago. Then I returned their matter and energy to the game of complexity playing out in my vegetable garden, and in me.

Pet smells: the outer animal

To conclude our tour of generic animal smells, a source that finally includes some pleasant ones: our dogs and cats! Not their cages or litter boxes but their surfaces, which we often press our noses to with pleasure. They're our companions and make us happy, and their smells are comfortingly close to ours. We're all mammals, with similar body structures and systems that emit the same volatile hum of warm-blooded aerobic metabolism, and that house similar surface microbiomes. Scientists haven't yet studied pet smells in much detail. But we can extrapolate broadly from what they have discovered about animals in general and the human animal in particular.

The body of a mammal is enclosed by its skin, the outer boundary and interface with the rest of the world. The skin is made up of several layers. The outermost layer, the epidermis, consists mainly of a tough protein, keratin, and oil-like lipids that help make it water-resistant. Embedded in the skin are keratin hairs that help retain body heat and shield the skin from water. And the skin and hair are kept flexible and moist by secretions from several different kinds of glands. Some produce waxy, oily molecules that lubricate the skin and hairs and help them retain moisture. Some secrete mainly cooling water along with traces of salts and sugars; in dogs, these sweat glands are localized on the nose and feet. And some glands secrete a richer mix of amino acids, urea, and lipid and other molecules; these are found mainly on a dog's head and on the base of the tail and the anal region.

So the skin and hair of mammals are coated with a film of bodily fluids. That film itself has a distinct but mild smell. It comes from traces of molecules from the body's basic metabolism that are small enough to be volatile, and from carbon-chain fragments created when the air's oxygen and sun's ultraviolet radiation break long-chain molecules apart. Judging from what's known about our own skin volatiles, most of these metabolic products and fragments are medium-chain aldehydes and ketones with pleasantly waxy, sweet, floral, solvifty qualities. These are what probably make up the bouquet of a freshly washed, unfragranced pet.

This initial bouquet quickly evolves into something less delicate as microbes grow and make their own contribution. The skin of our pets offers a number of different ecological niches to microbes, and the more moist and protected they are, the more populated and more odorous they're likely to become. The mouth is a microbial paradise, with a constant supply of moisture and food residues, so pets can develop bacterial bad breath just as we do, with the volatiles dominated by protein breakdown products—sulfides, branched acids, putrescine and cadaverine. Outside the mouth, ear canals and skin flakes and hair follicles offer nooks for both aerobic and anaerobic microbes, which feast on the nourishing sweat, the protein-rich saliva deposited when animals groom themselves, stray excremental residues from their anus and anal sacs (short-chain acids, trimethylamine), and whatever the animal may have been rolling in. As the microbes thrive, they generate the same kinds of volatile carbon-chain fragments that fill the animal interior and its excretions. And that's the direction in which the smells of pet skin and hair develop, especially on the head and feet: toward the stronger, less pleasant products of the surface microbiome. Dogs in particular have been found to accumulate branched methylbutyric acid, with its cheesy, sweaty qualities.

Pets are especially smelly when they get wet from the rain or a swim. Not only is moisture especially conducive to microbial growth and to general chemical reactions, but water molecules actually increase the volatility of volatile carbon chains by displacing them from the hair's protein fibers and pushing them into the air. So the smell of **wet dog** is the amplified smell of its myriad surface hangers-on, usually dominated by sulfides, methylbutyric acid, and phenol-cresol rings. It's familiar enough to have become a landmark smell. The term "wet dog" signifies the unpleasant mix of volatiles from uncontrolled microbial activity in moist materials and processes of all kinds, from carpets and drywall to laundry and dishwashers to wine and beer.

And then there's the smell of **dog paws**, which like our own feet offer microbes plenty of moisture and food and protective skin folds, and which therefore tend to generate strong smells. But those smells can be surprisingly pleasant. There are pages of internet postings devoted to them. The most common description of their quality is an odd and specific one: they're said to resemble corn chips or tortilla chips.

Not surprisingly, I've been unable to find any peer-reviewed studies of dog-paw volatiles! But this crowd consensus points to the likely presence of a volatile called **aminoacetophenone**, a six-carbon ring with amine and acetate decorations. Amino-acetophenone makes a major aromatic contribution not only to corn chips, but also to grapes, strawberries, linden blossom tea, and chestnut honey. It's also diagnostic for the presence of a specific bacterium, *Pseudomonas aeruginosa*, which can cause serious infections in both dogs and humans. However, this microbe tends to produce the volatile when it's in relatively innocuous survival mode rather than aggressively infectious. Species of *Pseudomonas* are common inhabitants of soils, waters, and even the air: they're thought to contribute to ice crystal formation in clouds. So while the corn-chip smell may be pleasant and amusing, it's also a reminder of the invisible multitudes clinging to those paws, some of them biding their time for a scratch or cut to infect. If the smell is strong, it's probably time to think about a bath.

SOME SMELLS OF DOGS

Component smells	Molecules	Sources
grassy, waxy, sweet, fruity, floral	hexanal, heptanal, octanal, nonanal, decanal, benzaldehyde, damascenone, nonanone	oxygen, UV breakdown of sebum
"wet dog," sweaty, cheesy, fecal, medicinal, earthy	methylbutyric acids, dimethyl trisulfide, cresol, phenol	microbes metabolizing skin & hair proteins, lipids
tortilla chip	aminoacetophenone?	*Pseudomonas* bacteria metabolizing skin proteins?

Pet smells offer some respite from the usual run of animal body volatiles and what they make manifest, the harsh material realities of life living on other life. Here at chapter's end you might wonder: Why spend so many pages on such unpleasantness? Why suggest how the smell explorer can get a whiff of death?

Some time ago I saw Roberto Calasso's intriguing explication of a passage in an ancient Sanskrit text, the Satapatha Brahmana, which says that the smell of death in cattle came originally from the *soma*, the gods' drink of immortality, so "one must not close his nose at that foul smell, since it is the smell of king Soma." Here the stench of death is somehow the smell of a state that transcends earthly existence, and must be acknowledged rather than shut out. A less paradoxical modern formulation might be: the animal volatiles that trigger revulsion in us are the actual, material manifestations of a fundamental aspect of animal life: its end. They're powerful evidence of what makes the living world possible and drives it, the relentless hunt for energy and matter that feeds animal life, and eventually consumes it. Though we instinctively hold our breath to shut out repugnant smells, we can also choose to open up, let them in, and engage with their meaning.

Enough about the animal body's common life and fate. There's a far less grim, far more variegated dimension of animal smells to explore. Many animals actively construct special volatile molecules to stand out from their generic body smells. Some of these signature smells are outstandingly repellent, while others find their way into foods and fine perfumes! The volatile signals of cats and skunks and beavers and bugs: next chapter.

Chapter 5 · ANIMAL SIGNALS

....

Last month I made a trip to the highlands north of our [Brazilian] province, São
Bento, in the headwaters of the Rio Negro. It produced quite a nice haul, but almost
exclusively of lepidopterological interest. Often there was the [swallowtail butterfly]
Papilio grayi, . . . whose male can be called "flower of the air" for its smell. The scent
that it trails is so strong and spicy that I carried the butterfly in my hand like a
flower, to smell it now and then.

> · Fritz Müller, "Blumen der Luft" (Flowers of the Air), 1878

S ome animals are as delightfully scented as flowers! The German biologist Fritz
Müller lived in southern Brazil for most of his life, corresponded with Charles
Darwin and other eminent colleagues of the day, and made lasting contributions to our understanding of insect behavior. This image of a butterfly as a flying
flower, *Blume der Luft,* had come originally from the visual resemblance, and Müller's
added sensory dimension highlights an aspect of animal smells beyond the residues
of molecular motors and innards. Creatures on the move need to be able to find and
be found by others of their species. They can do so—as well as warn off or repel
predators—by purposefully emitting volatiles as signals, just as flowers do. And those
signals need not have anything to do with motors or innards. They can be as acrid as
tear gas, but also flowerlike, fruitlike, spicy. In fact, animals may have helped nudge
plants into making floral and fruity and spicy volatiles in the first place.

Land-dwelling animals have long-distance senses to keep track of what's going
on around them as they move through the world. Smell and sight and hearing help
animals find food and mates and avoid becoming food themselves. Smell is powerful
enough on its own that many animals rely more on it than on sight. It can detect the

presence of food and friends and enemies when vision can't, at night or when they're hidden nearby or at some distance. Smell plays the role of sensory historian and futurist when it guides one animal from where another was some time ago, to where it will be.

Animal excretions are a kind of involuntary signal, and as such are a ready-made medium through which animals can communicate with each other. We'll start with them, and the surprising sophistication of cat urine, then move on to the signaling materials in such specialized bodily *secretions* as skunk spray and deer musk, goat milk and sheep lanolin. We'll conclude with the most diverse and influential of all animal scent makers—the flying flowers and toe-biters and their fellow insects.

Signaling excretions: garden repellents, hunting lures

When an animal voids excrement and urine from inside its body, these strong-smelling materials become dead giveaways of its presence. By the same token, excremental volatiles are ready-made means by which animals can advertise their existence and proximity, to get the attention of possible mates or encourage potential competitors to move on. They can carry meanings and messages. Many animals therefore practice *scent-marking*: intentionally laying down volatile materials to mark territories, to assert dominance, to broadcast their readiness to mate. We see this in action all the time when dogs lift their legs to urinate on pre-sniffed tree trunks and lampposts. In the wild, some animals leave and renew their excrement in conspicuous places known as *latrines*. Some take the trouble to get their scent marks well up from the ground, where they'll be more noticeable at a distance; red foxes and pandas and African bushdogs urinate or defecate while doing handstands! Among the more prolific excrementalists are otters, whose deposits can number dozens per day and have their own name: *spraints*.

Urine is a better medium for sending messages than solid excrement, which is generated mainly by food remains and microbes in the digestive system. Urine volatiles come directly from the animal's own body and its complex metabolism, unaltered by microbes until it hits the ground. So urine volatiles provide a relatively direct report of that metabolism and can carry information about an animal's sex, readiness for mating, age, health, and diet. Scientists have found that rodent urine can affect the behavior of cagemates in a variety of ways: it can alarm them, increase indicators of stress, help them recognize kin, speed up or slow down their sexual maturation, and prompt sexual activity. Something like a score of volatiles have been identified as likely *pheromones*—signals that evoke particular behaviors in members of the same species—and in a variety of different chemical classes, amines and sulfur compounds and steroid hormones among them. Chemical communication via urine has been studied in animals from mice to tigers and elephants, some of which—including the bull elephant and billy goat—go so far as to "anoint" themselves with their own urine when they're ready to copulate.

Gardeners often have to deal with destructive attentions of animals like deer, moles, raccoons, and gophers or other rodents. The most humane way to deal with them is to persuade them to go elsewhere, and animal repellents are formulated to do just that by way of their volatiles. A frequent ingredient is "putrefied egg," which offers the smells of animal death and decay, presumably appealing only to scavengers. For some repellents the formulation has been left up to the bodies of carnivorous predators: on the web you can find and buy the strongly urinous-fishy urines of the coyote, wolf, bobcat, mountain lion, and fox. It's likely that the active ingredient in all of them is a particular combination of a carbon ring and an amine, phenethylamine, fishy like most amines, which a 2011 study found to be substantially more abundant in the urine of carnivores compared to other animals, and which on its own induces rats and mice to avoid a corner of their cage dosed with it. A more unusual molecule in fox urine, a ring with both nitrogen and sulfur atoms called a thiazoline, has also been identified as a repellent for rodents. This pattern makes sense: carnivores get abundant protein in their diet and probably have higher nitrogen levels in their urine. And amine volatiles that activate the TAAR receptors may instinctively put vulnerable animals on alert, just as death-associated cadaverine does (see page 63).

Hunters want to attract animals, not repel them. They too can go online and find such useful excretions as "Doe in Heat" or "Doe Estrous," the urine of female deer in their sexually receptive phase, when they produce a different mix of volatiles than do males or unreceptive females. One volatile found in receptive deer urine, ethylphenol, is a close relative of the carbon ring cresol, which sexually responsive mares excrete in large amounts in their urine, and which induces erections in stallions. Not surprisingly, both cresol and ethylphenol are described as smelling like a horse stable or barnyard.

Landmark cat pee

Perhaps the world's most familiar animal signals emanate from the urine of its hundreds of millions of pet and feral cats. The very particular disagreeable smell of cat pee is significant beyond the feline: it's also a landmark for wine and food specialists, who invoke it to describe the aroma notes of Sauvignon Blanc wine and blackcurrant fruits. Cat pee is surprisingly well studied, largely because its distinctive volatiles aren't just the incidental by-products of protein metabolism. They're purpose-made for signaling. And they are indeed also found in wines and foods.

In a 2014 research paper, a group of chemists at the Swiss fragrance and flavor company Firmenich summarized earlier analyses of cat urine volatiles, added new details of their own, and noted where else we encounter them. Mature male felines are especially fond of spraying urine to mark their territories, and theirs has the strongest smell; neutered males and females spray less extensively and obnoxiously. Fresh cat urine is relatively inoffensive, but its smell gets stronger and stronger with time—a valuable feature in a territorial marker, since volatile materials usually get weaker as their molecules escape into the air. It turns out that cats have come up with a clever time-release system that constantly generates new volatile molecules over the course of several days.

To prime this system, the cat's body, probably its liver, uses protein building blocks to synthesize several complex but nonvolatile molecules, precursors to the volatiles. It then sends them into the blood, from which they're absorbed by the kidneys and excreted in the urine. In the kidney and urine a special enzyme trims the complex precursors into simpler molecules. The first-discovered and best-known of the simple precursors has been dubbed *felinine*. When the cat sprays and the urine is exposed to the air and ground, bacteria colonize the urine, multiply, and gradually break felinine and the other precursor molecules into small fragments—some of which are the volatiles that give the urine its persistently strong smell.

And what are the volatiles that make cat urine so strong? Almost all of them contain a sulfur atom, and many of them a sulfur-hydrogen thiol group, making them relatives of the simple and powerful methanethiol that forms in outer space and smells like rotting vegetables. When the Firmenich chemists isolated these volatiles and smelled them individually, **methyl sulfanyl butanol** and another were urinous and catty. But others suggested less obnoxious elemental sulfur, cooked meat and onions, cabbage, asparagus, tropical fruit, and citrus! And when they checked the chemical literature to see where else these sulfur volatiles have been identified, they found that all of them appear in foods of some kind—even the key urinous catty volatiles, which contribute to the aromas of mangoes, wines, hops, and other delicious things. At least one of them, the **"cat ketone,"** is synthesized for use as a flavor additive in the food industry.

SOME SMELLS OF CAT URINE

Component smells	Molecules	Also found in
urine, catty	methyl sulfanyl butanol (from felinine)	coffee, wine
meaty, fried onion, skunky	methyl thioethanol	wine
catty, fruity	methyl sulfanyl pentanone ("cat ketone")	blackcurrant, mango, citrus, Sauvignon Blanc wine, hops
sulfurous, cabbage, tropical fruit	methyl thiobutanone	cheese, meats
blackcurrant	methylbutyl sulfanyl formate	beer, coffee

Obnoxious though it may be, cat pee deserves the smell explorer's attention. It marks the ingenuity of animal metabolism in inventing a signal that gets released only when an excretion is released and exploited by microbes, a system shared by many animals, from mice to elephants. And it hints at the more nuanced roles and smells of sulfur volatiles built by living things, as well as the commonalities in life's substance and strategies. Cats make the cat ketone to get the attention of other cats; fruits make it and other aromas to get the attention of animals that can carry their seeds away. And as we'll see in the next chapter, the human animal has its own version of the cat-pee system, much as we try to obscure it.

Signaling secretions: skunk spray, musk, beaver paste

Excrement and urine are two media by which animals can send signals to kin or potential mates or competitors. Many, perhaps most, animals have come up with a third medium: special materials that they prepare and release for the primary purpose of communication. In mammals these *secretions* are waxy, pasty materials produced in specialized skin glands and released separately from excretions. Mammalian scent glands are commonly found near the anus, where they can add a layer of information to excrement as it exits, but they can also mark an object—or another animal—when one animal rubs its anal region against it. They help mark territories, which may be many square miles (bears) or a single tree (opossums), as well as frequently used paths or nest areas in nonexclusive "home territories" (members of the deer family). Many animals have several sets of scent glands, which are often found near the genitals, tail, feet, or face.

Most mammals, including our pet **dogs**, aren't especially creative with their glandular scents. They simply fill the glands with proteins and fatty materials for anaerobic microbes to feed on, and the microbes generate the usual animalic short-chain and branched-chain acids and amines and sulfidic volatiles. But there are a few fascinating exceptions to this rule.

The most offensive of these, and the one mammal whose scent glands many people have direct experience of, is the skunk, which has adapted them not so much for communication with fellow skunks as for defensive chemical warfare.

Nearly all animals have to worry about becoming the prey of other animals, and so they have evolved many different forms of defense. Some are physical barriers that protect vulnerable tissues—shells or thick skin. Some are piercing projections—claws and teeth and horns. Some are venoms. And some are volatile chemical repellents—offensive smells. Some animals borrow volatile chemicals from other creatures and actively anoint their own bodies with them. Squirrels and chipmunks chew on shed snakeskins and snake and toad remains and then lick themselves all over to disguise their own smells.

Skunks make their own repellent. They're a New World subfamily of the weasel family, and like their cousin weasels, polecats, and civets, they're nocturnal, nonsocial animals that are unlikely to see predators from a distance or be warned by others. Weasel family members all spray a defensive mixture from two anal glands. In skunks these are the size of ping-pong balls and can hit targets ten feet (three meters) away. On contact with the eyes or skin, the spray is intensely irritating, a natural version of tear gas. The smell alone can cause nausea when strong, and humans can detect it in tiny traces (around ten parts per billion).

SOME SMELLS OF ANIMAL SCENT GLANDS

Animal	Component smells	Molecules	Sources
dog, coyote	rancid, sweaty, cheesy	acetic, propionic, butyric, methylbutyric acids	microbial breakdown of animal protein, oil
skunk	sulfurous, acrid	butenethiol & methyl butanethiol & their acetate esters	animal synthesis
musk deer	musky, animal	muscone	animal synthesis
civet	musky, animal	civetone	animal synthesis
beaver	barnyard, leather, smoke	ethylphenol, cresol, guaiacol	animal synthesis from tree bark

Skunks secrete a dozen or so volatiles in their spray, the main ones being **butenethiol** and other thiols, sulfur-containing molecules based on four-carbon chains, and combinations of these with acetic acid, the acetate esters. These molecules smell sharper and more irritating than their primeval relative methanethiol, with its rotting-cabbage smell (see page 14). It's thought that the esters are responsible for the lingering skunkiness that returns when a victim's hair gets wet. They aren't as volatile as the thiols alone, nor are they very soluble in water, so they tend to cling to the skin and hairs. Then when the victim gets wet, the water breaks off the acetate group to release the thiol, and the stink returns.

Thiols are readily oxidized to much less smelly molecules, so the standard treatment for a skunked pet is to wash it in a mixture of hydrogen peroxide and baking soda, with some detergent to help spread it through the hair. On pavements and buildings, a simple remedy is dilute laundry bleach containing the good oxidizer hypochlorite.

Polecat opposites of the offensively defensive skunk are three mammals whose scent glands have been treasured for a thousand years, used in perfumery and even in cooking! Two of these have a similar and distinctive, pleasantly animalic quality described as *musky*. The term derives from the ancient Sanskrit word for "testicle," and it refers to the pouch-like scent gland of the **musk deer**, a native of Central Asia. The particular smell of true musk comes from an unusually large fifteen-carbon ring molecule that has been dubbed *muscone*. An unrelated group of animals in the weasel family, the catlike **civets** of southern Asia and Africa, secrete *civetone*, a similar-smelling seventeen-carbon ring, in their anal-gland paste. And then there are **beavers**, in the genus *Castor*, the dam-building rodents of northern Europe and North America, which eat tree bark. They transform its astringent phenolic molecules into ingredients of a secretion, **castoreum**, that's intriguingly leathery and smoky from phenols, cresols, and versions of the related ring compound guaiacol.

Because obtaining musk and civet and castoreum means killing, confining, or trapping wild animals, these days it's difficult for smell explorers to find and experience their true smells. Nevertheless they continue to be important influences in perfumery, so we'll take a deeper whiff of them in chapter 17.

Goat cheese, lamb meat, wool

Life in the modern urban world limits our exposure to the volatile plumes that rise from animal life, its metabolism and inner anaerobics and excretions and secretions. But we don't have to keep pets or make a field trip to the zoo or farm to get a good whiff of animality. The flavors of goat cheese and lamb meat, and the smell of wool, all come from secretions that carry chemical signals.

Goats and sheep are especially smelly animals. For the ancient Greeks and Romans, goatiness was a major olfactory benchmark, the essence of beastliness that human body odor approaches when at its worst. It was intriguing to the eighteenth-century pioneers of biochemistry, the French chemist Michel Chevreul and his colleagues, who chemically dissected goat body and milk fats into their building blocks and named four of the strong-smelling short-chain acids from the Latin for "goat," *caper* or *hircus* (see page 46). Sheep have a similar but milder smell that was invoked more recently, in March 2013, by the newly inaugurated Pope Francis in a special Mass to bless fragrant ceremonial oils. Francis exhorted his priests not to be distracted by duties that distance them from the down-to-earth lives of their congregation and prevent them from walking closely with the flock: "Be shepherds with the smell of sheep."

Goats and sheep are closely related animals, members of the genera *Capra* and *Ovis*, respectively, and related to another animal named for its strong smell, the Arctic musk ox (*Ovibos moschatus*). Goats and sheep had a common ancestor in Eurasia ten or twenty million years ago but adapted to two different ecological niches. The goat specialized in mountainous terrain with scrubby, sparse vegetation. Individuals stake out territories, defend them from one another, and evade predators by climbing. Sheep took to the grassy foothills, where they developed the habit of grazing together and protecting themselves by clustering into a dense flock.

The territorial male goat in particular is infamous for its body stench. Some of the intensity comes from the billy goat's habit of self-enuration, anointing itself with its own urine. But its special goatiness comes from the copious production of volatile secretions from sebaceous (oil-producing) skin glands around the base of its horns and elsewhere on the head. Rams—male sheep—produce similar secretions on their wool.

The distinctive notes in the smells of both goats and sheep are branched fatty acids, similar to the cheesy, sweaty short-chain products of protein breakdown that we've come across in animal smells. But instead of being the usual short chains three or four carbons long, the goat branched acids are eight- and nine-carbon chains, and the branches themselves can be either one carbon (methyl) or two (ethyl). This means that they're not simple by-products of the general breakdown of amino acids, but are custom-built by joining small molecules together into larger ones. They're there for a reason.

The molecules that smell the strongest to us are **ethyloctanoic acid** (aka goat acid), methyloctanoic acid (aka hircinoic acid), and methylnonanoic acid, whose qualities are similar to one another and described as goaty, generally animal-like, and sweaty. Male animals produce more of them than females, adults more than the young, and sexually intact males more than castrated males. A number appear to be interesting to the animals themselves. The aldehyde form of ethyloctanoic acid, ethyl octanal, has been identified as a goat pheromone that stimulates ovulation in females. On its own, this molecule doesn't smell sweaty to us at all: it's described as floral, waxy, and milky!

But once secreted onto hair and wool and exposed to the air and microbes, it's slowly oxidized to the goat acid, which no longer induces ovulation but still attracts females to the male's vicinity.

The **milk and meats of goats and sheep** are also distinctive, and thanks to some of the same molecules, especially the eight-carbon branched chains. Goat's milk is several times richer in them than sheep's milk, but goat body fat contains less of them than sheep fat. This is why goat milk and cheese tend to have a stronger animal smell than their sheep counterparts, but goat meat is milder than lamb and mutton.

Lamb meat has another dimension of animality that goat appears to lack. It contains significant amounts of two protein breakdown products: horse-stable cresol and fecal skatole. Both of these notes are heightened when lambs and sheep are raised on green pasture, which is relatively high in protein and low in carbohydrate, and encourages gut microbes to break down the protein for energy. Some of the cresol and skatole then end up stored in body fat. Meat chemists call this manure accent "pastoral" flavor, an aroma overlaid on the already distinctive branched-chain goaty, sweaty flavor. Their prominent animalic quality is one reason that lamb and mutton are far less popular than the more generically meaty beef (see page 504). Young lambs raised on low-protein hay and high-carbohydrate grain generally produce the mildest meat; mature sheep on fresh grass, the strongest.

Then there's **wool**, the special version of hair grown by sheep and by the goats that produce cashmere and mohair. Like the hair of other mammals, wool is protected by oily, water-repellent secretions from the skin's sebaceous glands. In sheep these secretions are called *lanolin*, which can account for as much as a quarter of the weight of newly shorn wool. Lanolin is a mixture of thousands of different molecules, mainly long-chain acids and alcohols combined to form waxy esters. Most of these molecules are too big to be volatile and odorous themselves. But they are good at holding and slowly releasing smaller, volatile molecules that give raw wool a strong animal smell until it's "scoured," or washed to remove as much dirt and lanolin as possible. The smell of raw wool is dominated by the special goat-sheep eight-carbon branched chains, and various sharp, rancid, sweaty volatile acids, the products of microbial metabolism in the nutrient-rich nooks and crannies of the animals' skin and wool. The wool fibers themselves are proteins that contain sulfur, and the combination of moisture and ultraviolet sunlight can also generate sulfidic volatiles from them. Scouring often leaves some lanolin and wool volatiles behind, and when woolens get wet and the fibers absorb water, they release volatiles that had been trapped in them, amplifying the smell.

SOME SMELLS OF GOATS AND SHEEP

Goat & sheep . . .	Component smells	Molecules	Sources
bodies	goaty, sweaty	ethyl- & methyl-octanoic acids, methylnonanoic acid	animal sebaceous glands
milks, lamb & mutton meats	goaty, sweaty	ethyl- & methyl-octanoic acids	animal metabolism
	stable, dung	cresol, skatole	rumen microbes metabolizing fresh pasture
wool	vinegary, cheesy, rancid	acetic, butyric, hexanoic acids	oxygen, light, microbes metabolizing wool lanolin
	sweaty, goaty	methylbutyric acid, ethyl- & methyl-octanoic acids	animal sebaceous glands; microbes metabolizing fiber protein
	sulfidic, rotten, ocean air	hydrogen sulfide, methanethiol, dimethyl sulfide & disulfide	moisture, oxygen, light, microbes acting on fiber protein

The smells of wet wool and lamb meat and goat cheese are pretty distinctive, but their custom-built branched carbon chains do show up elsewhere. We'll encounter them again in the next chapter—on ourselves.

Butterflies and bugs

When it comes to scent-gland inventiveness, the champions by far are the insects. They may only be an occasional annoyance in urban life, but their small size and unobtrusiveness belie their eminent position in the animal kingdom. They've been on the planet twice as long as the mammals have, number in the millions of species to the mammals' six thousand, and to date have been found to secrete on the order of a thousand different chemicals as pheromones—attractants, alarms, trail markers—or as chemical defenses to repel predators. In fact, the first pheromone to be chemically defined was bombykol, a sixteen-carbon alcohol made by the female silkworm moth. Insect volatiles include the various starter-set chains and their extensions (acids and aldehydes and alcohols and ketones), animal specialty rings like barnyardy cresol and fecal skatole, and cat-pee thiols, but also the petroleum component naphthalene, and many other molecules that we'll meet properly when we enter the plant kingdom: fruity esters, flowery kinked-chain terpenoids and benzenoid rings, earthy and green-bean pyrazines. These crossover volatiles explain why insect smells can remind us of herbs and flowers, and they're evidence of the ancient co-dependence of plants and insects (see page 140).

I grew up sixty years ago in a suburban subdivision of Chicagoland where there were still remnant patches of native prairie to play in. There my friends and I encountered all kinds of insects and their smells: solvent-like or sharp **ants** crushed between the fingers, the green-pea smell of harassed **ladybugs**, the barnyardy smell of what we called "tobacco juice" from **grasshoppers**, the stink of **stinkbugs**, the sweaty, bruised-apple smell of **swallowtail butterfly caterpillars**—repellent defenses for these vulnerable precursors of flying flowers! City dwellers have always had a good chance of smelling musty **flour beetles** in the pantry, ants and sharp **cockroaches** under the sink, and leafy-musty **bedbugs** between the sheets. The Greeks are said to have named the herb and spice coriander for the bedbug, *koris*, with which it shares a couple of volatile molecules. The honey and honeycomb wax of bees contain signal volatiles evident to the insects, but what we smell are complex mixtures dominated by flower and pollen volatiles (see pages 525, 471).

SOME VOLATILE SECRETIONS OF INSECTS

Insect	Component smells	Molecules
ant (many species); odorous ant (*Tapinoma*)	sharp, lemony, nutty; green citrus, musty coconut	formic acid, citronellol, citral, methylpyrazines; methyl heptenone, heptanone
ladybug	green peas	isobutyl methoxypyrazine
grasshopper	disinfectant, asphalt, barnyard	phenol, cresol
stinkbug	green leaf, vinegar, fruity	hexenal, acetic acid, decenal, hexyl acetate
swallowtail caterpillar; swallowtail butterfly	cheesy, sweaty; waxy, floral, citrus	methylpropionic & methylbutyric acids; nonanal, decanal, linalool
flour beetle	musty, mushroomy	octenone, decanal, octenal, methylbutanal
cockroach	sharp, green leaf, irritating	hexenal, ethyl acrolein
bedbug	green leaf, floral, waxy, musty	hexenal, octenal, nonanal, decanal, methyl heptenone
vinegaroon	vinegar, rancid	acetic, octanoic acids
green lacewing	fecal	skatole

Among insects actually prized for their smells, like the lemony ants enjoyed in much of South and Central America, perhaps the most remarkable is the **giant water bug**, *Lethocerus indicus*, an Asian relative of the North American "toe-biter." These insects are up to three inches (eight centimeters) long, big enough to eat frogs, and make a substantial mouthful when boiled or fried. The males are endowed with large scent glands that they use to mark their path and find their way back to the egg clusters they're responsible for tending. These bugs are especially prized in Vietnam and Thailand, where the glands are removed to make an aromatic essence that's added to sauces and soups. The gland volatiles include a couple of animalic acids and cresol, but also prominent fruity esters and two unusual sulfur volatiles, a six-carbon chain that smells like guava fruit, and an ester that smells catty (it's close to two of the cat-pee sulfur volatiles). Giant water bugs have proven difficult to farm, so they and their scent-gland essence are becoming rare and expensive. Asian groceries often stock a synthetic version of the essence, a simple blend of the major esters that tends to smell mainly like pear fruit.

Flavor chemists report that whole bugs have an overall fruity and floral smell suggestive of pineapple, banana, and green apple, with a slight fishy note. Thanks to my friend Pim Techamuanvivit and the restaurants that keep her commuting between Bangkok and San Francisco, I can attest to those qualities and to the deliciousness of Thai sauces made with *maeng da*. Smell explorers: seek them out!

SOME SMELLS OF GIANT WATER BUG SCENT GLANDS

Smell	Molecule
green banana	hexenyl acetate
cheesy, banana	hexenyl butyrate
green leaf	heptenyl acetate
mushroomy	undecenone
catty	sulfanyl hexyl acetate
cheesy, sweaty, feet	butyric & methylbutyric acids
horse stable	cresol
guava	sulfanyl hexanol

So the smells of purposeful animal signals are far more diverse than the smells of animal bodies, and some of them hint at the full-blown delights of flowers and fruits. Before we enter the plant kingdom, though, there's one last set of animals to sniff at: the odd ones that, instead of flaunting their smells, try their best to avoid and eliminate and disguise them. Namely ourselves.

Chapter 6 · THE HUMAN ANIMAL

....

*Why is it that those who have sexual intercourse or are capable of
it have an evil odor and what is called a goat smell, whereas children do not?
Why has the armpit a more unpleasant odor than any other part of the body?
Why is it that the mouths of those who have eaten nothing, but are fasting,
have a stronger odor, "the smell of fasting" as it is called . . . ?*

· *Problemata*

Though we ended our survey of animal smells with an oddly fragrant insect signal, the great majority of them were unpleasant by-products of the animal body's essential metabolism and its inner community of microbes. Now we come to ourselves, the human animal, also smelly, sometimes reminiscent of the billy goat and worse. These days many people live in cultures that encourage the elimination or disguise of smells from the body, but we're still animals, and we still pay attention to volatiles that once gave notice of nearby mates or kin or competition. In any case, our own personal smells are a part of who we are, part of our identity. Who among us doesn't have a private fascination with them, one that goes back to early childhood and our first explorations of our bodies? As the ancient Greek *Problemata* shows, thoughtful adults have been openly curious about their smells for a very long time. Other animals are alerted and informed by each other's smells. We can be as well, and informed more deeply than ever by way of perception at second hand, the accumulated observations and insights of many centuries.

Much of this legacy comes from medicine. The earliest physicians recognized that smelly breath and excretions carry information about what's going on inside the body, and they paid intimate attention to them. Around 400 BCE, the Greek founder of Western medicine, Hippocrates, noted that particular smells were associated with particular conditions. Foul breath indicated a malfunctioning liver. Strong-smelling urine indicated inner ulcerations. If coughed-up sputum from diseased lungs had a strong smell when poured on hot coals, then the case would be fatal. Around the same time, Vedic physicians in India described an illness associated with abnormally copious urine that attracted ants and was sweet to the taste. This was what we now call diabetes mellitus, *mellitus* meaning "honey-like" and applied by the seventeenth-century London physician Thomas Willis, who found the taste of diabetic urine "wonderfully sweet as if it were imbued with honey or sugar." Another symptom of diabetes is a particular smell on the breath, described a hundred years after Willis by the Englishman John Rollo as "nearly the same as that produced by the effluvia of decaying apples." German chemists later identified its primary component as the molecule acetone, familiar today in nail polish remover, and also a by-product of the impaired sugar metabolism that's at the root of diabetes.

Today we know plenty about the human *volatilome*, the set of volatile molecules typical of the human body in sickness and in health, and also in death. Biomedical researchers have cataloged hundreds, and correlated particular molecules with particular biological processes. To help analyze evidence from crime scenes and locate disaster victims hidden under debris, forensic scientists track how the body's volatiles evolve after death. A more troubling application is the effort to define fingerprint-like "odorprints" for the purpose of surveillance, an updated version of the notorious 1970s practice of the East German Ministry for State Security, which collected clothing or other materials from individuals brought in for interrogation. These *Geruchsproben*, "smell samples," could later be presented to dogs to track dissidents or identify people who had handled incriminating documents.

The most intriguing recent insights into the smells of the human body have come from the front line of the battle against them: from corporations that sell personal-care products! This cosmetic interest is at least as old as the medical. The *Problemata* mentioned perfumes and unguents, and according to the scholar Constance Classen, upper-class Romans bathed several times a day, shaved their underarms, and applied the mineral alum—still used today—to constrict their skin pores and slow perspiration. Today's personal-care companies, notably the multinationals Firmenich, Givaudan, and Unilever, are analyzing which volatile molecules cause body odors and where they come from, in order to control them more effectively. Thanks to the published research of their in-house scientists, we now know exactly what the human body has in common with the billy goat—and the tomcat—and the trouble it takes to create odorous effects that human culture then works so hard to suppress.

It's this ancient and many-sided interest in human smells that makes them well worth paying attention to. They may be unpleasant, but they're not insignificant. They can inform us about our lives as individuals and as members of a species whose body chemistry is actively shaped by our living together. We can recognize molecules from our insides that are doing our bodies good—and others that aren't. We can recognize molecules from our armpits and feet in our foods, and ponder what it might mean that we find those foods appealing. So let's set aside the air fresheners and deodorants, unleash our natural animal interest, and sniff around ourselves.

As we did with the rest of the animal kingdom, we'll begin with the indelicate smells of our own metabolism and that of our inner microbiome, but in greater detail, so that we can follow their variation with age and diet. We'll move on to the portal to our innards, the mouth and breath, the smells of fasting and of feasting. And we'll conclude with our outer envelope: the skin, its secretions, and the cryptic signals that they bring to the surface.

Volatile flows in the human body

Every single cell in the human body houses a plenitude of biochemical systems, and many of these systems spin off carbon chains small enough to be volatile. Among them are members of all the basic carbon-chain families listed on pages 47–49, the hydrocarbons and alcohols and aldehydes and acids, as well as close relatives like acetone. And just like the cells of all mobile animals, ours are rich in proteins and their building blocks, the nitrogen- and sulfur-containing amino acids, and in the nitrogenous purines that help build and fuel the protein machinery. Thus our own cells generate the same typical odorous fragments of these animal systems surveyed in chapter 4: the cheesy, sweaty branched-chain acids, the sulfidic, rotting-vegetable sulfides, the barnyard carbon rings, the fishy amines, the fecal carbon-nitrogen rings, and the putrid carbon-nitrogen chains.

Many of these volatiles never leave the cells they're made in; they're used to build other molecules. The waste by-products that can't be used are exported into the blood, from which some are taken up and processed for disposal by the liver, the largest of our inner organs and the coordinator of the body's collective metabolic activities. It's in the liver, for example, that waste nitrogen from protein and energy metabolism is converted from toxic, smelly ammonia into nontoxic, nonvolatile urea and uric acid.

The liver also deals with many molecules that come into the body from our breath and food and drink but aren't nutrients, can't be used by the body, and in fact often interfere with the body's finely tuned machinery. Such "foreign" molecules are called *xenobiotics*, from the same root as *xenophobia*, fear of the foreign, and among them are many volatile molecules of foods and perfumes—the very molecules that make them appealing! The body sometimes simply excretes xenobiotics as is, but the liver often processes them first to make them less toxic, commonly by binding the xenobiotic molecule to a simple molecule, a sugar or a short chain of amino acids. This traps a volatile xenobiotic in a nonvolatile combination molecule that has no smell when it's excreted, usually in the urine or the sweat. But volatiles that are bound like this can also be unbound once they leave the body. This is one way that body smells can change and intensify with diet and with time.

From the blood, both processed and unprocessed wastes are passed out of the body. It's at this point that volatile wastes can enter into the air for us to smell. There are four main outlets through which our blood dumps them. One is the breath. As the blood passes constantly through the lungs, inhalations resupply it with oxygen to fuel our metabolism, and exhalations purge it of carbon dioxide and many other by-products of that metabolism. A second outlet for our metabolic volatiles is urine, the fluid produced by the kidneys as they filter the blood, removing waste molecules and excess water. Much smaller amounts of blood volatiles end up in two other body fluids, saliva in the mouth and sweat on the skin.

So much for the flow of volatiles from our own cells and metabolism, but there's more to the human body than our own cells. Our digestive system houses what amounts to a second body, a four-pound (two-kilogram) mass of some hundred trillion microbial cells with its own collective metabolism and volatile by-products. As a 2006 report in the journal *Science* put it, "Humans are super-organisms whose metabolism represents an amalgamation of microbial and human attributes." We're just beginning to understand how much our health and even our state of mind may depend on our microbiome, the microbial partners that inhabit every nook and cranny of our body. They're often called our *commensal* microbes, from the Latin for "together at the table"—appropriately, as they're dining on our leftovers. Most of the volatile molecules that the gut microbiome generates accumulate and stay in the digestive system along with residual volatiles from our food, until they're excreted along with the solid wastes. But some do find their way into the bloodstream and thereby into other body fluids or the breath.

So the volatiles in our various excretions and secretions vary as our diet and metabolic systems and microbial partners change, and as time passes. This means that our smells reflect what's going on on us and inside us. An unusual smell can be a sign of our body or commensal microbes responding to an unusual meal, a period of fasting, a metabolic system gone awry, or an infection. Even our usual smells carry direct evidence of how our superorganism is doing, whether our habits are helping it thrive or courting trouble.

The signifying smells of excrement

We'll start our tour of human smells with the most powerful and least pleasant source: the solid wastes from our digestive system. But let's also start with the least unpleasant sample: the excrement of a healthy infant fed exclusively on breast milk. As many parents can attest, it's ambrosial compared to the excretions of infants who have made the transition to formula or solid foods. This kind of observation has been well supported by formal research! Reported a 2001 study in the *Journal of Pediatric Gastroenterology and Nutrition*: "Stools of soy-fed infants were significantly more often recorded as foul smelling (25% of stools) than were stools of breast-fed infants (0%) or of milk formula-fed infants (11%)." The mix of volatiles from breast-fed infants was dominated by acetic acid—the smell of vinegar—and sulfidic hydrogen sulfide. The mix from cows' milk formula included cheesy propionic and butyric acids, and largely replaced hydrogen sulfide with rotting-vegetable methanethiol. But the mix of volatiles from soy-fed infants was very high in both of these sulfur volatiles, the sulfidic and the rotten.

Why the differences? It's complicated but comes down to the fact that several components of breast milk favor the growth of benign commensal bacteria called *bifidobacteria*, which consume the milk sugar lactose and produce acetic and formic acids (along with nonvolatile lactic acid). Breast milk also provides fewer sulfur-containing amino acids and minerals for microbes to convert into sulfides and thiols. Bifidobacteria appear to have many positive effects on the developing infant digestive and immune systems, and the prominent smell of vinegar is an affirmative sign that they've taken up residence.

Of course, once we're weaned off mother's milk and onto foods from plants and animals, the community of microbes in our gut changes, and our excremental volatiles shift decisively in the direction of unpleasantness. But if we can get past our disgust and pay the briefest attention to what we smell of ourselves, we get clues to what our superorganismal partners are doing inside us, for better and for worse.

Whenever we exercise our essential animality and eat, our stomach and small intestine break down nutrients and absorb their building blocks for pickup by the blood and delivery throughout the body. The food molecules for which we don't have effective digestive enzymes instead pass unabsorbed into our large intestine, the colon. The most abundant of these are the carbohydrates in vegetables, fruits, and seeds that give them their physical structures. We call these indigestible plant carbohydrates *fiber*. Starch can also evade our digestion and enter the colon when it's uncooked, or when it has cooled for some time after cooking and becomes less digestible *resistant starch*.

The colon is a hollow tube about five feet (one and a half meters) long, whose primary job is to absorb water and salts from the slurry of digestive leftovers before they're excreted from the body. It's also the primary residence of our gut microbiome, which can include thousands of different species and contributes several million different genes to our superorganism, compared to our own body's measly twenty thousand. Its diverse families with diverse biochemical specialties feed on our leftovers, cooperate to scavenge as much energy and cell-building material as possible, and in the process provide us with additional energy and nutrients. Excrement is the residue of the microbiome's handiwork. It's around half microbial cells and half food residues, with traces of fat, proteins, minerals, pigments—the brown remains of red-blood-cell hemoglobin—and volatile molecules that mark what the microbes have been up to.

Excrement is malodorous because the colon is airless and the gut microbiome anaerobic. Its microbes break down large molecules incompletely, producing the short carbon chains and the nitrogen and sulfur volatiles that we've sniffed in the bodies and excretions of our fellow animals. Many gut microbes can break down resistant starch and plant fiber into their sugars and use them for energy. In the process they generate the familiar volatile short-chain fatty acids—acetic, propionic, butyric—vinegary and cheesy. Other microbes specialize in breaking down proteins and their amino acids, and are able to exploit materials from the lining of the colon itself, which secretes a protective protein- and sulfur-rich layer of mucus, and constantly sheds and regenerates its own cells. These protein specialists leave behind cheesy, sulfidic, ammoniac, fecal, barnyardy, and putrid fragments.

So the basic volatile formula for human excrement is: short carbon-chain acids, carbon-ring cresol, hydrogen sulfide and sulfurous thiols, and nitrogenous ammonia, amines, indole, and skatole. Collect a few of those chemicals and you have the makings of a standard stink bomb, a stench generator deployed in both practical jokery and in riot control.

SOME SMELLS OF HUMAN EXCREMENT

Component smells	Molecules	Sources: microbial breakdown of
sharp, vinegary, cheesy, sweaty, vomit	acetic, butyric, propionic, methylbutyric acids, branched acids	carbohydrates, fats, proteins
sulfidic, cooked vegetable	hydrogen sulfide, methyl sulfides	proteins, sulfur compounds from onion & cabbage families
rotten vegetable	methanethiol	proteins, sulfur compounds from onion & cabbage families
ammonia, fishy, urinous	ammonia, amines	proteins, purines
mothball, fecal	indole, skatole	proteins
antiseptic, tar, horse stable	phenols, cresols	proteins
putrid, rotting flesh	putrescine, cadaverine	proteins

The families of microbes that are most active in the gut microbiome, and the volatile molecules they generate, are largely determined by the particular menu of leftovers in the colon. If we eat a lot of vegetables, fruits, whole grains with their fiber-rich seed coats, or cooked foods with resistant starch, then a lot of carbohydrate fiber will make it to the colon, and the fiber specialists will produce plenty of the vinegary, cheesy short-chain acids. If we eat more meat, eggs, and dairy products than vegetables and whole grains, then the fiber specialists will produce less acids, the acid-averse specialists in protein breakdown will thrive, and their ammoniacal, sulfidic, fecal, barnyardy, putrid by-products will be more prominent. The protein specialists also dominate when we fast and don't supply any food materials at all, because they have the advantage of a continuous supply from the colon's own mucus and shed cells.

Because we're omnivores and eat a variety of foods, our gut microbiome is always producing a mixture of volatiles. But even an untrained nose can pick up on these two excremental accents, the vinegary-cheesy and the barnyardy-putrid. I'm always reminded of my years of diaper duty a day or two after I gorge on artichokes, which are rich in the soluble fiber that bifidobacteria are especially fond of. Indulging in a massive steak shifts things to the feedlot end of the spectrum. And these accents turn out to be direct signs of otherwise insensible processes that influence our long-term well-being. The acid accent is a good sign; the feedlot, not so good.

Biomedical scientists have found that the volatile acid products of gut fiber metabolism have a range of beneficial effects on the gut itself and the body in general. Acetic and propionic and butyric acids make the gut inhospitable to many harmful microbes, but also to the specialists in protein breakdown. They can be taken up by the colon cells and travel to different parts of the body. Acetate is sent throughout and helps to control inflammation, lower blood pressure, improve glucose tolerance, and reduce appetite. Much of the butyrate stays in the gut, provides energy for the colon cells, stimulates new cells to replace the cells that are constantly sloughed off, and at the same time suppresses the growth of abnormally active cells that can give rise to cancer. So a vinegary-cheesy cocktail of short-chain acids is a sign of good works, in the gut and elsewhere.

By contrast, the volatile products of protein metabolism appear more likely to cause trouble. Nitrogen-containing ammonia is toxic to cells. It and its related amines can lead to the formation of DNA-damaging nitrosamines. Barnyardy cresol has come under suspicion of causing kidney and skin damage. Sulfidic hydrogen sulfide is apparently an essential signaling molecule in the tiny quantities produced by our own cells, but when bacteria flood the gut with it, it breaks sulfur cross-links in the protective intestinal mucus and thins it, leaving cells in the colon wall more vulnerable to attack by disease microbes and to damage from various reactive molecules, including meat hemoglobin. This loss of protection predisposes the colon wall to inflammation and excessive cell growth. All of these particular effects are consistent with frequent findings that diets high in meat and therefore protein are associated with an increased risk of colon cancer.

Every time we empty our colon, we're greeted by volatiles that flow from our food choices, some of them the very molecules that are doing good for us, and some likely doing harm. These smells can remind us that when we eat, we're feeding both our animal selves and also trillions of other creatures whose lives affect ours, day in and day out, over years and decades. They can remind us to provide our superorganismal partners with the vegetables and fruits and seeds that they need to do their vinegary, cheesy good.

Sulfurous gas

The molecules that make excrement smell execrable are only minor products of the gut microbiome. The major volatile products of its collective metabolism are hydrogen, carbon dioxide, and methane, gases for which we have no olfactory receptors, and which therefore have no smell. Because these gases are produced in large amounts and accumulate, the colon releases them periodically to relieve the pressure. When it does so, they carry with them the trace volatiles that do smell. The rapid release of gases from the colon—the *fart*, an onomatopoetic word with roots in prehistory—disgusts with its odor and either offends or amuses with its rude sound. Among the first to propose the study of fart odor was Benjamin Franklin. In a satirical 1783 letter to a friend, he suggested it as a topic for a prize to be offered by the Royal Academy in Brussels, "for the serious Enquiry of learned Physicians, Chemists, etc. of this enlightened Age":

My Prize question therefore should be: "To discover some drug, holesome and not disagreeable, to be mixed with our common Food or Sauces, that shall render the Natural Discharges, of Wind from our Bodies, not only inoffensive, but agreeable as Perfumes." That this is not a chimerical Project, and altogether impossible, may appear from these Considerations. That we already have some knowledge of Means capable of varying that Smell. He that dines on stale Flesh, especially with much Addition of Onions, shall be able to afford a Stink that no Company can tolerate: While he that has lived some Time on Vegetables only, shall have that Breath so pure as to be insensible to the most delicate Noses; and if he can manage so as to avoid the Report, he may anywhere give Vent to his Griefs unnoticed.

Though Franklin was right to associate the worst smells with meat and sulfur-rich onions, a vegan diet is no panacea and no gas-perfuming drug has yet been discovered. But learned physicians did soon begin to study what they dignified with the term *flatus*, from the Latin for "wind." Probably the first to analyze its chemical composition was the Swiss physician Louis Jurine, who reported in 1789 on finding oxygen, carbon dioxide, nitrogen, and hydrogen sulfide in the colon of a man who had just died of exposure. A couple of decades later, the French physiologist François Magendie collected gas from four condemned men immediately after their executions, and added hydrogen and methane to the roster. Later researchers developed methods for monitoring the flatus of living subjects, some of them uncomfortably invasive, and essentially corroborated Magendie's findings for the major constituents.

One of the most recent and humane flatus protocols was inspired by bathtub bubbles. In the laboratory of the Dutch gastroenterologist Albert Tangerman, volunteers feeling pressure to "deflate" immersed their hindquarters in a warm bath, where the rising bubbles were trapped under an inverted beaker. Tangerman and others have found that their subjects produced between one and three pints (500 to 1,500 milliliters) of flatus per day, a single emission varying between a tablespoon and a cup and a half (15 to 375 milliliters). Of course they've detected many of the volatiles found in excrement, but the sulfur molecules dominate by far, with rotting-vegetable methanethiol the most important, sulfidic hydrogen sulfide next, and a few methyl sulfides also present. This ranking probably reflects the fact that hydrogen sulfide and methanethiol are more volatile than the short-chain acids and amines, held less tightly to the wet wastes, and faster to escape.

The potency of these molecules—that is, our sensitivity to them—is indicated by Tangerman's calculation that if the smeller's nose is three feet (one meter) away from the source of an emission, the volatiles it detects have been diluted by a factor of around fifty thousand. Recall that hydrogen sulfide and methanethiol are both primordial molecules, found in decomposing animal tissues, and toxic to oxygen breathers like us—a good reason for our sensory systems to be on high alert for reports of their presence.

Changeable urine

Urine is the human body's fluid excrement, largely composed in the kidneys and very different from the gut's semisolid mass of bacterial cells and food wastes. Fresh from a healthy body, urine is nearly microbe-free, filled with several thousand chemical by-products of our superorganism's metabolism, microbial partners included. (The yellow color, like the brown in excrement, comes from breakdown products of the blood's red hemoglobin pigment.) It's about as pristine as such a rich soup could be, and normally has a fairly mild and complex smell created by traces of many volatiles, some of them also found in foods. These contribute nutty, fruity, sweet aspects that mask the unpleasant notes of protein and purine metabolism. One notable urine volatile is the multi-ring steroid molecule androstenone, which we'll encounter in pigs and pork and which men excrete more of than women. Due apparently to genetic differences in smell receptors from one person to another, some people can't smell androstenone at all, while among those who can, some describe it as musky and urinous, and others as pleasantly flowery!

SOME SMELLS OF FRESH HUMAN URINE

Component smells	Molecules	Sources
sweet, coconut, peach	lactones (9-, 10-, 11-carbon chains)	our metabolism?
potato	methional (sulfur aldehyde)	?

continued

Component smells	Molecules	Sources
popcorn, basmati rice	acetyl pyrroline (nitrogen-carbon ring)	?
vanilla	vanillin (carbon ring)	?
horse stable	cresol	gut microbe metabolism of proteins?
fecal	skatole	gut microbe metabolism of proteins?
fishy, urinous	trimethylamine	our metabolism; gut microbe metabolism of urea & trimethylamine oxide
urinous, musky	androstenone (steroid)	our metabolism (sex hormone)
rotting & cooked vegetables	methanethiol, methyl sulfides	our metabolism of asparagusic acid in asparagus

Our foods and drinks can affect the smell of fresh urine. Some, like garlic and coffee, contribute their own volatiles. A more complex and notorious example is asparagus, whose influence has intrigued learned physicians and others at least since medieval times. Marcel Proust was unusually appreciative; his narrator in *In Search of Lost Time* recalls asparagus as "delightful creatures who had fun turning themselves into vegetables,

whose precious essence I would recognize when, all night long after a dinner at which I had eaten them, they played, with their poetic and coarse joking as in a fairy comedy of Shakespeare, at changing my chamber pot into a vase of perfume.

This distinctive quality of asparagus, precious or not, derives from a peculiar two-sulfur molecule, asparagusic acid, apparently unique to asparagus and concentrated in the growing tips, which it may help protect from microbes and insects. The body breaks down this foreign molecule and excretes fragments that include methanethiol, methyl sulfides, and related sulfur compounds that smell of rotten and overcooked vegetables. Within an hour or two of eating asparagus, these molecules end up in the urine at levels a thousandfold higher than normal. But not everyone notices a change. Recent studies indicate that some people don't excrete the stinky sulfur volatiles after eating asparagus, and some people aren't able to smell them when they do. Genetic differences in metabolism and olfactory receptors are likely to account for these variations.

It may sound like a frivolous exercise to study the genetics of urine smells, but in fact it has saved lives. Physicians throughout history have used these smells to diagnose disease, and in the twentieth century, medical researchers discovered that by smelling the urine of newborns, they could detect several serious genetic defects of basic metabolism early enough to treat them before they caused irreparable damage. Infant urines that smell like a mouse cage, maple syrup, sweaty feet, or fish are all scented by telltale by-products of impaired amino acid or amine metabolism.

Of course most of us wouldn't describe the usual smell of urine as mild and complex. It's strong and, well, urinous: ammonia-like and fishy, sulfurous, unpleasant. It's the smells of neglected toilets and urinals, skid-row sidewalks, nursing-home incontinence pads. These are the smells of *old* urine. They can begin to develop if urine is held for a long time in the bladder, and they develop quickly the moment it leaves the body and omnipresent microbes begin to feed on its biochemical riches, especially its animalic nitrogen wastes.

The body excretes much of the excess nitrogen from its protein and energy metabolism in the urine. Because simple ammonia, and small carbon-nitrogen molecules, the amines, are very reactive and therefore toxic to living cells, the body transforms these wastes into less reactive, nonvolatile urea, uric acid, and trimethylamine oxide (TMAO). In the quart (liter) or two of urine that each of us passes every day, there are all of these along with scant traces of trimethylamine (TMA), which is volatile and gives fresh urine its faintly urinous-fishy smell. Bacteria break down these molecules for energy, and in the process generate more TMA, other volatile amines, and ammonia. The accumulation of amines and ammonia has the additional effect of raising the pH of the urine and making these molecules increasingly volatile: quicker and quicker to escape from the liquid into the air. The result is that as it sits, urine gets more strongly ammonia-like and fishy.

SOME SMELLS OF STALE HUMAN URINE

Component smells	Molecules	Sources: microbial metabolism of
strongly urinous, fishy	trimethylamine, ammonia	urea & trimethylamine oxide
rotten vegetables	methanethiol	proteins, amino acids
overcooked vegetables, garlicky	methyl sulfides	proteins, amino acids
smoky, barnyardy	guaiacols	food by-products (coffee)
fruity, caramel	furaneol, maltol	food by-products (strawberry)
clove	eugenol	food by-products?
maple syrup	sotolon	food by-products (fenugreek)
cheesy	butyric, methylbutyric acids	proteins, amino acids

At the same time, other changes conspire to amplify the smell by adding sulfurous, cheesy, smoky, and spicy volatiles to the mix. Where do these come from? The cheesy and sulfurous molecules probably come from the microbial breakdown of proteins and amino acids, which urine also contains. But others are aroma molecules absorbed from our foods. The body can't use them, treats them as xenobiotics, and binds them to nonvolatile carrier molecules, often sugars or amino acids, for excretion. Once they're excreted, microbes in the urine break apart these combination molecules to use the carriers, and in the process liberate the food volatiles.

The breath and mouth

The most public and frequent release of volatiles from within the body occurs when we breathe. An ordinary breath takes in and then expels two to four quarts (two to four liters) of air from the lungs, where blood absorbs oxygen for our cells to generate energy, and releases carbon dioxide and hundreds of other volatile by-products of their biochemical machinery. The overall effect of an exhalation, the smokestack emission of healthy animal life, is a mild olfactory hum.

The most abundant volatiles in our breath are slightly rubbery isoprene and solventy acetone. Isoprene is a branched five-carbon molecule that serves in animals as a building block for making steroid hormones and cholesterol—and as we'll see (page 207) is also the major volatile exhaled by the leaves of green plants! Acetone is a two-carbon by-product of breaking down fats for energy. Because our cells burn fats primarily when they're running low on carbohydrates from recent meals, breath acetone levels rise significantly when we go for hours without eating, and are often detectable before breakfast.

Some volatile molecules that we take in with food and drink can get into our breath via the blood. Alcohol is one, and another is a particularly persistent by-product of eating garlic. Garlic volatiles are small sulfur compounds, but nearly all of them are so reactive that they never make it into the blood. The major exception is a three-carbon chain attached to a sulfur atom that circulates in the blood and persists on the breath for hours no matter how well we wash or brush our mouth.

In fact, it's the mouth that generally amplifies the smell of our exhalations and causes the strong and unpleasant smells of "bad breath." When we speak or otherwise exhale through the mouth, the lung gases pick up mouth volatiles and deliver them to nearby noses. And the mouth is the portal to the microbe-packed digestive system. Food and drink, breath mints and chewing gums, toothpastes and mouthwashes—all leave temporary smells, but the major contributors are microbes, for which the mouth is a warm, moist, protected, oxygen-poor refuge with no shortage of nutrients to feed on.

In healthy people with healthy gums, the major source of mouth volatiles is bacterial growth on debris from food and from the mouth itself. Food particles accumulate mostly between the teeth and on the upper rear surface of the tongue, which is rough, with countless pits and fissures in which anaerobic bacteria can evade oxygen and toothbrush and mouthwash. Species of *Porphyromonas*, *Prevotella*, *Fusobacterium*, and others thrive there by the billions and emit their usual foul brew: sour and cheesy and sweaty acids, sulfidic and rotten-cabbage sulfur compounds, fishy and putrid amines, including putrescine and cadaverine. Methanethiol often predominates.

You might think that not eating would starve mouth microbes and suppress their volatile production. Not so, as the ancient *Problemata*'s question about fasting forecast. Saliva is a rich source of nutrients, proteins in particular, as are sloughed-off cells from our cheeks and tongue. In fact, eating has the advantage of massaging some microbes out of the tongue, increasing saliva flow and washing its surface of microbes and volatiles alike. This is why our breath is usually at its worst first thing in the morning, after hours of fasting and low saliva flow.

SOME SMELLS OF HUMAN BREATH

Smells	Molecules	Sources
solvent	acetone	fasting metabolism of fats
sour, vomit	acetic, butyric acids	microbial metabolism of mouth proteins, carbohydrates, lipids
cooked vegetable, sulfidic	methyl sulfides, hydrogen sulfide	microbial metabolism of mouth proteins
rotten vegetable	methanethiol	microbial metabolism of mouth proteins
garlic	allyl methyl sulfide	food residues in mouth & circulation
cheesy	methylpropionic, methylbutyric acids	microbial metabolism of mouth proteins
fishy, spermous, ammonia	methylamines, pyrroline, ammonia	microbial metabolism of mouth proteins
putrid	cadaverine	microbial metabolism of mouth proteins

Remedies for bad breath go back to ancient times and have included debris-clearing chewing sticks, toothbrushes and powders and pastes, and mouthwashes based variously on salt water, beer, wine, even urine. In Han China two thousand years ago, members of the court freshened their breath with cloves, whose distinctive volatile, eugenol, is also an antibacterial. The modern era for mouthwashes came around the turn of the twentieth century, when Joseph Lawrence and Jordan Lambert began to sell their antibacterial mix of phenolic compounds—eugenol is one such compound—named Listerine for the English proponent of antisepsis, Joseph Lister. This is how volatile phenolic molecules from plants, including thymol, menthol, and eucalyptol, became canonical flavors of mouthwash. Today they're augmented with alcohol and detergent-like molecules that can disrupt the protective biofilm that mouth microbes form. Dilute hydrogen peroxide is also effective at oxidizing prominent sulfur malodors. A less drastic approach is to diminish the sulfur volatiles by chewing on foods rich in oxidizing enzymes or reactive phenolic compounds. These include raw fruits and vegetables that quickly turn brown when cut or crushed, a sign of both phenolics and active enzymes: apples and pears, lettuces, mushrooms, basil. Green tea can also be helpful.

The mouth's liberality

The mouth is much more than the gateway to our innards. From it emanates speech, the expression of our inmost thoughts, and song, emotion translated into sound. Like the eyes, it embodies feeling. It can be beautiful and seductive, an object of desire to touch and taste. In a passionate kiss, two mouths mingle breaths and experience them from within. No wonder that love poets across the millennia have found many delightful ways to idealize the breath of the beloved—and that iconoclastic Shakespeare would insist on the reality of "the breath that from my mistress reeks."

In reality, the mouth's microbiome can do more than malodorize: it also intensifies the delight of eating and drinking! More precisely, it liberates some volatile molecules that have been bound up in nonvolatile combinations with other molecules. This binding of volatiles with nonvolatiles may sound familiar: our bodies do it to excrete unwanted xenobiotic molecules in urine, and cats do it to seed their urine with time-release signals (see pages 94, 81). Well, food plants bind some of their aroma volatiles in the same way, and our mouth microbes do us the favor of unbinding them so we can enjoy them. The first clue of this came in the 1980s from the pioneering French wine chemist Émile Peynaud. In *The Taste of Wine*, he noted the mystery that "wine smells more of fruit than the grape itself," and that the mouth itself could also amplify aroma. The wine grape called Sauvignon Blanc

> has a very specific smell, floral, musky, smoky, with a slight raw herbaceousness suggesting bruised leaves. . . . When you bite into the thick skinned, golden Sauvignon grape you can smell this particular odor, albeit rather weakly; similarly the fresh squeezed juice of the grape has relatively little smell. . . . It is only 20 or 30 seconds after swallowing the juice that you suddenly experience a powerful aromatic rush at the back of the mouth as the Sauvignon

fragrance returns. Doubtless saliva reacts with and releases the Sauvignon essence which is present in the grapes in a relatively odorless form.

In 2008 scientists at the Swiss flavor and fragrance company Firmenich demonstrated that holding Sauvignon Blanc juice in the mouth does release a floral-musky-green aroma compound, a six-carbon version of the smaller sulfur molecules in cat pee, and that microbes did the releasing. They also showed a similar release of sulfur volatiles important to the flavors of green bell peppers (oniony-vegetal heptanethiol) and onion powder (cabbagey propanethiol). Other Firmenich chemists have found that cooking at temperatures high enough to cause browning can also generate bound volatiles with meaty and coffee-like qualities, and that enzymes of mouth microbes can liberate them. These discoveries help explain why the flavors of some foods seem to linger or even return after we've finished swallowing: microbes continue to release aromas from residues of their bound precursors. This may well have been why the flavor of my life-changing lunchtime grouse persisted well into the evening.

So the same mouth microbes that can embarrass us also enhance the pleasures of eating and drinking. The moral: Don't decimate them with mouthwash before a special meal. Munch an apple.

The skin, its secretions and denizens

Enough of exhalations and excretions from within! Now for the smells of our surfaces and composed secretions, some of them blessedly pleasant for a change, others attention-getting collaborations between animal and microbe—our body odors proper.

The human body's outer boundary and interface with the world is the skin, its largest organ, accounting for between 10 and 15 percent of our weight, and covering about twenty square feet (1.8 square meters). Skin determines what comes into us and what stays out, from heat to molecules to other life forms. Its many folds, creases, and pores are home to a diverse community of commensal microbes that helps exclude potentially harmful creatures. This skin microbiome also generates most of the volatiles that we refer to as body odor.

Our skin has several layers. There's an underlying support layer filled with blood vessels and various small glands that produce and store the skin's secretions. Above that is the dermis, a layer of actively growing skin cells, and finally the outermost layer, the epidermis, a layer of dead dermis cells that protects the living layers beneath. The epidermis consists largely of mechanically tough proteins. Hairs are inert protein fibers that project out of the skin proper; they're formed and pushed out from tiny organs called hair follicles embedded in the dermis layer. Skin hair is common among mammals, and serves many different purposes: retaining body warmth and repelling water, both camouflage and signaling, and extending the sense of touch well away from the skin. We humans have much less body hair than our primate relatives, and some of the patches we do have are probably there in part to amplify our body smells.

The smells of our skin start with several fluids that the skin secretes onto its surface to keep itself moist, flexible, and water-resistant. These fluids, sebum and sweats, are generated by three different kinds of glands in the dermis. Fresh from the glands, they're essentially odorless.

The sebaceous glands secrete *sebum*, a waxy, fatty material that provides a water-resistant film for both the skin and the hair, and also protects them from attack by oxygen in the air and ultraviolet radiation from the sun. Sebaceous glands usually empty directly into hair follicles, and are mainly found on the head and upper part of the torso, with smaller numbers on the eyelids, nose, and genitals, in the ear canal, and around the nipples. Sebum consists of the disintegrated remains of whole gland cells, so it's a complex mix of cell membranes and machinery along with their specialized products, which include fatty and waxy molecules and squalene, a long carbon chain (twenty-four chain carbons with six one-carbon branches) involved in the production of cholesterol and steroid hormones. Most of the carbon chains in sebum are too long to be volatile, but they can hold on to small volatile molecules and release them slowly over time; and with time they themselves can be broken into volatile fragments.

The familiar perspiration that we work up with physical exertion is *eccrine sweat*, produced by eccrine glands, the most numerous and widely distributed skin glands, numbering several million over the entire adult body, but at highest density on the forehead, palms of the hands, and soles of the feet, with hundreds in the area covered by a small coin. They release an odorless, watery solution that carries small amounts of salts, sugars, lactic acid, amino acids, urea, and proteins. The lactic acid helps control which microbes thrive on the skin by helping keep its pH at an acidic 5.5—about the same as brewed coffee.

Then there's a second kind of sweat called *apocrine sweat*, produced by glands that are few in number but powerful in their influence on body odors. There are only about two thousand apocrine glands on an adult, concentrated in the armpits, the areolas around the nipples, and the pubic area, just above the genitals, with a few on the head and cheeks, on the eyelids, and in the ear canal. They release a milky, thick fluid that includes fatty materials, steroid molecules related to the sex hormones, proteins, and—most notably for the smell explorer—special molecules whose primary function appears to be to release body odors. But not immediately, and only with the help of microbes: these body odor precursors are odorless.

There are plenty of microbes teeming on our skin and in its secretions, the external members of our superorganismal ecosystem. They include single-cell members of the fungus kingdom, relatives of baking yeast and mushrooms. Skin microbes that require or tolerate oxygen live in the moist flaky layers of the epidermis, while anaerobes grow in oxygen-poor hair follicles, in the tubular ducts that carry gland secretions to the epidermis, and in the glands themselves. (Even tiny insects called *Demodex* mites live in follicles and sebaceous glands.) The many different communities in the different niches produce distinctive odors. In general, exposed drier areas, like the forearm, have relatively low numbers of microbes. Areas like the nape of the neck and forehead have many sebaceous glands and support mainly bacteria and fungi that specialize in breaking down fatty materials into shorter-chain aldehydes and acids. Sheltered regions like the axillary and pelvic areas, where the arms and legs join the trunk, tend to be moister, and their plentiful apocrine and sebaceous secretions make them hospitable to bacteria that metabolize a wide range of molecules, including skin proteins and the special body-odor precursors.

Before we embark on a tour of our microbial communities and their volatiles, let's sniff at skin before any microbes get into the act.

Pristine skin: pleasant fatty fragments

Our skin cells and hairs and secretions have very little odor of their own, so the smell of freshly washed human skin is subtle and easily masked by soaps and other scented personal-care products. Scientists have detected nearly three hundred different volatile molecules escaping from skin, but they're present at low levels and combine to produce a background hum, the indistinct animal presence that we share with our freshly washed pets. Some of the predominant volatiles are medium-length carbon chains that are formed when oxygen in the air attacks longer chains and breaks them into pieces. The fatty sebaceous secretions give aldehydes eight to ten carbon atoms long, octanal, nonanal, and decanal, with waxy and citrusy qualities. Also prominent are solvently acetone, fresh floral geranyl acetone, and citrusy methyl heptenone, all ketones that probably arise from the oxidation of squalene, that long branched carbon chain in sebum and apocrine sweat.

The skin's long carbon chains can also be broken into smaller volatile fragments by high-energy ultraviolet rays in sunlight. Here's a simple five-minute experiment for a sunny day: Scrub your hands with scentless soap, rinse and rub them dry with an unscented towel, then hold one out flat and palm downward in the sunlight for a few minutes while keeping the other palm up. Now smell the backs of both hands side by side. The shaded hand will smell pretty neutral, while the sunned hand will have that characteristic sunskin smell: a little metallic, a little mushroomy, and a little pungent, all from UV-fragmented skin lipids. A 2006 study of sun-exposed skin identified two- and three-carbon acetaldehyde and propionaldehyde, pungent and earthy, along with the sweet-smelling two-carbon hydrocarbon ethylene.

The UV that breaks skin lipids and scents us is the same radiation that can penetrate down to our living skin cells, break apart their DNA and other critical molecules, and cause sunburn and sometimes cancer. And aldehydes themselves are reactive molecules that damage cells. As we'll see, aldehydes are the same family of carbon-chain volatiles that are produced when we expose plant oils and animal fats to air and the high energy of a stovetop—that is, when we fry foods. The smell of sunned skin is a sign that we're beginning to fry ourselves.

SOME SMELLS OF (MOSTLY) PRISTINE HUMAN SKIN

Smells	Molecules	Sources
waxy, sweet, citrus	octanal, nonanal, decanal (aldehydes 8–10 carbons long)	oxygen breaking sebum fats (14–20 carbons long)
solvent, fresh, floral, citrus	acetone, geranyl acetone, methyl heptenone (ketones 2–11 carbons long)	oxygen breaking 30-carbon squalene from sebaceous & apocrine glands
sharp, pungent	acetaldehyde, propionaldehyde	UV sunlight & oxygen breaking skin lipids
kareishu: fatty, waxy, green, floral	nonenal, nonanal, diacetyl	oxygen breaking skin lipids; microbes metabolizing lactic acid

Skin aldehydes have also been fingered as volatiles that help give older people a particular smell, what the Japanese call *kareishu*. Various studies have confirmed that untrained sniffers can distinguish individuals over and under age forty by their skin smells. Japanese research teams attribute *kareishu* to higher levels of 2-nonenal, a nine-carbon aldehyde with a double bond that contributes fatty, green notes in melons and cucumbers, along with the buttery ketone diacetyl, which species of *Streptococcus* bacteria generate from the lactic acid in sweat. A group at the Monell Chemical Senses Center in Philadelphia instead found higher levels of the straight-chain nonanal, with waxy and floral qualities, and suggested that differences in Japanese and Western diets might account for the discrepancy (nonenal likely comes from unsaturated fats, nonanal from saturated). It seems reasonable that the body's general metabolism and skin microbiome might shift with age, and that these could affect overall body odor.

I myself may now smell detectably old, but I vividly remember the smells of my two grandmothers from years of lap-sitting and hugs: and they were very different from each other, one fresh and sweet, the other strong and musky. I think their smells had mostly to do with their homes and diets and activities. One was a lifelong Midwesterner and Christian Scientist, strict in her habits, who stayed out of the kitchen when she visited. The other was Indian, living in England, cooking and eating the strongly flavored curries and pickles that she and my grandfather loved. We don't know much about the effects of diet on skin volatiles, but one well-documented example in addition to garlic's persistent sulfides happens to be fenugreek seed, a common ingredient in Indian curries. The primary volatile in fenugreek is sotolon, a carbon-ring molecule that has a characteristic smell of maple syrup (see page 172). The human body alters this molecule into others, so far un-identified, that have a similar smell and make their way into our skin secretions, so we're flavored by it too: we give off a faint whiff of that spicy, caramelly odor.

Peachy scalp and hair

Nuzzling the hair of a loved one is one of the great pleasures of our animal life, calming or stimulating depending on the loved one. Cool silken strands of hair have their smell, and through them the hidden warmth of the scalp radiates and lifts its volatiles. Shampoos and hair conditioning treatments all cover the intrinsic faint smells of our head and hair, but these become more assertive with time. The scalp is one of the body areas rich in sebaceous glands and the fatty materials that oxygen and ultraviolet light break apart into small volatile acids and aldehydes and alcohols, as well as ketones and lactones. The few published surveys of head vola-tiles report dozens and dozens of these, with no particular molecules predominat-ing; so the main smell of a clean, healthy head is a low-level cocktail of fatty breakdown products.

There's another aspect of that smell that comes from a community of microbes well adapted to the relatively dry environment of the scalp. One dominant mem-ber of the healthy scalp flora is a species of *Malassezia* (aka *Pityrosporum*), a kind of fungus that's technically a yeast but a closer relative of plant diseases and even table mushrooms than to bread yeast. When scientists at the University of Penn-sylvania grew the commonest scalp species, *Malassezia furfur*, in petri dishes that included human sebum and monitored the volatile molecules it produced, they detected large amounts of three lactones, unusual chain-ring hybrids (see page 161) that are the characteristic aroma molecules in coconuts and peaches, and are described as having fatty, waxy, creamy aspects. Sure enough the scientists re-ported that their scalp-yeast cultures smelled like "canned peaches." Because they're not simple breakdown products of energy metabolism, these lactones likely serve some purpose for *Malassezia* cells, perhaps as signal molecules.

SOME SMELLS OF THE HUMAN HEAD

Smells	Molecules	Sources
metabolic medley	medium-chain hydrocarbons, alcohols, aldehydes, acids, ketones	sebum lipids oxidized by oxygen, UV light
fatty, waxy, creamy, peach	g-nonalactone, g-decalactone, g-undecalactone	yeast transformation of sebum lipids

So the appealing native aroma of the head carries hints of pleasant foods. Of course the longer a head goes unwashed, or a comb or brush uncleaned of sebum and protein-rich scalp debris, the more we smell the stronger, ranker aromas of general carbon-chain breakdown. In individuals and cultures that wear their hair long and don't wash it every day or two, this amplified smell is common. According to the historian Charles D. Benn, in the Tang period of China a thousand years ago, when long hair was the norm, government officials were allotted a day off in every ten to wash their hair, and their salary was termed "a subsidy for clothing and hair washing."

Cheesy feet

At the other end of the body, a very different volatile world! Feet can be powerfully odiferous body parts, especially when we confine them in shoes and create a warm, moist, airless paradise for microbes. But even people who wear sandals or go barefoot have hospitable pockets between the toes and between the nails and skin. Of the various slang terms for the smelly residues that collect there, "toe cheese" is especially apt, because foot odor can be strongly reminiscent of cheeses—and vice versa. The Dutch have the equivalent word, *tenenkaas*, and in 1996 the olfactory resemblance led Dutch scientists to discover that they could decoy the malaria-carrying mosquito, which tends to feast on human feet and ankles, with Limburger cheese.

The personal-care and shoe industries have long had a vested interest in understanding what foot odor is and how to control it, so we know a fair amount about it—and why the resemblance to cheese is so strong.

A human foot has about a quarter of a million eccrine sweat glands that deposit abundant moisture and minerals and even a bit of glucose sugar on the surface, mainly on the sole. There are few sebaceous and no apocrine glands, so foot microbes get most of their energy from breaking down the epidermis proteins. The upper foot surface has around a thousand bacteria per square centimeter, while the sole has about a hundred thousand and the protected toe clefts about ten million. Several kinds of bacteria dominate foot ecosystems, and they make different contributions to foot odor. A number break down the various remains of skin cells to the usual simple carbon chains, the short-chain fatty acids, and especially two- to four-carbon acetic, propionic, and butyric acids, the first two sharp and vinegary, the third like Parmesan and other long-aged cheeses.

The distinctive smell of unwashed feet comes from other short-chain fatty acids, the rarer ones that have a one-carbon methyl branch off the main chain. Their production starts with species of the uncommon bacterium *Kytococcus*, named from a Greek root meaning "skin," which are especially good at attacking the fibrous skin proteins, breaking them down into smaller fragments and eventually to building-block amino acids, which they and other bacteria then use for energy. Several different members of the foot microbiome—species of *Kytococcus*, *Staphylococcus*, *Brevibacterium*, and *Micrococcus*—are able to metabolize the amino acids with branched chains (leucine, isoleucine, valine), and in the process produce branched-chain fatty acids. And it's these volatiles that produce the distinctly sweaty-foot, cheesy smell.

SOME SMELLS OF HUMAN FEET

Smells	Molecules	From microbial breakdown of
sharp, vinegary, aged-cheesy (Parmesan)	acetic, propionic, butyric acids	skin oils, proteins
sharp, sweaty, cheesy (Swiss washed-rind cheeses)	methylpropionic, methylbutyric acids	skin proteins

Why the connection between cheese and feet? Cheese is a coagulated mass of milk proteins and fats, not so different in chemical composition from our skin, and the family of "washed-rind" cheeses is regularly moistened during months of maturation with a brine—like a concentrated version of our eccrine sweat—to encourage the growth of surface bacteria rather than molds. Washed-rind cheeses often carry species of *Brevibacterium*, one of the major foot bacteria, on their surface. But *Brevibacterium linens* on cheeses breaks down sulfur-containing amino acids to the rotten-vegetable compound methanethiol, which produces a different kind of funkiness, the room-filling smell of Limburger or French Époisses or Italian Taleggio (see page 566). The cheeses that smell most like feet tend to be those whose rinds are washed for only a brief period and then matured with a relatively dry rind, which must somehow favor the metabolism of branched-chain amino acids over sulfur-containing amino acids. These include Swiss cheeses like Appenzeller and Tête de Moine.

Sexual smells, microbial and spermous

Our sexual organs develop at the junction of our trunk and legs, surfaces well endowed with sweat and sebaceous glands, but their own particular smells come from within their specialized anatomies. A woman's vagina supports a distinct community of microbes that generate familiar metabolic volatiles, while male semen is relatively microbe-free and has a set of volatiles and a smell all its own.

The vagina is the main passageway of a woman's reproductive tract; through it the male's sperm passes inward to reach the uterus, and the fetus passes outward during birth. Secretions from the vagina wall keep it moist, flexible, and hospitable to a community of beneficial bacteria that prevent harmful microbes from invading. This vaginal microbiome varies from woman to woman but usually has a prominent representation of lactic acid bacteria, often *Lactobacillus* species related to the ones that help turn milk into tart, spoilage-resistant yogurt (see page 548). Their metabolism keeps the vaginal secretions acidic and inhibitory to the growth of most undesirable microbes, and releases similarly protective peroxide and other compounds.

Volatiles from a healthy vagina include the usual medley of animal metabolic by-products. Studies suggest that some vaginal microbiomes produce acetic and lactic acids and little else, while others produce a stronger, cheesier range of straight and branched acids. When the normal microbial community is disrupted by antibiotics or other treatments—including spermicides and lubricants—undesirable microbes can invade and cause infection. These are mainly anaerobic bacteria and yeasts, which tend to metabolize proteins and amino acids in the secretions and produce ammonia, fishy-smelling trimethylamine and other amines, and both putrescine and cadaverine. Vaginal infections are often diagnosed by the presence of an unusual strong smell, and borderline cases can be clarified by means of the "whiff test," in which the physician adds an alkaline solution to a sample of the secretion to increase the volatility of the alkaline amines. Some healthy women also develop a strong amine smell during menstruation, apparently because the hormonal changes temporarily lower their bodies' ability to convert trimethylamine from their gut microbes into nonvolatile trimethylamine oxide.

The smells of semen are a very different and simpler story. Semen is the collective fluid secretion produced by a number of male organs to carry and support sperm cells on their journey up the woman's vagina to fertilize an egg cell. Since semen is held within a man's body until it's ejected during sex, microbes have little opportunity to grow in it and transform it. In addition to the usual faint volatile hum of animal metabolism, semen also carries the steroid hormone molecule androstenone, a relative of testosterone, and a handful of distinctive nitrogen volatiles. Many people can't detect androstenone; those who do usually describe it as musky and urinous. It's the nitrogen molecules that give semen its more particularly "spermous" smell.

Remember putrescine, which we first met along with cadaverine on page 62 as a breakdown product of proteins in animal cadavers? It's also a constructive molecule. Both animal and plant cells intentionally make its four-carbon, two-nitrogen chain as a building block for larger molecules called spermidine (seven backbone carbons and three nitrogens) and spermine (ten backbone carbons and four nitrogens). Both spermine and spermidine have a number of important functions in cells, including the stabilization and proper functioning of DNA and RNA, which sperm are jammed full of to deliver the male's genes to the egg. Together with putrescine, these molecules are readily oxidized to form pyrroline, the four-carbon, one-nitrogen ring that gives semen its distinctive smell, and is therefore described as semen-like or spermous itself. It's likely pyrroline that perfumes many city streets with the incongruous smell of semen! A number of plants scent their flowers with it to attract pollinating insects—not bees, but flies (see page 215).

SOME SMELLS OF VAGINAL FLUIDS AND SEMEN

Smells	Molecules	Sources
sharp, vinegar	acetic acid	microbial metabolism of secretion carbohydrates
alcohol, solvent	alcohol, acetone, etc.	microbial metabolism of secretion carbohydrates
cheesy	short & branched acids	microbial metabolism of secretion carbohydrates, amino acids
fishy, urinous	trimethylamine	microbial metabolism of secretion amino acids
semen, musky	pyrroline, pyrrole, ammonia	our metabolism + oxidation
musky, urinous	androstenone (steroid)	our metabolism (sex hormone)

So the smell of semen is a rare direct whiff of some of the innermost agents working in all the cells in our body, while vaginal secretions offer the body's general metabolic hum along with sharp traces from its protective commensal microbes. Not the most intrinsically appealing smells, either on paper or on skin, but in the right context they can captivate. According to recent classical scholarship, this is what the Roman poet Catullus may have hinted at two thousand years ago in the famously startling final image of his poem 13, "Cenabis bene, mi Fabulle, apud me." The poet addresses his friend Fabullus and says that he will dine very well chez Catullus as long as he brings along his own food and drink—and his own girl. And what would Catullus and his empty purse provide?

> In exchange you will receive pure love
> Or whatever is more sweet or elegant:
> For I will give an ointment that to my girl
> Venus and Cupid have given,
> Which, when you smell it, you will ask the gods
> To make you, Fabullus, all nose.

Armpit riches: goats, shallots, duck with olives

The smells we usually mean by the term *body odor* come from the armpits. Two thousand years after the authors of the *Problemata* asked why this is, we have a few hard facts and a reasonable theory. The most important fact: Our body goes out of its way to *compose* a special set of volatile molecules there. Why bother, when our skin microbiome generates plenty of volatiles from sebum and sweat? One idea is that specialized armpit smells are a relic of a stage in the distant past when our primate ancestors relied more on their chemical senses to detect and recognize friend and foe and family. A strong and particular smell emanating when the arms are raised may have been a way for individuals to assert their dominance within a group, a visual-olfactory version of a peacock's tail or lion's mane or ape's fire-engine-red buttocks, and closer to the noses of upright hominids than the anal and genital scent glands of other mammals would be. This general interpretation is consistent with another fact noted in the *Problemata*, that children lack strong body odors and develop them during adolescence, as they reach sexual maturity.

Whatever their original function may have been, armpit smells now have a very different significance: millions of people spend billions of dollars trying to hide them! And so though they might seem an unlikely subject for scientific inquiry, some of the major corporate makers of deodorants and fragrances have underwritten decades of research—and allowed their chemists to publish much of it. That research, together with the complementary work of sensory psychologists and geneticists, has brought a number of surprising insights into our nature as social animals, a growing appreciation for the importance of body odors in everyday life, and hints of an end to the long chemical warfare waged against them.

The *axillae* or *axillary vaults*, as the armpits are impressively known in the medical and scientific world, are said to be the richest microbial habitat on the human skin. They're our body's equivalent of the tropics, irrigated with secretions from abundant sweat and sebaceous glands, warm and humid because they're sheltered from the open air, and sprouting a jungle of hairs that wick nourishing secretions up from the skin surface and trap moisture. Moreover, the profusion of hairs means many follicles and ducts in which oxygen-sensitive anaerobes can thrive. Bacterial populations are estimated to be up to a million per square centimeter.

A number of these microbes are also found elsewhere on the body, notably species of *Staphylococcus*, *Corynebacterium*, and *Propionibacterium*, and it's no surprise that they generate many of the same metabolic by-products of protein-loving animal commensals: vinegary-cheesy short-chain acids, and sweaty-foot and cheesy short branched-chain acids. Oxygen in the air attacking sebum lipids is probably responsible for another couple of armpit volatiles, an eight-carbon alcohol and a ketone, which have metallic and mushroomy notes.

However, there's more to ripe armpit sweat than oxidation and microbial business as usual. Two other groups of molecules, unusual branched-chain acids and sulfur volatiles, make the smell distinctive and especially strong. These molecules are larger and more complex than the short-chain acids, purpose-built rather than waste products. Their source is the apocrine sweat glands, which mature and begin active secretion in adolescence. The glands make these key volatile molecules, but they also bind them to nonvolatile molecules, often sugars or amino acids, that prevent them from being released into the air. (As we saw above, our liver similarly binds xenobiotic food volatiles to dispose of them; see page 94.) These odorless precursors then generate odors when the volatile-nonvolatile bond is broken and the volatile portions are liberated.

What liberates the volatiles from the precursor molecules? Microbial confederates of the apocrine glands, their collaborators on the outside. A small group of bacteria—species of *Corynebacterium*, *Staphylococcus*, and *Bacillus*—have just the right enzymes required to break the volatile-nonvolatile bond (and be rewarded with the sugar or amino-acid portion). Sound familiar? In the previous chapter we saw that cats have a similar system for generating volatiles from their territory-marking pee (see page 80). The human system likewise provides a continuing slow release of body-marking volatiles even when we're not actively sweating.

What do these gland-created volatiles smell like once they're liberated? One set includes sterol relatives of testosterone, musky androstenone and urinous androstenol, but about half of U.S. adults tested can't smell these molecules, so they're not thought to be significant in body odor. Our ancestors may well have been better at detecting them. More significant are some unusual branched-chain acids six and eight carbon atoms long. **Hydroxymethyl hexanoic acid** and others

contribute rancid, animal, cumin-like, pungently sweaty notes, and are the source of the goatish smell that the Greeks and Romans often invoked, as the Roman poet Ovid did when he wrote that the women readers of his *Art of Love* don't need to be warned: "no wild goat in the armpits." The methyl- and ethyloctanoic acids in human armpit sweat are the very same molecules that scent goat and sheep meats, milks, cheeses, and wools (see page 85).

Another intriguing group of slow-release armpit volatiles is a set of sulfur molecules. Two of them are straight five- and six-carbon chains, and two others are branched four- and six-carbon chains. The branched four-carbon chain (**3-sulfanyl-2-methyl butanol**) has its branch just one carbon atom over from the otherwise identical molecule in cat pee. So we and our cats are probably using very similar metabolic machinery to make these molecules. And what do they smell like? Of the straight-chain sulfur volatiles, one smells like cooked meat, the other like grapefruit, tropical fruit, and onion. One of the branched-chain volatiles smells meaty, the other oniony.

Chemists first identified these sulfur-containing sweat volatiles about twenty years ago, but the food echoes were recognized long before. Around 1900 the Belgian novelist Joris-Karl Huysmans, a connoisseur of smells, wrote a brief sketch about armpits ("Le Gousset"), in which he noted "the harsh fumes of the buck" rising from the shirts of Parisian workmen in the summer, and described the smell of haymakers in the countryside as "the musky lingering odor of wild duck cooked with olives, and the sharp odor of shallots"!

Of course the resemblance of sweat volatiles to foods works both ways. It means that meat and onions also smell like aspects of sweat: an odd fact with very practical applications. Three different industrial laboratories independently discovered the meaty-oniony-fruity sulfur molecules in human sweat and published their findings in 2004. But the laboratory at the Swiss company Givaudan, which makes both fragrances and flavors, had already found them several years before, and kept their discovery under wraps while they prepared patents for the use of sweat volatiles— synthesized in the lab—as food additives for enhancing meaty flavors.

SOME SMELLS OF HUMAN ARMPITS

Smells	Molecules	Sources
vinegar, cheesy	acetic, propionic acids	microbial metabolism
sweaty feet, cheesy, rancid	methylpropionic, methylbutyric acids	microbial metabolism of proteins

continued

Smells	Molecules	Sources
animal, goaty, cheesy	ethyloctanoic acid	microbial liberation from human precursor
metallic, mushroomy	octenol, octadienone	oxidation of skin lipids
sweaty, sharp, cumin	methylhexenoic acid, hydroxy methylhexanoic acid	microbial liberation from human precursor
meaty	methyl sulfanyl butanol, sulfanyl pentanol	microbial liberation from human precursor
grapefruit, tropical fruit, onion	sulfanyl hexanol, methyl sulfanyl hexanol	microbial liberation from human precursor

So that's the basic recipe for armpit sweat: touches of vinegar, cheese, metal, mushrooms, rancid goat, meat, grapefruit, and onions. Rich! And variable from one person to another. Huysmans rhapsodized that underam odor is "diverse as hair color . . . no aroma has more nuances, it plays across the entire keyboard of smell." It's thought that differences arise from variations in apocrine-gland activities and skin microbiomes. There seem to be some broad trends: for example, a 2015 study found that European women produce more of the meaty-oniony-fruity sulfur volatiles; men, more of the sweaty-goaty-cheesy branched acids, and more volatiles overall. Men also tend to support more *Corynebacteria*, which are the most active volatile producers.

Limiting body odors, West and East

Underarm smells were important enough in the lives of our distant ancestors that their bodies evolved an intricate system for producing them. With the development of human culture came new ways to assert dominance, and increasingly crowded and indoor living conditions in which body smells would have become inescapable, perhaps therefore less meaningful, and eventually annoying. In the ancient Mediterranean and China, strong undisguised smells were signs of a low position in the social hierarchy, a lack of the resources necessary to wash regularly or mask smells with perfumes. The Romans learned to use alum to block the sweat gland pores; and historian Charles Benn reports that courtiers in medieval China hung bags in their armpits that contained aromatics like cloves and frankincense, and also mineral lime, whose alkalinity limits the volatility of short-chain acids. In modern times, George Orwell famously observed in *The Road to Wigan Pier* that leftist intellectuals seldom had anything to do with the lower classes that they championed, for a simple reason, his emphasis: *"The lower classes smell."*

Nowadays strong body odors are less evident than ever thanks to the cosmetic sciences, which have come up with effective chemicals for reducing sweat-gland secretions (especially aluminum chlorohydrate, a knockoff of the Roman alum), and suppressing the skin microbes that transform them. Unnatural as these chemical deodorants are, there's good evidence that the biological evolution of our species has been moving in the same direction. Modern science has confirmed what East Asian cultures have known for many centuries: that Asian peoples emit far fewer goaty-meaty-oniony volatiles than Europeans do. It's evident from historical documents that when the Chinese first encountered travelers from the West, they were shocked and repelled by their smell. In medieval Tang China (roughly 600–900 CE), which had regular contact with people from Central Asia and India, strong body odor was called *hu chou*, or "barbarian stench." Because *hu* could also mean "fox," animals noted for their smelliness and wile, fables were told of fox-people from foreign lands who hid their telltale smell with perfumes.

Why do Westerners have stronger body odors than East Asians? Genetics. There are specific human genes that direct the apocrine sweat glands to construct both the odor precursors and the cellular machinery for delivering them to the skin surface. Scientists have identified a gene that is essential for secreting the odor precursors, and have found various inactive versions of this gene that arose via mutation. We all carry two copies of this ABCC11 gene, one from each parent. Nearly all Africans and Europeans carry two active genes, while nearly all native Chinese and Koreans carry two inactive genes. South and Central Asians and Native Americans tend to carry one of each. The same gene is responsible for a parallel difference between East Asians and other peoples; while Africans and Europeans secrete a yellowish "wet" sticky earwax that has a sharp, cheesy smell thanks to short-chain acids, East Asians and Native Americans tend to secrete whitish "dry" earwax: not much of it, powdery rather than sticky, and with only a faint smell.

The geographic distribution of the inactive ABCC11 genes suggests that the mutation arose during a migration of humans out of Africa and Europe and into Asia around fifty thousand years ago. Its strong prevalence in Asia suggests that it has somehow been advantageous. One possibility is that along with minimizing physical signs of female fertility and tucking body smells under the arms, minimizing those smells may be another of the modifications to our ancestral primate body that have tended to reserve sexual signals for relationships that are already close, reduce sexual competition, strengthen monogamous bonds, and encourage social cooperation.

Acknowledging our emanations

Of all the smells we emit, including the standard wastes of our own cells and our gut microbiome and the peach-coconut notes from our scalp yeasts, armpit odor is the most significant. The human body directs and amplifies the production of those volatiles, and like the identical or related molecules produced by goats and sheep and cats, they get the attention of others in the species—and not just when they're offensively strong. There's intriguing evidence that our brains pay attention to human body smells from the very beginnings of our lives. Researchers in Europe have found that the odorless sweat precursors are present in the amniotic fluid that bathes a fetus in the womb. They're also in colostrum, the fluid that the mother secretes from the breast just before and after giving birth, and in the initial breast milk. The volatiles might be released from amniotic fluid by microbes in the birth canal, or when colostrum or milk encounters bacteria on the mother's skin or in the infant's mouth. If they are, then hints of those animal-oniony smells could contribute to the newborn's well-documented ability to orient to the nipple, recognize her mother's body odor, and bond with her.

There's more to human body odor than the obvious strong smells and volatiles that scientists have been able to catalog to date. Many studies have shown that untrained people can often smell a difference between everyday armpit sweat and the sweat of volunteers who have been induced to feel stressed or fearful or anxious or sad, or whose immune system has been revved up by the injection of traces of a microbial toxin, which simulates an infection. These emotional and physical states cause complex changes in body chemistry. Untrained smellers also can often distinguish the body odor of strangers from blood relatives and romantic partners. What are the volatile clues? No one yet knows, in part because we don't consciously notice them and so can't describe them. But these discriminations indicate our olfactory alertness to the identity and state of other significant creatures close by.

There's a flip side to this heightened attentiveness to body odors: namely the relaxing pleasure that we get from the smells of loved ones. The psychologist Donald McBurney and others have found that romantic partners, parents, children, and siblings commonly smell each other's clothes and bedsheets for "olfactory comfort," an extension of the well-established phenomenon of *contact comfort*, the calming, reassuring effect that bodily contact has on infants. Here it seems that our attention is drawn to the generic smells of human presence as well as the particular volatiles that distinguish our loved ones, both their own and those they borrow from soaps and scents.

McBurney gave his 2012 essay the title "Olfactory Comfort in Close Relationships: You Aren't the Only One Who Does It." In the modern world, body smells are seldom openly sniffed at or enjoyed or discussed. It's impolite to assert our presence with our volatiles, so we do what we can to suppress them and not force others to notice them. When we do encounter them, they come as intrusions into our mostly smell-less personal airspace.

But there are hints that body smells may make something of a comeback. Some modern scholars of body odor, acknowledging the intricate cooperation between our apocrine glands and commensal skin microbes, have suggested that abolishing it may not be wise. Researcher Yoshihiro Hasegawa and colleagues at the Kao Corporation in Tokyo, which markets Ban brand antiperspirant among many other similar products, concluded their pioneering 2004 study of the human sulfur volatiles by saying:

> Although it is controversial, we think that axillary odor is not just malodor. It probably does have a function, maybe pheromonal or emotional effects, and only became undesirable in the modern age. . . . It might now be possible to develop less aggressive methods for a more targeted deodoration that respects the symbiotic relationship between humans and microorganisms.

And Johan Lundström, a sensory psychologist at the Monell Chemical Senses Center in Philadelphia, wrote in 2009 that "it is reasonable to begin pondering ways to enhance the positive emotional and informational signals concealed within

body odors, while simultaneously reducing the conscious negative odor percept."
These ways could involve shifting the balance among our skin microbes so that
strong odors are reduced and other, more subtle signals become more prominent.
Perhaps one day we'll be able to apply the olfactory equivalent of an audio equal-
izer to ourselves, modulate and refine our smell identities, lower the goat and raise
the fruit.

Finding ourselves in foods

In the meantime, we're already exercising our entirely natural interest in body
smells in a more socially acceptable way than sniffing ourselves and one another:
namely by enjoying them in our food and drink.

If you go back and scan the table of armpit smells above, you'll see that the
individual volatile components have qualities that the researchers describe as like
cats, goats, cheese, cumin, meat, onions, grapefruit, blackcurrants. They refer to
these animals and foods because the same or very similar molecules occur in those
animals and foods. We identify the smell qualities of individual volatile molecules
through our prior experience of them, and the memories and associations that
come from that experience.

But remember the example of Alex Atala and his lemongrassy Amazonian
ants: to Amazonians who don't grow up with it, lemongrass tastes like ants. Because our description of smells is contingent on what we experience them in first
or most frequently, it's interesting to pause and notice when and where human
beings would first have encountered particular body volatiles. Clearly the smells
of our ancestors' bodies would have been familiar to them from their earliest days
in Africa, long before they encountered Eurasian goats and sheep and onions, long
before they cooked meat or made cheese, even before they were officially *Homo
sapiens*. So body smells would have been their reference point for the same or
similar volatiles showing up in other animals, in onions and roasted meat, fruits
and coconuts and spices.

What might that mean? For early humans, perhaps encountering whiffs of
themselves in roots and fire-scorched foods and aromatic seeds made those initially strange materials more approachable as foods: a primal version of "If you like
that, try this." Maybe early agriculturalists even came to *cultivate* human smells in

some foods. Cheeses, for example. Cheese is a solid concentrate of proteins and fats, the same major ingredients of our body, and the surface of freshly made cheese is readily colonized by some of the same protein- and fat-eating bacteria that live on our skin. Some traditional cheeses are ripened by the laborious application of a sweat-like brine over the course of days or weeks, and those cheeses develop a smell very close to the smell of toe cheese. Why go to all that trouble—or to the similar trouble it takes to make the many other cheeses with strong animalic smells—when coagulated salted milk will keep very well without it? Those smells must have given pleasure—as they still do, at least to fans of Limburger and Époisses and Appenzeller. Early cheesemakers discovered that they could transform bland coagulated milk into some of the most sensorially stimulating materials in their world. Maybe a few also discovered that they could coax cheeses into smelling a bit like themselves, and so give them a human note, a touch of their author.

If that sounds far-fetched, consider the testimony of connoisseurs in China, where mild body odors had evolved long before the invention of cheese, where frequent bathing has long been common, and where dairying has been rare. The Chinese enjoy strong-smelling foods, including notorious rotted-fermented "stinking" vegetable and soy products (see pages 553, 556), but many can't stand Western cheeses. To clarify why, Fuchsia Dunlop, an English authority on Chinese food, brought a set of cheeses to Shaoxing, the capital of stinking preparations, and presented them to a number of prominent Chinese chefs. As she reported in the *Financial Times* in 2011, the chefs told her that "the cheeses had a heavy *shan wei* (muttony odor), an ancient term used by southern Chinese to describe the slightly unsavory tastes associated with the northern nomads. Another said that the selection 'smells like Russians.'" To at least some Chinese, cheeses smell like the goaty foreigners who make them, and not at all like their own stinking vegetables, which involve little protein breakdown and develop only the generic short-chain acids and sulfur volatiles.

Today the smells of the human body may be socially embarrassing, but for children, and privately for adults, they're often irresistible. Cheeses that smell like feet, goaty lamb and wool, onion and garlic and sweaty spices, wine and blackcurrants with a touch of cat pee: all of these offer volatiles that remind us of ourselves and one another, but in something other than ourselves, at unembarrassing remove, and at moments of our choosing. (The animalic ingredients in perfumes do something similar; see page 465.) It's not exactly that we like these smells the way we like the smell of a strawberry or rose, and many people actively dislike them. But they catch our attention, momentarily make us all nose, because they were once central to our animal life, and fleetingly restore that lost sensory dimension to it.

Part 3

LAND PLANTS

Independence, Immobility, Virtuosity

Chapter 7 · SWEET SMELLS OF SUCCESS

....

[The archangel Raphael] now is come
Into the blissful field, through Groves of Myrrh,
And flowering Odours, Cassia, Nard, and Balm;
A Wilderness of sweets; for Nature here
Wantoned as in her prime, and played at will
Her Virgin Fancies, pouring forth more sweet,
Wild above Rule or Art; enormous bliss.
Him, through the spicy Forest onward come,
Adam discerned, as in the door he sat
Of his cool Bower.

> • John Milton, *Paradise Lost*, book 5, 1674

A t last, things that smell *good*! Flowers and spices and forests, actual sniffable things worthy of perfuming an imagined paradise, and all are creations of the plant kingdom. This and the following several chapters are
devoted to land plants and their many happy contributions to the osmocosm.
(We'll get to the very different seaweeds and their relatives in chapter 15.)

Why postpone botanical delights until after mineral and animal malodors? In
part to appreciate the contrast more fully. And in part because plants advanced
the possibilities of the volatile far beyond the mineral and animal worlds. Though
animals are sophisticated avatars of Hero Carbon, able to sense their surroundings
and move at will, their volatiles are relatively few, simple, and incidental. Carbon's
land-plant avatars may not move much, but they are brilliant chemists. Their volatiles are numerous beyond counting, and they're purpose-built—often to exploit

the senses and mobility of their animal neighbors. Any delight that plants inspire in us also serves their own interests.

In John Milton's version of biblical creation, sweet plant smells are traces of the world's original innocence, the freedom from evil and death that it enjoyed before Adam ate the fruit of the Tree of Knowledge. Well, according to the knowledge that Adam's descendants have harvested from that tree, plant smells are actually the *product* of life-and-death constraint, the struggle of living things rooted in the ground to survive on a planet crawling with hungry animals. For all the beauty of Milton's poetry and the tradition that underlies it, this scientific account inspires its own sense of wonder and gratitude. Sweet plant smells are all the more precious for being so hard-won.

Plants that live on land are both vulnerable and resourceful. They can't run away from predators, and they can't seek each other out to reproduce. But as descendants of the microbes that invented photosynthesis eons ago, they have access to abundant energy from the sun and carbon from the air. And they've invested these resources in the unparalleled invention and manufacture of carbon chains and rings, thousands upon thousands of different kinds, and in amounts sufficient to build massive trees. Some of this material wealth they use to deter roving animals from feeding on them, and some to invite animals to help them reproduce. Their obvious inventions include defensive structures like thorns and bark, and beckoning flowers and fruits. Their invisible but smellable inventions can act as missile-like chemical weapons and warnings, and as come-hither floral and ripening perfumes. Plants generate volatiles to persuade animals both to spare them and to serve them.

The volatile repertoire of plants goes way beyond life's standard starter set and the protein fragments of animal bodies. It's a cornucopia overflowing with themes and variations on carbon chains and rings and smells. Many of our favorites have been successful twice over. They helped their makers to survive in the natural world, and then persuaded our species to adopt their makers from the wild and coddle them in gardens and farms.

On then to the blissful fields! Here's a brief map of the pages ahead. This chapter sketches the success story of the plant kingdom, and addresses a basic puzzle: Why do human animals find its anti-animal weapons sweet and blissful? The following chapter introduces the main families of plant volatiles: why and how they're made, some of their outstanding members, and the smells they contribute to the osmocosm. Chapters 9 and 10 then trace a path through a virtual botanical garden, pointing out individual plants of interest and the volatiles that contribute to their scents. And chapters 11 through 13 proceed through imaginary markets stocked with edible plants: kitchen greens and herbs, seeds and spices, fruits.

It may all seem a little daunting at first. So many molecules, so many plants! But it's our good fortune that Hero Carbon has poured forth such an abundance of both. There's no need to read these chapters straight through. Feel free to wander and jump and backtrack and pause, as your own nose and mind and experiences lead you.

The greening and scenting of Earth

In chapter 1, I invited you to imagine yourself in the role of assistant to the Chef of the cosmos, following the smells as it simmered and our planet evolved. We paused that story shortly after the Great Oxygenation, when the success of the first photosynthetic microbes, ancestors of the plant kingdom, cleared Earth's atmosphere of volcanic and anaerobic hydrogen sulfide. Let's pick up where we left off, and catch the emergence of plants and their smells.

Don your white jacket and lean over the Earth once again, watching and sniffing as time rolls on. For a long stretch it's likely that not much changes. You and the Chef see a planet mostly blue from its vast oceans of water, and various barren landmasses, pocked with volcanoes and glittering with lakes, that slowly grow and shrink and grow and eventually occupy about a third of the surface, constantly shifting position. From volcanoes and vents rise occasional puffs of hydrogen sulfide and ammonia and other primeval volatiles, and from the waters, milder methyl sulfides, amines, and chlorine-bromine seaside volatiles.

Then, after a couple of billion years, come changes that for the first time make you breathe deeper and bring smiles to your faces. They arrive in three overlapping phases.

Around 500 million years ago, the landmasses gather to form three main continents near the South Pole. You see their edges and their lakes and rivers take on an outline of green, and you pick up the smells of green leaves, at first very faint, then stronger and stronger as the green coloration spreads.

Then the separate continents gradually rotate and merge into a landmass that spans one side of the planet from pole to pole: the supercontinent known as Pangaea. Its empty interior fills in with green, especially along the equator. And up comes a burst of woody smells, cedarwood and pine.

Then Pangaea begins to break up, this time into masses that are vaguely suggestive of the continents we know: a caricature of South Asia, the eastern slice of North America, the main bulk of South America, Antarctica, and Australia. A wedge off the east coast of Africa slowly moves north, collides with southern Asia to form the vast peninsula of India. And up from these masses comes a kaleidoscope of smells, flowery and fruity, minty, spicy, getting more and more complex as the continents settle into their current configuration.

What you and the chef have witnessed is Hero Carbon assuming the mantle of plant life on Earth's land and jumping to a new level in the cosmic game of complexity. If the past billion years had just passed for you in the space of an hour, then the first whiffs of green leaves would have come about thirty minutes ago, the woodsy smells fifteen minutes after that, flowery-fruity-spicy-minty after another five minutes, and a burst of variations on those themes just two minutes ago.

Now let's zoom in for a closer look and sniff at the plants themselves.

Life grows up . . . and up

The story of the plant kingdom's volatile riches begins with its occupation of the land and the air above it. The Earth's solid surface was barren when early forms of photosynthetic algae washed up onto its edges five hundred million years ago. There was nothing like nice loamy soil, just patchy remains of whatever pioneer microbes had managed to survive on wet minerals. Multi-celled algae could thrive as simple sheets, all of their cells bathed in water and exposed to sunlight, and the earliest land plants may have been not much different, something like puddle scum. But life on land encouraged the pioneers to develop a third dimension. Water and minerals collected in gouges and crevices and weathered rock debris on the ground, while the sun shone from above. These essential resources pulled on plant architecture from opposite directions, and over time drew it into an elongated form, with roots probing the land for water and minerals, green photosynthetic leaves seeking the sun, and a set of connecting tubes in between.

The earliest of these land forms were ancestors of today's ground-hugging mosses and liverworts. Then Hero Carbon developed a molecular innovation that ultimately lifted plant tissues hundreds of yards skyward: a material named *lignin*, from the Latin word for "wood," which gives the wood of tall trees its mechanical strength, chemical stability, and resistance to attack by microbes and animals. Lignin isn't a single definable molecule. It's the aggregation of countless small building blocks into a strongly bonded three-dimensional network that can be enlarged indefinitely. The building blocks are rings of six carbon atoms variously "decorated" with short carbon chains and oxygen atoms. Plant cells deposit lignin in the walls that surround them and hold them together.

Lignin raised plants off the ground in two different ways. It forms a continuous, rigid molecular skeleton that can support structures weighing many tons. And it repels water and therefore makes an excellent rigid lining for the tubes that conduct water from roots to leaves. Lignin made it possible for plants to develop large areas of light-gathering tissues—canopies of leaves that can span hundreds of square yards—and raise them above nearby obstructions, including other plants. The first tentatively woody plants were club mosses, ferns, and horsetails. Over the course of a few hundred million years, they spread across the landmasses, created soils by dissolving rock with their acids and dropping their own debris, grew to fifty yards high, and formed dense forests. Of the plant families we're familiar with today, conifer and ginkgo trees arose about a hundred million years after those early forests, and the first flowering plants and broad-leaved trees a hundred million years after that, a hundred million years before the present.

Lignin has been such a successful invention that it now accounts for a third of all the biomass on the planet—and much of the industrialized world is powered by energy extracted from the fossilized remains of the first lignified forests, the carbon deposits we call coal. As we'll see, though lignin and its building blocks aren't volatile molecules themselves, they are the source for many familiar smells today, both nasty and nice.

The boon of growing up: energy and carbon chains galore

The bountiful smells of the plant kingdom wouldn't exist without the bountiful resources it has for spinning off carbon chains and rings. As descendants of aquatic microbes that had invented photosynthesis billions of years before, land plants can harvest vast amounts of energy from sunlight, and carbon from the air's carbon dioxide. The most awe-inspiring evidence of this power is the massiveness of trees. The record holders, redwoods in the coastal and mountain forests of California, weigh many thousands of tons, and rise more than twenty stories high. Nearly all of that mass is Hero Carbon in the guise of the supporting wood's rings and chains, with only a tiny fraction in the photosynthesizing cells of the leaves and the actively growing branch tips. This extravagant construction has been called "the debris of photosynthesis": as long as the sun shines and the temperature is right for all the chemical machinery to work, new carbon chains keep streaming out of the leaves, far in excess of what's needed to keep the living cells running, and the plant has to do something with them. Trees turn many of them into wood, which is so durable that it accumulates, year after year, ring after ring.

This plenitude of energy and carbon also gives plants leeway to divert some of it into making other molecules, to let Hero Carbon explore new structures, new bits of machinery that might come in handy for one purpose or another. One way this can happen is for plants to make extra copies of their genes, have one copy of one gene become mutated, and have the altered product of that mutation ultimately lead to some improvement or useful addition. Some plants have double the number of genes that we do, and most plants synthesize a far greater variety of molecules than we do—despite having no brain or muscles.

Among those diverse plant molecules are the volatiles that give trees and herbs, flowers and fruits their smells. They aren't part of the essential photosynthetic machinery, or the cell-constructing machinery, or the support materials, but they're produced through detours in the pathways that make these essentials. The detours persist because their volatile products have been adaptive: they have helped plants thrive on land by meeting the land's challenges.

The challenges of plant life on land: self-defense

The greatest threat to the success of early land plants was the presence of other life forms that could take advantage of their productivity and immobility and consume them as food. Opportunistic microbes would have accompanied them from the waters, and there's fossil evidence that some of the first terrestrial animals, ancestors of the insect tribe, were chewing on leaves and piercing fluid-conducting tubes very early in the history of plant life, when it was still confined to wet coastal areas.

Plants overcame their physical vulnerability by investing some of their considerable resources in anti-microbe and anti-animal defenses, both physical—lignin-toughened tissues—and chemical. They've been spectacularly inventive with chemical defenses that react and interfere with their enemies' essential systems: plant toxins are thought to number in the tens of thousands. Many of these aren't volatile. Tannins, the molecules that cause the mouth-drying, rough sensation that we feel when we drink tea and red wine, are ancient nonvolatile toxins; newer ones include bitter alkaloids like curare and strychnine and nicotine, and molecules that release cyanide.

Volatile defenses, however, offer the advantage of going airborne like missiles, being detectable by animal sensory systems at some distance and in trace amounts, and so possibly deterring predators before they can do much damage. And because many insects rely on volatiles to recognize their own friends and foes, plant volatiles that mimic their signals can confuse and misdirect them. In fact, most of the volatiles emitted by plant leaves and seeds, and the wood and bark of trees, are warnings and weapons that help defend against animals and disease microbes. The smells of freshly cut grass, mint and lavender, vanilla and cloves, pine and cedar and cinnamon: all the fumes of chemical warfare!

The combination of tough lignin and small-molecule chemical defenses has worked well for land plants. In the oceans, most of the wild plant biomass produced every year is consumed by animals, but on land, less than a fifth is eaten. In the oceans, animals are estimated to outweigh plants by a factor of thirty; on land, plants outweigh animals by a factor of a thousand. Sweet smells contribute to the difference.

Land-plant challenges: sex and the flower

A second major smell-inspiring challenge for plants stuck in place on land: How could they convey their fragile reproductive cells to each other to fuse and give birth to the next generation? Aquatic plants could simply release bits of themselves into the surrounding water, but cells exposed to air quickly lose moisture and die. Early wetland mosses and ferns made do by entrusting naked sperm and egg cells and fertile spores to standing water and drops of rain and dew. Later plant forms developed armored transport vehicles for spores and then for sperm cells, the structures we call seeds and pollen grains; and with these and the help of the wind, ginkgos and cycads and ancestors of the conifer trees colonized drier regions and came to dominate the planet-spanning continent Pangaea. Then Hero Carbon turned its attention to the egg cell and came up with the remarkable structure we call the flower. As Pangaea broke apart, the flowering plants began their rise to dominance in the plant kingdom and made their special contribution to the Earth's smells.

The essential feature of the flower is the *ovary*, an envelope of plant tissue that surrounds the egg cells and the seeds they form. Whereas conifers and cycads protect egg cells in the recesses of rigid cones, flower ovaries are exposed and un-hardened. They provided a new reproductive platform that plants could shape and accessorize and thereby explore new ways of combining genes from different parents.

For us it's the accessory structures that define a flower, above all the petals, whose name comes from an ancient root meaning "to spread out." The petals are a cluster of leaflike sheets that cover the flower tightly while it matures, then un-fold to reveal the delicate pollen-bearing stamens, stalks called stigmas that re-ceive pollen and transmit it to the egg cells within the ovary—and the inner surfaces of the petals themselves, which are often beautifully colored, patterned, and scented. These visual and olfactory features that appeal so powerfully to our sense of the beautiful have a very practical function. They recruit special agents to carry pollen from plant to plant in a far more efficient and selective way than wa-ter and wind can.

Those agents were mostly insects, which had been flying over the land since the era of the earliest seed-bearing plants. The first insect pollinators may have been beetles and flies; later recruits, as Pangaea was breaking up, were the ances-tors of modern-day moths and butterflies and eventually early bees. Visual pat-terns and volatiles help get the attention of insects while they're still some distance from a flower, guide them to it, and help them recognize and seek out reliable sources of sugary nectar, their main inducement for visiting and incidentally carry-ing pollen grains. There are many variations on this theme of recruiting insects as go-betweens, and many flowering plants—notably the grasses—still rely on the wind and so are neither showy nor scented. But the flower platform gave plants access to the largest animal group on the planet—there are something like three million different insect species—and with it an unprecedented ability to explore new partnerships and reproductive strategies.

The results of that exploration are impressive. Today there are about 1,000 species of nonflowering conifers and close relatives, but 250,000 species of flowering plants. The smells of flowers are both sweet and fascinating; we'll delve into dozens in chapter 10.

Land-plant challenges: offspring and fruit

A third problem for land plants to solve: How could they get their progeny out into the world? Aquatic plants don't have this problem; the reproductive cells of algae come together in the waters near their parents and develop independently from them. But the embryonic offspring of land plants develop while firmly attached to their grounded parents. If seeds simply detach and drop when they're mature, they compete with one another and their parent for the same small patch of soil. So land plants developed a number of different seed transport systems to help their offspring leave home. Flowering plants were especially inventive. Some came up with wind-catching fluff, others with fur- and sock-catching burrs. Still others take advantage of animality in full, its mobility *and* sensitivity *and* hunger: they chaperone their seeds with showy masses of nourishment for animals to detect, carry away and consume, and in the process scatter an intact seed or two. These masses are fleshy fruits, by far the most flavorful foods in the natural world thanks to their endowment of tasty sugars and acids and aromatic volatiles.

Botanists define a *fruit* as any structure that derives from a flowering plant's ovary and carries seeds—including fluffs and burrs, nuts in shells, and wheat and corn kernels. Fleshy fruits—just plain *fruits* in everyday usage, and henceforth in this book—became especially popular seed chaperones with the relatively recent emergence of birds and mammals. The fruits aimed at these animals tend to start out harsh-tasting and inconspicuous, green like the surrounding foliage, until the seeds are mature. At this point they hail the local animal services by *ripening*: by changing color, emitting volatile molecules, and becoming soft and sweet inside. At its most effective, this system induces animals to consume the fruit flesh without damaging all the seeds; those that are swallowed intact even have the benefit of being deposited in a nourishing pile of dung.

Plants developed sets of flower smells to establish pollination relationships with insects, and fruit volatiles to establish seed-dispersing arrangements with larger animals, our ancestors included. "Flowery" and "fruity" are broad categories for describing smells, and they do overlap, but the distinction is rooted in biology and is useful. Floral smells have no immediate biological significance for us. Fruit smells and flavors, on the other hand, are our original template for the sensory richness of foods, for deliciousness, and therefore for the possibilities of creating deliciousness ourselves. Much more on flowers, fruits, and cooking in later chapters.

Flying flowers and crawling spices

Remember Brazilian chef Alex Atala's Amazonian cook Dona Brazi, who thought that his Asian ginger and lemongrass tasted like her local ants? To anyone who shops in a supermarket, the direction of her comparison sounds incongruous. But step back from the happenstance of personal experience, ask whether ants or plants should be the point of reference, and it turns out that Dona Brazi's perspective is truer to the primal relationship between plants and insects. The insects may well have made many spicy- and flowery- and fruity-smelling volatiles first, long before grasses and lemons and ginger came to exist. At the very least, insects goaded plants into making many of the volatiles that they do, and deserve a share of the credit for the pleasure that plants give us.

Insects are descendants of shrimp-like creatures that were the first animals to crawl out of the waters onto land. When they did, these pioneers and the plants that preceded them had to adapt to each other's presence. Both developed systems of airborne signals to communicate with their own kind and with each other.

Immobile plants mainly needed means to discourage the animals from feeding on them, and they could and did do this with frankly toxic molecules, volatile and nonvolatile. But they could also prevent the first bite by emitting the pests' own alarm or repellent volatiles to confuse them, or volatiles that would attract other animals to feed on the pest species. And an advantage may have come to early plants that managed to attract animal companionship. Their moss and liverwort descendants are prolific volatile chemists, capable of synthesizing members of all the major families of plant volatiles. Some of these volatiles attract tiny "microarthropod" insect relatives, springtails and mites, whose presence appears to facilitate the transfer of reproductive cells from one plant to another. Sound familiar? This is a primitive version of insect pollination that may predate the evolution of the flower by hundreds of millions of years.

The Austrian biologist Florian Schiestl has surveyed which plant and insect lineages are capable of making which volatile molecules, and suggests that "plants use the insects' own chemical language to influence their behavior." He finds especially clear evidence for this in the case of flowers and the insects that specialize in pollinating them. The ancestors of moths and butterflies developed the chemical machinery to make characteristic flower volatiles called benzenoids long before the flowering plants did. (Lowly moss-tickling springtails also make benzenoids.) Less charmingly, some relatives of the philodendron and dieffenbachia emit the same volatiles that attract beetles and flies to animal dung and carrion, and thus enlisted them as pollinators and even seed dispersers (see page 214).

With these remarkable relationships in mind, flip back to chapter 5 to appreciate the roster of insect smells—stinky and musty, but also herbaceous, fruity, flowery—and especially Fritz Müller's description of the aromatic Brazilian butterfly. Flying flowers and crawling Amazonian spices: fine emblems of the shared authorship of the plant world's volatile delights.

Getting acquainted with plant volatiles: smells and associations

In the next chapter I'll introduce a selection of the volatile molecules that plants make to survive on a planet crowded with predators and potential partners. Scientists have isolated many thousands of different volatiles from plants, though only a few dozen of them are responsible for most of smells we would recognize. A few dozen is still a lot of molecules and smells, though they fall into a handful of distinctive families, and members of those families often share a family resemblance.

When it comes to assigning particular smell qualities to particular molecules, you'll notice that most molecules are assigned more than one smell. And often the several smells aren't just slight variations on a basic smell; they can be very different. The eight-carbon alcohol octanol, for example, is described as orange, mushroom, and melon. Similarly, one smell quality can be applied to several different molecules, as melon is to octanol, nonenol, and nonenal. Don't worry: the confusion is in the very nature of plants and in the way we perceive smells. Plants deposit countless different molecules in their wood and leaves and flowers and fruits, because they can. The smell of any particular item is a composite of many different volatiles, of which perhaps a dozen or two predominate. So several different molecules can remind us of the same item. And then on our end, we encounter many of the same molecules in different plants: so one molecule can remind us of several different items. In fact, that's part of the interest of paying close attention to flavor. When we do, we notice echoes and rhymes in very different things.

With any perception of smells, there's a subjective element involved: different people have different sensitivities to molecules and different associations for them. Few of us, apart from professional flavorists and perfumers, ever have the chance to smell pure versions of the molecules we'll meet. In the tables that follow, I've relied for broad characterizations of molecules on perflavory.com, a superb online database, which lists smell qualities that have been reported in the professional fragrance and flavor literatures.

"Chemical" smells in natural materials

Most of the plant volatiles we'll meet have smells that remind us of their familiar plant sources, and they're often named for those sources. Vanillin, for example, is the primary volatile in vanilla; menthol, in mint; thymol, in thyme; eucalyptol, in eucalyptus leaves—and they smell like vanilla, mint, thyme, and eucalyptus, respectively. However, some plant volatiles—including thymol and eucalyptol— have smells that can remind us of less pleasant "chemical" or "medicinal" sources: household and industrial treatments like turpentine, paint thinners, sanitizers, and insect repellents; and personal-care products like antiseptics, mouthwashes, and pain-relieving ointments.

There's a good reason for these associations: we humans put plant volatiles to use in many different ways, and we've been doing so from before recorded history. Above all, volatile-rich plants help repel or kill microbes, insects, rodents, and other pests that would otherwise infest our dwellings and storehouses and bodies.

Think, for example, of keeping clothes in moth-repellent cedarwood chests. Even more effective than the plants themselves are concentrated extracts of their volatiles. We know from the Roman writer Pliny that the Egyptians extracted volatiles from cedarwood by heating it and collecting the "oil" that flowed out, a version of what we know today as turpentine. Three thousand years ago, the Egyptians used cedar oil in mummification to preserve human and animal bodies against decay and pests (they had used raw tree resins two millennia before that). At some point the oil-like properties of turpentine led to its use as humankind's first good solvent for greasy and tarry materials. Pine turpentine eventually became a standard ingredient in cleaning fluids of all kinds—to which it also lent a pine smell that came to connote cleanliness.

The volatile oil of eucalyptus trees arrived much later, with commercial-scale manufacture beginning in Australia around 1850, but it has been used ever since as an antiseptic and insect repellent. Around 1880 a St. Louis doctor combined eucalyptus oil with alcohol and the volatile oils of mint, wintergreen, and thyme to make a general-purpose antiseptic that was later marketed as the first commercial mouthwash (see page 106). To this day, manufacturers routinely use the key volatiles in these oils—eucalyptol, menthol, methyl salicylate, and thymol—to formulate a wide range of personal-care and cleaning products.

So certain plant volatiles simply *do* smell like manufactured solvents and cleaners and medicines. Most people today are more likely to encounter these molecules first and most often in bathrooms and hospitals and malls, not in the wild and garden where humankind first found and cultivated them. To smell the plant in the product and the product in the plant is to be reminded that these molecules were invented by plants eons ago, have a central role in plant life, and are a legacy that we continue to rely on today, even if we now manufacture some of them ourselves.

Non-smell volatile qualities: cooling, warming, irritating, soothing

Plant volatiles are some of the best antimicrobial agents we have, but there's another, fascinatingly twisted reason that we put them into medicines. Just as they attack microbes, they can attack us: we too are creatures that plants defend themselves from. We in turn have several systems that alert us to potentially toxic chemicals, and the smell receptors in our nose are part of just one of them. The others are deployed more widely over our exposed surfaces—our skin, eyes, and the lining of the mouth and nose and passageways to the lungs. They send signals to the brain that we experience as sensations of irritation, pain, physical touch, and heat and cold. So the smells of some plant volatiles are accompanied by non-smell sensations that also become an important part of our experience of them. And as folk medicine has recognized for thousands of years, these alerting responses to potential toxins can actually be beneficial to us.

The most familiar of these smell-plus volatiles are the ones that we encounter in nonprescription medicines for relieving cough and pain. Menthol, the characteristic volatile in peppermint, creates the sensation of coolness, and has been shown to reduce coughing and give the impression that irritated airways are more open. Camphor, a volatile found in several trees and herbs, creates the sensation of warmth and even irritation, but when camphor is applied in an ointment to the skin, those sensations help relieve muscle pain. Eucalyptol, thymol, and methyl salicylate have similar effects, and in our nose and airways as well. So when flavor and perfume chemists describe the sensory qualities of volatiles like menthol and camphor, they often supplement the smell with sensations like cooling, warming, irritating, stimulating, and soothing.

This phenomenon of plant chemicals triggering sensations other than smell and taste is called *chemesthesis*—*esthesis* coming from the Greek for "feeling." It's still not well understood, but since pain is a central problem in medicine, it's the subject of much active research. One key element is that some of the same skin molecules that detect plant volatiles also detect unusual changes in temperature or physical pressure—touch—or actual damage to the cells around them. When these multi-functional detectors are triggered, they in turn trigger sensations to make us aware that something significant has changed. The signal may be ambiguous—either the air is cooler, or we just inhaled menthol—but either way we're invited to pay more attention to the movement of air through our nose and airways.

There seem to be two keys to the soothing and pain-relieving effects of plant volatiles. The first is that while some volatiles trigger specific detectors, they can also overwhelm and inactivate them, or block other, entirely different detectors from working. The second key is that the many different detector systems influence one another, so that one system can trigger others, or prevent others from being triggered. The particular plant volatiles with noticeable chemesthetic effects have somehow ended up being exquisitely effective at triggering particular detectors and systems, and some of them help mask or reduce discomfort caused by injury or illness. So it is that ointments meant to reduce muscle pain have long included camphor, eucalyptol, menthol, thymol, and other detector activators.

In the tables of plant volatiles below, I've included some chemesthetic effects that perfumers and flavorists attribute to these special molecules. But don't take them too literally or absolutely. For example, scientists have shown that camphor triggers the same detector as capsaicin, the nonvolatile molecule in capsicum peppers that we experience in the mouth as "hot." Yet some fragrance experts describe camphor as cooling. Given what we know about the complexity of the detecting systems, it's not surprising that our sensations and perceptions are sometimes ambiguous. For the smell explorer, the appearance of non-smell, chemesthetic terms should simply be suggestive, a general alert: this molecule will make you feel your breath. And remember that they're being applied to single volatile chemicals, which plants actually emit in rich mixtures. When we encounter them in the world, we may or may not notice these qualities of warmth or freshness. If we do, we're getting a hint of what they sometimes do more clearly and powerfully in medicaments based on them.

The sweetness of plants and their chemical weapons

My first experience with a single plant volatile in pure form came nearly thirty years ago, when I ordered a sample of thymol, the main volatile of thyme, from a chemical supply company. It arrived in a small bottle that was cushioned inside a tightly lidded metal can, with a label that carried a prominent black X on a red-orange background, and the following warnings:

HARMFUL BY INHALATION, ON CONTACT WITH SKIN
AND IF SWALLOWED
IRRITATING TO EYES, RESPIRATORY SYSTEM AND SKIN
WEAR SUITABLE PROTECTIVE CLOTHING

But handling and inhaling and swallowing are exactly what we do with thyme and other herbs and spices! Nothing could make clearer the defensive nature of plant volatiles and our paradoxical relationship with them: many are hazardous materials, yet we enjoy them and use them to scent our foods and even ourselves.

Why don't these weapons wound us and warn us off? One part of the story is that we encounter them in small doses, as whiffs in the air or accents in our foods, and not by the mouthful as a bug or goat would. When we cook or compose fragrances, we imitate ripe fruits, which mix and dilute many plant volatiles so that they serve mainly as signals, not weapons. Chew on a sprig of thyme loaded with thymol, and you will indeed feel the pain. But of course even pain doesn't necessarily deter; many people love black and capsicum peppers precisely for the hurt. That's another part of the story: our sensory systems exist to be stimulated, and when we're not on the alert for danger, strong or complex sensations can be more engaging and enjoyable than their absence.

Yet another reason that we find most plant volatiles pleasant may be that they're simply different from the usual animal and microbial volatiles. Instead of those mostly small, incidental, funky waste products of metabolism, plant volatiles are generally larger molecules, six to fifteen carbon atoms long, actively built by the most prolific synthetic chemists on the planet. They're tokens of invention and growth rather than consumption and depletion. Some support for this idea comes from studies at both the Weizmann Institute in Israel and Rockefeller University in New York, where researchers surveyed the perceived qualities of volatile molecules by asking people to sniff a wide range of them one by one, and describe what they smelled like. The larger the molecule, the more likely the sniffers were to describe it as pleasant.

Many of the plant volatiles we'll get to know have a pleasant, soothing quality that will be named with the descriptor *sweet*, just as John Milton described the trees and flowers of Eden as a "Wilderness of sweets." Today the word *sweet* refers mainly to the taste of sugar, not to smells, and many flavor scientists consider its application to smells a misnomer arising from the association of certain smells with sweet tastes. Nevertheless, from its earliest days in Old English, *sweet* has meant "pleasing," and has been applied not only to edible sweets but to just about everything in the world that's somehow agreeable: sights, sounds, people, their characters, circumstances. There are April's sweet showers in the first line of Chaucer's *Canterbury Tales*, and many dozens of instances in Shakespeare beyond

Romeo's rose by any other name. Today, fragrance and flavor specialists routinely apply it to single molecules. In the perflavory.com database, hundreds of different molecules are described as sweet.

The plant kingdom is a rich source of sensations and sweetness and significance, and they all arise from its carbon chains and rings, some of our best molecular friends. It's time to get to know them.

Chapter 8 · PLANT VOLATILE FAMILIES: GREEN, FRUITY, FLOWERY, SPICY

Early apples begin to be ripe about the first of August, but I think that none of them are so good to eat as some to smell. One is worth more to scent your handkerchief with than any perfume which they sell in the shops. The fragrance of some fruits is not to be forgotten, along with that of flowers. Some gnarly apple which I pick up in the road reminds me by its fragrance of all the wealth of Pomona. . . .

There is thus about all natural products a certain volatile and ethereal quality which represents their highest value, and which cannot be vulgarized, or bought and sold. No mortal has ever enjoyed the perfect flavor of any fruit, and only the god-like among men begin to taste its ambrosial qualities. For nectar and ambrosia are only those fine flavors of every earthly fruit which our coarse palates fail to perceive—just as we occupy the heaven of the gods without knowing it.

· Henry David Thoreau, "Wild Apples," 1862

Whereas the English poet John Milton imagined Eden as a long-lost paradise of exotic aromatics, the American essayist Henry David Thoreau found heaven's scent in the everyday here and now, in fallen roadside apples. For Thoreau, the smells of fruits and flowers are earthly pleasures at their finest, the source of whatever ideas we have of paradise or heaven. His gnarly apple is an apt tip for the smell explorer: there's plenty to savor in the simplest things.

It was during Thoreau's lifetime that the first specific molecules of Earth's ambrosia were identified by chemists in Germany and England. These and thousands discovered since then have become the basis for a highly lucrative chemical industry, and synthetic scents of manufactured foods and fragrances—bought and sold and vulgarized despite Thoreau's idealism. It's largely thanks to this industry that we know as much as we do today about the volatile molecules of the plant kingdom as a whole, from the tiniest of mosses to giant redwoods. By drawing on this body of knowledge to inform our experience of the plants that we live alongside and enjoy in food and drink and perfumes, we can make the effort that Thoreau implicitly suggests, to cultivate our palates and better appreciate our earthly paradise.

In this chapter I first describe the major families of plant volatile molecules and how they're made. Then I introduce particular members of each family that figure prominently in the osmocosm and the chapters to come. I briefly give their names, their smell qualities, and a few of the objects and materials that they contribute to.

It shouldn't be Too Much Information for a smell explorer, but it is a Lot of Information packed into a few pages, and more than anyone can really absorb by reading straight through. I suggest that you first skim through it to get a general idea of the families and the qualities their members share—some "green," some spicy, some fruity, some "exotic." Then browse to find a few particular qualities or aromatic things that interest you, and notice the names of the molecules involved in them. Don't worry about hanging on to specific chemical names. But as you read through later chapters and become curious about particular molecules, remember that you can find them in this chapter. Come back whenever you'd like to be reintroduced.

The highways of plant metabolism

Let's begin with a bird's-eye view of where volatile molecules fit in a plant's overall metabolism. Plants get their energy from the sun and their raw materials from the soil and air. Soil provides water and minerals, including the nitrogen and sulfur that plants use to make their protein machinery. Air provides carbon dioxide for building carbon chains, and oxygen to help make that process efficient. Because plants lack the anaerobic digestive system and muscle machinery that incidentally scent the animal body, their volatiles and smells are dominated by the molecules that plants themselves purposefully make.

A plant has a handful of basic systems for building and maintaining itself. I think of them as the highways of plant metabolism, the major routes on which

Hero Carbon's many forms travel in order to transform soil, air, and sun into the defining materials of plants. The chart below shows these metabolic highways in schematic form. Let's take a quick look at the overall map.

The central vertical arrow represents the defining process of photosynthesis, in which plants assemble single carbon atoms from the air's carbon dioxide into six-carbon, six-oxygen sugar molecules. These sugars serve as the plant's raw material for building most of its other myriad carbon chains and rings. That's why sugars are at the center of the diagram.

Radiating outward from Sugar Central are lines that represent the highways of plant metabolism. Each represents many individual biochemical reactions that gradually transform the sugars flowing from photosynthesis into the various destination materials that make up a living plant. I've grouped these essential materials into four basic kinds: light-collecting pigments, structural supports, protective sheathings, and the active machinery that keeps cells and the plant as a whole functioning.

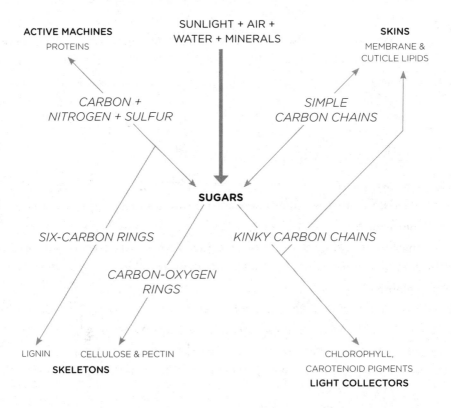

PLANT METABOLIC HIGHWAYS

Begin with the highway that runs to the upper right. It leads to the plant's physical **skins**, the sheaths that enclose individual cells and separate them from one another, and those that coat the whole plant and separate it from its environment. Both sets of skin molecules are lipids, among carbon's simplest structures, chains with an occasional oxygen atom or two, and so a common source of life's starter set of volatiles. Simple carbon chains don't mix well with water (see page 12), and that makes them ideal materials for isolating one watery cell from another and whole organisms from rains and drying winds. Combined to make the materials we call oils and fats, they're also an efficient form of stored chemical energy.

Now return to Sugar Central and follow the highway heading to the lower right. Carbon atoms on this route are destined to become either accessory sheath molecules or **light collectors**, the pigment molecules involved in photosynthesis. Green chlorophyll molecules capture a portion of the energy in sunlight, and orange and yellow carotenoids and xanthophylls absorb and dissipate excess light energy that would damage the chlorophyll machinery. These molecules are more complicated than the barriers, still essentially long carbon chains, but with kinks and twists.

Back to Sugar Central, and now heading to the lower left: the carbon atoms that travel this highway are destined to form **skeletons**, the structural supports that give plant cells and whole plants the mechanical strength to resist the forces of gravity and wind. One branch produces celluloses and pectins, long strings of sugar molecules, which we generally refer to as "fiber" and which the plant deposits in the walls of its cells to glue them to each other. The other branch produces tough lignin, indefinitely expandable masses of especially rigid six-carbon rings, which line the plant's inner circulatory system and bind celluloses and pectins together to make wood.

The last highway from Sugar Central, heading to the upper left, is the road that carbon atoms take to become amino acids and then proteins, the **active machines** that do all the work of making and breaking the plant's molecules. This is the highway on which carbon is joined by nitrogen and sometimes sulfur atoms, the two elements that elevate the stink of protein-rich animal bodies.

And volatile molecules? They're mostly produced on detours off the main highways. As sugars travel along this or that highway, being modified atom by atom and step by step, some of the intermediate structures are diverted into secondary assembly lines that produce different sets of molecules, less essential than the destination materials but still useful. It's on these byways that Hero Carbon is especially free to try out variations and novelty structures without compromising a plant's basic functioning, and it has done so to the tune of many thousands of small molecules. This flexibility helps plants make the special adjustments necessary to survive in a vast range of environments. The distinctive smells of different plants arose from the gradual selection of volatile molecules that have benefited their ancestors.

Volatile byways

Let's refresh the metabolic highway map, and this time overlay onto it the byways that generate significant volatiles and smells. Start by moving along the upper-right highway to plant and cell skins. We actually sampled the first byway back in chapter 3: it includes simple carbon chains from life's starter set, the short-chain alcohols and aldehydes. Look back to the table on pages 47–49 and you'll see the smells of cut grass and of cucumber and melon, fruits that share its fresh quality. Further along the same highway are diversions that produce two different sets of fruit smells: peach and coconut, and then apple, pear, strawberry, and banana. And then there's a flowery turn, a whiff of jasmine.

A delightful range of smells already—and from just one corner of the map!

Now circle back to Sugar Central and set out on the highway toward the lower right and the light-collecting chlorophyll and carotenoid pigments. Take the first left turn, a branch that leads to plant lipids related to animal cholesterol and steroid hormones. There you'll see a byway that includes woody and spicy smells, and grapefruit. Now double back to the main highway and you come to a byway with pine and eucalyptus, and also mint and thyme, lemon and rose: a gamut from herbaceous to citrusy to intoxicating! And then, past the pigments to a couple of their fragments, the intense and exotic floral smells of violet and saffron.

Backtrack again to Sugar Central, and start this time along the upper left highway, with amino acids being formed along the way to serve as the building blocks for destination proteins. Remember how protein-rich animals emit some pretty unpleasant breakdown products containing nitrogen and sulfur? Plants use the same atoms to construct potent volatiles of their own. One byway spins off sulfurous garlic and onion smells; a second, sulfur-nitrogen mustard and horseradish and wasabi. But these are real outliers in their aggressiveness. More consistent with other plant volatiles are the byways that feature nitrogen in green-vegetable smells, and in distinctive berry and flower smells.

Now take the middle branch from the protein highway that leads to the lower left. It begins with one particular ring amino acid, phenylalanine, and leads to lignin, the woody support material that consists mainly of carbon rings. Early on that branch the nitrogen is removed from phenylalanine, so the volatile diversions that come further on lack nitrogen. In addition to another set of flowery and fruity smells, they generate molecules that define some of our favorite spices: cinnamon, clove, anise, vanilla.

Finally we return to Sugar Central and the highway leading to cellulose and pectin, structural materials that support all plants, woody or not. It has what might look like a relatively skimpy byway, but the volatiles it produces are key to

PLANT VOLATILE BYWAYS: SAMPLE SMELLS

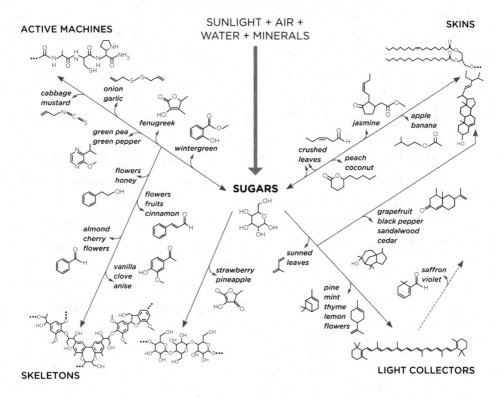

two of our most intensely flavorful fruits, pineapple and strawberry—and, quite remarkably, these wonderful molecules are created in fruit-free cooking! We'll encounter them often.

So the smells of the plant world come from families of similar molecules that are assembled on byways of plant metabolism. Now that we have this bird's-eye view of their origins, let's take a closer look at each of these families and meet some of their members.

Simple carbon chains: green, fresh aldehydes and alcohols

We'll start with the skins highway and its mostly simple straight carbon chains. We've already met many of them as members of life's starter set. "Green" and "fresh" are unusual descriptions for smells. We see green colors with our eyes and feel freshness on the skin. But these are common and appropriate terms for the smells produced by straight-chain alcohols and aldehydes that are six to ten carbon atoms long. The six-carbon chains are known as the **green-leaf volatiles**, or **GLVs**, because they're molecules that the living leaves of most plants emit: so we associate them with the green color of those leaves and with their vitality. Old leaves look faded or brown, lack the GLVs, and smell very different, like hay or the forest floor. We encounter GLVs often, from recently mown grass, from a meadow on a sunny summer day, from a green salad.

Despite our pleasant associations, for plants the GLVs are signals of damage and danger. They emit these molecules to repel or poison attacking insects and microbes, and to trigger the beefing up of chemical defenses in nearby leaves on the same plant, and even in neighboring plants. Most living plant cells, but especially the cells in green leaves, are packed with membranes made up of carbon-chain lipids. In the same cells, but isolated from the membranes in separate compartments, are plant enzymes that can break apart lipid chains into small fragments and oxidize them. When leaf tissue is damaged and the compartments are breached, enzymes mix with the lipid chains, and small volatile fragments begin to escape, like smoke from a fire. This process can be triggered by a caterpillar's bite, a snail's rasp, a fungus's invasion, a crushing human foot, or even strong summer sun and heat.

The particular volatile fragments liberated from damaged leaves depend to some extent on the plant. But the vast majority of plants produce fragments that are six carbons long, a portion of them with a double bond between two of the chain carbons, and with oxygens on one end that make them members of either the alcohol or the aldehyde families. The first green-leaf volatiles formed by freed enzymes include the six-carbon aldehydes **hexenal** and hexanal. Aldehydes are reactive molecules, especially toward proteins. The leaf aldehydes interfere with the digestive proteins that insects and disease microbes use to eat into plant tissues, and attack the attackers themselves, sometimes fatally.

While the leaf aldehydes do their work, other plant enzymes transform some of them to their corresponding six-carbon alcohols, **hexenol** and hexanol. Alcohols are less reactive than aldehydes and diffuse into the leaf and even other parts of the plant, where they trigger the healthy tissues to ramp up their other preformed chemical defenses. The volatile alcohols also escape from the leaf into the surrounding air, where they can alert nearby plants to the presence of insects, and attract predator insects like wasps to come attack the attackers. The leaf alcohols also readily combine with the common two-carbon acetic acid to form another green-leaf volatile, **hexenyl acetate**, a member of the fruity family of "esters" (in the next section). It too primes the defenses of nearby plants and attracts insects.

The various GLVs share a fresh, green quality but have their own nuances, **some grassy, some leafy, some solvent-like, some like unripe apples or bananas**, which probably reflect their predominance in these different plant materials. Hexenyl acetate also has **sweet, tropical-fruit** facets. And thanks to the actions of the plant enzymes and oxygen in the air, the proportions and the overall smell of a given leaf or fruit can change quickly, from simple and fresh to complex and heavier. This is why the flavor of newly chopped or pounded or pureed greens isn't the same after just a few minutes.

THE SIX-CARBON GREEN-LEAF VOLATILES, GLVs, AND THEIR SMELLS

Carbon chain	Aldehyde	Alcohol
6 carbons, no double bonds	hexanal fresh, green, grassy	hexanol green, fruity, solvent
6 carbons, 1 double bond	hexenals sharp, green, grassy, fruity	hexenols fresh, green, leafy, fatty

In addition to the six-carbon GLVs, we frequently encounter some of their longer relatives. **Eight- and nine-carbon aldehydes** are also produced by enzymes when plants are wounded; they smell fresh, like green citrus peel and like **cucumber**, a fruit that we enjoy as a green vegetable. The **eight-carbon alcohols** take us in a different direction, toward the **earthy** smell of **mushrooms**. The longer chains tend to evoke **flowers** and **citrus peels**, which are often described as fresh-smelling, but in their own non-grassy ways.

SOME 8-, 9-, AND 10-CARBON CHAINS AND THEIR SMELLS

Carbon chain	Aldehyde	Alcohol
8 carbons, no double bonds	octanal waxy, citrus peel, green	octanol orange, mushroomy, melon
8 carbons, 1 double bond	octenals fresh cucumber, green	octenols mushroomy, earthy, green
9 carbons, no double bonds	nonanal waxy, rose, citrus peel	nonanol fresh, floral, orange
9 carbons, 1 double bond	nonenals green, cucumber, melon	nonenols waxy, green, melon
10 carbons, no double bonds	decanal sweet, waxy, citrus peel	decanol waxy, floral, orange
10 carbons, 1 double bond	decenals cilantro, citrus peel, floral	decenols floral, rose, fresh

Fruity esters and lactones

Ethereal and ambrosial as a gnarly apple may be, its fragrance—along with the essence of fruit smells in general—begins off the skins highway with some of the least ethereal volatiles on Earth. These are the short carbon-chain acids, vinegary acetic acid and cheesy butyric acid and others, supplemented by the sweaty, toe-cheesy branched-chain acid residues of protein breakdown (see pages 40, 59). Plants make these simple molecules not mainly as metabolic waste products, but as building blocks for larger molecules called **esters**. Esters are hybrid molecules that result when one acid molecule and one alcohol molecule are jammed together at their oxygen-bonded ends, knocking out a molecule of water, H_2O, in the process. Fruits are the plant kingdom's main ester-making factories. They ramp up production to signal their ripeness when their seeds are mature.

Since plants make dozens of different acids and alcohols, they can make hundreds of different combinations of the two. Most fruits tend to make a blend of a dozen or so of the smaller esters. This means that different fruits emit many of the same esters, but in varying proportions that give them their own overall smells. That's why when flavorists evaluate individual ester molecules on their own, they tend to describe their smells as generically fruity, with suggestions of the fruits in which they're especially prominent.

Here I've compiled a table of the esters made from simple alcohols and acids up to eight carbons long—here, and henceforth, abbreviating the number of carbons in a chain as C1, C2, and so on. Give the table a quick scan, and you'll get a sense for the possible variations on the theme of fruitiness—and how different the esters are from their acid parents! Esters are named for their two parent chains, alcohol first and acid second, with the acid term ending in the suffix *-ate*. One of the most common combines one-carbon methyl alcohol and four-carbon butyric acid to make **methyl butyrate**, an ester that contributes to the smells of apple, pineapple, and strawberry. Replace the methyl alcohol with branched-chain methylbutyl alcohol (also called isoamyl alcohol) and butyric with acetic acid, and you get **methylbutyl (isoamyl) acetate**, more evocative of banana. Off this chart: the combination of ethyl alcohol and ten-carbon decadienoic acid gives **ethyl decadienoate**, the specific note of European pears.

Acids ➤ Alcohols ▼	C1 formate (pungent, fruity)	C2 acetate (vinegar)	C3 propionate (vinegar, cheesy)	
C1 methyl (alcoholic)	fruity, plum	green, solvent	fresh, solvent, fruity, strawberry, apple	
C2 ethyl (alcoholic)	lemon, strawberry	solvent, fruity, grape, green	ethereal, fruity, sweet, apple, grape	
C3 propyl (alcoholic)	ethereal, sweet, green, berry	fruity, solvent, banana, pear	sharp, sweet, fruity, pineapple	
C3 isopropyl (alcoholic)	sweet, solvent, fruity	ethereal, fruity, banana	fruity, sweet, ethereal, pineapple, banana	
C3+1 methyl-propyl (isobutyl) (fruity, apricot)	sweet, solvent, fruity	sweet, fruity, banana, pear	sweet, fruity, banana, green	
C4 butyl (whiskey)	fruity, plum, solvent	solvent, apple, pear, ripe banana	sweet, earthy, banana, floral	
C4+1 methyl-butyl (isoamyl) (whiskey, fruity)	solvent, green, plum	sweet, banana, apple, solvent	sweet, banana, ripe, tropical	
C5 pentyl (amyl) (whiskey)	sweet, fresh, solvent, fruity	solvent, fruity, banana, pear, apple	sweet, apricot, pineapple	
C6 hexyl (green, fruity, apple)	solvent, sweet, green banana, green plum	fruity, green banana, apple, pear	pear, green, musty	
C7 heptyl (fresh, floral, lemon)	green, waxy, floral	green, fresh, ripe fruit	rose, apricot	
C8 octyl (citrus, mushroom)	fruity, rose, cucumber	green, earthy, mushroom	fruity, sweet, jammy	

COMBINATIONS—ESTERS—AND THEIR SMELLS

C4 butyrate (cheesy, vomit)	C5 pentanoate (cheesy, sweaty)	C6 hexanoate (cheesy, rancid)	C7 heptanoate (waxy, cheesy, dirty)	C8 octanoate (fatty, rancid, cheesy)
pineapple, apple, strawberry	sweet, fruity, sweaty	fruity, pineapple	sweet, fruity, green	green, fruity, waxy, citrus
banana, pineapple, strawberry	apple, strawberry, pineapple	pineapple, green banana, pear	pineapple, banana, strawberry	fruity, waxy, apricot, fatty
fruity, sweet, pineapple, sweaty	ethereal, fruity, pineapple, animal	fruity, sweet, pineapple, green	ripe fruity, apple, pear, pineapple	coconut
ripe fruity, green	fruity, pineapple	fruity, pineapple, berry	—	fruity, banana, coconut
sweet, fruity, pineapple, apple	ethereal, fruity, apple, strawberry	sweet, fruity, pineapple, green	green, herbal, fruity	fruity, green, oily, floral
fresh, banana, pineapple, pear	sweet, pineapple, banana	fruity, pineapple, berry	grassy, marigold, apple, coconut	buttery, solvent, dank
fruity, green, apple, pear	ripe apple, strawberry	fruity, green, pineapple, banana	herbal, grassy, unripe fruit, banana	sweet, waxy, soapy, pineapple, coconut
sweet, fruity, banana, pineapple	ripe apple	green, waxy, apple, pineapple	solvent, banana, coconut	floral, sweet, solvent
green, fruity, waxy	oil, fruity, green	fruity, green, tropical	green, dank	green, apple, fruity
fruity, tropical, floral	fruity	green, bruised leaves	green, grassy	waxy, green, tropical fruit
fruity, green, creamy	—	fruity, herbaceous	fruity, fatty	coconut, fruity

There are all kinds of interesting patterns to be seen in the ester table, and one in particular highlights an important fact about fruit smells and smells in general. Fruits like apple, banana, pineapple, and strawberry appear in a number of different boxes because individual esters contribute to the smells of more than one fruit, and individual fruits contain more than one ester. Most smells are pixelated, like a mosaic, with a number of different volatiles combining to create an overall identity, and fruit lovers have recognized for centuries that very different fruits can share important qualities. Take the example of strawberry and pineapple, which share four boxes in the ester table. The most widely grown strawberries are varieties of the hybrid species *Fragaria* x *ananassa*, named from the Latin word for "strawberry" and the Latinized version of a South American word for "pineapple." Why that combination? As the French strawberry devoté Antoine Duchesne wrote in 1766:

> One of the merits of the odor and taste of the pineapple, is that it shares in the odor and taste of strawberries: they in turn seem to approach the pineapple, more than any other fruit, with their delicious perfume: from which came the name of *fraise-ananas*.

Such olfactory echoes and rhymes are great fun to discover and share.

Not all esters are pleasantly fruity. Some formate and acetate esters also smell like solvents—nail polish remover, varnishes, glues—because we encounter them in these products of human industry. The solvent overtone of **hexyl acetate** (lacking the double bond in the GLV hexenyl acetate) may contribute to the heavy, not-so-fresh quality that develops over time in bruised and crushed green foliage. In addition, because they are combination molecules, esters can be broken apart into their parent alcohol and acid, or their formation may not keep up with the production of their parents, so some fruits can develop vinegary and cheesy notes when they're overripe or stored tightly wrapped.

There's a separate subfamily among the esters that's responsible for a distinctive set of fruity smells. The **lactones** are technically esters but don't look anything like the straight chains we've been sniffing at. Plants make lactones by starting with a single long-chain acid that has an alcohol-like oxygen-hydrogen group projecting from one of the middle carbons. They bend the molecule onto itself to bring the oxygen-carrying end next to that group, and jam the two together, creating a molecule with a ring and a straight tail and again releasing a molecule of water. The most common lactones have eight to twelve carbon atoms and share a fatty quality, individual lactones having particular notes of coconut, peach, or apricot, as well as nuts and milk products and animal fats (*lactone* comes from the Latin for "milk"). Serve peaches with cream and you double down on lactonic richness.

SOME RING ESTERS, OR LACTONES

Lactone	Smells
g-octalactone	coconut, fruity, green
d-octalactone	coconut, dairy
g-nonalactone	coconut, creamy
g-decalactone	fresh, peach, creamy
d-decalactone	coconut, dairy, peach
g-undecalactone	peach, apricot, fatty
g-dodecalactone	dairy, fruity, nutty
d-dodecalactone	coconut, dairy, fruity, nutty

The initial letters indicate the number of carbon atoms in the ring, g meaning 4 and d meaning 5; chemists use the Greek letters γ (gamma) and δ (delta). The remainder of the 8 to 12 carbons in the original acid form the tail.

The jasmonoids: signals, defense, floral delight

In addition to the green-leaf volatiles and fruity esters, there's one more notable family of volatiles that plants create from simple carbon chains along the skins highway. It's small but potent, and serves as a surprising link between the smells of flowers and tea leaves. To make these molecules, plants start with a common component of their membranes, eighteen-carbon linolenic acid. Enzymes trim off six of the carbons and generate a five-carbon ring with two short chains projecting off it. This modified acid and an ester of it were first discovered in jasmine flowers and so were given the names **jasmonic acid** and **methyl jasmonate**. There's another slight variation called **jasmone**, as well as a lactone-like version, **jasmo-**

lactone. Jasmonic acid isn't very volatile, but methyl jasmonate, jasmone, and jasmolactone are. These three **jasmonoids** are all flowery and reminiscent of jasmine, but with different nuances that reflect their particular structures: the ester fruity and green, the lactone peachy, the jasmone spicy.

We most commonly encounter the jasmonoids in jasmine, gardenia, honeysuckle, and other flowers, but they're also a hidden presence in green tissues, where they serve as general hormones and chemical defense signals for whole plants. Normal low levels of jasmonic acid and methyl jasmonate help regulate the growth of plant tissues and the processes of flowering, fruit ripening, and leaf dropping. High levels are generated when plant leaves are stressed or damaged. Then the nonvolatile jasmonic acid circulates beyond the damaged area, inducing nearby tissues to boost their chemical defenses, while volatile methyl jasmonate carries the same warning signal to nearby leaves and other plants.

THE JASMONOIDS

Jasmonoid	Smells
methyl jasmonate	jasmine, green, fruity
jasmone	floral, herbal, woody
jasmolactone	fatty, coconut, peach, floral

This warning system turns out to be the secret to how makers of oolong tea manage to coax intoxicatingly floral aromas from green leaves of the camellia bush; see chapter 19 for the ingenious details. Whenever we notice the smell of jasmine, in flowers or in tea, we're detecting some of the molecules at the very heart of plant life.

Prolific kinky chains: introducing the terpenoids

Along the skins highway we've met mainly fresh and fruity smells, with a few flowery ones. Now we return to Sugar Central and embark on the highway of plant metabolism that manages to spin off by far the most diverse set of volatiles and smells, flowery and fruity and much more. A few of its individual molecules are even well known by name to connoisseurs of cannabis and beer! This metabolic highway was pioneered early in the history of life, and today all living things rely on it to make molecules for various essential functions. In plants it has two main forks, one leading to molecules that stabilize cell membranes, the other leading to the light-collecting pigments of photosynthesis. Its molecules are assembled from **five-carbon building blocks** that include double bonds and branches, so unlike the simple starter-set chains, they're kinked and bristling rather than straight. They can twist and double back on themselves and be tweaked into all kinds of intricate structures—chemists have cataloged tens of thousands. Of this vast number, the molecules built from two or three building blocks are small enough to be volatile and so have smells.

These kinked-chain molecules are named **terpenes** and **terpenoids**, after the material from which chemists first isolated them. That was turpentine, the strong-smelling, thin, multi-use liquid that can be cooked out of some tree resins, these days especially from pines. Turpentine in turn had gotten its name from the resinous terebinth tree, a relative of the pistachio that was prominent in the ancient Mediterranean. Terpenes strictly defined are hydrocarbons, made up only of hydrogen and carbon atoms, while terpenoids are variations on the theme that include a few other atoms, usually oxygen. I'll use *terpenoid* for all members of the family.

It appears that volatile terpenoids may well have been developed for signaling by early land insects, then adopted as insect-confusing defenses by early land plants (see page 140). It's shared terpenoids that make ginger and lemongrass resemble Amazonian ants. Plants also make some small terpenoids as hormones to control their own growth and development, and as means of dealing with the stress of drought and high heat and light. The smells of conifer trees, pines and firs and cedars, are defined by their terpenoid volatiles, but it was the flowering plants that explored the terpenoid highway's many possible byways most thoroughly. Today volatile terpenoids dominate the smells of many plant materials, from roses and mints to marijuana and hops to citrus peels and lychee fruits. Along with the green-leaf volatiles, they're the plant world's most common chemical defense. And they're effective at poisoning a variety of essential systems in their enemies. They work so well against insect pests—and spoilage and disease microbes—that our ancestors enlisted them thousands of years ago to care for their dwellings and their own bodies.

Plants tend to produce terpenoids in groups, and many of the same terpenoids show up in different plants and different plant parts. As we've seen with esters and fruits, this molecular promiscuity explains why a single terpenoid molecule can suggest several different qualities to the flavorist trying to describe it, or seem generically herbaceous or flowery. It's also why a lime can remind us of a pine tree, a lychee of a rose, why we name lemongrass after a fruit. Terpenoids give us a lot to smell and a rich web of echoes and resonances to work with. Some also trigger a range of different chemical detectors in our airways, sensations of cooling and warming, irritation and soothing, in addition to their smells. Familiar examples are menthol, which smells minty and has a cooling quality, and camphor, which smells medicinal and seems warming.

The following two sections offer sniffs of the two main classes of terpenoid volatiles: the lighter *monoterpenoids*, ten-carbon molecules made from two building blocks, and heavier fifteen-carbon *sesquiterpenoids* made from three. Within each class I've grouped the individual volatiles roughly by similarities in smell quality.

Monoterpenoids: trees, herbs, citrus fruits, flowers

The monoterpenoid volatiles are constructed on a byway early on the light-collecting highway, and they run an impressive gamut of qualities and source materials with which we associate them. Here are a handful of groups: woody, herbal, floral, and fruity.

Let's start with the monoterpenoids that are mostly typical of and suggestive of conifer trees. They may be truly primeval, going back to the early history of seed plants before flowering plants came on the scene. Flavorists describe these molecules as piney, woody, resinous, and "terpy," a generic turpentine quality reminiscent of solvents and liquid fuels, including the petroleum products that have largely replaced turpentine in the modern world. These volatiles have self-explanatory names like pinene, terpinene, and terpineol. **Pinene** is especially common, and conifer forests release it in vast amounts into the air, where it reacts with sunlight and ozone and volatiles from human activity to produce a complex, odorous, visibly hazy mixture of molecules. In addition to their primary sources in resinous conifers, these terpy terpenoids are also found in and lend their notes to a wide range of herbs and fruits. The resin exuded from cannabis leaves is rich in **myrcene** and pinene. Myrcene is also an important contributor to the smells of a number of flowers and fruits, and **menthatriene** is largely responsible for the woody quality of fresh green parsley.

SOME PINEY-WOODY MONOTERPENOIDS

Terpenoid (C10)	Smells	Contributes smells to
pinene	piney, woody, terpy	pine, cypress, citrus fruits, herbs, spices, mastic
terpineol	piney, terpy, floral	pine, cypress, eucalyptus, many herbs, spices, fruits
terpinenol	woody, earthy, cooling	cedar, marjoram, thyme, lavender
terpinolene	woody, terpy, citrus	lime peel, ginger & galangal, spices
cymene	woody, terpy, harsh	eucalyptus, lavender, oregano, thyme
myrcene	woody, resinous, green, citrus	conifer trees, eucalyptus, marijuana, citrus peel
phellandrene	green, terpy	conifer trees, dill, fennel, chrysanthemum, wild strawberry
thujone	cedar leaf	arborvitae, cedar, wormwood, sage
menthatriene	woody, terpy, camphor	parsley, citrus peel, bay laurel

Another group of monoterpenoids is also mainly found in trees and shrubs and described as woody, but these have warming or irritating qualities as well. **Camphor** is the standout here, the rare plant volatile that we can experience in close to pure form because it's sold in pain- and cough-relieving medicaments. It's used on its own in Southeast Asian cooking and lends its qualities to cinnamon, other spices, and some flowers and herbs. **Borneol** brings a similar woody warmth to a number of conifers, as well as spices and citrus peels. **Umbellulone**, which seems to trigger sensations of heat, cold, and irritation, is largely specific to the California bay tree. **Carvacrol** and **thymol** are familiar from the kitchen: they protect the shrubby herbs oregano and thyme.

SOME WARMING, PUNGENT MONOTERPENOIDS

Terpenoid (C10)	Smells	Contributes smells to
camphor	medicinal, woody, warming/cooling	cinnamon, lavender, citrus, herbs, spices
borneol	woody, warming	pine, cypress, ginger, lavender, citrus peels, spices
sabinene	woody, pine, warming	cedar, thyme, oregano, marjoram
umbellulone	minty, irritating/cooling	California bay laurel
cineole (1,4-)	mint, pine, warming	citrus, bay laurel, rosemary, lavender
carvacrol	oregano, warming	oregano, summer savory, thyme
thymol	thyme, warming	thyme, oregano, savory, marjoram

A third cluster of monoterpenoids share minty and/or cooling qualities and are found in plants that we cultivate as herbs specifically for these qualities. They include **menthol**, a familiar flavoring in peppermint candies, gums, and cigarettes, and **mint carvone**, which gives spearmint its different version of mintiness. Carvone comes in two different structures, each the mirror image of the other, one with the quality of spearmint, the other especially prominent in caraway seeds. (Their chemical nomenclature is confusing; I'll refer to them as the mint and caraway carvones.) **Fenchone** gives a similar fresh quality to fennel, **perilla aldehyde** to the shiso leaves found most often in sushi, **dill ether** to dillweed, and **eucalyptol** to its namesake trees and a number of herbs. **Cuminaldehyde** gives cumin seed its very individual quality. And umbellulone reappears in this cluster because it's both irritating and cooling.

SOME MINTY-HERBAL MONOTERPENOIDS

Terpenoid (C10)	Smells	Contributes smells to
menthol	minty, cooling	peppermint & other mints
menthone	minty	mints, some geraniums
pulegone	minty, pungent	pennyroyal mint, yerba buena

Terpenoid (C10)	Smells	Contributes smells to
mint carvone	spearmint	spearmint & other mints, lavender, citrus
caraway carvone	caraway, dill	caraway, dill, field mint, lavender
fenchone	cooling, minty, earthy, camphoric	fennel, lavender, cedar
ocimene	green, woody, tropical	mint, many flowers, lavender, tarragon
perilla aldehyde	fresh, green, woody, citrus	shiso leaf, peppermint, citrus peels
cuminaldehyde	cumin, green, sweaty	cumin, cinnamon, some citrus peels, wormwood, yarrow
dill ether	dillweed, green	dill, grapefruit
eucalyptol (1,8-cineole)	eucalyptus, cooling, fresh	eucalyptus, bay laurel, California bay, sage, chrysanthemum
umbellulone	minty, pungent, irritating	California bay

Terpenoid group number four: the florals. These volatiles are far more common in flowers than the jasmonoids derived from the simple carbon chains. Most of the floral terpenoids are alcohols, including **citronellol**, cousin to the citrusy aldehyde citronellal, and **geraniol** and **nerol**, cousins to the lemony aldehydes geranial and neral. All three tend to be dominant terpenoids in flowers and lesser presences in citrus fruits. **Rose oxide** is pretty specific to roses and rose-scented geranium leaves, but it's also what makes Asian lychee fruit so flowery.

The most important of the floral terpenoids, one we'll encounter often, is the oddly named **linalool**, which can be pronounced several different ways. I prefer lin-AL-low-ol, which best reflects its nineteenth-century derivation: an alcohol (the ending -ol) first identified in an extract of a Mexican aromatic wood (Latin *lignum*) similar to an Asian aromatic wood (aloeswood; see page 199). Linalool is present in something like 75 percent of all flowers analyzed (geraniol and nerol trail at about half that), and thus confers a floral quality on anything it's prominent in, including many household and personal-care products and manufactured foods. It also contributes to the prominent floweriness of papaya fruits. Despite linalool's pleasantness for us, its origins in chemical defense are evident from its use as an active agent in pet treatments for ticks and fleas. A less evident sign: scientists have found that strawberries respond to mold infections by boosting their emission of linalool, and that this increased florality does measurable damage to the mold cells' membranes and energy-generating systems.

Linalool comes in two different structures, mirror images of each other, both floral but with different secondary qualities: one is dominant in coriander seed, the other in lavender.

SOME FLORAL MONOTERPENOIDS

Terpenoid (C10)	Smells	Contributes smells to
linalool (coriandrol, licareol)	floral, sweet-citrusy; lavender-woody	many flowers, coriander seed, lavender, herbs, lychee, papaya
citronellol	floral, rose, geranium, green	rose, geranium, citronella, ginger, citrus peels, many flowers
geraniol, nerol	sweet, floral, rose, citrus	rose, geranium, many flowers, lemon, palmarosa, citronella grass, citrus
rose oxide	rose	rose, geranium, lychee

Now a fifth and beloved group of terpenoids: the ones that largely define the characteristic quality of citrus fruits. Unlike apples and strawberries and bananas, the citrus fruits—lemon, lime, orange, grapefruit—are relatively poor in esters, but they're rich in terpenoids, especially in their peels. Despite its name, **limonene** isn't especially lemony; it's often the most abundant volatile in citrus fruits and provides the generic fresh background note. **Geranial** and **neral** are mirror images of each other and provide the true lemon note in lemons as well as a variety of herbs. **Citronellal** is lemon-like but with a flowery aspect. And what about orange and lime? Their characteristic smells come from additions to the citrus base of simple-chain aldehydes and piney terpenoids, respectively.

SOME CITRUSY MONOTERPENOIDS

	Terpenoid (C10)	Smells	Contributes smells to
	limonene	citrus, herbal, terpenic	citrus peels; many fruits & flowers
	geranial, neral	lemon	lemongrass, citrus peels, lemon verbena, eucalyptus, ginger
	citronellal	citrus, floral, rose	lychee, citronella, makrut lime, citrus peels

A last monoterpenoid group is a handful of esters, molecules that result from the combination of a terpenoid alcohol with an acid. The most important terpenoid esters are all formed from acetic acid. They aren't just fruity; they're found in and smell of a range of terpenoid-rich materials, from trees to herbs to flowers and citrus fruits. The most distinctive among them is **linalyl acetate**, the characteristic note of lavender and of Earl Grey tea, which is scented with peel oil from the bergamot orange.

SOME MONOTERPENOID ESTERS

Terpenoid ester	Smells	Contributes smells to
terpinyl acetate	herbal, bergamot, lime, lavender	citrus fruits, cedar, bay laurel, nutmeg, sage, juniper, coriander
bornyl acetate	camphor, woody, pine	pine, larch, fir, spruce, cypress
linalyl acetate	lavender, bergamot, sage	lavender, bergamot (Earl Grey tea), citrus fruit peels, jasmine, mints
geranyl acetate	fruity, floral, rose	citrus peels & juices, orange flower, lemongrass, citronella, geranium, eucalyptus

Sesquiterpenoids and terpenoid relatives: woods, spices, violets, apricots

Now we leave the crowded monoterpenoid byway for two other groups of kinky-chain volatiles. The sesquiterpenoids are made on the kinky-chain branch that leads to membrane stabilizers similar to the cholesterol in animal membranes. Because they're more massive molecules than the monoterpenoids, they're generally less volatile and slower to leave their source, but also more persistent in the nose. Many share a woody quality, and several are largely specific to particular trees: **cedrol** and **himachalenes** to cedars and similar conifers, **santalols** and **santalene**, with their interesting milkiness, to exotic sandalwood. **Caryophyllene** may be the most common sesquiterpenoid and lends a woody quality to a number of spices, herbs, and even flowers. **Nootkatone** was named for a kind of cedar but also contributes to the unusual citrus smell of grapefruit. **Humulene** is named from the Latin word for the beer ingredient hops and also lends its particular woodiness to marijuana, a member of the same plant family. **Rotundone** is notable for giving some of the woody-peppery aroma—not pungency—to black and white pepper, to a couple of Mediterranean herbs, and also to Syrah wines and to roasted chicory. And **zingiberene** and **zingiberenol** provide the fresh sharpness of ginger and its relative turmeric.

SOME COMMON PLANT SESQUITERPENOIDS, ROUGHLY GROUPED BY SMELL QUALITY

Sesquiterpenoid (C15)	Smells	Contributes smells to
cedrene, **cedrol**	common cedar wood	N. American cedar, cypress, juniper
himachalenes, himachalols	woody, cedar	Lebanon/Mediterranean cedar
nootkatone	cedar, grapefruit	grapefruit, blood orange, cedar
santalols, santalene	sandalwood, milky, sweaty	sandalwood
caryophyllene	woody, camphoric, peppery	cloves, cinnamon, marijuana, sassafras leaf, tulip
germacrene	woody	honeysuckle, stock
cadinene	woody	avocado
humulene	woody	hops, marijuana
turmerone	woody, turmeric	turmeric
rotundone	peppery, woody	white & black pepper, marjoram, rosemary

Sesquiterpenoid (C15)	Smells	Contributes smells to
farnesene	citrus, green, herbal	many flowers
nerolidol	fresh, floral, woody	orange flowers, citrus peels, strawberry
zingiberene, zingiberenol	fresh, sharp	ginger, turmeric

Back now to the light-collecting highway and a handful of appealing volatiles that arise on much smaller byways, none of them true terpenoids, but all of them close relatives. One, **isoprene**, is a close five-carbon approximation of the terpenoid building blocks and has a faint rubbery smell because it's the chemical building block for natural rubber, which emits it. Isoprene is made by many plants, from mosses to oak trees, and has been called a kind of gaseous sweat: leaves store it in their photosynthetic membranes and release it as a way of dealing with overheating and oxygen stress. It's the major cause of the summer haze often seen over deciduous forests; like pinene over conifer forests, it reacts with other molecules to form particles of complex mixtures, and contributes to the formation of clouds.

Another handful of terpenoid relatives emerge only at the end of the light-gathering highway, as small fragments broken off of the destination carotenoid pigments. The **ionones** are largely responsible for the smell of violets, and along with **dihydroactinidiolide**, **damascone**, and **damascenone** contribute to the smells of fruits, especially berries. **Safranal** dominates the smell of saffron crocus flowers after they've dried out, which is the smell of saffron the spice. And **geranyl acetone** is remarkable for its floral-fruity presence both within the plant world and without: it contributes to the smell of our own skin! (See page 111.)

SOME VOLATILE TERPENOID RELATIVES

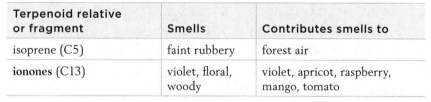

Terpenoid relative or fragment	Smells	Contributes smells to
isoprene (C5)	faint rubbery	forest air
ionones (C13)	violet, floral, woody	violet, apricot, raspberry, mango, tomato

continued

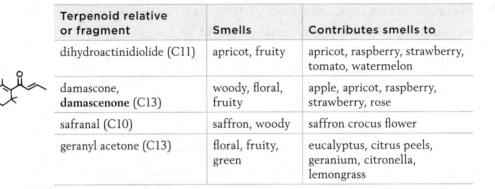

Terpenoid relative or fragment	Smells	Contributes smells to
dihydroactinidiolide (C11)	apricot, fruity	apricot, raspberry, strawberry, tomato, watermelon
damascone, **damascenone** (C13)	woody, floral, fruity	apple, apricot, raspberry, strawberry, rose
safranal (C10)	saffron, woody	saffron crocus flower
geranyl acetone (C13)	floral, fruity, green	eucalyptus, citrus peels, geranium, citronella, lemongrass

Carbon-oxygen rings: caramel furanones

So far, two highways down, and a large range of smells sniffed: green, citrus, fruity, flowery, woody. Now let's start over at Sugar Central and follow the skeleton highway leading to celluloses and pectins, the carbohydrates that firm plant cell walls. There's a minor turnout that leads to just a couple of volatiles, but they're wonderful molecules indeed. **Furaneol** and mesifuran are both members of a group called **furanones** (the Latin root *furfur* refers to the bran of grains). Plants make them by starting with sugar molecules, which are sweet but not volatile, and using enzymes to rearrange them into volatile molecules that have a central ring of four carbons and one oxygen atom, with an added oxygen decoration: a structure resembling the lactones. Their smell is suggestive of sweet fruits and also of caramel—browned sugar—because high heat can also rearrange sugar molecules into furanones.

It's primarily furaneol and mesifuran that give a shared caramel note to some of our favorite fruits, including strawberries, pineapples, raspberries, and mangoes. A third important furanone is **sotolon** (named from *sotō*, the Japanese word for unrefined sugar, which includes caramel-rich molasses). Sotolon has exactly the same chemical formula as furaneol, but it's made on a different byway, from an amino acid rather than a sugar, and has a different placement of the oxygen decoration that makes it both a furanone and a true lactone. Sotolon is suggestive enough of maple syrup that it's often added to corn-syrup imitations of the real thing. In the plant world, it's the major volatile in fenugreek seeds—which is why high concentrations smell more spicy and curry-like—and also shows up in pineapple. And as we'll see in later chapters, sotolon and furaneol provide fruity-caramel echoes in many cooked and fermented foods.

SOME SWEET-SMELLING PLANT FURANONES

Furanone	Smells	Contributes smells to
furaneol (dimethyl hydroxyfuranone)	caramel, fruity	strawberry, pineapple, raspberry, Concord grape, mango, tomato
mesifuran (dimethyl methoxyfuranone)	caramel, musty	strawberry, pineapple, raspberry, mango
sotolon (dimethyl hydroxyfuranone)	caramel, maple syrup, fenugreek	fenugreek, candy cap mushroom, pineapple

Rigid carbon rings: introducing the benzenoids

Back now to Sugar Central, and we start out on the last major highway of plant metabolism, where carbon chains join up with nitrogen and sulfur atoms to make amino acids, building blocks of the active molecular machines called proteins. But first follow the branch that breaks off from the protein straightaway toward skeleton molecules and has wood-strengthening lignin as its destination. Lignin is built up from countless rigid six-carbon rings from the amino acid phenylalanine, minus its nitrogen atom. And byways on the lignin highway are the source for many of our favorite spice smells.

The six-carbon ring is a major motif in plant metabolism, and its name reflects the importance of aromatic plants in human culture. It's called a **benzene** ring, and variations based on it **benzenoids**, after **benzoin**, a prized resin collected from certain Southeast Asian trees, which sixteenth-century alchemists studied by heating it and collecting the fragrant vapors. The word *benzoin* in turn comes from the Arabic *luban jawi*, "incense of Java"; to this day benzoin and its carbon rings remain important components of the incense burned in Catholic and Orthodox churches (see page 195).

It's impossible to overstate the importance of the **benzene ring** in the life of plants. Each of its six carbon atoms shares some electrons with its immediate neighbors and some communally, an arrangement usually depicted as alternating double bonds. This unusual electron sharing gives the ring a physical rigidity that straight and kinky carbon chains lack, as well as the ability to absorb ultraviolet radiation from the sun—the makings of a chemical sunscreen that was probably essential for algae to move from the waters to the unshielded surface of the land. Small aggregates of several rings have the property of absorbing visible light radiation and act as pigments, giving leaves and flowers and fruits some of their yellow, red, purple, and blue colors.

To construct the benzene ring and its many variations, plants start with a sugar molecule, modify it in a couple of dozen steps to form the ring, and incorporate it as a side chain into the amino acid phenylalanine for protein building. Then, to make essential sunscreens and lignins, plants remove the benzene ring from the nitrogen-containing portion of phenylalanine, and then go on to "decorate" the ring with various groups of atoms, including other rings. It was along this post-amino, nitrogen-free highway that plants developed detours for making many useful volatiles.

The benzenoid volatiles form several families that are defined by their decorations. Their names can be confusing, and in fact each has several aliases. I'll stick with the names most commonly used in the flavor and fragrance world, and explain briefly where they come from.

Pre-amino benzenoids: wintergreen and musky fruits

We'll start with two unusual volatiles that plants actually make from newly minted benzene rings *before* they're incorporated into the amino acid phenylalanine. I don't include benzene itself because, while it's found in interstellar space and in petrochemicals (see page 420), it's not a significant plant volatile. (A good thing for us: it's toxic.) Both pre-amino benzenoids are esters formed with methyl alcohol. **Methyl salicylate** is a volatile relative of aspirin (acetylsalicylic acid), whose smell is usually called "wintergreen" after the evergreen North American shrub whose leaves and flowers release it when they're damaged: clearly a defensive function. Few of us ever encounter the shrub, but its volatile is familiar from candies, root beer, and breath fresheners—and in Britain, from household cleaners. Its name fortuitously suggests a cooling quality, and in fact it triggers receptors for both cold and general irritation.

The second pre-phenylalanine benzenoid volatile is **methyl anthranilate**, which is decorated with a nitrogen atom. It has a musky kind of fruitiness. It's an important note in small and intense "wild" strawberries and in citrus flowers, and it helps give Concord, muscadine, scuppernong, and other native American grapes the smell that distinguishes them from table grapes of European origin.

SOME PRE-AMINO BENZENOID RINGS

Benzenoid ring	Smells	Contributes smells to
methyl salicylate	wintergreen, root beer, warming	wintergreen, tuberose, carnation, tulip
methyl anthranilate (methyl amino benzoate)	orange flower, grape	Concord grape, strawberry, citrus flowers & peels, tuberose

Phenyl, cinnamic, and benzyl rings: honeyed, fruity, balsamic

Once the benzene ring is incorporated into the amino acid phenylalanine and then cleaved free of the nitrogen-containing chain, a short byway leads to a small group of volatiles known as *phenyls*. They're named from the Greek for "light" because a nineteenth-century French chemist isolated some benzenoids in residues of the coal gas that was used to illuminate early streetlights (hence also *phenyl*-alanine). Phenyl acetic acid, phenyl ethanol, and phenyl acetaldehyde are benzene rings decorated with the two-carbon starter-set chains. All of them are found in flowers and have a floral smell. The alcohol **phenylethanol** is a very common volatile in flowers, prominent in many aromatic roses, and along with the similarly ubiquitous terpenoid linalool helps define the general quality of floweriness. **Phenylacetic acid** and **phenylacetaldehyde** have a weightier sweetness suggestive of honey, which of course bees make from flower nectar (see page 525).

SOME PHENYL RINGS, FLORAL AND HONEYED

Phenyl ring	Smells	Contributes smells to
phenylacetic acid	floral, honey, animal	orange flower, rose, mango
phenylacetaldehyde	honey, floral	many flowers & fruits
phenylethanol	rose, floral	rose, orange flower, narcissus, geranium, many fruits

The benzene rings that don't get diverted into the phenyl family proceed down the highway, and some are diverted to form a larger group of volatiles with a less obscure family name, the **cinnamyls**, first isolated from cinnamon. Cinnamon is the bark of Southeast Asian trees in the genus *Cinnamomum*, and it owes much of its character to **cinnamaldehyde**, a benzenoid ring with a three-carbon aldehyde chain decorating one corner, a molecule that now effectively advertises the presence of baked sweets in many malls and airports. Cinnamaldehyde also triggers our cold and pain sensors, so at high levels it's cooling and pungent. Its more widespread cousin molecules, **cinnamic acid** and **cinnamic alcohol**, and their esters, suggest benzoin tree resins, flowers, and fruits, especially spicy cranberry.

SOME CINNAMIC RINGS, SWEET, SPICY, FLORAL, FRUITY

Cinnamic ring	Smells	Contributes smells to
cinnamic acid	balsamic, sweet	benzoin resin
cinnamaldehyde	cinnamon, pungent	cinnamon
cinnamyl alcohol	floral, green, balsamic, cinnamon	cinnamon, benzoin resin, cranberry
methyl, ethyl cinnamates	fruity, cinnamon, honey	benzoin resin, clover flowers, cinnamon basil, cranberry, clove
cinnamyl acetate	honey, balsamic, floral	cinnamon, guava
acetophenone (methyl phenylketone)	almond extract, floral, powdery	alyssum & clover flowers
aminoacetophenone	grapey, sweet	Concord & muscadine grapes

A short byway from cinnamic acid leads to two other flowery-fruity volatiles with names based again on "phenyl." The ketone **acetophenone** helps give the popular flower alyssum its rich aroma, as it does for clover flowers. The same ketone with an additional nitrogen-hydrogen group decorating one corner of the ring is fruity, musky **aminoacetophenone**, a molecule that, in combination with the similarly nitrogen-decorated methyl anthranilate, makes North American grapes—and their juices and jellies and wines—immediately recognizable.

A second byway from cinnamic acid takes benzene rings to the family with which they share their name, the **benzyls**. Their namesake tree resin *benzoin* also gives us the more common words *balm, balsam,* and *balsamic,* which suggest the soothing, calming, pleasant qualities of the resin and its individual benzyl volatiles. These molecules consist of the basic benzene ring decorated with a one-carbon group on one corner. They form the usual series of benzyl alcohol, aldehyde, and acid, and various esters. Not surprisingly, **benzyl alcohol** and **benzoic acid** smell like tree resins, not piney like the terpenoid-rich conifer resins, but aromatic in their own distinctive sweet, "balsamic" way. The **benzyl esters**, like most esters, have a fruity quality but are also often flowery. They're prominent volatiles in flowers, where they seem to be especially important in attracting butterflies and moths.

SOME BENZYL RINGS, RESINOUS, FLORAL, FRUITY

Benzyl ring	Smells	Contributes smells to
benzoic acid	resinous, sweet	benzoin resin
benzaldehyde	almond extract, cherry	bitter almond, cherry, peach, apricot, heliotrope, petunia
benzyl alcohol	floral, resinous	many flowers, cherry, cantaloupe melon
methyl, ethyl benzoates	fruity, floral	many flowers, guava
benzyl acetate	jasmine, fruity	jasmine, narcissus, hyacinth, heliotrope, cantaloupe melon
benzyl benzoate	balsamic, fruity	carnation, petunia, tuberose

Of all the benzyl volatiles, the most intriguing is **benzaldehyde**. It has the distinctive smell of almond extract, and is a noticeable component of the smell of cherries. But it smells nothing like ordinary almonds! That's because the almonds from which the natural extract is made are "bitter almonds": bitter because in addition to generating aromatic benzaldehyde, they generate bitter—and deadly—hydrogen cyanide. Nibble on the tender kernel inside a hard cherry or apricot or peach pit—these are close almond relatives—and you'll get a sample of both the aroma and the bitterness. They're generated together when the kernel is damaged as a two-pronged chemical defense: the cyanide toxin accompanied by a volatile warning signal.

In nature, almond-essence smell betokens danger and possible death. Chemically separated from cyanide in almond extract, or present in a small number of bitter almonds that flavor traditional marzipan, or contributing in trace amounts to the flavors of stone fruits, it pleases.

Spicy rings: hay, anise, clove, vanilla

Back now on the highway to wood, and a couple of byways take us to coumaric acid, the base structure for a last group of benzenoid volatiles. The decorations on their rings are pretty miscellaneous, so they don't really constitute a closely knit family. But they give us some of our all-time favorite smells.

Begin with **coumarin**, from which coumaric acid got its name. Its structure is similar to the lactones, and it provides the warm smell of crushed clover and hay, which often mixes with the green-leaf volatiles when lawns and meadows are mowed. Several other Eurasian plants, including "lady's bedstraw" and "sweet woodruff" (species of *Galium*), are prized for emitting the same volatile and smell, and are used as insect-repellent bedding and in cooking (sweets, teas, flavorings for beer and wine). The tropical American tonka bean (*cumaru* in the Tupi of the Brazilian Amazon) is a concentrated source of coumarin and used as a spice—and sometimes as a cheap approximation of vanilla. Coumarin can cause liver and kidney damage in moderate amounts, however, and some countries ban its use in foods.

Among the other individualist benzenoids, several share a similar quality that we experience most purely in the seed spices anise and star anise. **Anethole** and estragole are identical molecules except for the position of one double bond on a three-carbon decoration, and each contributes a generally anisey but slightly individual quality to a different set of spices and herbs, anethole mainly to seeds like anise and star anise, **estragole** mainly to herbs like tarragon (French *estragon*) and basil and fennel. **Anisole** lacks that side chain altogether, and has a harsher, solvent-like quality along with some anise character, while another variant, **anisaldehyde**, is sweeter and lends a slight anise note to some flowers.

Eugenol is the key volatile in cloves, the dried flower buds of a tree in the Asian tropics (genus *Eugenia*). Concentrated sources like cloves and clove oil are also pungent and numbing because eugenol triggers several different pain receptors. In traces it has a supporting role in other spices, in ripe bananas, and in carnation and other flowers. Its cousin **methyl eugenol** (not an ester but an *ether*) smells like a blend of cinnamon and clove, important in carnations and allspice. **Myristicin** gets its mysterious name from the tropical Asian nutmeg tree (*Myristica*) and carries much of that woody seed's distinctive aroma, which it also lends to green parsley and dillweed. **Safrole** is the main volatile in sassafras root and was the original flavoring of root beer until it came under suspicion as a liver toxin. Today it's hard to smell on its own unless you can find the herb hoja santa, "holy leaf" (see page 261).

Vanillin is the key volatile in vanilla, one of the most popular spices in the world. Warm and sweet, not unlike coumarin, it was one of the first flavor molecules to be manufactured industrially. It's found naturally in a few fruits and flowers, and in some kinds of wood, especially when heated; this is part of the reason for the ongoing use of wood barrels to age wines and spirits.

SOME MOSTLY SPICY RINGS FROM COUMARIC ACID

Coumaric acid derivative	Smells	Contributes its smells to
coumarin	sweet, clover, hay	clover, alfalfa, tonka bean, woodruff, sweetgrass, cassia
anethole	anise, medicinal	anise, star anise, fennel
anisaldehyde	sweet, floral, balsam	anise, star anise, basil, flowers
anisole	gasoline, medicinal, anise	flowers
estragole	anise, green, herbal	tarragon, basil, fennel
eugenol	clove, warming	clove, basil, cinnamon, nutmeg, oak wood, carnation, stock, rose, banana

continued

Coumaric acid derivative	Smells	Contributes its smells to
methyl eugenol (ether, not ester)	cinnamon, clove	allspice, basil, carnation
myristicin	resinous, woody, nutmeg	nutmeg, mace, parsley, dill
safrole	sassafras root, root beer	sassafras, nutmeg, hoja santa
vanillin	vanilla	vanilla, oak wood, cherry
raspberry ketone (hydroxyphenyl butanone)	sweet, berry, floral	raspberry, loganberry

And a last volatile that derives from coumaric acid, one totally unlike the other spicy molecules: **raspberry ketone**, a benzene ring decorated with one oxygen-hydrogen group and a four-carbon ketone group, named for the fruit that it marks so clearly.

That's it for the major benzenoid volatiles. They expand the smell universe to the non-coniferous tree resins, contribute a new dimension to the realms of flowers and fruits, and largely create the realm of spices. Pleasant though they can be, most or all of the benzenoids are primarily chemical weapons meant to repel and deter. Concord-grape methyl anthranilate is a proven bird repellent. Benzoates are common food additives because they're good at killing microbes, as are cinnamaldehyde and eugenol. Benzaldehyde isn't acutely toxic in itself but is a warning signal that cyanide is in store.

In sum, the benzenoid branch of the protein highway lifts plants up to the sun, shields them from sunburn, brings their enemies down—and has made countless human friends.

Accents from nitrogen and sulfur

Let's return now to the direct highway to active protein machines: just a few more byways and we'll have completed our tour. And these bring new dimensions to plant smells. All the volatiles that we've met up to this point have been constructed from just three elements—carbon, hydrogen, oxygen—and their smells have been generally pleasant. Now on the protein highway, nitrogen and sulfur enter the mix. As we've seen, animals and their microbes break down their copious proteins into smelly nitrogen- and sulfur-containing fragments. Plants are happily different. They don't carry nearly as much protein machinery and don't need to tap it for energy. Their nitrogen and sulfur volatiles aren't wastes; they're built molecules that augment the volatile arsenal for dealing with animals, and incidentally augment our pleasure in eating.

Let's start with the relatively few volatiles that include nitrogen. One is not pleasurable at all. "Stinking goosefoot" is the common name for a weedy spinach relative that makes fishy-smelling **phenylethylamine** by decorating a benzene ring with a short chain ending in a nitrogen atom. But put the nitrogen directly on one corner of the benzene ring and you get **methyl anthranilate** and **aminoacetophenone** (see above among the benzenoids), which are found in fruits and flowers and have an unusual musky aspect to their generic fruity florality. And remember pyrroline, a ring of four carbons and a nitrogen that gives human semen its smell? Plants add a two-carbon chain to pyrroline to make **acetyl pyrroline**, the molecule responsible for a distinctive floral-musky, popcorn quality in basmati rice and in leaves of the tropical pandan plant, and a popcorn-like note in cooked foods.

Some plants construct **pyrazines**, six-cornered rings with two nitrogen atoms opposite each other. Plant pyrazines with simple methyl and ethyl decorations on the ring tend to have earthy, musty smells, while **isobutyl** and other **methoxypyrazines**, with oxygen atoms in the side chains, have both earthy and heavy green-vegetable qualities that they give to peas, green bell peppers, and asparagus. Methoxypyrazines also lend vegetal notes to some grapes and wines, sometimes appreciated in the case of Sauvignon Blanc, usually regretted in Cabernet Sauvignon and Merlot.

SOME PLANT VOLATILES CONTAINING NITROGEN

Nitrogen volatiles	Smells	Contributes smells to
phenylethylamine	fishy	stinking goosefoot
methyl anthranilate	orange flower, grapey	citrus flowers & peels, gardenia, tuberose, Concord grape, strawberry
aminoacetophenone	grapey, sweet	Concord & muscadine grapes, chestnut honey
acetyl pyrroline	basmati rice	aromatic rices, pandan leaves
methyl, ethyl pyrazines	earthy, musty, nut skins	asparagus, nuts, potatoes
methoxypyrazines	green, earthy	fresh peas, green capsicum peppers, asparagus, lettuce, wine grapes

A last group of nitrogen-carrying plant volatiles also includes sulfur. **Allyl** and other **isothiocyanates** are a particular specialty of the large cabbage family, and make the strongest impression on us whenever we inflict on ourselves the sting of strong mustard or horseradish or Japanese wasabi. In addition to that pungency, they give a characteristic sulfurous aroma to those condiments as well as to cabbage and its other family members, among them mustard greens, arugula, and radishes. A few more distantly related plants also make traces of thiocyanates and so have similar sulfurous notes without the pungency; these include the buds and fruits of the caper bush, nasturtium leaves and flowers, and—oddest of all—fruits of the tropical papaya tree.

Sulfurous accents are fairly common in plants, which insert sulfur atoms into many different molecular structures. Among the simplest are **sulfides**, a group that includes primeval and sulfidic hydrogen sulfide and the two-carbon dimethyl sulfide, which become prominent when many vegetables are cooked. The cabbage and onion families both make these, along with a number of other sulfides with more than one sulfur atom and various longer carbon chains. Particular sulfides have their own particular smells and strike us as more or less cabbagey, more or less oniony or garlicky, depending on which plant they're more prominent in—**diallyl disulfide** is garlic's signature—but all smell vaguely sulfurous. Durian, the notorious tropical Asian fruit (see page 333), also accumulates unusual sulfides, and has a distinctly oniony-garlicky aroma, as does the Central Asian spice asafetida.

A diverse set of other sulfur volatiles provides more unusual accents. **Thiols** are carbon chains with a sulfur-hydrogen couple at one end. Methanethiol is the primeval thiol with just one carbon atom, formed in outer space and bacterial metabolism, and unsubtly smelly. But plants can add sulfur-hydrogen groups to longer chains, straight and kinky, to make a variety of thiol alcohols, ketones, esters, and terpenoids—just as we've seen cats, skunks, and our own bodies do! Thiols like **menthenethiol** provide what are often described as "exotic," sometimes "animalic" notes to blackcurrants and several tropical fruits, notably grapefruit, guava, and passion fruit. They're also important in the aroma of wines made from the Sauvignon Blanc grape.

Then there are the specialty sulfur volatiles of the remarkable *Allium* tribe, named from the Latin word for "garlic" and also including onions, shallots, leeks, and chives. Alliums are the plant kingdom's sulfur virtuosos, and among their specialties are mixed chains called **sulfinates**, which include oxygen atoms. They give onion, garlic, and their relatives their irritating, tear-inducing effects and add their particular versions of sulfurousness to the sulfide family.

SOME PLANT VOLATILES CONTAINING SULFUR

Sulfur volatiles	Smells	Contributes smells to
isothiocyanates (also contains nitrogen)	sharp, mustard	cabbage family (mustard, radish, horseradish, wasabi, arugula . . .), capers, papaya, nasturtium
sulfides	cabbagey, cooked vegetable, oniony	cabbage family (broccoli, cauliflower, turnip . . .), onion family (leek, garlic, chive), durian, asafetida
thiol alcohols, ketones, esters	tropical fruit, sulfurous, sweaty, catty	grapefruit, passion fruit, guava, blackcurrant, capsicum pepper, hops, boxwood, durian
menthene (terpenoid) thiol	grapefruit, sulfurous	grapefruit, pomelo, orange
sulfinates	oniony, garlicky, pungent	onion, leek, garlic, chive

Of all the plant volatiles we've surveyed in this chapter, the oniony sulfinates and garlicky sulfides and mustardy thiocyanates are the most obviously defensive. They burn our mouths and eyes and make us cough. So it's ironic that they're all such valued ingredients in the foods of people all over the world! We enjoy them exactly because they provide dimensions of aroma and taste and feeling that other foods don't, because their aggressive qualities can be modulated by combining them with other ingredients, and because cooking transforms them into yet other aroma and taste molecules.

With the nitrogen and sulfur volatiles we conclude our tour of plant metabolism and its smell-making byways. Riffle back and notice how various and expansive the plant repertoire is compared to the fumes of animal and microbial engines, their simple chains and sulfides and thiols. Green, resinous, woody, herbaceous, flowery, fruity, sweet, spicy, exotic, aggressive: the ingredients of earthly nectar and ambrosia.

It's been many pages, many highways and byways, many molecules! And all of this organization and dizzying traffic, and unimaginably more, is contained in the specks that are individual plant cells, barely visible to the eye. Now that we have an idea of what Hero Carbon has been up to in there, it's time to zoom back out to our own everyday experience of trees and herbs and flowers and fruits. Armed with this inside knowledge, let's see how much more of their fine flavors we might perceive.

Chapter 9 · MOSSES, TREES, GRASSES, WEEDS

Their country is called "Arabia the Blessed": a natural sweet aroma pervades the whole country, because almost all the plants that excel in fragrance grow there pell-mell. Along the coast grows the balsam tree, and cassia. . . . Inland are thick forests with big trees yielding frankincense and myrrh, alongside date, kalamos, cinnamon, and all other such aromatics. . . . The aroma strikes and stirs the senses of every visitor: it seems divine, beyond the power of words. Those coasting by, even at some distance off, have their share of the pleasure, because in summer, when there is an offshore breeze, the aromas exhaled by the myrrh and other such trees will carry that far: these are not the products known to us, dried and stored and stale with age, but fresh living plants at the peak of their vigour.

> • Agatharchides, *On the Erythraean Sea*, about 150 BCE

I doubt if there is any sensation arising from sight more delightful than the odors which filter through sun-warmed, wind-tossed branches, or the tide of scents which swells, subsides, rises again wave on wave, filling the wide world with invisible sweetness.

> • Helen Keller, *The World I Live In*, 1908

Those tiny plant cells we've just escaped from, with their molecule-crowded highways and byways: they scent the wide world. They scented the Arabian breeze described by the Greek geographer Agatharchides, a real-world model for what John Milton imagined for his earthly paradise. They sweeten the breeze of Helen Keller's American countryside. When we manage to escape from our dwellings and cities and their smells and take in a deep breath of the open air, we're enjoying the collective exhalations of Hero Carbon's green avatars unobscured, their life-giving oxygen along with faint traces of the volatiles that help keep them alive.

We'll start our appreciation of the plant kingdom with denizens of forests and meadows that contribute to the wide world's sweet tides. I've arranged this chapter and the next as a virtual botanical garden. Imagine a few acres of land planted with aromatic specimens from all over the world, a collection that would take a few hours to tour at a leisurely pace. That's what we'll proceed through here. In the next chapter we'll pass into the flower garden with its more intimate, intense, short-lived smells. Later we'll move to produce markets, then indoors to the kitchen and its global collection of aromatics. Feel free to fly through all these chapters like Raphael through Eden for a quick overview of the different settings. Or stroll, stop when you come to something that you happen to love or hate or find interesting, and look in the accompanying table to find the component smells that help make it what it is.

Of course imaginary gardens and kitchens are no substitute for real plants and molecules and sensations. I hope these chapters will prompt you to finger real pine needles or blades of grass, to stop to smell the roses, try different varieties of mint and apple, and seek out things you've never experienced before. As Thoreau said of heaven, these pleasures are available all around us. We can find them today in parks and forests, florist shops and plant nurseries, supermarkets and groceries, in the lumber section of home improvement stores. Unusual items are often obtainable online. So browse in the virtual gardens and markets below, find something interesting, then sally forth into your world and sniff.

When you do sniff, don't worry if you can't make out any component smells right away. Plants are profligate chemists. Any leaf or flower or fruit emits dozens to hundreds of different volatiles, and our sensory systems aren't designed to notice each one. The overall blend of volatiles tells us what we need to know. Most leaves smell "green" when fingered or crushed thanks to the several green-leaf volatiles; most flowers smell "floral" thanks to a handful of terpenoids and benzenoids, most fruits "fruity" thanks to bunches of esters.

Usually, though, there are a few volatiles that rise above the generic hum, sometimes because they're present in unusual proportions, sometimes because they're unusual ingredients for a leaf or flower or fruit. These are the volatiles that

I emphasize, drawing whenever possible on published studies of living or fresh materials, not altered by drying or distillation into an essential oil. To keep the tables as simple as possible, I'll name the generic green-leaf volatiles only when they're a dominant or major contributor to a plant's smell.

One or more of a plant's component smells may jump out at you at first sniff; others may become perceptible after several more, or months later when you're not even hunting for them. Or they may always remain too integrated in the mix to pick out. You'll have all these experiences. What makes each valuable is what underlies them all: the attention to what you're perceiving, your awareness of its complexity, and the expansion of your database of perceptions and reference points.

Not to mention the pleasure of recognizing subtleties and patterns and themes. The volatiles and smells of cedar differ from those of pine, ditto for spearmint and peppermint, damask and tea roses. But trees and greens and seeds also share volatiles, and their smells often echo and suggest each other. Keep that in mind and you'll begin to notice and enjoy whiffs of the forest in a sprig of rosemary or a cannabis bud—and vice versa. Similarly, there are flowery and fruity notes in a number of greens and roots and seeds. Some volatiles that strike us as attractively flowery were invented as defensive weapons long before the first flower appeared on the Earth, by diminutive plants whose descendants today are mostly underfoot and ignored.

Curious and informed smelling can lift the dull shroud of familiarity so that we perceive plants and their creations afresh. It can also remind us of their significance in human culture: the scents of tree resin that inspire thoughts of deep time and immortality; ceremonial flowers that represent life's intensity, fragility, and brevity; fruits that, as carefully prepared and flavorful nourishment for creatures of a different kingdom, provide a model for cooperation—and the gold standard for cooking.

So begin by getting a sense for the range and richness of plant creativity. Stop and linger when you see a smell or source that resonates with your experience or interest. Just follow your nose.

Mosses, liverworts, horsetails, ferns

Let's enter the garden on a path that takes us into a stand of tall trees whose canopies filter the sunlight; its rays sparkle off a gently flowing stream. We'll begin not with the trees, but with the easily ignored plants underfoot, far from nose level, coating patches of rock and soil and tree trunk. These are mosses and liverworts, modern versions of very early land plants. Their clan never developed the circulatory system that carries water from roots through stems to leaves: they don't have true roots, and can't survive without plentiful moisture. Nor could they have survived over the eons without effective defenses against insects and microbes. Sure enough, lowly mosses and liverworts make many of the same volatiles that we associate with fragrant trees and flowers and fruits.

That they do is a token of the deep history of plant volatile virtuosity. Smells that we prize today were likely first emitted by ancient plants to deal with their enemies, and they persist today because they continue to be effective defenses. Those plants developed something like the volatile equivalent of sound synthesizers, and found the notes that confused or annoyed or stunned or killed their enemies. Ever since, their descendants have been adjusting both notes and chords to suit their evolving, increasingly complex relationships with animals and microbes. **Liverworts**, whose direct ancestors go back nearly 400 million years, developed specialized oil droplets for accumulating volatiles in their cells, and in them chemists have identified more than seven hundred different terpenoid molecules and two hundred benzenoids. Liverwort specialists report that they can smell of wood, turpentine, carrots, mushrooms, and seaweed. **Mosses** generally don't have particularly strong smells, but they're recognizably planty, with green-leaf volatiles and other simple carbon chains, selections of terpenoids and benzenoids, and dimethyl sulfide, which we'll see is more common among seaweeds than land plants. Their eight-carbon chains have been found to repel insects, and their nine-carbon chains suppress fungal growth.

SOME PRIMITIVE LAND PLANTS (GLVs = green-leaf volatiles)

Plant	Component smells	Molecules
liverworts (family Hepaticae)	green, mushroomy, woody, spicy, turpentine, carrot, seaweed	GLVs, octenol, terpenes (pinene, camphene, limonene, phellandrene, menthenol)
mosses (family Musci)	green, mushroomy, piney, earthy	GLVs, octenol, nonenal, nonanal, terpenes (pinene, limonene, humulene, ionone, bornyl acetate, geosmin), benzenoids (benzaldehyde)
horsetails (*Equisetum*)	green, mushroomy, fruity, floral, spicy	GLVs, octenol, terpenes (ionones), benzenoids (phenylethanol, vinylguaiacol)
ferns (*Polystichum, Dryopteris, Phegopteris, Pteridium*)	green, sweet, floral, spicy, hay	GLVs, nonanal, decanal, terpenes (pinene, terpineol, limonene, linalool, ionone), benzenoids (benzaldehyde, acetophenone, coumarin)
maidenhair fern (*Adiantum capillus-veneris*)	green, plastic, soapy, stinkbug	GLVs, decenal, decadienals

Rising above the mosses and liverworts in this moist shady spot are **horsetails**, erect green leafless quills, and **ferns**, with their lacelike incised fronds. Their families are more advanced than the mosses and liverworts, still lacking true roots but with vascular systems for water transport and the ability to grow quite tall—the planet's first forests were largely horsetail ancestors as high as a hundred feet (thirty meters). They, too, have generically green smells, though a number of ferns are notable for sweet, haylike, floral smells that help define a whole category of perfumes, the *fougères* (see page 478). Among the ferns, the common houseplant **maidenhair fern** is notable for being decidedly un-perfume-like. It emits the same simple-chain aldehydes that give the herb cilantro its characteristic controversial smell, which can suggest both soap and bugs (see page 258).

Conifer trees

Now let's raise our eyes and thoughts to the trees, various groupings of needle-leaved conifers: cedars, pines, firs, redwoods, and their ilk. Trees are familiar ornaments of parks and yards and city streets, but try to imagine that you're encountering them afresh as the strange and marvelous creatures they are. Towering, hulking, sometimes centuries or millennia old, they exist on an altogether different scale from ours. Two species of California redwood include the tallest and most massive living things on today's Earth. Trees have helped awe human cultures into imagining a higher realm of powerful beings that they point toward and represent. In the version of the *Epic of Gilgamesh* composed by a Babylonian priest over three thousand years ago from much earlier Sumerian sources, the hero and his companion Enkidu reach the Cedar Mountain, "dwelling of gods, throne-dais of goddesses" and "stood there marveling at the forest, / Observing the height of the cedars":

> The cedar was scabbed with lumps of resin for sixty cubits' height,
> resin oozed forth, drizzling down like rain,
> flowing freely for ravines to bear away.

The fragrances emanating from trees and their resins have often been taken as ways of communicating with higher realms while getting a whiff of them here on Earth. Strange, that creatures stuck in the mire and moving only with the wind should inspire ideas of the ethereal and the transcendent—ideas that become part of the experience of their smells.

Our first grove of trees includes clusters of cedar and its resin-oozing relatives, all members of the ancient cone-bearing tribe whose heyday came 100 million years before the dominance of flowering plants and their broad-leaved forests. The smells will be familiar to anyone who's come within sniffing distance of a fresh Christmas tree or gone for a hike in the mountains. Many conifers produce the kind of resin that Gilgamesh saw and smelled on the Cedar Mountain, a viscous, sticky, fragrant fluid that defends them against attack by microbes and by wood-boring beetles and other insects, and that repairs wounds to their tissues.

Resins are a complex mixture of defensive chemicals. The primary components are terpenoids, the carbon chains that are ten, fifteen, or twenty carbons long, ill-suited to interacting with water molecules and so mostly water-resistant. There are diterpenoid molecules, with twenty carbons too large to be volatile, but which give the fluid substance and stickiness. There are also monoterpenoids and sesquiterpenoids, smaller and volatile, which dilute the diterpenoids and give the resin its fluidity and its smell. When a tree is physically damaged, stored resin flows from nearby ducts or "blisters" to the damaged surface, where it traps or repels invaders and creates a seal that slowly hardens as the volatile terpenoids evaporate into the surrounding air. Even trees that don't ooze resins still store defensive terpenoids in their wood to make it less susceptible to insect attack and to *rot*, the name we give to damage caused by fungi that specialize in attacking lignin.

Tree resins are ancient, ingenious, effective materials that provide both physical and biological protection. They were appreciated as such by the earliest human civilizations, from Gilgamesh's Sumer to the Indus valley farther east, which used resins scraped from various conifers or cooked out of wood to treat wounds in human flesh, to mummify the dead, and to waterproof containers, roofs, boat hulls, and even cloth. Later on, as cities, states, and trade developed, countless conifers were felled to protect the wood of oceangoing ships, and *turpentine*—an extract of resin and wood, usually pine, enriched in the smaller terpenoids, excellent for thinning and dissolving resin as well as other oily, greasy materials like paints—became an important commodity for human industry in general (see page 417). Turpentine still finds some use, and its continuing presence far removed from the forest explains why, fresh and pleasant as the smell of tree resins may be, they can also remind us of paint thinners and floor cleaners.

Though there's a broad family resemblance among the conifer resins, they all have their own typical terpenoid mixes and smells. Gilgamesh's cedar didn't smell quite like the aromatic, moth-repelling cedar chests to which we entrust our special clothes. His renowned but now rare **cedar of Lebanon** belongs to a different genus altogether than the North American trees that give us cedar oil and cedar wood, which are actually species of **juniper** and **arborvitae** (*Thuja*). Both the true cedar and the juniper called **eastern red cedar** contain woody-smelling sesquiterpenoids, but they're different sets of molecules. The arborvitae, or **western cedar,** contains yet another set of sesquiterpenes, but they're outnumbered by monoterpenoids that add a distinctively fresh piney, turpentiney note.

SOME CONIFER TREES

Tree	Component smells	Molecules
Lebanon (aka Atlantic) cedar (*Cedrus*)	resinous, woody, sweet	pinene, himachalenes, himachalols, atlantone
Texas, Mexican, eastern red cedar (*Juniperus*)	woody-camphor, green	cedrene, cedrol
western red cedar (*Thuja*)	camphor-minty, woody, piney	thujone, sabinene, terpinenol
pine (*Pinus*), juniper (*Juniperus*), fir (*Abies*), spruce (*Picea*), redwood (*Sequoia, Sequoiadendron*)	piney, turpentine, fresh, woody	pinene, limonene, carene, phellandrene, myrcene, terpinyl & bornyl acetates, sabinene

That lighter pine impression is common to many familiar conifers, **pines** and **firs** and **spruces** and **redwoods**, and we come across it in lumber, plywood, and Christmas trees. The needle- and scale-like conifer leaves share background green-leaf volatiles but carry different mixes of their characteristic terpenoids and sometimes other volatiles. For example, redwood foliage includes the woody benzenoid safrole and a green, nitrogen-containing methoxypyrazine. Conifer greens are especially aromatic when young and actively growing.

As we emerge from the edge of the conifer grove, notice the outward-facing pine trunks where patches of resin bake in the sun for hours at a time. The smell they give off isn't just piney and fresh; it has a rich quality, warm like the sun. Years ago that smell caught the attention of Roman Kaiser, a distinguished chemist at the Swiss fragrance company Givaudan, as he hiked along the Mediterranean coast of Liguria in western Italy. Kaiser has studied and written eloquently about the volatiles and smells of the plant world for decades. He analyzed a sample of sun-warmed pine resin, and found that it harbored unusual fifteen-carbon chains that are greatly prized in perfumes! So conifer resins scent the forest air themselves, and also provide the ingredients for new volatiles that may be incidental for the trees but give great pleasure to attentive passersby. More about Kaiser's discovery in chapter 17.

Hardwood tree resins: frankincense, balsams, copals

We move on now from the impressive shade of the conifers to a more open grove of mostly short trees with broad leaves. They belong to the younger and much more diverse tribe of flowering plants, whose trees are called *hardwoods* to distinguish them from the (often) softer-wood conifers. Like conifers, the trees in this grove also secrete aromatic resins when they're wounded. They represent our more comprehensive version of the spicy forest in Eden that the poet John Milton had Raphael traverse on his way to talk with Adam. Milton planted the forest with myrrh, cassia, and balm because, along with frankincense, they figure in the Bible as materials for anointing and incense, rare gifts of the Creator whose incomparable smells transport us from ordinary life to a more spiritual, contemplative plane. Some of these materials are still incense ingredients in the Catholic and Orthodox churches, and they're also readily available online for aromatherapy and as unusual ingredients for foods and drinks. To learn how censing intensifies resin and wood aromas, jump to the section on incense (see page 444).

Frankincense and myrrh are resins that come from related shrubby trees native to the arid Arabian Peninsula and northeastern Africa. Prized for millennia throughout much of the Old World, they're harvested by gashing the tree trunks to stimulate resin production, then scraping off the resin that collects on the wound when it's hard enough to handle.

Frankincense, also known as olibanum, is the defining material of Western church incense. It comes from trees in the genus *Boswellia*, one major species marked by pine and spicy notes, another by fresh and floral notes. The distinctive frankincense smell that they share comes from unusual **olibanic acids**, eight-carbon chains with a three-carbon ring at one end. The chemists who discovered their role in 2016 described their smells as the incense's typical "balsamic, old church-like endnote," and found that they persist on surfaces for months: so when launched from the censer's glowing coals during religious services, they "settle on walls, furniture, and drapes and then allow a continuous diffusion of their odor."

Myrrh is often mentioned in the same breath with its relative frankincense, and has been considered the finest of an entire subfamily of resins from related species that includes balm of Gilead and such verbal standouts as bdellium, opo- ponax, and gum guggul. It's sweeter and earthier than frankincense, with unusual mushroom and leather facets contributed by modified sesquiterpenoids, including **furanoeudesmadienes.**

Less exotic and liturgical than frankincense and myrrh is **mastic**, the resin from a relative of the pistachio tree growing on the Greek island of Chios and nearby regions. It's used as a flavoring in Greek and Turkish cooking and as a persistent, tooth-gluing chewing gum—as pine resin has been in northerly regions. Its vola- tiles are dominated by the terpenoids pinene and myrcene.

SOME TREE RESINS

Resin	Component smells	Molecules
frankincense, olibanum, *al-luban* (*Boswellia sacra* & *carteri*; Oman & Somalia)	incense, old church, woody, piney, peppery	olibanic acids, pinene, myrcene, linalool, cresol, mustakone, rotundone
frankincense, olibanum, *al-luban* (*Boswellia papyrifera*; Sudan, Ethiopia, Eritrea)	incense, old church, fresh, green, earthy, mushroomy, floral	olibanic acids, octyl acetate, octanol, limonene, geraniol, linalool
myrrh (*Commiphora myrrha*)	warm, sweet, leathery, mushroom, undergrowth	furanoeudesmadiene, lindestrene, curzerene (3-ring furan-sesquiterpenoids)
mastic (*Pistacia lentiscus*)	piney, woody, fresh, floral	pinene, myrcene, limonene, linalool
benzoin, Siam & Sumatra (*Styrax tonkinensis* & *benzoin*)	Siam: balsamic, fruity, floral, sweet, vanilla Sumatra: balsamic, floral, fruity, plastic, sweet	Siam: benzoic acid & benzoates, vanillin Sumatra: cinnamic acid & cinnamates, benzoic acid & benzoates, styrene

Resin	Component smells	Molecules
sweet gum "styrax," storax, Oriental & American (*Liquidambar orientalis* & *styraciflua*)	sweet, plastic, piney, woody, floral, fruity	styrene, pinene, caryophyllene, cinnamate esters
copals (species of *Bursera, Hymenaea, Protium, Pinus*)	woody, spicy, piney, resinous	copaene, germacrene, pinene, limonene
diesel tree resin, copaiba (*Copaifera*)	woody, camphor, balsamic	caryophyllene, germacrene, selinene

The aromatic material that John Milton called "balm" is a defensive tree secretion like conifer resins and frankincense and myrrh, but its chemical makeup and smells are very different. *Balm* now means something soothing and healing. It came via the Greek *balsamon* from Hebrew and Arabic words meaning "spice" or "perfume"—aromatic plant materials also used to *embalm*, to prevent the decay of the human or animal body and replace the smells of death with the smells of pleasure and ceremony. These overtones of healing and preservation hover in the name of "balsamic" vinegar, traditionally made by slowly fermenting a sweet wine vinegar in barrels made from aromatic woods (see page 583). Today plant biochemists use *balsam* to refer specifically to tree resins that are dominated not by piney-woody terpenes but by sweet, mild, soothing, "balsamic" benzenoid rings, which include benzoic and cinnamic acids and their esters, and vanillin.

Balsams have been important incense ingredients for many centuries and in many cultures, from Asia through the Middle East and into Europe. **Siam and Sumatra benzoins** come from sister Asian tree species, the Sumatra giving more prominence to fruity-floral cinnamic volatiles. A distinctive kind of balsam comes from **sweet gums**, ornamental trees of the Northern Hemisphere genus *Liquidambar*, whose hand-shaped leaves turn deeply red in the autumn. They're common street and park trees and often have beads of resin on their trunks. Sweet-gum resins contain significant amounts of balsamic-smelling vinylbenzene, a molecule made from a six-carbon benzene ring plus a two-carbon side chain. Since it was originally discovered in sweet-gum styrax balsam, vinylbenzene was dubbed **styrene**, a name that most of us would know from polystyrene plastics, the synthetic

material of light spongy packaging materials and takeout food and drink containers. Polystyrenes often retain traces of the styrene building block and its smell, so
styrene-heavy resin straight from a tree can seem to have an artificial, plastic
quality. Given the tree's priority, we should really call the plastic's smell balsamic.

The civilizations of the New World also treasured tree resins, and some of their
most important sources are subtropical and dry-land members of the same botanical family as Old World frankincense and myrrh (species of *Bursera*, *Hymenaea*,
Protium). They are known as **copals**, from a Nahuatl root, and were and continue
to be used as preservatives, folk medicines, fumigants, and incense for honoring
gods and powerful men, as well as for making a fine varnish. Except for those collected from pines, copals are dominated by woody-smelling sesquiterpenoids, notably woody-spicy-honey copaene.

The last resin-making tree we'll smell comes from the South American genus
Copaifera, a name that means "copaiba-bearing," **copaiba** being the name for this
particular kind of copal. But it's also called the diesel or kerosene tree and is
known less for unusual volatiles than for its prodigious yield of flammable woody-
smelling resin. It's usually taller than the other resinous hardwoods, and single
trees in a single year can give more than ten gallons (forty liters) of copaiba. The
diesel tree is a living monument to the astonishing productivity of these avatars of
Hero Carbon that we call plants, turning sunlight and air into tangible, smellable,
useful matter.

Aromatic hardwoods: cinnamons and laurels, eucalypts, incense woods

The cassia that the archangel Raphael caught a whiff of in the spicy forest grows
in the next grove of trees in our garden, a group of hardwoods that don't secrete
resins from wounded tissue but do protect their woody parts and leaves with
abundant defensive volatiles. Along with cassia, most of them are used as spices,
to bring interesting smells to our foods. The first five span the planet from Asia to
the Mediterranean to California, and yet all belong to the same plant family, the
Lauraceae, named after the Mediterranean member, the bay laurel, which gives us
our spice-rack bay leaves. The laurel clan has been traced back to the very early

days of the flowering plants, and it's likely that the production of volatiles—their general aromatic quality—has been one of the keys to their longevity.

SOME AROMATIC HARDWOOD TREES IN THE LAUREL FAMILY

Tree	Component smells	Molecules
cinnamon (*Cinnamomum verum*)	cinnamon, honey, flowery, woody, clove	cinnamaldehyde 5–15%, cinnamyl acetate, linalool, caryophyllene, eugenol
cassia, Indonesian cinnamon, Saigon cinnamon (*Cinnamomum cassia, burmannii, loureiroi*)	cinnamon, pungent, sweet, hay	cinnamaldehyde 15–20%, 20–50%, 55–70%; cinnamic acid, coumarin
camphor laurel (*Cinnamomum camphora*)	camphor, fresh, medicinal, floral, woody, warm, spicy	camphor, eucalyptol, linalool, nerolidol, safrole, borneol
bay laurel leaf (*Laurus*)	camphor, lavender, pine, woody	eucalyptol, terpinyl acetate, pinene, sabinene
California bay leaf (*Umbellularia*)	fresh-harsh, medicinal, woody, pine	umbellulone, eucalyptol, terpineol, thymol
sassafras root bark, leaf (*Sassafras*)	bark: root beer, camphor, nutmeg, clove; leaf: lemon, pine, woody	bark: safrole, camphor, myristicin, eugenol; leaf: neral, geranial, pinene, caryophyllene
avocado leaf (*Persea americana*, Mexican race)	anise, woody, pine, cinnamon	estragole, caryophyllene, pinene, methyl eugenol

Though cassia was once as rare and exotic as myrrh and balm, it's now very much an everyday ingredient: it's a kind of cinnamon, the inner bark of medium-tall Asian trees in the genus *Cinnamomum*. The spice trade distinguishes among several different kinds, though product labels usually don't. The flavor differences are mainly due to the proportions of the benzenoid cinnamaldehyde, the main volatile and also a trigger of heat and pain receptors, and the source of flavor and heat in Red Hot candies. **True cinnamon** has a mild and complex smell, with a fruity ester and flowery terpenoid complementing moderate, not especially pungent levels of cinnamaldehyde. **Cassia** is more intensely cinnamony and pungent from as much as double the cinnamaldehyde, with a sweet, haylike note from coumarin, another benzenoid. And **Indonesian** and **Saigon cinnamon** take the cinnamaldehyde up to red-hot levels. A third species of *Cinnamomum* has been known as **camphor**, and gave its name to and was the main source for that distinctive monoterpenoid, at the same time fresh and medicinal, cooling and warming. Camphor the volatile has long been extracted from the tree wood and leaves by heating them and collecting the vapors, from which it solidifies in the form of waxy white crystals. It has been used since ancient times as a medicine and as an ingredient in cooking, mainly in Southeast Asian and Arabic traditions. Six variants of the camphor species actually specialize in some other volatile instead—there are linalool trees and eucalyptol trees, for example—and are cultivated to make fragrance extracts.

The **bay laurel** is a midsize Mediterranean tree, easily trimmed to be a garden-scale bush, whose tough leaves carry a wonderfully versatile terpenoid mix of eucalyptus, pine, lavender, and woody notes that infuse an added dimension to stews. The so-called **California bay** is a distant relative geographically and aromatically, much less versatile because its leaves and bark are usually dominated by the harsh pungent/cooling terpenoid umbellulone, which is probably responsible for this bay's nickname, the "headache tree." Another member of the laurel family that made its mark in North America is the Eastern native **sassafras**, which grows as a tree or shrub. Sassafras root bark once flavored root beer until its major volatile, the benzenoid safrole, was suspected of being a carcinogen and banned from commercial foods. But the distinctive three-lobed or mitten-like leaves still flavor Louisiana gumbos with lemon and pine terpenoids in the form of filé powder. And the Mexican race of the **avocado** tree, which thrives at cooler high elevations than other tropical races, protects its leaves with the anise-like benzenoid estragole, along with a couple of piney, woody terpenoids, while its bark includes the prominent copal sesquiterpenoid copaene.

Now we walk along to a handful of aromatic trees that are unrelated to the laurel family and to each other. The first, the **eucalyptus**, with weeping, brittle branches and elongated leaves, is familiar to anyone who has visited Australia, its home, or California, where it was widely planted to replace native redwoods and oaks. Eucalypts are prolific terpenoid producers, with more enzymes devoted to that function than any other known plant. There are several hundred species with different forms and volatile mixes. Most can grow to be quite tall, and most feature the eponymous monoterpenoid **eucalyptol**, which has a fresh, cooling quality in addition to its distinctive smell. It's not much used in cooking but distilled eucalyptus volatiles have been an important commercial product since the nineteenth century, and both eucalyptus oil and eucalyptol are ubiquitous ingredients in household and personal products.

Next to the eucalypts are two clusters of small to medium-size Asian trees that have been prized from ancient times for their aromatic woods. **Sandalwood** is thought to derive its name from a Sanskrit root meaning "glowing" or "shining," reflecting the use of its older heartwood and root wood as incense; but its wood is also traditionally ground with water into a fragrant paste for use in ceremonies, medicine, and food, and distilled for an essential oil used in perfumery and to aromatize tobacco. The tree is parasitic, like its relative mistletoe, and taps the root systems of other trees; it's now cultivated in plantations, notably in Australia. It owes its uniquely soft, milky, somewhat animalic, urinous smell to unusual sesquiterpenoids, the **santalols**.

SOME OTHER AROMATIC HARDWOODS

Tree	Component smells	Molecules
eucalyptus leaf (*Eucalyptus*)	eucalyptus, camphor, pine	eucalyptol, pinene, cymene
sandalwood (*Santalum album*)	woody, creamy, powdery, sweaty, smoky, urinous	santalols, santalals (sesquiterpenoids)
agarwood, aloeswood, oud, *jinkoh*, *kyara* (*Aquilaria* species)	rich, woody, sweet, floral, vanilla, animal	unique & myriad sesquiterpenoids, benzopyrones, furans
palo santo (*Bursera graveolens*)	sweet, minty, floral, woody, pine	mint carvone, mint lactones (benzo-furanones), pulegone, terpineol, bisabolene

The second cluster of Asian trees is devoted to **agarwood**, aka aloeswood, *jinkoh*, *kyara*, and oud—the last word Arabic for "wood" itself, *agar* deriving from ancient Sanskrit. They're mostly Southeast Asian species in the genus *Aquilaria*. When they're wounded and their inner tissues exposed to fungi and bacteria, these trees (perhaps with the help of their resident microbiome) respond by producing a protective terpenoid-rich resin that gradually permeates the wood surrounding the wound. Agarwood is the most remarkable incense material, the rarest and most expensive, so sought after for so long that it has become endangered. Fine specimens cost hundreds of dollars per gram, and efforts are under way to induce resin production artificially.

The smell of agarwood is unlike any other wood or resin, with so many complementary facets, from woods and resins to flowers, spices, and leather, that the fragrance chemist Roman Kaiser describes it as "the mother of all scents." Its richness reflects its chemical complexity, which perhaps defeats microbial attack by deploying too many different chemical weapons for any microbe to overcome. Chemists have cataloged dozens of sesquiterpenoids found nowhere else, along with two-ring benzopyrones that are close relatives of sweet-hay coumarin.

The last tree in this aromatic grove is a South and Central American relative of the trees that produce copal and frankincense. Less renowned for incense than sandalwood and agarwood but well worth knowing, the heartwood of **palo santo**, "holy wood," is cut into sticks, ignited, and the flame blown out so that it smolders. The smell of palo santo is remarkable for its unique and surprising note of mint: it shares a carvone, pulegone, and mint lactones with members of the herbaceous mint family.

Mild hardwoods: oak, maple, cherry

We come now to the last grove of trees, mature and majestic, more common and familiar in Europe and North America than any we've sniffed so far, and not as distinctively aromatic: namely the oaks and maples and others that populate the broadleaf forests of the Earth's temperate climates. These trees emit similar and fairly generic smells when their wood is cut, initially with fresh and green starter-set aldehydes from breakdown of various carbon chains, and later vinegary acetic acid and bready furfural (see page 345) as the more complex cell-wall materials dry out and break down.

The best studied of the temperate hardwoods is the **oak**. The English word *tree* and many of its European cognates come from the proto-Indo-European word for "oak," an index of how important a presence it has been in human life. There are several hundred different species of oak across the Northern Hemisphere, and they share with most of their non-coniferous forest companions a lack of character-defining volatiles. They don't bother much with terpenoids. Recall that the molecular backbone of wood is lignin, a complex mass of interlinked benzenoid and other carbon rings. The volatiles of oak and many hardwoods turn out to include aldehyde forms of these building blocks, along with a smattering of volatile benzenoids, including vanillin, clove-like eugenol, and guaiacol, which smells smoky because it's a lignin fragment produced from wood when it burns (chapter 16). Oak in particular also carries a modified eight-carbon lactone, the "oak" or "whiskey" lactone, which shares with other lactones a sweet, coconut-like quality. **Cherry** wood carries a different lactone, along with some of the same benzyl volatiles that also flavor cherry fruits. Many of these carbon-ring volatiles are boosted when "green" woods are dried, and become even more prominent when the dry wood is lightly singed, charred, or burned. They bring welcome aromas to foods that are cooked above smoldering wood, and to drinks—wine, whiskey, rum—stored in wood barrels. They even help give the wood-pulp pages of old books and newspapers their characteristic sweet smells, with vanillin, benzaldehyde, and furfural dominating a few starter-set acids and aldehydes.

SOME MILD HARDWOODS

Tree	Component smells	Molecules
ash (*Fraxinus*), maple (*Acer*)	fresh, green, bready, vinegar	pentanal, hexanal, furfural, acetic acid
oak (*Quercus*)	bready, fresh, woody, sweet, clove, smoky, coconut	syringaldehyde & other lignin aldehydes, vanillin, eugenol, guaiacol, oak lactone (methyl g-octalactone)
cherry (*Prunus*)	almond-essence, balsamic, coconut-creamy	benzaldehyde, benzyl alcohol, massoia (C10) lactone

As we leave the oaks and maples and cherry trees behind, we come to a very different scene: several beds of shrubs and low plants, and beyond them, a small meadow. An open vista that invites us to lower our gaze, amble, pluck and sniff.

Leafy plants

The shrubby, grassy, and herbaceous members of the plant kingdom may not inspire the awe that trees do, but they're good at delighting. They're our everyday
companions, closer to us in size and sturdiness and life span, easy to handle,
sources of nourishment. Their aboveground mass is mostly foliage, thin, pliable
green leaves. We see them everywhere, from the city to the countryside; we walk
over them, snooze on them, carpet parklands with them, farm them en masse,
plant them around homes and in gardens, bring them onto windowsills, into our
kitchens, into us.

The same traits that make leafy plants appealing and useful push them to
make major investments in chemical defense. They're tender, low to the ground,
vulnerable to microbes and hungry animals of all kinds, from slugs to humans.
And their photosynthesizing leaves are essential assets. Some plants arm their
leaves with nonvolatile molecules that are toxic to animals, or armor them with
prickles and spines. But for nearly all, the first lines of defense are the green-leaf
volatiles that are formed and released when the leaf cells are damaged and that
give green leaves their characteristic smell (see page 154).

In addition to releasing GLVs and similar simple-chain volatiles, many plant
leaves defend themselves as the trees do, with premixed cocktails of kinky terpenoids and/or hex-ring benzenoids. Some stockpile these weapons in special internal canals, and others in external, hairlike structures, the visible fuzz on the
leaf surface of mint and basil, tomato plants, and cannabis. These external glands
are fragile, and break and release their contents on the slightest physical contact,
even before the leaf itself is damaged.

When it comes to our experience of their smells, leafy plants fall into three
groups. There are wild plants and plants that we use for landscaping, whose volatiles are much diluted in the open air. There are plants whose leaves we eat as
vegetables, which rely mainly on the GLVs and have a relatively narrow range of
green-vegetable flavors. And there are the plants we call herbs, which augment the
GLVs with terpenoids, benzenoids, furanones, and other volatiles, which have a
wide range of strong smells, and which we use in small quantities to add flavor to

our foods. We'll sample each of these groups and see that, unimposing though they may be in stature, the green plants are chemical virtuosos. Wild and landscaping plants coming up; vegetables and herbs in later chapters.

Aromatic shrubs, terpenoidal and stinky

The shade of oaks behind us, we pass now toward an area of mostly bare ground dotted with groups of low, scruffy plants, representatives of some common and assertively aromatic shrubs. The first several emit mainly terpenoids, which ally their smells with conifer trees. The plant genus *Artemisia* includes dozens of terpenoid-rich species. Many of them are weedy denizens of dry soils and variously named mugwort, wormwood, sagebrush, and sagewort. **Common mugwort** smells mainly of pine, eucalyptus, and camphor (there's also an edible mugwort, see page 245), while **California sage** has a solvent quality and leans in the direction of cedar and the kitchen herb. **Yarrow** is a decorative garden shrub in the daisy family that mixes camphor, eucalyptus, and pine terpenoids.

The next several plantings have downright unpleasant smells that are reminiscent of animals and ourselves. **Boxwood** is a fine-leaved European shrub that lends itself well to trimmed hedges and topiary, but has always been notorious for an unpleasant edge to its sun-warmed smell. Chemists have identified the responsible volatile as the "cat ketone," the sulfur-containing molecule also found in cat urine (see page 81). The several species of **rue** are Mediterranean shrubs with a strong solvency, medicinal quality from acetone-like ketones, and some with a host of sulfur volatiles, including the cat ketone and several others identical or closely related to molecules found in human sweat (see page 121). Lower lying than boxwood or rue is **stinking goosefoot**, a relative of quinoa and spinach native to Europe and western Asia but now widespread. Its leaf is shaped like a webbed foot—hence *Chenopodium*—and it has a fishy, ammonia-like smell due to an unusual stockpiling of trimethylamine—hence *vulvaria*, "vulva-like," for the supposedly similar smell of female genitalia (see page 116).

SOME AROMATIC SHRUBS

Plant	Component smells	Molecules
mugwort (*Artemisia vulgaris*)	turpentine, fresh	pinene, camphene, camphor, eucalyptol
California sagebrush (*Artemisia californica*)	eucalyptus, fresh, cedar, sage, solvent	eucalyptol, thujone, camphor, artemisia ketone
yarrow (*Achillea*)	camphor, eucalyptus, turpentine	camphor, sabinene, eucalyptol, borneol, pinene
English boxwood (*Buxus sempervirens*)	green, woody, sweaty, cat pee	phellandrene, humulene, methyl sulfanyl pentanone
rue, fringed rue (*Ruta graveolens* & *chalapensis*)	solvent, penetrating, rancid, sweaty	undecanone, nonanone, nonyl acetate, sulfanyl hexanols, methyl sulfanyl pentanone
stinking goosefoot (*Chenopodium vulvaria*)	fishy	trimethylamine
creosote bush (*Larrea tridentata*)	turpentine, pungent, solvent, sweet, seashore, musty, medicinal, smoky	mixed monoterpenoids, methyl butenone & butanone, methacrolein, methyl acetate, acetone, dimethyl sulfide, phenol, guaiacol, chloroanisole, carbon tetrachloride
sweet or pink everlasting, rabbit or ladies' tobacco, cudweed (*Pseudognaphalium*)	maple syrup, curry, fenugreek	sotolon?

Then, standing apart from the rest in a patch of crusty dry soil is the remarkable **creosote bush**, a denizen of vast swaths of the Mojave Desert and similar arid regions down into South America. It's named for the tarlike residue of combustion once used to treat wooden railroad ties (much more about creosote in chapter 16). The Spanish name is *hediondilla*, from *heder*, "to stink." Its stems and leaves are coated with a shiny, tacky resin that protects against water loss and damage from the sun's ultraviolet rays, and that, when washed by rare rain, is said to saturate the desert air with its smell. Creosote bush resin emits a hodgepodge of volatiles that indeed include some of the same molecules found in tars and asphalts. Terpenoids, several; benzenoids, the medicinal phenol and smoky guaiacol; solvent-like ketones; methyl acrolein, the relative of an acrid aldehyde that characterizes overheated cooking oil; sulfurous dimethyl sulfide; *and* a number of chlorine-containing volatiles, musty chloroanisole, and carbon tetrachloride and chloroform, two important industrial solvents. A nasty collection of molecules! And apparently a very effective one. A sparse ring of creosote bushes south of Barstow, California, sixty feet in diameter and first spotted from the air, marks the clonal extensions of a long-dead plant at their center whose remains have been dated to eleven thousand years ago.

Of course I had to make a pilgrimage to the home of King Clone. In July 2014, I found myself on the lonely Bessemer Mine Road east of Lucerne Valley, gazing at nothing but creosote bushes scattered from horizon to horizon. It was starkly evident that they've managed to defend themselves against all comers. To me their smell seemed "chemical" rather than plantlike, with the harsh halogenic chlorine/bromine quality of an oversanitized swimming pool. In the desert!

To purge that smell and prepare for the pleasantness of the meadow ahead, stoop to sniff at a different California wild plant, one found in sunny exposures on Bay Area hills that teased me for years with its elusive scent of maple syrup. When I finally identified it, it turned out to be the sister species of a weed known in the eastern United States as **sweet everlasting** or **rabbit tobacco**. These members of the daisy family have glands on their leaf surfaces that exude a sticky aromatic resin. So far the few species studied emit mostly terpenoids; I'm betting that in ours there's also some of the furanone sotolon (see page 525).

Meadow grasses and clovers, green and sweet

Try now to lend our imagined botanical garden a hint of that great pleasure of the natural world: walking on a sunny day through a meadow green with countless blades of grass and the leaves and flowers of their weedy companions, or *forbs*. The moist air and the smell it carries, grassy and haylike and sweet, are part of that experience, which we approximate with parklands and golf courses and home lawns. In fact, the delightful smells of sun-warmed meadow and freshly mown lawn are products of the physical and chemical damage that the plants are suffering. There's the mower's blade, of course, and the sun's double-edged gift: enough energy to fuel life and growth, but also enough to damage their fragile machinery.

Common turf **grasses** are the quintessential green of the humanized landscape, an ideal carpet because they spread horizontally along the ground and their growing tips can be trimmed close without killing them. Grassiness is also the quintessential green smell, the product of green-leaf volatiles released when grasses are cut or chewed on, and also when they wilt in strong sun. Unlike the rest of the leaves in the table below, turf grasses don't have other major defensive volatiles to augment the green-leaf volatiles. This may be because their growth habit actually allows them to thrive despite being nibbled at by grazing animals, so there's no great need.

Various species of **clover** are a frequent companion to grasses in meadows and hayfields and lawns, and some of them, notably sweet clover, are responsible for the presence of the benzenoid ring coumarin and its wonderfully sweet, vanilla-like scent. Coumarin is a chemical defense that plants stockpile in inert form and release when their tissues are damaged by crushing or by drying. It's often described as haylike, because even when sweet clover is a minor component of mixed animal forage crops, it aromatizes them when they're piled together and dried for storage as hay. Pleasant though it is, the scent of coumarin in forage signals potential danger for animals; mold growth in damp hay converts coumarin into extremely toxic dicumarol, which can cause hemorrhage and death (Coumadin is a coumarin-based blood-thinning drug for humans).

SOME LEAFY GRASSES AND FORBS (GLVs = green-leaf volatiles)

Plant	Component smells	Molecules
grass (*Festuca, Lolium, Poa,* etc.)	green, grassy	GLVs
sweet clover (*Melilotus*)	sweet, fresh hay, vanilla	coumarin, GLVs
sweet woodruff (*Galium*)	sweet, fresh hay, vanilla	coumarin, GLVs
sweet vernal grass (*Anthoxanthum odoratum*)	sweet, fresh hay, vanilla	coumarin, benzoic acid, GLVs
sweet grass, vanilla grass, holy grass, bison grass (*Anthoxanthum nitens, Hierochloe odorata*)	sweet, fresh hay, vanilla	coumarin, benzoic acid, GLVs

The sweet smell of coumarin is pleasant enough that a number of leafy plants are cultivated to bring it to gardens and indoors. So at the far edge of the meadow we encounter a few separate plantings. **Sweet woodruff** is a coumarin-stockpiling forb now common through Europe and Asia, grown in lawns and gardens and dried to scent sachets, rooms, perfumes, and drinks. Then there are a couple of similar-smelling clumping grasses: Eurasian **sweet vernal grass**, and Northern Hemisphere **sweet grass**, aka vanilla or holy or bison grass, used by North American peoples in basketry and ritual, and in northern Europe to flavor foods and drinks.

The volatile tides of forests and grasslands

In this chapter we've sniffed closely at some of the plants that dominate unculti-vated lands, taking note of the volatile chemical weapons that they deploy against plant eaters: terpenoid mixtures in conifers, terpenoids and benzenoids in broad-leaved trees, the green-leaf volatiles in grasses and forbs. All of these plants and their defensive volatiles help generate the airy tides of scent that Helen Keller, deprived of sight and hearing from early childhood, counted as among the greatest of sensory delights. Her "invisible sweetness" is something like a bouquet of all plants in the neighborhood, their volatiles mixed and diluted in the atmosphere to the point that no particular smell stands out. All contribute to the general sensa-tion of a living presence, of active, productive, companionable creatures that make the air breathable and the planet nourishing. A walk in the woods—what in Japan has recently been called *shinrin-yoku*, "forest bathing"—immerses us in that pri-mal pleasure.

Terpenoids and benzenoids and GLVs are certainly in the scent-tide mix, but some less distinctive volatiles also contribute. When chemists studying the Earth's atmosphere turned their attention from man-made pollutants to pristine environ-ments, they discovered that microbes and plants fill the air with volatile emissions that dwarf those of human activity emissions by a factor of ten or more. The most voluminous of plant emissions by far is isoprene, the five-carbon chain related to terpenoid building blocks (see page 171). Isoprene doesn't appear in any of the tables above because it has only a faint rubbery smell that's usually masked by other plant volatiles. It suggests rubber to us because it's a building block for the milky plant latex that hardens into natural rubber. But isoprene is produced by most green plants apart from the conifers, in the largest quantities by both temperate and tropical trees. Its function is to deal not with specific predators, but with the stress of the sun's heat and high-energy ultraviolet radiation, and the highly reactive by-products of photosynthesis, oxygen included—all of which can damage the chemical machinery of plant cells. Isoprene is a molecule that absorbs solar energy, reacts preemptively with damaging molecules, and when it's released, cools leaves the same way that evaporating perspiration cools our skin.

The conifers apparently achieve similar protection with their major terpenoid emissions, pinene and limonene. (A few pines, loblolly and ponderosa among them, also emit significant amounts of the tarragon-like benzenoid estragole.) And both isoprene and terpenoids have another property that makes their initially invisible tides visible, as the haze that gives the world's various Blue and Smoky mountains their names. As they react with other chemicals in the atmosphere, their by-products form clusters, some of which attract water molecules and develop into water droplets or ice crystals. These suspended particles, or aerosols, both absorb and scatter light—which is why they're visible as haze—and thus deflect some of the sun's energy from reaching the leaves. Much like the aerosols formed by dimethyl sulfide over the oceans (see page 378), these isoprene and terpenoid aerosols encourage the formation of clouds, reduce the amount of solar energy reaching the Earth's surface, and so have significant effects on local climate and possibly the global climate as well.

After isoprene, the most abundant plant volatile in the air is one-carbon methanol, the simplest alcohol, which smells like its two- and three-carbon relatives in vodka and rubbing alcohol. Methanol is released from nearly all plants as a by-product of strengthening their cell walls. A number of other starter-set carbon chains are emitted as by-products of other metabolic activities: among them acetaldehyde, acetic acid, acetone, and ethanol. These are the typical volatiles in the air above grasslands until they begin to suffer from strong sun, high temperatures, high winds, hailstorms, or grazing, when they emit copious amounts of the green-leaf volatiles. Mow grasses and let the cut leaves wither in the sun, and the GLVs fade in favor of the more generic starter-set molecules along with sweet, warm, haylike furfural, a six-carbon sugarlike molecule generated by the breakdown of cell-wall cellulose (see page 345).

All of these carbon chains, variously defensive, stress-relieving, and ancillary to growth and metabolism, contribute to what we perceive as the plenitude of forest and meadow air. Physically, these molecules are unimaginably small, and they're sparsely distributed, but plants emit so many of them, and there are so many plants on the land, that they add up to an equally unimaginable mass: on the order of a billion tons every year, or the equivalent of a hundred cubes of lead with the dimensions of a football field.

What's the fate of these countless airborne carbon chains and rings? The more unstable among them, isoprene and the terpenoids and GLVs, react with one another and with pollutants to form new molecules that eventually fall to the ground in raindrops or snow. The less reactive starter-set chains probably end up broken apart completely and oxidized all the way to carbon dioxide, the basic one-carbon molecule that plants absorb and use to build their complex chains. Those chains and rings in turn feed the land's microbes and animals, us included.

So when we become aware of the fullness of the open air, we're noticing bits of the substance of life passing from Earth's teeming land and waters back to the harsh simplicities of its upper atmosphere. As they make that passage, those bits leave their mark on the atmosphere, and eventually reenter the pool of primal materials from which life is built.

The medical term for calm, restful inhalation and exhalation is "tidal breathing." It's one likely and apt result of sensing and contemplating the scent tides of the green world.

Chapter 10 · FLOWERS

....

Man alone, so to speak, among animals perceives and takes pleasure in the odors of
flowers and such things. . . .

Odors agreeable in their essential nature, e.g. those of flowers . . . have been
generated for human beings as a safeguard to health. . . . The odor arising from what
is fragrant, that odor which is pleasant in its own right, is, so to say, always beneficial
to persons in any state of bodily health whatever.

> • Aristotle, *Sense and Sensibilia*, about 325 BCE

Cato has recommended that flowers for making garlands should also be cultivated in
the garden. Their delicate variety is impossible to express, since no individual has
the power for describing them that Nature does for coloring them. In flowers Nature
shows her playful mood, her delight in showing off the profusion of her creativity.
The other plants she has produced for our use and our food, and to them accordingly
she has granted years and more of life: but to the flowers and their scents she gives
only a day—a mighty lesson, to teach us that the most beautiful and attractive
things in life are the very first to fade and die.

> • Pliny, *Natural History*, about 75 CE

Small, intricately shaped and colored and scented, flowers are among Hero
Carbon's most exquisite creations. It seems plausible that these fragile and
not especially nourishing plant parts might have been the first things to
nudge the human animal into enjoying smells for their own sake, not as signs of
food or friend or foe. At some point our ancestors made flowers a regular part of
their lives. They've been with us ever since, often as exemplars of intense yet brief
existence, natural symbols for human beauty, desire, love, and life.

The cultural history of flowers is long and variegated. By four thousand years ago, Egyptian wall reliefs show women holding water lily flowers up to their nose, and collecting terrestrial white lilies and pressing them for perfume. The ancient Chinese paid special attention to a few flowers representing different seasons and qualities: in the spring, peach tree blossoms and the tall tree peony; in the summer, the Indian lotus, associated with the Buddha and enlightenment because it rises from smelly sediments to lift a beautiful blossom to the skies; in the autumn, the modest chrysanthemum, its smell more forest-like than floral. In ancient Greece and Rome, and in India to this day, garlands of flowers crown rulers and heroes, brides and grooms, nearly anyone and anything to be celebrated. The Persians and their Arab conquerors, poised between Europe and Asia, had pleasure gardens scented with lilies, roses, jasmine, and orange blossoms. For many centuries in Europe and elsewhere, people strewed common flowers and aromatic herbs to relieve the stale dankness of their dwellings.

And today? For a few dollars we can buy once exotic Asian natives, "improved" in Europe and grown in South America, as cut flowers in the supermarket. And they're often scentless. Our love for flowers and our ability to produce them at will have conspired to drain them of both significance and smell. The modern global flower industry is the product of growing wealth and city markets for cut flowers, the professionalization of gardening and plant breeding, commercial flower shows, and plant-collecting expeditions. Its business is driven by the competitive breeding of visually striking new varieties that have a long vase life. This program has often meant a steep decline in floral scents: in part because volatile and pigment molecules share biochemical resources, so that more color means less scent, in part because some scent volatiles are also plant hormones that shorten vase life. And in today's largely deodorized world, buyers often prefer scentless flowers for their unobtrusiveness.

Happily, this isn't the end of the story. Breeders have recently become more interested in flower smells, so aromatic varieties are making a comeback, and amateur and small-scale gardeners have kept older varieties alive. Smell explorers intrigued by flowers can easily rediscover them as improbably delightful gifts of the soil, from underfoot and from the four corners of the Earth.

The meanings of flowers and their smells

Why do we humans take such pleasure in plant organs that are otherwise mostly useless to us? A good and ancient question! Aristotle simply asserted that the pleasantness of their odors signifies a healthful influence on our bodies. Pliny extrapolated our delight to the goddess Nature and saw in flowers an expression of her playful creativity, which anticipates the philosopher Immanuel Kant's definition of free beauty as "purposefulness without purpose." There are durable kernels of truth in these early ideas. The smells of flowers probably aren't healthful in themselves, but the feelings of pleasure they can generate may well be. And though there actually is a serious purpose underlying the variety of flower tints and shapes and smells, that variety is also a manifestation of Hero Carbon's aimless exploration of complexity.

As we understand it today, Nature's flower making is a matter of life and death over the long haul, of adaptation and extinction. The diversity of flowers demonstrates the survival value of sexual reproduction, in which two individual organisms mix the genes that define them to make offspring that are different from both of them—so that over time their descendants can try out new traits and become better adapted to their circumstances. Flowers are organs of sexual reproduction that evolved from flowerless ancestors and turned out to be effective and versatile structures for moving genes between creatures that can't themselves move. Flowers are adaptations whose survival value to their plants is to expedite the process of adaptation.

Some flowers rely on the wind and chance to carry pollen grains to potential mates, and most of their pollen is wasted. Flowers that rely on the sensory systems and mobility of animals can target their pollen more reliably and efficiently. To recruit an animal, flowers have to get the animal to notice them, attract it to them, prevent it from simply eating them, induce it to pick up some pollen grains, then nudge it to leave and visit other flowers of the same species and drop off some pollen grains near those flowers' ovaries. That's quite a set of specifications! And it's why flowers are the complex structures they are, with many possible combinations of size and shape, color and pattern, surface texture, nectar and other food rewards, opening and closing times—and emissions of volatile molecules, both attractive and defensive.

Most pollinators are insects, and no wild plant depends on us to move its pollen. So we enjoy the shapes and patternings and scents of flowers as disinterested bystanders, free to experience the qualities of flowers purely as features of the natural world and attribute to them whatever associations and meaning we want. Shapes and colors and patterns and scents may capture our attention and interest simply because they're unusually rich sensory inputs for our brain to make sense of. Part of the pleasure might come from what the biologist E. O. Wilson named *biophilia*, an innate sense of connection with the living world; such intricate constructions are manifestations of vitality, Dylan Thomas's "force that through the green fuse drives the flower." Whatever factors might be most responsible, our pleasure in flowers has led to a less disinterested, more involved relationship with some of them, not just as their cultivators, but as *elicitors* of their hidden potential to be themselves, but differently.

Plant are living systems with astounding inner resources for adapting themselves to the challenges of survival, but most of these resources remain untapped in any given generation or plant. What breeders have done since the beginning of agriculture, first by selecting which seeds to sow, later by controlling pollination, today by directly altering plant DNA, is to elicit potential qualities in plants that had remained unexpressed: larger seeds, sweeter fruits, faster growth. Flower breeders have done the same for purely aesthetic qualities, and have developed untold thousands of varieties with forms and colors and scents that bear little resemblance to the original wild parents. Like the crazy range of dog breeds that have emerged from a single wolf-like progenitor, the endless variety of edible and ornamental plants—including many modern scentless wonders—testifies to the malleability of Hero Carbon's living avatars.

There are many sides to the appeal of flowers. They suggest pleasure and healthfulness and playful creativity, beauty and love and shared fragility. Modern biology helps us appreciate them as superb emblems of relationship, as key nodes in the intricate web of resources, services, and communication through which Earth's living kingdoms manage to coexist and flourish.

And flowers smell good! Most of them, anyhow. The exceptions are instructive.

Not so flowery: corpse and cabbage, beetles and flies

If you start to think about flowers and pollinators, what probably come to mind are butterflies and bees flitting and humming through the roses and lilies. Well, consider what is probably the most celebrated flower of 2016, the year in which I'm writing this chapter. The **titan arum** is a member of the plant family Araceae, which includes the familiar calla lily and indoor-tolerant ornamentals, philodendrons and dieffenbachias. It is native to the Indonesian jungle and was first brought to a Western botanical garden in the late nineteenth century. The plant grows for years relatively inconspicuously. When it matures it develops a titanic structure for a flower, as tall as a good-size human.

In 2016, nearly a dozen titan arums happened to bloom in botanical gardens all across the United States, drawing thousands of people to wait in line for the opportunity to see them—and above all, to smell them and revel in communal disgust. Another name for the titan arum is **corpse flower**, and it emits what can be an overwhelming stench of rotting flesh. What are the animal partners it's beckoning? Beetles and flies, crawly, buzzy, unclean. There are something approaching a half million different species of these insects on the planet, far more than there are butterflies and bees. It would be botanical malpractice if no plant recruited them for pollination duty!

Beetles are among the oldest of extant animal lineages and probably among the earliest flower pollinators, beginning more than a hundred million years ago when their diet was mainly decaying plant matter, long before the first honeybees and butterflies flew onto the scene. It's likely that plants first attracted them with simple starter-set volatiles that imitated musty, sour compost, along with terpenoids and benzenoids that early animals were already developing to attract and repel and communicate with each other. Later, when large fleshy reptiles and mammals evolved, their dung and corpses became newly nourishing food sources for insects. Today some beetles specialize in animal dung—including the Egyptian scarab—and certain choosy flies will lay eggs in the dung of vegetarian cattle but not of omnivorous pigs or of carnivores. Carrion beetles and blowflies feed and lay their eggs in animal corpses.

SOME SMELLS EMITTED BY FLOWERS THAT ATTRACT BEETLES AND FLIES

Smells	Molecules
feet, cheese	methylpropionic, methylbutyric acids
fishy, urinous	trimethylamine
barnyard	cresol
fecal	indole, skatole
rotting animal flesh	dimethyl di-, tri-, tetra-sulfides

All of these insects find animal remains by detecting their telltale volatiles, and hundreds of modern plant species from a range of families are now known to construct and emit the same volatiles from their flowers in order to attract beetles and flies for pollination duty. Biologists have found that these plant emissions are dominated by various sulfides, the ring molecule cresol, and nitrogen-containing rings indole and skatole, all of which microbes commonly generate as they break down animal flesh and its rich mine of proteins. **Dimethyl disulfide** and cresol appear to be the floral keys to imitating the dung of vegetarian horses and cattle, while the addition of indole or skatole suggest the dung of carnivores and omnivores (cats, dogs, pigs), and **dimethyl trisulfide** sketches the stench of carrion.

The titan arum has plenty of unpleasant company in the Araceae, including the Mediterranean **dead-horse arum**, which actually mimics rotting seagulls, the less exotic North American **skunk cabbage**, and the **devil's tongue**, or **konjac**, which I've found makes an undemanding and entertainingly weird houseplant. There are other temperate-climate flowers whose malodorousness is tempered by floral notes. The **Callery or Bradford pear**, a widely planted street tree originally from China, produces copious springtime blossoms that are notoriously redolent of semen, probably due to the nitrogen-containing pyrroline documented in similar-smelling flowers. Various species of **barberry**, *Berberis*, share pyrroline and related volatiles, so that even the plant-loving Henry David Thoreau described one as having a "sickening buttery odor, as of an underdone batter pudding, all eggs but no spice." And many **hawthorn** flowers have a fishy undertone from trimethylamine.

SOME MALODOROUS FLOWERS IN THE ARUM FAMILY

Arum species	Smells	Molecules
titan arum, corpse flower (*Amorphophallus titanum*)	rotting flesh & fish, urinous	dimethyl disulfide & trisulfide, trimethylamine, methylbutyric acid
dead-horse arum (*Helicodiceros*)	rotting flesh	dimethyl sulfide, disulfide & trisulfide
skunk cabbage (*Symplocarpus foetidus*)	cabbage, sulfurous, skunky	dimethyl disulfide
devil's tongue, konjac (*Amorphophallus konjac*)	rotting flesh	dimethyl di- & tri-sulfides

Why begin our exploration of flowers with these undesirables? Because they hint at what the very first flowers might have smelled like, they show that modern flowers can deploy very unflowery volatiles, and they demonstrate that flower volatiles can be attractive to some animals and repulsive to others. In fact, this appears to be the rule rather than the exception. Even the very volatiles that make flowers smell sweet to us repel some insects.

SOME COMMON TREES AND SHRUBS WITH MALODOROUS FLOWERS

Plant	Smells	Molecules
Callery, Bradford pear (*Pyrus calleryana*)	semen	pyrroline (probably)
barberry (*Berberis*)	"sickening," semen	pyrroline, acetyl pyrroline
hawthorn (*Crataegus*)	fishy	trimethylamine

Mixed signals

As flowering plants and insects evolved side by side over tens of millions of years, flower structures became more complex, and insect bodies and behaviors changed to take advantage of them. For example, flowers developed sugary nectar as a food reward, and some hid it in an ever-deepening recess so that only certain long-tongued partners could get to it, and sometimes only by hovering in the air. The first such sippers appear to have been butterfly-like ancestors of modern lacewings, with moths and butterflies themselves (and hummingbirds) coming much later.

Flowers also enlarged their volatile vocabulary. Instead of simply imitating the smells of decaying organic material, they developed their own distinctive smell— the generic smell of flowers, floweriness—to advertise their special offerings, nectar and spare pollen chief among them. To do so they used many of the same volatiles from the terpenoid and benzenoid pathways that we've seen trees and other plants rely on for chemical defense, to repel insects rather than attract them. So how do flowers manage to turn them into attractants? Perhaps by emitting them as they do, in trace amounts and diluted in complex cocktails of a dozen or more.

Two volatiles show up in so many flowers and so few other plant organs that they smell generically "flowery" to us: the terpenoid alcohol linalool, and the benzenoid alcohol phenylethanol. Perfumers call molecules like these *floralizers*. Phenylethanol tends to suggest roses in particular because it's so prominent in them. Of the other terpenoids, citrusy limonene, green-woody ocimene and myrcene, and pinene are found in about 70 percent of surveyed flowers. Of the other benzenoids, almond-extract benzaldehyde and wintergreen methyl salicylate are found in more than half. They and hundreds of other less common volatiles modulate generic floweriness and give different flowers their distinct scents, which can help pollinators identify and visit members of the same species.

Attracting pollinators is an important role for floral volatiles, but there are others. A flower is a complex, often fragile structure, with a variety of soft tissues rich in moisture and sugars and other nutrients. What's to stop an insect from detecting its smell, finding it, and chewing it to pieces? Or hanging out and drinking all the nectar, or packing up all the pollen, so the next visitor goes unrewarded and is less likely to visit another flower like it? What prevents bacteria or molds or yeasts from souring or spoiling nectar, or infecting the exposed sexual organs?

Volatiles, of course! Take the two preeminent floralizers, linalool and phenylethanol. Both have been found to attract bees, moths, butterflies, and beetles. But phenylethanol repels ants. Linalool repels crickets, and like many terpenoids it suppresses the growth of microbes.

So individual flower volatiles can simultaneously be signals of welcome to some animals and warnings to others. They can be prophylactic antibiotics. Some attract wasps and other predatory insects that attack the insects that attack the plant. There's also evidence that volatiles help keep pollinator visits short and therefore more numerous, perhaps by directly affecting insect feeding reflexes, perhaps by distracting or intoxicating or numbing, or simply making the in-flower air quality barely tolerable. That's what occurs to me when I imagine myself bee-size, flying toward an Asiatic lily or jasmine flower and landing on volatile-emitting petals that can be smelled from yards away! It's because we're literally above it all, much larger than insect or flower and always diluting the volatiles as we inhale, that we can take the pure pleasure in flowers that we do.

Clearly there's far more to flower smells than meets our nose, and plenty of further research for botanists and entomologists and evolutionary ecologists. For the nonspecialist smell explorer, it's intriguing just to know that floral scents are complex chemical mixtures, that their individual components play a number of different roles, and that we still understand very little about why particular flowers have developed their characteristic scents.

A few decades ago, biologists laid out a handful of pollination "syndromes" that associate particular flower characteristics with the pollinators they attract. The expert consensus now seems to be that promiscuity rules: most flowers are visited by a variety of insects, and a given insect group can pollinate a variety of different flowers having different scents. Beetles and flies are huge groups and aren't limited to dung and carrion lovers, for instance; some of their members are the prime pollinators for the heavenly scented Egyptian water lilies. And there are thousands of different species of bees.

Still, even generalizations with copious exceptions can help open our eyes and nostrils to what's going on in the flower garden. Here are a few.

- The most common pollinating insects are: beetles, the most ancient; flies, which may well pollinate more flowers than they've been given credit for; bees and butterflies, the familiar day laborers; and moths, which are mainly crepuscular—appearing briefly in the twilight of early evening—and nocturnal.
- Scented flowers that appeal to a wide range of pollinators tend to produce a mix of simple-chain, terpenoid, and benzenoid volatiles, in effect cover-

ing all the bases, although one category usually dominates the other two somewhat.

- Honeybees seem to have a broad palate. They show some preference for terpenoids but are also attracted by spicy and fruity benzenoids and the complex mixtures in roses.
- Butterfly flowers tend to be brightly colored and to have a mild, fresh, sweet scent coming mainly from a couple of floral-honey benzenoids, phenylethanol and phenylacetaldehyde, and the common floral terpenoid linalool along with fresh-woody ocimene.
- Moth flowers form a distinctive group and include many of the flowers we think of as exotic and intoxicating, jasmine and gardenia and the like. Moths tend to be active at twilight and night, when smells are likely to be more important signals than colors and shapes. Moth-pollinated flowers are often white, perhaps to maximize whatever visibility they may have and not waste resources on pigments. They're often strongly scented, perhaps to make their scent trails as far-reaching and long-lasting as possible. The so-called white flower scents of these evening and nocturnal blooms often supplement the generically floral mix found in butterfly flowers with intensifying benzenoid ingredients: fruity benzoate esters and methyl anthranilate, almond-extract benzaldehyde, and wintergreen methyl salicylate.

"Dirty" notes in white flowers: ambiguous indole

There are several odd and intriguing volatiles that sometimes show up in floral cocktails, most notably in those exotic white flowers. We first met them in the malodorous animal kingdom. Cresol, a carbon ring derived from an amino acid, is reminiscent of the horse stable. Indole and skatole are nitrogen-containing rings that are prominent in animal feces, so their smells are often described as fecal, even "intensely" so. Yet some professional noses also describe indole much more neutrally, as smelling like mothballs—that is, musty and penetrating and even a bit chemical-like. When I've sniffed vials of nearly pure indole, it seems far more like mothballs than otherwise. Decorate the carbon-nitrogen ring of indole with just one more carbon, to make its near relative skatole, and the smell does become quite fecal (see page 61).

It's no surprise that cresol, indole, and skatole would be common components in flowers that mimic animal remains to attract beetles and flies, including the common housefly, *Musca domestica*. There's also indole in edible squash blossoms, where biologists have found that it specifically attracts pollinating cucumber beetles. But these animalic volatiles, indole in particular, also show up in flower smells that are described as exotic, sexy, sometimes bordering on the unpleasant: among them jasmine, narcissus, tuberose, orange blossom, wisteria, and hyacinth. Indole is often singled out by perfumers as being responsible for their arousing, animalic, "dirty" facets, the quality that can make them so alluring but can also go too far. (I think cresol is also a likely culprit; see the entry for ornamental jasmine on page 233.)

What's indole doing in flowers that aren't frankly angling for beetles or flies? Jasmine is the prime example: its volatiles can be as much as 10 percent indole. From what little we know about the pollination of wild jasmines, which depends mainly on wild bees and flies, it could well be serving as insurance, to attract flies as backup pollinators during the day. (Jasmine volatile emissions aren't constant; they're richer in linalool during the day, in benzenoids at night.) Indole might also appeal to some butterflies or moths or help them identify and patronize the flowers that emit it. It might also have roles beyond pollination. Whereas in microbes it's commonly a by-product of breaking down proteins—as in the guts of carnivores—in plants indole is deliberately constructed, in part to serve as a precursor to an important growth hormone. Plants release indole when they're damaged, and it signals neighboring parts of the same plant, and neighboring plants, to ramp up their various chemical defenses. It also seems to be a chemical weapon itself, limiting the growth of fungi.

Whatever indole might be doing for flowers and pollinators, what is it doing for us, for disinterested animals who experience indolic flowers as unusual and intoxicating? Even if it doesn't really smell downright fecal itself, the musty note of indole is out of place in the standard bouquet of floral volatiles, and could simply hint at the full animal context it's usually found in. It introduces associations with a different living kingdom, and might even put us on a subliminal alert the way other frankly animalic smells do.

In fact, neurobiologists at the University of Oxford have reported evidence that this is exactly what's going on. They asked volunteers to smell two versions of an artificial reconstitution of jasmine volatiles, one in which indole was included and the other in which it was left out, scanned their brain activity as they did so, and asked them to rate the pleasantness of the two jasmine scents as well as indole alone. The subjects rated indole alone as unpleasant, but the jasmine scent with indole as either similarly pleasant to the indole-free scent, or even *more* pleasant. When the researchers scanned the brains of the subjects as they smelled, they detected a stronger and more prolonged activation when the scent included indole. These results led them to suggest that an unpleasant element in a complex odor mixture promotes "attentional capture"—gets the brain to commit more resources to processing the sensation—and this strengthens and prolongs the overall sensation. What makes indole deserve this special attention, even if it's subliminal? Perhaps the fact that it usually signals animals and decay, both potential threats.

The Oxford researchers suggest that a similar effect might be at work with other volatiles that are unpleasant alone but enhance complex smells. It's a promising perspective on the various cat-pee and sweaty sulfur molecules in fruits and wines (see pages 300–301), and the prized animal ingredients in traditional perfumes (see page 465).

Approaching the flower garden

Now that we have a sense for the general nature of flowers and some of their volatile motifs, it's time to sniff our way through some of the world's best-loved blooms. Rather than organize them strictly by botanical family or geography or scent quality, I've grouped sixty-odd specimens in a dozen plantings that I hope will convey something of their histories and broad affinities, botanical and volatile. We'll start with samples of ancient inspiration, the flowers of Egypt and China; move on to the common strewing, garlanding, and garden flowers of Europe; then to the very different flowers of Asia, from the eastern edge of the Mediterranean to the Pacific; and finally to flowers from the New World, especially South America. Unless you're an avid gardener, you'll probably have scant experience with the parent plants, and many of the flowers themselves may not be familiar. So this section of our virtual botanical garden is more challenging to imagine than trees and meadows. If you like, follow along with online images; better yet, take an actual field trip or two.

Before we begin, a word of clarification about the smells and volatiles that I assign to each flower. I want the next few pages to inspire you to get out to the garden or florist or nursery and smell the flowers you know and don't. When you do put your nose into a real flower, it's likely that sometimes what you smell won't match what I describe. When that happens, trust your own nose, and let the discrepancy help you sharpen your perception of the actual smell. Understand that there are several possible reasons for the discrepancy. One of the most likely is that you're smelling an atypical variety of the same flower: plant breeders are constantly developing new versions of old favorites, and they can smell quite different. Flower volatiles can also vary depending on how they're grown and handled, their age, even the time of day. In addition, plant names are confused and confusing: lots of very different flowers are called lilies and chrysanthemums.

It's also true that any given flower emits on the order of dozens of volatile molecules, they're not easy to identify, and scientific reports on a given flower usually differ from one another, sometimes in major ways. So the descriptions I give below are very much sketches, not full portraits. I've tried to find some kind of consensus for the most prominent or abundant volatiles in each, one that accords with my own experience.

Whenever possible, I've relied on reports of the volatiles emitted from living flowers. When these aren't available, I use analyses of flower extracts, which often differ from their source because the extraction process alters some volatiles. Extracts have their own interest and are important ingredients in perfumes, as we'll see in chapter 17.

Flowers of the ancients

We begin with flowers whose smells captured the attention of early civilizations. Two of the most remarkable were prized in Egypt thousands of years ago. In our garden they hover above a murky pool of water, not soil, and represent an ancient lineage that predates those of most other flowering plants. These two species of **water lilies** have several dozen sister species from across the globe, all of them rooted in shallow-water sediments, floating their sometimes very large green leaves on the surface, and supporting striking, many-petaled flowers. Of the two Egyptian favorites, the **blue water lily** opens in the morning and closes in the afternoon, while the **white water lily** opens in the evening and closes in the morning. (Both actually come in a variety of hues.) Night flowers typically attract moths with more complex volatile cocktails, but both of these appeal mainly to beetles, and it's the day-blooming blue lily that has the richer aroma. It was the one included in the burial chamber of King Tutankhamun.

Growing in the ground alongside the pool is another flower prized in ancient Egypt and the Mediterranean more widely. The **white lily**, also called the **Madonna lily**, is entirely unrelated to the water lilies and looks nothing like them—it's borne on a long leafy stalk and has six curved petals. It smells more generically flowery, with the two main floralizing volatiles plus a honey note from phenylacetaldehyde. *This* lily is the namesake for what botanists define as the lily family, which includes a hundred-odd other "true" species of lily, along with tulips. Confusing!

Beyond the white lilies, there's a patch of small, low-lying plants with rings of thin leaves and central blossoms of white and purple and gold. Around the same time that the Egyptians were enjoying water lilies, the Minoan civilization on the island of Crete recorded women collecting these **crocuses**, no doubt to dry their stigmas to make the powerfully scented spice and dyestuff **saffron**. The smell of dried saffron comes from the terpenoid fragment safranal, but the fresh flower is delightful in a completely different way, with violet overtones from a different terpenoid fragment, and a fruity note from the benzenoid acetophenone.

SOME FLOWERS OF THE ANCIENT MEDITERRANEAN

Flower	Component smells	Molecules
blue water lily (*Nymphaea caerulea*)	floral, fruity, anise, violet	benzyl alcohol & acetate, anisaldehyde & alcohol, cinnamyl alcohol & acetate, ionone
white water lily (*Nymphaea lotus*)	fresh, green, solvent, waxy	dimethoxytoluene, methylmethoxymethyl butyrate & valerate, cembrene
white lily (*Lilium candidum*)	floral, honey, citrus	phenylethanol, phenylacetaldehyde, linalool
saffron crocus (*Crocus sativus*)	floral, violet, fruity	linalool, ionone, oxoisophorone, acetophenone

A last planting of flowers prized for thousands of years comes from China, whose vast geography encompasses a more varied set of ecological niches than the Mediterranean. Its early rulers celebrated several very different flowers. First another shallow pool, this one home to the **lotus**, a plant revered throughout much of Asia, with a relatively mild, fresh smell, not especially rich. Similar in appearance to the unrelated water lilies, lotuses are an incongruously beautiful inhabitant of stagnant waters that became associated with the Buddha and enlightenment. The small white **peach tree** blossoms of spring are dominated by benzenoids, almond, fruity, and floral. Confucius cited as exemplary the inconspicuous **cymbidium orchid**, these days a familiar houseplant: just as its scent fills a room and unobtrusively delights without being noticed, the virtue of a good person fills a gathering until everyone is permeated by it. The large flowers of shrubby, woody-stemmed **tree peonies** can have a number of different scents, but those from the region long renowned for them, Henan Province in the central plateau, often smell like roses. And then there's the autumn-blooming **chrysanthemum**, from a familiar group of ornamental garden plants traditionally associated with quiet countryside retirement, whose scent is more reminiscent of tree resin than flowers, with camphor and other terpenoids typical of conifers.

SOME FLOWERS OF ANCIENT CHINA

Flower	Component smells	Molecules
lotus (*Nelumbo nucifera*)	fresh, medicinal, woody, floral	dimethoxybenzene, sabinene, eucalyptol, jasmone, terpineol
peach blossom (*Prunus persica*)	almond, floral, resinous, rose, citrus	benzaldehyde, benzyl alcohol, geranyl acetone, acetophenone, methyl heptenone, nonanal
orchid (*Cymbidium goeringii*)	floral, jasmine, lemon	jasmonoids, nerolidol, farnesol
tree peony (*Paeonia suffruticosa*)	rose (also woody, medicinal varieties)	citronellol, phenylethanol (ocimene, trimethoxybenzene)
chrysanthemum (*Chrysanthemum indicum*)	camphor, fresh, pine, woody	camphor, pinene, bornyl acetate, caryophyllene

Flowers of Europe

The next area of the garden is divided into five different beds, the last partly shaded by a small open grove of trees. These are representatives of Europe, wildflowers and weeds first, then mostly native garden flowers, then rambling climbers and a notable flowering bush and a tree. A number of these plants are found throughout the Northern Hemisphere, not just in Europe.

The first bed is planted with wildflowers. They're generally small, grow in numbers large enough to scent fields and woods, are free for the picking, and would have been among the candidates for strewing to freshen the indoors in medieval and Renaissance times. **Meadowsweet** was said to be Queen Elizabeth's favorite strewing flower. **Violets** are more retiring, growing close to the ground in shaded woods, the small, strikingly purple flowers usually opening underneath the leaves that support them. Violets have a landmark intense and intoxicating smell created largely by the terpenoid fragment ionone; courtiers were said to carry and sniff them to cover the nastier smells of city life.

Lily of the valley is also a woodland native, not a true lily; its light, fresh, "watery" scent comes from an unusual cluster of alcohols. **Dandelions** and **clovers** are common suburban weeds today, with woody and fruity adjuncts to the standard floralizer volatiles. **Alyssum** is an adaptable low-lying plant that produces copious clusters of small white flowers and has become a favorite for garden bedding and ground cover. It's one of the nicest-smelling members of the cabbage family, with a sweet, honey-like aroma. As with similar-smelling clover flowers, its main volatile is acetophenone.

SOME COMMON EUROPEAN WILDFLOWERS

Flower	Component smells	Molecules
meadowsweet (*Filipendula*)	fruity, sweet, vanilla-like	methyl benzoate, benzaldehyde, hexenyl acetate, anisaldehyde
violet (*Viola*; N. Hemisphere)	violet, intense floral, woody	ionones

continued

Flower	Component smells	Molecules
lily of the valley (*Convallaria*)	fresh, floral, watery	benzyl & cinnamyl alcohols, geraniol, farnesol, phenylacetonitrile
dandelion (*Taraxacum officinale*)	light, fresh, watery	phenylethanol, benzaldehyde, nerolidol, hexenol, phenylacetonitrile
clover (*Trifolium; Melilotus*)	honey, fruity; hay	phenylethanol, acetophenone, methyl cinnamate; coumarin
sweet alyssum (*Lobularia*)	honey	acetophenone

The next bed has three clusters of plants that bear a family resemblance, all with strappy, elongated leaves growing from ground level and bearing stalks of intensely aromatic flowers. They grow from bulbs and first became popular around 1600. We've already met the **white lily** proper, known to the ancients and generically floral. The others are various species in the genus *Narcissus*, including the flowers commonly called narcissus, as well as paperwhites and jonquils and **daffodils**: another set of confusing names! Daffodils tend to smell of the florist's shop, fresh and medicinal, while jonquils and paperwhites are some of the most intensely floral flowers there are, with the usual terpenoids and benzenoids augmented by fruity esters and musty indole.

EUROPEAN LILY, NARCISSUS, DAFFODIL

Flower	Component smells	Molecules
white lily (*Lilium candidum*)	floral, honey, rose	phenylethanol & phenylacetaldehyde, linalool
narcissus, jonquil, paperwhite (*Narcissus jonquilla, tazetta*; SW Europe, N. Africa)	floral-fruity, green, woody, jasmine, animal	narcissus, jonquil: methyl benzoate, ocimene, linalool, indole, hexenol paperwhite: ocimene, benzyl acetate, linalool, cineole, indole
daffodil (*Narcissus pseudonarcissus*)	green, fresh, woody	ocimene, dimethoxybenzene & dimethoxytoluene, farnesene

The third bed in the European garden brings together a few members of the very large aster or daisy family. What looks like a single daisy flower is actually a mass of tiny flowers surrounded by larger petal-like flowers. The European favorites have in common yellow and orange colors and a not especially floral smell, with little or none of the usual floralizers. The **daisy** is the most flowery among them, dominated by a honey-like benzenoid. The **marigold** is dominated by woody terpenoids, and the European version of **chrysanthemum** by camphor and resinous myrcene, much like the Chinese, though it also has a touch of minty, shiso-leaf perilla aldehyde. **Chamomile** is more complex than the other asters, with fresh minty terpenoids and a number of esters that give it a distinctly fruity quality, a combination that has made dried chamomile flowers a favorite ingredient in herbal teas.

SOME EUROPEAN GARDEN ASTERS

Flower	Component smells	Molecules
daisy (Paris or federation), marguerite (*Argyranthemum frutescens*)	honey	phenylacetaldehyde
common/pot marigold (*Calendula officinalis*)	woody, spicy, violet	cadinene, copaene, caryophyllene, humulene, ionone, thujene
garland chrysanthemum (*Glebionis coronaria*)	camphor, green, woody, minty	camphor, myrcene, ocimene, perilla aldehyde
chamomile, German & Roman (*Matricaria, Chamaemelum*; Europe, Asia)	terpy, woody, fresh, minty, fruity	cymene, eucalyptol, artemisia ketone, ethyl & propyl methylbutyrate

The fourth European bed collects a miscellaneous group of garden flowers. **Snapdragon** has a fresh and fruity quality. **Lavender**, which got its name from its being used to perfume bathing waters, is a shrubby member of the mint family whose leaves are also aromatic, as we'll see. Three different species are cultivated for use in perfumery and are defined by different terpenoid mixes: English lavender offers the prototypical scent of distinctive acetate esters; Spanish lavender is more minty and medicinal with camphor; and French lavender lacks linalool and smells more herbaceous than floral.

Three different European garden flowers share the spicy note of clove imparted by the benzenoid eugenol. Together with **sweet William** and other **pinks** (the flower gave its name to the color), **carnations** belong to the genus *Dianthus*. Carnations are some of the most common cut flowers in the West, third most important in commerce, but they're now largely scentless or smell faintly of green-leaf volatiles and wintergreen. This is in part because breeders have emphasized red colors and color patterns that draw on the same benzenoid pathway that produces eugenol, and perhaps also because modern tastes run toward unobtrusively scented flowers for wearing in button-holes. Still sweetly and frankly redolent of cloves is **stock**, like alyssum an unusually pleasant member of the cabbage family, and also known as **gillyflower**, from the French word for "clove," *girofle*. And notorious for their sweaty, cheesy smell are the clustered tiny flowers of **baby's breath**, a popular visual accent in florists' composed bouquets.

VARIOUS EUROPEAN GARDEN FLOWERS

Flower	Component smells	Molecules
snapdragon (*Antirrhinum*; Europe, N. America)	floral, fruity, woody, green	methyl benzoate, myrcene, ocimene
lavender: English, French, Spanish (*Lavandula angustifolia, stoechas, latifolia*; W. Europe to India)	English: fresh, floral, lavender, green, woody; French: medicinal, pine, lavender, minty; Spanish: floral, fresh, minty, medicinal	linalool, linalyl acetate, ocimene, terpinenol, lavandulyl acetate; fenchone, camphor, pinene, lavandulyl acetate; linalool, cineole, camphor
sweet William (*Dianthus barbatus*)	green, waxy, floral, woody	C6–10 aldehydes, esp. nonanal, caryophyllene, linalool
carnation (*Dianthus caryophyllus*; Europe to Asia)	heirloom varieties: clove; new varieties: green, fresh, wintergreen, fruity, resinous	heirlooms: eugenol; new: C6 alcohols & aldehydes, methyl salicylate, hexenyl benzoate, benzyl benzoate, benzyl alcohol
stock (*Matthiola*; Europe, Asia)	clove, woody	eugenol, isoeugenol, methyl eugenol, phenylethanol
baby's breath (*Gypsophila*)	woody, cheesy, sweaty	ocimene, methylbutyric acids

The fifth and final bed of European flowers supports three long-stemmed climbing plants on several flowering shrubs and trees. **Sweet pea** is a relatively restrained and tender annual climber grown for its strongly floral scent from terpenoids found also in roses. Common **honeysuckle**, woody and rampant, is intensely floral thanks to its mixture of linalool and modified versions of linalool, with a touch of musty indole. It has sister species in Asia and North America.

Then there's the European **rose**, whose thorny stems can grow high into trees. The rose has a rich history of associations with love, devotion, martyrdom, chivalry, and the English royal houses, and today it's the preeminent flower in world trade. At the end of this chapter I'll devote a few pages to the story of its changing smells. Here it's enough to note that wild roses were noted as much for their red color and thorns as their scent, but that scent is another landmark smell. The benzenoid floralizer phenylethanol and the terpenoids citronellol, geraniol, and nerol are its prominent volatiles; they define rosiness, and rosiness defines them. Conspicuous by its absence is the terpenoid floralizer linalool.

Giving support to these three climbers is the shrubby **lilac**, a native of the Balkan region that has a number of sister species throughout the Northern Hemisphere. Its conical clusters of flowers and intensely floral scent have helped make the lilac a favorite garden shrub in North America. That scent comes from uniquely high doses of oxidized forms of linalool known as lilac aldehydes and alcohols, together with fresh-smelling ocimene and dimethoxybenzene, and musty indole.

Behind the lilacs, taking the honeysuckle and rose vines to new heights, are a couple of saplings and a soaring mature specimen of what's known as the linden or lime tree; the North American species are called basswood. They do well in city parks and streets and so are often found there—Berlin has a street named Unter den Linden. **Linden flowers** have an unusually complex mixture of resiny, floral, and herbaceous volatiles, including a distinctive hybrid benzenoid and furan called the **linden ether**. Linden flowers dry well and are used to make an aromatic, honeyed tea; in the famous passage in Marcel Proust's *In Search of Lost Time*, the narrator takes his madeleine with linden tea, and it's the tea whose aroma would have been most distinctive.

SOME EUROPEAN VINE, SHRUB, AND TREE FLOWERS

Flower	Component smells	Molecules
sweet pea (*Lathyrus*; temperate worldwide)	floral, woody, rose, honey	ocimene, linalool, nerol, geraniol, phenylacetaldehyde
honeysuckle (*Lonicera*; N. Hemisphere)	floral, woody, citrus, pine, musty	linalool & its oxides, germacrene, farnesene, terpineol, indole
European roses (*Rosa gallica, rubiginosa*)	classic rose, intense, floral	phenylethanol, citronellol, geraniol, nerol
lilac (*Syringa*; SE Europe to E. Asia)	floral, lilac	lilac aldehydes & alcohols, ocimene, dimethoxybenzene, indole
linden tree (*Tilia*; N. Hemisphere)	fresh, terpy, floral, fruity, minty	limonene, terpinene, terpinolene, linalool, rose oxide, benzyl alcohol, phenylethanol, damascenone, linden ether

Flowers of Asia and Australia

In the next four sections of our garden we travel eastward from Europe into Asia, the original source for many of the flowers whose smells we tend to call exotic, tropical, intoxicating. By and large, these flowers—Asian lilies, jasmine, gardenia—emit complex mixtures of terpene alcohols, benzenoid esters, and an ambiguous molecule or two, much as the European species of *Narcissus* do.

We start in western Asia with a relatively subdued bed of two lily relatives and one true family member. Low-growing **hyacinth** and **grape hyacinth** bear stalks with many small flowers, often purple. They lack a significant dose of linalool and are not quite as floral as some of their relatives. Upright single-stemmed **tulip** flowers are familiar from spring and early summer gardens and named from the Turkish and Persian words for "turban." They usually have little or no scent: shapes and colors have been more prized in the six thousand varieties to come out of the seventeenth-century Dutch tulip craze, when single bulbs could cost the same as a house, and in subsequent breeding. The scented tulips that do exist are variable; apparently the genus doesn't have a primary set of volatiles.

Farther east, and in the next bed, are two notable groups of garden flowers from China and Japan. Somewhat sprawling **daylilies** are cultivated in China for their edible flower buds and medicinal roots, and in the West mostly for their multiple sunny yellow-orange flowers. Some carry fruity and floral benzenoids and terpenoids, but the landscape varieties are usually scentless.

The largest plants in this bed, ranging from knee- to head-high, are **Asiatic** and **Oriental lilies**, true sister species to the European white lily, and volatile powerhouses. With their great range of flower sizes and colors, and their upward-facing attitude, they're much used in breeding hybrids for the cut-flower trade. These new varieties often inherit their copious and complex volatile mix, which can include clove-like eugenol and animalic cresol and indole. Their smell is distinctively strong, aggressive, and penetrating, so much so that one Japanese report called it "an unfortunate quality" because it limits their use in restaurants and other confined spaces. A Chinese study notes that Oriental hybrids smell strongest at night, and suggests that "nocturnal emission should be considered when placing, as many people suffer from intolerance to strong fragrances." Chemical treatments and selective breeding can tone down the scent of Oriental lilies and will likely expand the market for them.

SOME ASIAN HYACINTHS, TULIPS, AND LILIES

Flower	Component smells	Molecules
hyacinth (*Hyacinthus*; E. Mediterranean to C. Asia)	floral-fruity, citrus, mothball/animal, honey	benzyl acetate, farnesene, indole, octenol, phenylacetaldehyde
grape hyacinth (*Muscari*)	green, woody, wintergreen, balsamic	ocimene, methyl salicylate, isoeugenol, benzyl benzoate
tulip: General de Wet; Montreux; Jan van Nes; Tender Beauty (*Tulipa*)	fruity; rose; medicinal & spicy; wintergreen	ionone & decanal; phenylethanol; dimethoxytoluene; methyl salicylate

continued

Flower	Component smells	Molecules
daylily (*Hemerocallis*)	floral, fruity, balsamic, woody, animal	benzenoid esters, linalool, myrcene, pinene, indole
Asiatic & Oriental lily (*Lilium* & hybrids; N. Hemisphere)	floral, sweet, fresh, green, woody, clove, horse stable	benzyl alcohol, benzaldehyde, benzenoid esters, eugenol, linalool, ocimene, cresol, indole

The next section of Asian flowers provides some contrast with its immediate neighbors; these plants are relatively restrained in their smells. We've already encountered the Chinese **chrysanthemum** among the flowers appreciated in ancient times. Modern large-flowered hybrid "mums" are currently the second most important flower in commerce, in part because they have a long vase life. Like the European species, Chinese chrysanthemums have a smell dominated by camphor and woody notes. China came to appreciate its several native **rose** species much later than it did the chrysanthemum and lotus, and they don't smell much like European roses. The mild benzenoids dimethoxytoluene and trimethoxybenzene, and woody sesquiterpenoids, are more prominent than their traces of linalool and ionone. But Chinese roses played an important role in the smells of modern garden roses, as we'll see in a few pages. The last plant in this miscellaneous section is the woody **wisteria** vine, with cascades of purple or white flowers hanging from its branches. Single wisterias have been trained on supports to cover as much as an acre with leaves and blossoms. These are the most typically floral of the flowers in this section, with a pleasing mix of terpenoids and benzenoids and an accent of indole.

ASIAN CHRYSANTHEMUM, ROSE, WISTERIA

Flower	Component smells	Molecules
chrysanthemum (*Chrysanthemum* x *morifolium*; Asia, NE Europe)	medicinal, pine, woody, fresh	camphor, pinene, safranal, myrcene, eucalyptol, phellandrene, camphene
Chinese roses (*Rosa chinensis*, *R. chinensis* x *odorata* var. *gigantea*)	fresh, earthy, spicy, green tea, violet	dimethoxytoluene, trimethoxybenzene, caryophyllene, germacrene, linalool, ionone
wisteria (*Wisteria*; Asia & NE America)	floral, green, fruity, musty	linalool, ocimene, methyl benzoate or benzyl alcohol, indole

Now we come to a section of the garden with three beds, each home to a dense clutch of small-leaved vines bearing many small white blossoms. These are among the most distinctive and prized of all Asian flowers: **jasmine**, whose vines originated in the Himalayas and were probably first cultivated in Persian gardens and later brought to Spain by the Arabs; the name comes from the Persian. The best-studied species are "Spanish," grown primarily to provide extracts for the perfume industry, and "Arabian," the type grown in Indian plantations and home gardens for garlands, and used in China to make jasmine-scented tea. Jasmine flowers generally open at dusk and close around midnight, but their volatiles accumulate in the closed flower during the night and are at their maximum early in the morning, so that's when they're usually picked. Spanish jasmine is intensely floral and rich, with a mix of benzenoids, terpenoids, jasmonoids, and a strong animalic undercurrent from indole and cresol, while Arabian or sambac jasmine is sweeter and lighter thanks to a lack of jasmonoids and cresol, and the presence of green and woody sesquiterpenoids.

The common ornamental species, which came from China, carries only a trace of indole, but its volatile output is more than 10 percent horse-stable cresol. In my experience, that cresol long outlasts the more floral ingredients. Years ago I had a prolific jasmine vine cascading over my backyard fence; when it was past full bloom, its flowers beginning to decay, it smelled disconcertingly urinous. In a 1997 study, Danish researchers reported that some people found even its fresh scent "'overpowering,' with the prolonged sensory stimulation by the volatiles emitted causing headaches and nausea." Best kept outdoors!

SOME SPECIES OF JASMINE

Jasmine species	Component smells	Molecules
Spanish jasmine (*Jasminum grandiflorum*)	floral, fruity, animal, horse stable, jasmine	benzyl acetate, linalool, indole, cresol, jasmonoids
Arabian, sambac jasmine (*J. sambac*)	floral, citrus, musty, resinous, jasmine	linalool, farnesene, indole, benzyl alcohol & acetate
ornamental, pot jasmine (*J. polyanthum*)	floral, fruity, spicy, horse stable	benzyl acetate, cresol, isoeugenol, eugenol, linalool

The final section of Asian flowers is planted with a couple of shrubs and a small tree, all with shiny dark-green leaves, all from China but with sister species elsewhere in Asia. One shrub bears white, medium-size, large-petal, transportingly fragrant flowers. These are **gardenias**, a frequent component of Western corsages; other species native to the Pacific islands are included in necklace-like leis. The most common species, *Gardenia jasminoides*, shares many of jasmine's volatiles while omitting the ambiguous aspects, so its smell is more purely floral. The second shrub is a fragrant species of osmanthus, which belongs to the same family as jasmine and lilac but gives its tiny, white or orange flowers an entirely distinctive set of volatiles. **Osmanthus** flowers have a distinctly fruity, apricot-like smell thanks to a simple-chain ester and the lactone that helps define apricot, along with a heavily floral violet note from the terpenoid ionone. They're used to scent teas, wine, and various sweets. Rising above the shrubs is a citrus tree, the species that bears the sour and pleasantly aromatic bitter orange. These fruits start out as **orange flowers**, which are highly valued in their own right but aren't at all citrusy. They offer their own unique mix of floral and resinous terpenoids, fruity anthranilate esters, and musty indole. In medieval Persia and the Arab and Ottoman cultures that came later, both orange and rose flowers were extracted into aromatic waters to perfume people and foods alike; these waters (or artificial imitations) are still much used in western Asia.

A last flower from the Pacific: the little yellow fuzzballs of the **silver wattle** tree, or **mimosa**, a species of *Acacia* from Australia. It's widely planted in southern France and also in California, where it unforgettably scented the moist spring air as I cycled between home and library during my first months in the Bay Area: fruity, vanilla-honeyed, hints of anise and mushroom from a mix of esters and benzenoids and alcohols.

ASIAN GARDENIA, OSMANTHUS, BITTER ORANGE; AUSTRALIAN WATTLE

Flower	Component smells	Molecules
gardenia (*Gardenia jasminoides*; Asia, Pacific, Africa)	floral-fruity, jasmine	benzoate esters, linalool, jasmolactone
osmanthus (*Osmanthus*; E. Asia)	sweet floral, violet, apricot, fruity	linalool, ionone, g-decalactone, hexyl butyrate

Flower	Component smells	Molecules
bitter orange (*Citrus*; SE Asia)	floral, animal, grape, woody	linalool, myrcene, linalyl acetate, methyl anthranilate, indole
silver wattle, "mimosa" (*Acacia dealbata*)	fruity, honey, mushroomy, vanilla, anise	ethyl propanoate & butyrate & hexanoate, phenylacetaldehyde, octenol, vanillin, methyl & ethyl anisate

Flowers of the Americas and Africa

This penultimate section of the flower garden takes us beyond Europe and Asia to a selection from the rest of the planet, mostly the Americas. The plants themselves often range over wide areas, so I'll group them by the relative prominence of benzenoid and terpenoid volatiles.

Terpenoid showcases first: an area with several large trees, a cluster of shrubs, two sets of erect bedding flowers, and small patches of green lining the pathway. The southern United States is home to a number of species of **magnolia**, trees that are direct descendants of some of the most ancient flowering plants. Their large aromatic flowers are often pollinated by beetles and tend to emit the generically floral linalool or rosy-floral geraniol, along with a few other terpenoids and sometimes a fruity ester. Past the magnolias are shrubby **frangipani**, originally from the Caribbean, whose small and pleasantly intense, rosy-nuanced blossoms are now a frequent ingredient in Hawaiian leis. Then come a couple of bedding flowers familiar from gardens and florists. **Freesias** are one of Africa's few contributions to the scented flower trade; they come from the far south and are now found in many different colors. Along each flower stalk there are several blossoms that supplement their floral terpenoids with a couple of fruity esters. Similarly fruity, as its name suggests, is the diminutive and charming **pineappleweed**, a common American wildflower hugging the ground along the garden path. It's more often stepped on than sniffed, but it's endowed with a delightful mix of terpenoids and esters and worth stooping for.

Rounding out the American terpenoid specialists is the garden stalwart **marigold**, familiar and yet a volatile maverick. Like its camphoraceous fellow member of the aster family the chrysanthemum, it isn't especially floral. But instead of camphor, marigolds synthesize the rare terpenoid **tagetone**, which gives them a fresh but medicinal, somewhat harsh quality, unlike any other flower.

SOME TERPENOID-RICH FLOWERS FROM THE AMERICAS AND AFRICA

Flower	Component smells	Molecules
magnolia: southern, sweet bay, saucer (*Magnolia grandiflora, virginiana, M.* x *soulangeana*; Americas & E. Asia)	southern: floral, rose, green, woody; sweet bay: floral, fruity; saucer: green, woody, pine, floral	geraniol, ocimene, myrcene; linalool, methyl decanoate; ocimene, pinene, linalool
frangipani (*Plumeria rubra* var. *acutifolia*; C. & S. America)	floral, rosy, green, fruity	linalool, geraniol, nerolidol
freesia (*Freesia*; SE Africa)	floral, fresh, fruity	linalool, terpinolene, hexenyl & phenylethyl acetates
pineappleweed, wild chamomile (*Matricaria discoidea*)	fresh, citrus, sweet, fruity, pineapple	farnesene, geranyl isovalerate, myrcene
marigold (*Tagetes*; S. America)	fruity-harsh, medicinal, fresh	tagetones, ocimene, ocimenone, phellandrene

In the next bed, the American benzenoid bearers, in four groupings: low leafy plants, small velvet-leaved shrubs, a thick-leaved plant cradled in a shaded nest of moss, and finally a tangle of thick green spiny stems and branches. The **petunia**, a native of South America and common garden ornamental, is the first flower I remember smelling as a child, and flats of them can dominate the air in plant nurseries. It has a distinctive scent coming from benzenoid alcohols, aldehydes, and esters. The rich-smelling Mexican native **tuberose**, whose numerous leaves emerge from the ground as lily leaves do, is a relative of the desert-dwelling agave; it's especially rich in floral-fruity benzenoid esters and also carries a touch of musty indole. Pumpkins and squashes originated in Central America; their large-leaved spreading vines bear sizable bright yellow **squash** or **zucchini** blossoms, which we can find battered and fried in Italian restaurants or raw in markets and gardens. They emit a balanced, relatively subdued mix of fresh, medicinal benzenoids and floral terpenoids. They also contain indole, which attracts pollinating cucumber beetles.

Garden heliotrope is a shrubby Peruvian member of a genus found worldwide. Its clusters of tiny, usually purple flowers are also benzenoid dominated, but with a very different sweet, vanilla-like quality from prominent fruity and balsamic benzaldehyde and anisaldehyde. Reposing in the heliotropical shade are several varieties of the showy Brazilian **cattleya orchid**, another familiar nursery plant, whose smells move in the direction of spiciness, with wintergreen and clove benzenoids carried on the floralizers linalool and phenylethanol.

The tangle of green stems is a patch of **night-blooming cereus**, a cactus native to the dry American Southwest and northern Mexico, whose populations manage to open their white multipetal blooms in synchrony and mostly on one night in early summer, attracting hawkmoth pollinators and human admirers of their intense, sweet, balsamic scent.

SOME BENZENOID-RICH FLOWERS FROM THE AMERICAS

Flower	Component smells	Molecules
petunia (*Petunia*; S. America)	floral, fruity, almond extract, honey	methyl & benzyl benzoate, benzaldehyde, benzyl alcohol, phenylacetaldehyde, phenylethanol
tuberose (*Polianthes*; Mexico)	floral-fruity, wintergreen, sweet, Concord grape, citrus, musty	methyl & benzyl benzoates, methyl salicylate, methyl isoeugenol, methyl anthranilate, farnesene, indole
squash, zucchini (*Cucurbita pepo*; C. America)	fresh, green, floral, pine, musty	dimethoxytoluene, trimethoxybenzene, linalool, pinene, indole
heliotrope (*Heliotropium arborescens*; S. America, Europe, Asia)	almond extract, jasmine, balsamic, green	benzaldehyde, benzyl acetate, anisaldehyde, ocimene

continued

Flower	Component smells	Molecules
cattleya orchid (*Cattleya labiata*; Brazil)	floral, wintergreen, clove	linalool, methyl benzoate & salicylate, phenylethanol, eugenol
night-blooming cereus (*Peniocereus greggii*; U.S. Southwest, Mexico)	wintergreen, sweet, balsamic, fruity, almond extract	methyl benzoate & salicylate, benzyl benzoate & salicylate, benzaldehyde

The volatile history of roses

To conclude our stroll through the world's better-known flowers, let's pause at the best-known of them all. If there's a single canonical flower in the world, prized from earliest times and capturing all that a flower can be, it must be the **rose**. There are around a hundred species in the genus *Rosa*, but ten to twenty *thousand* different varieties. This last section of the flower garden offers a few thorny bushes to sketch the vicissitudes of rosiness.

The first two bushes, the **Gallic rose** and the **sweetbrier**, are wild plants of species native to Europe and western Asia. They have flowers with five or more small pink petals that are borne once yearly, and the rose-defining mixture of monoterpenoid alcohols and the benzenoid phenylethanol. A third plant, with white blossoms, is the wild **musk rose**, which has a less typical rose smell, clove-like and spicy from other benzenoids. Rose breeders have developed the thousands of varieties by selecting and mating these and other wild species, usually to elicit their potential for expressing striking shades of red and other colors, many-petal blossoms, and the ability to flower repeatedly. Scent has generally been of secondary interest.

The great aromatic exception to this looks-first rule arose as an apparently natural hybridization among three different species native to Europe and the Near East, probably in what is now the northern area of Iran. This chance genetic mixing created the **damask rose**, borne on the fourth of our bushes, pink and many-petaled. Its name comes from the ancient Syrian city of Damascus, where European crusaders encountered its especially fine, strong scent in perfumers' rose water and rose essential oil (see page 452). One of the parent species, the Gallic, likely contributed the basic rose volatiles; the musk rose perhaps the metabolic byway that generates terpenoid-pigment fragments; and the third parent, a little-studied western Asian species, is said to smell of brown bread and blackberry jam and so might have contributed fruity esters. In their joint offspring damask rose, the basic Gallic mixture is deepened with the addition of two terpenoid-derived fragments, violet-like ionone and cooked-apple damascenone—so named because it was first isolated from damask roses—along with several esters. Today the finest rose oils for perfumery come from Bulgarian, Turkish, and Iranian damask roses, and from a damask descendant, *Rosa* x *centifolia*, that's especially associated with the perfume center Grasse in southeastern France.

Damask roses can seem worlds apart from modern rose varieties, which are often barely scented or scentless. The post-damask phase began in the eighteenth century with the arrival in Europe of **Chinese roses**, some of them already hybrids, which had many desirable characteristics for breeding that European roses lacked, including the ability to flower repeatedly and petal shades of yellow and orange. But the yellow multipetal flowers on the next bush, a Chinese specimen, are nothing like the European roses: less floral and powerful, instead fresh, earthy, slightly spicy.

The characteristic mild smell of Chinese roses comes from the dominance of a couple of unusual benzenoids, both slightly medicinal—dimethoxytoluene, fresh and earthy, and trimethoxybenzene, spicy and lightly animalic. The green-leaf ester hexenyl acetate also contributes a green note that's unusual in flowers. For Westerners, the smell was reminiscent of another Chinese import. The *American Flower Garden Directory* published in Philadelphia in 1832 includes a listing for "Rosa odorata, or **Tea-rose**, celebrated in this country for its fragrance being similar to fine Hyson [green] tea. It justly deserves the preference of all the China roses for the delicacy of its flavour."

SOME ROSE SCENT FAMILIES

Rose varieties	Component smells	Molecules
European/Middle Eastern species: (*Rosa gallica*, *rubiginosa*; R. *moschata*)	*gallica*, *rubiginosa*: classic rose, intense, floral	phenylethanol, citronellol, geraniol, nerol
	moschata: clove, violet, woody	eugenol, ionone, caryophyllene

continued

Rose varieties	Component smells	Molecules
damask natural hybrid rose (*Rosa* x *damascena*) (from *R. gallica, moschata, fedtschenkoana*)	classic rose + heavy, rich, fruity, violet, woody	phenylethanol, citronellol, geraniol, nerol; + damascenone, ionone, rose oxides, hexenyl & hexyl & phenethyl esters, rotundone
Chinese tea roses (*Rosa chinensis, R. chinensis* x *odorata* var. *gigantea*)	fresh, earthy, spicy, green tea, violet	dimethoxytoluene, trimethoxybenzene, caryophyllene, germacrene, linalool, ionone
hybrid tea, floribunda roses (complex Eurasian/ Asian crossbreeds)	often faint tea type	dimethoxytoluene, trimethoxybenzene, hexenyl acetate, geraniol, citronellol
moss damask rose (*R.* x *damascena*, Perpetual White Moss)	damask rose + pine, woody, fresh	damask rose volatiles + pinene, myrcene, sabinene, phellandrene

The frankly rosy scent of European roses began to fade in the nineteenth century, when French plant breeders pioneered the crossing of Chinese tea roses with European roses to obtain **hybrid tea roses**. For many decades, most of the popular garden roses you could sniff at a plant nursery would have been handsome and productive but largely scentless. Happily, many older pre-tea varieties survived, and beginning in the 1960s, the English rose breeder David Austin worked to overcome the olfactory deficit in modern varieties, with great success. Aromatic roses have made a comeback.

We've covered quite a swath of the osmocosm in this chapter, from corpse flowers to water lilies to dandelions and jasmine, and roses several ways. The universal visual appeal of flowers means that they're easy to find in everyday life, in front yards and in parks and markets, nurseries and botanical gardens. Many municipalities in many countries have rose gardens, some with hundreds of varieties growing side by side. All of these offer ready opportunities for the smell explorer to experience what flowers are capable of, while marveling at the astoundingly beautiful scent-emitting structures that Hero Carbon and human minds have collaborated to create.

Before we depart the flower garden, two last whiffs in the rose patch. Go back to the second plant, the wild European *Rosa rubiginosa*, and rub a leaf between your fingers. This plant has been known in English as **eglantine** or **sweetbrier**—"brier" for its thorns, "sweet" not for the smell of its flower but for its leaves, apple-scented with acetaldehyde and sesquiterpenes! Greenery too can be delightfully aromatic. Then come back past the tea and hybrid teas to another damask bush with several unopened buds. Take a close look: this is one of the varieties of damask and centifolia roses, the **moss roses**, that cover their green buds with thousands of tiny hairs, which give them a fuzzy appearance. Now lightly rub the fuzz and smell your fingers, then smell one of the open blossoms. The blossom emits the usual damask volatiles, but the hairs release an entirely different set of terpenoids when they're brushed against: the smell is piney, woody. These hairs are clearly protective weapons, bristling volatile-loaded spikes to discourage chewing insects from trying a bite.

As we depart the flower garden and its many sweet smells, the sweetbrier leaf and moss rose are reminders that the pleasant volatiles in flowering plants are mostly defensive, and greenery often smells more than just green. Volatile hairs are an interesting oddity on a flower, but fuzz is what matters most on mint and basil leaves. On now to the herbs.

Chapter 11 · EDIBLE GREENS AND HERBS

In the Composure of a *Sallet*, every Plant should come in to bear its part, without being over-power'd by some Herb of a stronger Taste, so as to endanger the native *Sapor* and Vertue of the rest; but fall into their places, like the *Notes* in *Music*, in which there should be nothing harsh or grating: And tho admitting some *Discords* (to distinguish and illustrate the rest) striking in the more sprightly, and sometimes gentler Notes, reconcile all Dissonancies, and melt them into an agreeable Composition.

It being one of the Inquiries of the Noble Mr. *Boyle*, what *Herbs* were proper and fit to make *Sallets* with, and how best to order them? we have here (by the assistance of Mr. *London*, His Majesty's Principal Gard'ner) reduc'd them to a competent Number, not exceeding *Thirty Five*; but which may be vary'd and inlarg'd, by taking-in, or leaving out, any other *Sallet*-Plant, mention'd in the foregoing List.

• John Evelyn, *Acetaria: A Discourse of Sallets*, 1699

Thus far in our exploration of the green world, we've sniffed our way through forest, meadow, scrubland, desert, and flower beds. Now our path brings us home to the dozens of plants whose smells we know most intimately: those that we put in our mouths and eat, freeing their inner volatiles as we chew and pulling them up into the nose as we exhale. We leave the virtual botanical garden behind and proceed into an adjoining open-air produce market, shaded to keep the just-harvested materials fresh. It's the most extensive and organized and unprofitable market imaginable, with stand after stand of greens and herbs, roots and tubers, nuts and grains, spices, fruits—nothing for sale, all free for the sniffing and tasting.

In this and the next couple of chapters, imagine strolling along these market stands and sampling the items you're interested in, calling their smells up from memory and comparing their sets of components. Better yet, leave imagined stands for your kitchen and check the real things. Explore further afield by adding less familiar ingredients to your shopping lists, and bring home a few at a time for tastings. Even a modestly provisioned fridge and spice rack amount to a trove of aromatic plant materials from all over the planet, many of them grown specifically for their volatile molecules.

Some of the items in this market and these chapters will be most familiar in their dried or cooked form. Here, though, we'll focus on the smells of the raw plant materials. Both drying and cooking drive off volatiles and generate new ones, usually obscuring the original, natural qualities of the materials. The one important exception is the group of materials we call *spices*, which are seeds and other plant parts whose strong flavors persist even after they're dried. We'll sniff at spices in the next chapter, cooked foods in chapter 18, and dried herbs in chapter 19.

This chapter begins our survey of edible plants with a dozen stands bearing piles of leaves: some large, some tiny, some bunched together, some decorating long stalks. These are things that we do often eat raw, in salads and side dishes and garnishes. *Kitchen greens* include large-leaved vegetables, lettuces and chicories and kales and the like, whose smells are mostly dominated by the green-leaf volatiles. *Kitchen herbs* are a diverse set of small-leaved plants that carry a variety of additional volatiles and stronger smells—mint, rosemary, parsley—and are used by cooks more as flavor accents than as major ingredients.

These days our experience of kitchen greens and herbs tends to be fairly standardized, limited, and easily taken for granted. So it's salutary to remember the pioneering salad sage John Evelyn, the interest taken in his studies by noble colleagues at the Royal Society and by the royal gardener, and his final list of thirty-five suitable salad ingredients—trimmed from an initial set of eighty! The finalists included many familiar fixings, but also such forgotten greens as burnet, samphire, scurvy grass, and trip-madam, a dozen of which might find their way into a given salad. Clearly there are more flavorful salad ingredients in the world, and underfoot, than are dreamt of in our modern kitchens. And more interesting challenges to the cook in composing with them.

In this chapter we'll sniff nearly as many leaves as Evelyn's eighty, nearly all of them cultivated and widely available. They span a good range of smells, from lettuce's generic green-leaf volatiles generated by cutting and chewing, to the specific terpenoids and benzenoids of mint and lemongrass and the other herbs, premade and stockpiled in hairtrigger leaf fuzz or hidden interior canals. Along with spices, kitchen herbs enable the cook to bring heightened flavors and interest to the relatively bland foods that supply most of our nourishment. Their chemical defenses, aimed at microbes and insects and other animal enemies, fail to repel *us* only because we're large and can control how much we take in: enough to be stimulated but not enough to be irritated or injured.

Despite being ineffective as defenses against us, these weapons have been highly effective as inducements to assist in their plants' defense. They lead us to replace the individual plants we consume, to care for them, multiply them far beyond their natural homeland and numbers, and breed types that would never survive in the wild: all so that the species will continue to bring olfactory interest to our lives. We've transformed plant weapons into favors, potential injury into pleasure, individual sacrifice into collective triumph. Expansive thoughts for the salad course!

Kitchen greens

The first stands in the market are piled with familiar leafy greens, lettuces and others, which are tender and mild enough for us to eat raw or lightly cooked. Most of these vegetables come from the temperate regions of Europe and western Asia, and all are domesticated, meaning they've been selected and bred over millennia to be productive and pleasing, with toned-down chemical defenses and milder flavors than their wild ancestors.

Of course the flavor of most salad leaves is "green"! When we eat raw leaves, chewing damages the tissues and stimulates production of the grassy, fresh green-leaf volatiles (see page 154) in our mouth. That's the dominant quality that most green vegetables share—at least when we enjoy them raw. Cooking kills the GLV-producing enzymes and forms other molecules, so cooked greens have different flavors (see page 514). And each particular vegetable carries its own mix of preformed non-green volatiles.

Three of our most common salad leaves occupy the very first stand, all large-leaved and members of the populous daisy family. **Lettuce** is the mildest and most generically grassy of the leafy vegetables, but it can have woody and earthy bell-pepper notes from a terpenoid and a nitrogen-containing pyrazine. **Endive** and **radicchio** are sister species to each other and are more complex than lettuce, valued in part for the bitter counterpoint they can bring to a salad. Endive is touched with almond, cucumber, and flowery notes; radicchio, with consonant honey, waxy, flowery, and sometimes minty notes.

On the stand next to the familiar salad fixings are two piles of related but very different-looking greens, long stalks with many small divided leaves. Rub a few and sniff your fingers: you're in a different olfactory realm, dominated by terpenoids. These are two other members of the daisy family better known in Asia than the West, both species of *Artemisia*, the genus that gives us the inedible sagebrush and common mugwort (see page 203) and the barely edible wormwood (see page 259). **Garland chrysanthemum** is a vegetable cousin of the chrysanthemum species that are grown for their flowers; it's popular in China and Japan, its raw leaves scented by woody and resinous terpenoids. And vegetable **mugwort**—distinct from the common weed—is a species of *Artemisia* that the Japanese are especially fond of, with terpenoids suggestive of eucalyptus, cedar, and camphor.

SOME LEAFY VEGETABLES IN THE DAISY FAMILY

Plant	Component smells	Molecules
lettuce (*Lactuca sativa*)	green, vegetal	GLVs, isopropyl methoxypyrazine, caryophyllene
endive (*Cichorium endivia*)	green, almond, cucumber, floral	GLVs, benzaldehyde, nonenals, ionone
radicchio (*Cichorium intybus*)	green, honey, floral, waxy, minty	GLVs, phenylacetaldehyde, ethyl dodecanoate, methyl pentanone
garland chrysanthemum, shungiku, tong ho (*Glebionis coronaria*)	green, woody, resinous	GLVs, ocimenes, myrcene, farnesenes, germacrene
mugwort, yomogi (*Artemisia princeps*)	green, eucalyptus, cedar, camphor	GLVs, eucalyptol, thujone, bornyl acetate, borneol

The next stand shifts the smell register again, to the overtly sulfurous. Its half-dozen piles of leaves large and small are representatives of the cabbage family: **cabbage** itself, **kale**, **rocket** or **arugula**, **mustard greens**, **garden cress**, and **watercress**. All of these are Eurasian plants, all usually combining a strong green smell with sulfur aromas and volatile mustardy pungency from nitrogen-sulfur defensive molecules, all formed when the leaves are cut or chewed. The ornamental flowering **nasturtium**, *Tropaeolum*, is a South American cousin to the cabbage family, with a similar if more muted flavor; its common name, which is the same as watercress's botanical name, is said to come from the Latin words for "nose" and "twist": an allusion to the pungent smells.

SOME LEAFY VEGETABLES NOT IN THE DAISY FAMILY

Plant	Component smells	Molecules
kale, cabbage, rocket, mustard greens, cresses (*Brassica oleracea, B. juncea, Eruca sativa, Lepidium sativum, Nasturtium officinale*)	green, sulfurous, pungent	GLVs, methyl & ethyl sulfides, nitrogen-containing nitriles, sulfur- & nitrogen-containing isothiocyanates, thionitriles, thiocarbamates
spinach (*Spinacia oleracea*)	green, fatty, earthy, sulfurous	GLVs, C8 aldehydes, isopropyl & butyl methoxypyrazines, methanethiol, dimethyl trisulfide
rhubarb leaf stalks (*Rheum* hybrid)	green, grassy, apple, cucumber, floral, fruity	hexenals, nonadienal, ionone, hexenol
corn salad, mâche (*Valerianella locusta*)	green, fruity, floral, horseradish	GLVs, methylbutyrate esters, phenylethanol, citronellol, phenylethyl isothiocyanate
borage (*Borago officinalis*)	green, citrus, seaweed	hexenol, octanal, decadienal
oyster plant (*Mertensia maritima*)	cucumber, mushroom, geranium, melon	nonenal, octadienol, octadienone, nonadienal

Now comes a stand with three unrelated items for sampling, two of them familiar, one worth getting to know. Salad favorite **spinach** shares membership in the amaranth family with both stinking goosefoot and kerosene-like epazote (see pages 203, 261). Happily, it doesn't smell like either of those, but it is more complex than lettuce, with fatty and sulfurous notes added to the green and earthy. Next to its hand-size bunches is a handful of large, thick, celery-like stalks, some green and some red: the edible portion of **rhubarb** leaves, sour with oxalic acid and mostly vegetal in aroma, but with floral and fruity notes that make it sweetenable into an honorary fruit. Then there's a cluster of small spinach-like rosettes: **corn salad** or **mâche**, also known as lamb's lettuce and a number of other names. It's a diminutive European relative of two plants with notably aromatic roots, the exotic nard included in John Milton's garden of Eden (see page 463), and valerian, the original source of cheesy-sweaty five-carbon valeric acid. The tender leaves of mâche emit an intriguing mix of green, fruity, flowery, and faintly sulfurous notes, unlike any other salad green.

The last stand of kitchen greens provides a kind of transition to the more distinctive herbs; they're usually accents rather than primary ingredients. On it are the dark green leaves of two members of the borage family, which includes the flowering heliotrope (see page 237). Both smell startlingly like raw oysters! One is **borage** itself, a Mediterranean native that makes both aromatic leaves and small blue-purple flowers; the other is the **oyster plant**, a sister species to the Virginia bluebell, which can be found throughout the Northern Hemisphere. Their similarity to oysters and to cucumbers comes from shared volatile cocktails, a dozen or so aldehydes and alcohols from eight to ten carbon atoms long, in addition to the generic green-leaf volatiles.

Kitchen herbs: introducing the mint family

As we walk away from the market stands of kitchen greens toward the herbs, we come to a table with several virtual-reality headsets. Put one on, adjust it, and look around. What you see is an almost blindingly white span of fractured rock stretching to the horizon: rough, jumbled, and sparsely populated with wiry-stemmed shrubs. It's a simulation of the homeland of the Mediterranean mint family (and of the laurel tree that produces bay leaves; see page 198). Though mints, thyme, rosemary, and their cousins flourish in well-tended gardens and farms, they developed their defining strong smells as a means of surviving in their homeland. When they're pampered in cultivation, they tend to lower their defenses and become a little less themselves. To appreciate them fully, you need to know where they came from.

Over the years, I've had several chances to walk through uncultivated lime-stone areas of the Mediterranean, near Minervois in southern France, in Puglia in southern Italy, near Erice in Sicily, on Crete, and in the foothills of Mount Olym-pus. Dry areas of these terrains are called *garrigue*, *gariga*, and *phrygana* in French, Italian, and Greek; moister areas are *maquis* or *macchia*. In all of them I've recog-nized, plucked, and tasted many of the herbs that commonly flavor our stews and meats and dressings and sweets, that scent our sachets and soaps.

Thyme, oregano, sage, savory, rosemary, lavender, and mints are all members of the same family—for our purposes, the mint family—and still grow wild in their native habitat. What becomes clear when you see them there is that many are sitting ducks, easy targets, obvious swatches of green on the white limestone canvas—like the creosote bushes in the American desert (see page 204). These plants will certainly be chewed on. They're especially exposed in the garrigue, sometimes growing out of a mere crack in the rock, with no soil or nearby plants for even a bit of camouflage to hide from hungry snails, insects, rabbits, goats. To have a chance of surviving long enough to produce seeds, they have to make them-selves as unpalatable as possible.

They do so by loading their small, tough leaves with chemical defenses, mainly terpenoids, that supplement the basic green-leaf volatiles. To get a sense for their power, try what I did on my first visit to the garrigue: pretend to be a grasshopper or rabbit, and chew on a sprig of thyme or rosemary. It stings! Biologists have dis-covered that their volatiles deter competitors as well as predators: when these scrappy plants drop their leaves or are rained on at the start of the growing sea-son, some of the volatiles end up in the soil and inhibit the seeds of other plants—and sometimes their own—from growing there and taking up scarce water and minerals.

Now, since we have VR conveniently at our disposal, activate the headset's voice control and speak the trigger word "rosebud." The scene shifts from rocky garrigue to a spring-green background bristling with a thicket of translucent stalks, each tipped with what looks like a tiny droplet. This is a microscopic view of the fuzzy-looking buds of the moss rose, which we fingered at the end of the previous chapter. Then activate the voice control again and say "rosemary." The rosebud close-up fades and gives way to a darker green scene with similar but stubbier pro-jections. Of course this is a close-up of a rosemary leaf. Like the moss rose, rosemary and other members of the mint family synthesize and store defensive terpenoids in tiny glands that project from the plant surfaces. These glands give the plants a dull or fuzzy or hairy appearance, and they break and release their contents even when the plant is barely brushed against, not actually bitten into. They're a first line of deterrence that can discourage more damaging attacks. The green-leaf volatiles, by contrast, are generated only when the leaf tissue itself is breached.

This segregation of the mint family's volatile defenses makes it possible for bartenders to give a minty aroma to drinks without the generic green leafiness. Instead of crushing mint leaves with a muddler, they slap them gently between their flat palms, releasing only the terpenoids. Because slapping deposits some of the terpenoids on the hands, it's more effective, if less showy, to rub the undersurfaces of two leaves together—the lower surface has more glands—and drop both into the drink.

Time now to take off the VR headset, return to our low-tech imaginary market, and meet the mint family. Because it's their preformed gland volatiles that distinguish these herbs from other leafy plants, I'll simplify the tables below by omitting mention of the ubiquitous green-leaf volatiles.

The basic set of mint-family herbs

The first stand of herbs has only three piles on display. These are the most familiar species of the genus *Mentha*, the mints themselves, which lend their name to a family that includes nearly three hundred other genera. The mints are atypical of the mint-family herbs because they grow throughout much of the Northern Hemisphere and prefer the moister macchia and similar habitats to the garrigue. They contribute two landmark volatiles and smells to our everyday lives. Menthol, fresh and cooling and minty, is a terpenoid that has been flavoring endless confections and personal-care products since the late nineteenth century. It's abundant in two species: **field mint**, which is grown on an industrial scale for the production of natural menthol, and **peppermint**, which is a naturally occurring hybrid between two parents that smell nothing like it. Piled next to the bunch of peppermint is one of those parents. **Spearmint** is a favorite cooking ingredient in the eastern Mediterranean and the source of the distinctive aroma and flavorings named after it, which is generated by a specific and otherwise uncommon terpenoid, the mint carvone. Its mirror-image molecule is the caraway carvone, and I find that there's often a hint of caraway aroma in spearmint leaves.

SOME COMMON MINT SPECIES

Herb	Component smells	Molecules
field/menthol mint (*Mentha arvensis*)	cooling, minty, fresh	menthol, menthone
peppermint (*Mentha* x *piperita*)	cooling, minty, fresh, earthy	menthol, menthone, menthyl acetate, menthofuran
spearmint (*Mentha spicata*)	spicy-minty, herbaceous, cucumber, cooked apple	mint carvone, dihydrocarvone & dihydrocarvone acetate, eucalyptol, nonadienal, damascenone

The next market stand is crowded with the many mint-family herbs that are denizens of dry garrigue areas from the Mediterranean to Central Asia. Perhaps for that reason, they retain their flavor unusually well when dried and are usually too strong to cook with in anything but small amounts. A lesser-known close relative of the *Menthas*, **nepitella**, has an aggressive minty quality from the strong resinous terpenoid pulegone. Several more familiar Mediterranean herbs share a similar landmark quality—a medicinal, tarry smell and pungent taste. These come from two nearly identical terpenoids, thymol and carvacrol, that take the form of a six-carbon ring with decorations and closely resemble phenol, a powerfully reactive molecule often used as an antiseptic and disinfectant (see page 426). Some mouthwashes include thymol and carvacrol not because they are pleasant volatiles, as minty menthol is, but because they help control the bacteria that cause tooth decay and bad breath. **Thyme** and **oregano** are dominated by thymol and carvacrol, while milder **summer savory** and **winter savory** blend them more evenly with woody, turpentiney terpenoids.

SOME MEDITERRANEAN HERBS IN THE MINT FAMILY

Herb	Component smells	Molecules
nepitella, mentuccia (*Calamintha nepeta*)	sharp, minty, resinous, herbaceous	pulegone, menthone, piperitenone, menthol
thyme (*Thymus vulgaris*)	medicinal, tarry, turpentine, woody, floral	thymol, terpinene, cymene, linalool

Herb	Component smells	Molecules
oregano (*Origanum vulgare*)	medicinal, tarry, woody, turpentine	carvacrol or thymol dominant, cymene, terpinene
summer savory (*Satureja hortensis*)	medicinal, tarry, turpentine, woody, green	carvacrol, terpinene, cymene, myrcene
winter savory (*Satureja montana*)	medicinal, tarry, woody, turpentine	thymol, cymene, terpinene, carvacrol
marjoram (*Origanum majorana*)	woody, pine, citrus, peppercorn	terpinenol, sabinene, rotundone
rosemary (*Rosmarinus officinalis*)	eucalyptus, pine, woody, camphor	eucalyptol, pinene, borneol, camphor
sage (*Salvia officinalis*)	cedar, camphor, fresh, pine	thujone, camphor, eucalyptol, pinene
lavender leaves (*Lavandula* species)	eucalyptus, camphor, earthy, pine	eucalyptol, camphor, borneol

A number of other Mediterranean herbs bring conifer and other tree smells to our food and drink. **Marjoram** carries lighter pine notes along with rotundone, the woody, spicy terpenoid that helps define the aroma of peppercorns. **Rosemary** provides an intense but balanced mix of eucalyptus, pine, and camphor. **Sage** highlights the cedar note of the terpenoid thujone at the top of a similar mix. **Lavender** leaves, very different from the flowers (see page 227) and the lavender scents made from them, mix fresh eucalyptus and camphor notes with woody borneol.

Pass along now to a stand with just four heaps: these are farther-flung members of the mint family. Though we normally associate it with Italy, common **basil** originated in Asia and only later found its second home in summer gardens around the Mediterranean. It mixes terpenoid and benzenoid volatiles and has smells that are reminiscent of anise and clove and cinnamon. **Perilla**, an Asian species known in Japan as **shiso** and most reliably encountered at the sushi bar, shares in the mild, refreshing quality of the mints, but with a distinctive quality from modified terpenoids, **perilla alde-hyde,** and perilla ketones, found nowhere else in the family. As its name suggests, **anise hyssop** smells of anise; its home is the plains of northern North America. And a last example, not so much culinary as historical: the low, spreading **yerba buena** ("good herb") from moist coastal forests of California and the Pacific Northwest gave its name to the Spanish settlement at Mission Dolores that later became known as San Francisco. It mixes medicinal and cooling camphor with minty terpenoids.

SOME ASIAN AND AMERICAN HERBS IN THE MINT FAMILY

Plant	Component smells	Molecules
basil (*Ocimum basilicum*)	tarragon/anise/clove more or less dominant, flowery	estragole, eugenol, linalool
perilla, shiso (*Perilla frutescens*)	perilla, minty, citrus	menthadienal (perilla aldehyde), perilla & egoma ketones (C10 furan rings)
anise hyssop (*Agastache foeniculum*)	anise, tarragon, woody	estragole, cadinene, limonene, caryophyllene
yerba buena (*Clinopodium douglasii*)	camphor, minty, pungent	camphor & camphene, sometimes + pulegone or mint carvone or isomenthone

These last three market stands span quite a range of smells, minty and medicinal and woody and spicy. Clearly the mint family is a bunch of volatile virtuosos. Many are unusually promiscuous in their production of terpenoids and benzenoids, probably so they can fine-tune their volatile cocktails to adapt to particular local microbes or animal predators in the wild. This means that individual plants can have very different smells there and in the garden and on the farm: there are many versions of "mint" and "thyme" and "basil" out there to explore.

To give just a taste of this diversity, each of the next three stands is devoted to one of these herbs and offers samples of the different smells it can carry. If you're a fan of the Italian pasta sauce pesto alla genovese, the basil stand may be revelatory.

Volatile variations in mints and thymes

The first stand is devoted to the mints themselves. Spearmint and peppermint are among our most familiar herbs, but there are more than a dozen other species. Spearmint is the canonical cultivated mint of the eastern Mediterranean, where it's much used in cooking and as a tea, and it's the standard garnishing mint elsewhere. But there are populations of spearmint that still grow wild in the Greek countryside, and they can have very different smells, floral or camphor- or peppermint-like. Peppermint itself, with its high levels of the cooling terpenoid menthol, is a hybrid between spearmint and the atypical musty/woody **water mint**. One particular variety of water mint is much more pleasant than the others: the variously named **lemon mint** or bergamot, lavender, or eau de Cologne mint isn't especially minty either, but it contains floral linalool and its acetate ester, two terpenoids that are indeed found in lemon and bergamot and lavender. The species known as **pennyroyal** has fallen out of favor over concerns about its toxicity; its strong minty-resinous terpenoid pulegone is known to be an effective insecticide. And the oddly named **apple mint** mixes the smells of pennyroyal and spearmint.

SOME UNCOMMON MINTS

Mint species	Component smells	Molecules
wild Greeks mints (*Mentha spicata*)	spearminty	mint carvone + dihydrocarvone
	floral/citrusy	linalool
	minty/ camphoraceous	piperitone + piperitenone oxides
	pepperminty/ camphor/sulfurous	menthone + isomenthone + pulegone
water mint (*M. aquatica*)	musty, earthy, eucalyptus, woody	menthofuran, eucalyptol, ocimene, limonene, caryophyllene
lemon/bergamot/ lavender/eau de Cologne mint (*M. aquatica* var. *citrata*)	floral, citrus, fruity, fresh	linalool, linalyl acetate, eucalyptol, myrcene
pennyroyal (*M. pulegium*)	minty-resinous, sharp, fresh	pulegone, menthone, menthol
apple/pineapple mint (*M. suaveolens*)	minty-resinous, spearmint, camphor	pulegone, mint carvone, piperitone oxide

Now to the thyme stand. Common garden thyme has a very particular identity in Western cooking thanks to its major and somewhat medicinal volatile thymol, but it came to the garden from the Mediterranean garrigue, and like the Greek spearmints, its still-wild brothers and sisters express a range of other identities. Surveys of thyme populations in the countryside of southern France have identified groups that smell variously like a cross between camphor and mint, like pine resin, like eucalyptus, like lavender flowers, and like roses—all thanks to different terpenoid cocktails. Surely some of these are also worth bringing into the garden! Two other sister species have already made it there. **Lemon thyme** produces the truly lemony terpenoids neral and geranial, and so has a more genuine lemon character than lemon mint. Some strains of **caraway thyme** smell like caraway seed thanks to their production of the terpenoid carvone, others lean into mintiness with a mint-carvone derivative, and a half dozen more offer yet other variations on the thyme theme.

SOME UNCOMMON THYMES

Thyme species	Component smells	Molecules
wild thymes, S. France (*Thymus vulgaris*)	medicinal, tarry, turpentine, woody	thymol or carvacrol, terpinene, cymene
	minty, turpentine, camphor	thujanol, terpinenol, myrcenol
	pine, floral, lime	terpineol, terpinyl acetate
	eucalyptus, minty, floral	eucalyptol, thujanol, linalool
	floral, lavender	linalool, linalyl acetate
lemon thyme (*T. citriodorus* or *pulegioides*)	floral, rose	geraniol, geranyl acetate
	floral, lemon	geraniol, geranial, neral
caraway thyme (*T. herba-barona*)	caraway, minty	caraway carvone, dihydrocarvone (2 of 8 types)

Basil variations, pesto particulars

This last stand of mint-family variations overflows with ten different versions of basil. Whereas the potential for diverse aromas in mint and thyme remains mostly in the wild, basil diversity is already well represented in cultivated forms. The basil genus is native to Asia and Africa, and it's unusual in the mint family for producing both terpenoids and benzenoid volatiles, most frequently clove-like eugenol and tarragon/anise estragole. Indian **holy basil**, **Thai basil**, and **African basil** are different species of the genus *Ocimum*, readily available as garden plants, and respectively dominated by the smells of clove, anise, and thyme (African basil is one of the rare non-thyme thymol producers). The basils most popular in the West are varieties of *Ocimum basilicum*, the species that includes Thai basil. It appears to have been carried from Asia to the Mediterranean by Arab traders and cultivated in Spain during the Moorish period around nine hundred years ago. Thereafter, European and American enthusiasts developed it into a number of very different varieties. Among these are **Mexican** or **cinnamon basil**, with cinnamon and fruity notes from benzenoid cinnamates; **lemon basil**, a hybrid that produces lemony terpenoids; and **African blue basil**, a medicinal hybrid that produces camphor. The large-leaf **ruffled** or **lettuce basil** varieties smell strongly of tarragon and anise.

The **standard basil** varieties in the West today mainly produce varying proportions of a couple of terpenoids, flowery linalool and fresh eucalyptol, and the clove- and anise-smelling benzenoids eugenol and estragole. But when it comes to a dish in which basil stars—pesto alla genovese, the Ligurian pasta sauce of pounded basil, garlic, nuts, and cheese—Italians are more particular.

SOME VARIETIES OF BASIL

Basil variety	Component smells	Molecules
holy, tulsi (*Ocimum tenuiflorum*)	clove or cinnamon + clove	high eugenol or methyl eugenol
African, clove (*O. gratissimum*)	thyme or clove	high thymol or eugenol
Thai (*O. basilicum* var. *thyrsiflora*)	tarragon, anise	high estragole
Mexican, cinnamon (*O. basilicum*)	fruity, strawberry, cinnamon	high methyl cinnamate
lemon or sweet Dani (*Ocimum* x *citriodora*)	lemony, floral	high neral & geranial

continued

Basil variety	Component smells	Molecules
African blue (*O. basilicum* Dark Opal x *O. kilimandsharicum*)	camphor, medicinal	high camphor
ruffled, anise types (*O. basilicum*)	tarragon, anise	high estragole
standard; small-leaf bush types (*Ocimum basilicum*; *O. minimum*)	floral, eucalyptus, clove, tarragon	linalool, eucalyptol, eugenol, estragole
Genovese types (*O. basilicum* cv. Genovese gigante)	floral, eucalyptus, clove	linalool, eucalyptol, eugenol

The European Community has bestowed an official Protected Designation of Origin on the basil variety *Ocimum basilicum* cv. Genovese gigante, aka *basilico genovese*, because it has a specific smell, arising from a specific set of volatiles, which is considered proper for pesto alla genovese. This volatile set omits estragole and emphasizes floral linalool. As plant breeders at the University of Bologna put it, estragole gives "a typical mint/anise flavour that is considered anomalous and thus undesirable in 'Genovese,' and not appreciated by the Italian consumers." So if you're a stickler for Genovese authenticity, sniff your basil before you buy!

A different line of research has confirmed that the volatile composition of basil changes as individual leaves and whole plants grow and mature. Young leaves have a higher proportion of scent glands to green tissue, and more volatiles in those glands (they gradually lose volatiles to evaporation). Moreover, leaves on young plants synthesize their volatiles in different proportions from those of new leaves on mature plants. Standard Italian practice for pesto basil is to harvest whole Genovese plants while they're still quite young, after just a few weeks' growth and with only three sets of leaves.

Kitchen herbs: the celery family

After six stands devoted to the many-sided mint family, we come to a single spread of celery-family herbs. They're fewer and less virtuosic but perhaps more broadly useful. In addition to its namesake, popular members of the celery family include cilantro, dill, fennel, and parsley, all of them natives of temperate Eurasia. The family resemblance shows in their general preference for moist soils, a more erect growth habit than the ground-hugging mints, tiny flowers borne aloft on showy umbrella-like "umbels," and a short life span, typically one or two years. They store their volatiles in tubular ducts that run through their leaves and stems, not in surface hairs, and most of them fill their ducts with more benzenoids or simple chains than terpenoids. The defining smells of celery itself and a couple of other family members come from unusual volatiles that fuse a modified benzene ring with an oxygen-containing furan ring. Celery-family species are pretty well fixed in their volatile identities and so are more predictable ingredients than basils and mints.

Celery aroma arises from several distinctive benzofuran hybrid molecules called **phthalides**, notably **sedanenolide**. Green-leaf volatiles are important in the leaves but less so in the thick stalks, which are substantial enough to serve as a vegetable. Along with carrots and members of the garlic family, celery is classified as an aromatic vegetable, used to flavor dishes as well as a primary ingredient. When it is cooked, its aroma changes drastically, developing the sweet-fenugreek furanone sotolon and floral terpenoid fragments (see page 510). **Lovage** is far less common than celery but sometimes grown as a leafier, more herblike approximation; it, too, stores a kind of phthalide but also a few flowery, piney terpenoids. **Dill**, like celery acknowledged with the species name *graveolens*, or "heavy-scented," gets its very particular smell from yet a different benzofuran dubbed the **dill ether**, in partnership with the terpenoid phellandrene; it also has a detectable caraway note from the terpenoid carvone.

SOME HERBS IN THE CELERY FAMILY

Plant	Component smells	Molecules
celery (*Apium graveolens*)	celery, sweet, green, woody	phthalides (benzofuranones), GLVs, myrcene, myristicin
lovage (*Levisticum officinale*)	floral, celery, pine, grassy	terpinyl acetate, ligustilide (benzofuranone), phellandrene
dill (*Anethum graveolens*)	dill, caraway, fresh, minty	phellandrene, dill ether (C10 furan), caraway carvone, myristicin

continued

Plant	Component smells	Molecules
fennel (*Foeniculum vulgare*)	anise, tarragon, minty, citrusy-fresh	anethole, limonene, estragole, phellandrene, fenchone
chervil (*Anthriscus cerefolium*)	tarragon, clove, citrus	estragole, methyl eugenol, limonene
cilantro, coriander (*Coriandrum sativum*)	cilantro, soapy, waxy, melon, hay	dodecenal, decenals, decanal (seed mainly linalool)
culantro, recao (*Eryngium foetidum*)	cilantro, soapy, musty, waxy, green, pungent	dodecenals, dodecanal
parsley (*Petroselinum crispum*)	woody, green, metallic, earthy, nutmeg	menthatriene, myrcene, methoxypyrazine, myristicin, decenal

A few herbs in the celery family share important volatiles and so can be suggestive of each other. The sturdy, feathery-leafed yard-tall **fennel** plant has a strong, sweet anise smell from two similar-smelling benzenoids, anethole and estragole, balanced by fresh citrusy limonene. Delicate **chervil**, hand-high and spreading, offers a more subtle anise quality with low levels of estragole alone. The divided-leaf Eurasian native **cilantro**, also known as **coriander**, is largely defined by ten- and twelve-carbon aldehydes, as is the tough strap-leafed New World native **culantro**, which probably got its Spanish name from the aromatic resemblance to cilantro. (Coriander seeds have an entirely different set of volatiles; see page 282.)

Cilantro and culantro are very popular in much of the world, notably southern Asia and Mexico, but they're despised by significant and vocal numbers elsewhere. The early English botanist Nehemiah Grew wrote that coriander leaves "stink so basely that they can hardly be endured," and modern haters often describe their flavor as "soapy." This difference in taste is likely due to genetic differences in sensitivity to the volatiles, cultural differences in exposure to them, and also to their unusual nature. They're neither terpenoids nor benzenoids, the usual herb volatiles, so they don't immediately suggest aromatic plants. Instead, these aldehydes are commonly encountered in everyday life as breakdown fragments of the long-chain lipids in soaps and cosmetics: materials that are inedible and distasteful. They're also emitted by various insects, including stinkbugs. No wonder they can take some getting used to! Fortunately for people who dislike cilantro, and unfortunately for those who like it, aldehydes are intrinsically reactive molecules, so they disappear soon after they're released by pounding or cooking. A tip for gardeners: the aldehyde content of cilantro plants rises as they develop, so the leaves smell mildest before flower buds appear, and strongest as the small green fruits are maturing.

A last bunch of celery-family herbs lies slightly apart from the rest to mark its special prominence in Western cooking. **Parsley** appears on many plates because its deep green color appeals to the eye and its leaves are sturdy and tolerate chopping and heat. It's also popular because its scent stands out among the plant smells that we choose to give added flavor to our foods. It's pronounced and at the same time generic: a green, fresh, woody quality that can contribute a welcome and unobtrusive accent to nearly any savory food, regardless of what other herbs and spices it might contain. That quality is created by a diverse blend of terpenoids, benzenoids, simple-chain aldehydes, and nitrogen-containing pyrazines, a volatile cocktail that both reinforces the defensive work of the green-leaf volatiles and meshes with their green smells.

Kitchen herbs: the daisy family

Now we pass to one last stand devoted to the herbs of a single plant family. It's the daisy family, and relatives of the green-vegetable artemisias, the strong-smelling chrysanthemums and mugworts (see page 245). Of the five bunches of herbs, only one is at all familiar, and it's an artemisian aberration that smells nothing like the others. **Tarragon** has a fresh anise scent that comes mainly from the benzenoid estragole with some support from another benzenoid, spicy methyl eugenol. The *Artemisia* species that gives us tarragon grows across the Northern Hemisphere and often has little or no scent; the culinary herb is a particular subspecies.

Wormwood is more in the artemisia mainstream, a Eurasian plant rich in woody and turpentine terpenoids, including cedar-sage thujone. It's too bitter for cooking but was valued in traditional medicines as a treatment for parasitic worm infections—hence its name, which it in turn gave to the herb-infused wine called vermouth (via the German form *Wermut*). Wormwood remains an important ingredient in some vermouths and in the distilled liquor absinthe.

Despite a name that suggests an Indian origin (see curry *leaf* on page 263), the

curry plant is a Mediterranean cousin of wormwood and tarragon, with a blend of terpenoids that mysteriously happens to approximate the smell of the base spice mix of many Indian dishes. Unfortunately, that blend fades quickly in the heat of cooking.

SOME HERBS IN THE DAISY FAMILY

Plant	Component smells	Molecules
French tarragon (*Artemisia dracunculus* var. *sativa*)	anise, clove, green, woody	estragole, methyl eugenol, ocimene, terpinolene
wormwood (*Artemisia absinthium*)	resinous, sage, cedar, woody, pine	myrcene, thujones, sabinene, pinene
curry plant (*Helichrysum italicum*)	pine, rose, cedar, thyme, fruity, lavender	pinene, neryl acetate, cedrene, thymol, pentanoate esters, bergamotene
pericón, Spanish or Texas tarragon (*Tagetes lucida*)	anise, herbal, resinous, woody	estragole, myrcene, germacrene
huacatay, anisillo (*Tagetes minuta*)	fresh, minty, woody, solvent	limonene, piperitenone, terpinolene, tagetone, ocimenone

The other two daisy-family herbs are American plants, both species of the marigold genus *Tagetes*. **Pericón or Spanish tarragon**, from Mexico and Central America, is a good approximation of true tarragon, with its dominant benzenoid estragole and supporting woody terpenoids. **Huacatay or Mexican tea**, a native of South America, has its own unique blend of terpenoids, fresh and mintlike, but accompanied by the medicinal tagetone characteristic of marigolds.

Other herbs from the Americas and Asia

It's remarkable that the several dozen herbs we've sampled so far come from just three of the four-hundred-plus families of flowering plants, and mostly from Europe and western Asia. The final three stands of herbs in our market display a selection from the rest of the plant kingdom, one cluster from the Americas, one from eastern

Asia, and the very last a handful plants that long ago spread across the Northern Hemisphere.

On the first stand are a couple of New World counterparts of Eurasian herbs, and a couple of more unique contributions. The familiar smells come from sister shrubs in the verbena family. The plant known as **Mexican oregano** spans the region from the southwestern United States to Central America, and it does indeed accumulate both carvacrol and thymol, the medicinal terpenoids found in oregano and thyme. The leaves of **lemon verbena**, a woody shrub native to southern South America, contain the same lemon-defining terpenoids as lemon thyme and lemon basil.

SOME HERBS FROM THE AMERICAS

Plant	Component smells	Molecules
Mexican oregano, orégano cimmarón (*Lippia graveolens*)	thyme, oregano, fresh	thymol, carvacrol, cymene, eucalyptol
lemon verbena, cedrón (*Aloysia citrodora*)	lemon, citrus, green	geranial, neral, limonene, methyl heptenone
epazote (*Dysphania ambrosioides*)	kerosene, pine, turpentine	ascaridole, pinene, myrcene, terpinene
hoja santa (*Piper auritum*)	warm, spicy, anise, pine	safrole, pinene, terpinene
tomato leaf (*Lycopersicon esculentum*)	green, citrus, turpentine, camphor, eucalyptus, clove	GLVs, limonene, phellandrene, caryophyllene, eucalyptol, eugenol

Far more unusual is **epazote**, whose name is the Spanish version of an Aztec word meaning "skunk sweat." It's a weedy plant in the amaranth family, which also includes stinking goosefoot (see page 203), and its smell is kerosene-like from an unusual terpenoid, **ascaridole**, that owes its name to its prowess at killing ascarid tapeworms. Epazote is an important herb in Mexican cooking, notably in bean dishes, as well as in folk medicine. **Hoja santa**, "holy leaf," comes from a New World sister species of Old World black pepper. Its large heart-shaped leaves are used to wrap various foods before cooking and impart the aromatic, near-anise warmth of safrole (see page 179).

Another distinctive aromatic native to Latin America is the **tomato** vine, whose hairy foliage leaves a strong smell on skin or clothes that brush against it. Tomato fruits are beloved around the world, but the leaves had long been thought to contain toxic alkaloids. Recent research has shown that their alkaloids not only are not toxic, but may even be beneficial. The volatile-rich leaves offer another variation on the green and terpenoid themes, with a chemical-medicinal mix of turpentine, camphor, and eucalyptus notes and a touch of the clove benzenoid eugenol. They can contribute another dimension when added at the last minute to sauces made from the tomato fruit.

We move on now to the next stand and a cluster of six East Asian herbs, very diverse in form and flavor. The first two are strongly citrusy. The fibrous base and long leaves of **lemongrass** are endowed with the truly lemony terpenoids neral and geranial, along with touches of pine and lavender. The **makrut or kaffir lime leaf** comes from a small citrus tree that may be in the ancestry of the trees that give us lime fruits (see page 320). Its leaves have strong green, citrus, and roselike floral qualities from the terpenoids citronellal and citronellol, which are named for citronella oil, an extract from sister species to lemongrass (see page 460). Makrut leaf lacks the specifically lemony terpenoids, so it and lemongrass are similar but different. They're often combined in Thai dishes, complementing each other with their largely non-overlapping sets of terpenoids. More delicately citrusy are **kinome or sansho leaves**, tender new leaves from the citrus-family tree that gives us tingly Japanese sansho pepper (see page 287). When the leaves are crushed they're dominated by GLVs and a grassy aroma; when they're slapped sharply but left intact for a garnish, the terpenoids are more evident.

SOME HERBS FROM EAST ASIA

Plant	Component smells	Molecules
lemongrass (*Cymbopogon citratus*)	lemon, green, lavender	neral, geranial, myrcene, linalyl acetate
makrut, kaffir lime (*Citrus hystrix*)	citronella, lemon, floral, rose, woody	citronellal, linalool, citronellol, geraniol, pinene, caryophyllene
kinome, sansho leaf (*Zanthoxylum piperitum*)	pine, citrus, woody, floral	pinene, limonene, phellandrene, citronellol

Plant	Component smells	Molecules
rau ram, Vietnamese coriander/mint, laksa plant (*Persicaria odorata*)	green, soapy, citrus, pungent, metallic	GLVs, dodecanal, decanal, undecanal, octadienone
curry leaf (*Murraya* or *Bergera koenigii*)	sulfurous, resin, citrus-floral, pine, eucalyptus	phenylethanethiol, pinene, linalool, eucalyptol
pandan leaf (*Pandanus amaryllifolius*)	basmati rice, popcorn, sweet, caramel, fruity, floral	GLVs, acetyl pyrroline, ethyl sotolon (maple furanone), nonanal

The smell of **rau ram or Vietnamese coriander**, a trailing weedy plant in the buckwheat family, is indeed very close to coriander leaves and culantro due to shared aldehydes and their distinctive fresh-green-soapy-stinkbug qualities. A more broadly appealing but less familiar flavor, nutty and woody, comes from the **curry leaf**, commonly used as a briefly toasted garnish on southern Indian dishes. It's borne on a small tree in the extended citrus family, though it's not a true citrus species itself. Curry leaf emits a rare benzenoid-sulfur volatile with a sulfurous, roasted/burnt quality accompanied by familiar piney, floral, and eucalyptus terpenoids. And the bladelike **pandan leaf** comes from a shrubby species of screw pine native to tropical Asia. It breaks from the usual terpenoid/benzenoid pattern with two unusual volatiles, one a caramel-like furanone more typical of fruits than leaves, the other nitrogen-containing acetyl pyrroline, slightly musty and spermous like its component pyrroline (see pages 181, 215), which is the distinctive note in basmati and jasmine varieties of rice and contributes to the smell of popcorn. Pandan leaf is widely used in India and East Asia to scent rice dishes and sweets.

Cosmopolitan herbs

To round out our worldwide tour of herbs, we finish at a stand with just three plants, all widely distributed across the Northern Hemisphere, one quite common in the kitchen and the others not.

Chives are relatives of onion and garlic, and their long, thin tubular leaves provide a mild version of the defining family sulfurousness when chopped and strewn on many dishes. They more closely resemble onions, while the **Chinese or garlic chive**, a flat-leaved species native to Asia, is indeed more garlicky. (Much more on onions and garlic in the next chapter.)

Various species of the low woody **wintergreen or teaberry** bush, a relative of the cranberry and blueberry, grow throughout North America and Asia. Most of them fill their small leaves and flowers with the benzenoid methyl salicylate, the specific smell of wintergreen, which finds its way not only into infusions of the dried leaves, but also into chewing gums, candies, mouthwashes, liniments, and household cleaning products.

Hops are climbing relatives of cannabis with species all over the Northern Hemisphere. They concentrate volatiles not in their main green leaves, but in terpenoid-rich glands on the leaflike bracts that cluster around their female flowers. Hops became important in the Middle Ages when European brewers, who typically made beer with *gruit*, a mixture of herbs that slowed spoilage, discovered that hop "cones" were effective on their own and contributed a pleasant aroma and bitterness. Today there are dozens of hop varieties for flavoring beers with different cocktails of terpenoids (humulene is named for the hop genus), and in some cases with sulfur volatiles as well. They contribute aromas that are variously resinous, flowery, and fruity. Hop varieties from Eurasia are said to be dominated by woody humulene; those from the New World, by resinous myrcene.

SOME COSMOPOLITAN HERBS

Plant	Component smells	Molecules
chive (*Allium schoenoprasum*); Chinese chive (*A. tuberosum*)	oniony, woody; garlicky, floral	methyl & propyl di- & tri-sulfides, farnesene; dimethyl & allyl di- & tri- & tetra-sulfides, linalool
wintergreen (*Gaultheria procumbens*)	wintergreen, fresh, cooling	methyl salicylate, limonene, pinene
hops (*Humulus lupulus*)	resinous, woody, floral, fruity, blackcurrant/cat pee, grapefruit	myrcene, humulene, caryophyllene, linalool, sulfur molecules (e.g., methyl sulfanyl pentanone, sulfanyl hexanol)

Of course, the plant family that includes hops is named for **cannabis**, or **marijuana**, a native of Central Asia and India whose renown has little to do with its smells! Nevertheless, THC and the other active cannabinoid molecules are produced in the same fuzzy-looking leaf glands that produce defensive volatiles, and they're constructed from terpenoid and benzenoid building blocks. The strong terpenoidal smell of cannabis isn't much appreciated in cooking, but we'll get to it in chapter 17 as it's going up in smoke.

Here endeth our survey of leafy greens and herbs, whose defenses provide an extensive arsenal for cooks to deploy in the service of our pleasure in eating. Of course the raw plant leaves we eat are also nourishing: they provide vitamins, minerals, fiber to feed our gut microbes, perhaps some useful microbes as well, and antioxidants and other phytochemicals that help optimize our metabolism. But humans can't live on salad alone. Other plant parts are far less fragile than leaves and far more concentrated sources of energy, protein—and stimulating defenses. On to the roots and seeds and spices.

Chapter 12 · EDIBLE ROOTS AND SEEDS: STAPLES AND SPICES

It is remarkable that the use of pepper has come so much into favor. In the case of some commodities their sweet taste has been an attraction, and in others their appearance, but pepper has nothing to recommend it. To think that its only pleasing quality is pungency and that we go all the way to India to get this! Who was the first person who was willing to try it on his food, or in his greed for an appetite was not content merely to be hungry? Both pepper and ginger grow wild in their own countries, and nevertheless they are bought by weight like gold and silver.

· Pliny, *Natural History*

Moving on in our virtual produce market, we pass from stands of green leaves to a less vibrant prospect: dirty, dun-colored swollen roots in various sizes and shapes, pile after pile of nuts and grains large and small. Despite their unimpressive looks, these underground organs and seeds are among the most valuable plant materials we have. While some are as dull in the mouth as they are to the eye, others are explosively flavorful and aromatic: special attractions for any smell explorer.

Plant seeds and starchy underground growths provide some of humankind's most nourishing foodstuffs. Wheat, rice, corn, coconuts, potatoes, and sweet potatoes are all concentrated sources of energy and body-building materials. Their cultivation made possible the first large human settlements and the rise of civilizations. These mainstays generally have little smell when raw and remain mild even when cooked, perhaps a desirable characteristic for any food eaten in quantity. Imagine making a winter's breakfast not with oatmeal but with a steaming bowl of black-peppercorn porridge!

But peppercorns ahoy, just a few tables down from the oats: they're also mostly seed. Like the previous chapter's green herbs, peppercorns and other spices are flavorful plant materials, but they dry out naturally on the plant or can be dried without losing their intensity, and they keep well enough to have been items of long-distance trade millennia ago—the term *spice* comes from the Latin for "goods." Most spices are small, seedlike dry fruits with the actual seed within (black peppercorns), but some are true seeds (mustard), some are rootlike (ginger), some are tree bark (cinnamon), some are flower parts (saffron), and some are fruit pods (vanilla). In this chapter, we won't worry about botanical exactitudes and discrimination by organ. Here *root* and *tuber* and *seed* are used as they are in common, loose parlance, and our spice market will carry all kinds.

The story of spices is vast and fascinating. It involves ancient civilizations, trade routes between Africa and China and throughout the mainland and islands of Southeast Asia, the voyages of discovery led by Columbus and Vasco da Gama, subsequent explorations and depredations by the European sea powers, and world-changing breaches of geographical, biological, and cultural boundaries. It touches on millennia of religious practices, medical theories, cooking traditions, and their mutual influences.

In light of all this rich history, the Roman natural historian Pliny's rhetorical question sounds small-minded and moralistic: why pay princely sums for mere irritation? In fact, two thousand years later, it still gets to the heart of what spices are all about. There's no mystery about the value of life-sustaining grains and tubers, but who was it who first doctored food not with appealing sweetness but with pungency? And why?

We know today that whoever they were, the first spice eaters lived thousands of years before there was a Rome—and their likely reason was and remains a good one.

In 2013, archaeologists reported finding clay pots in northwestern Germany and Denmark from about six thousand years ago, right around the beginnings of settled agriculture, that contain remains of meat, fish, starchy foods—and wild mustard seeds. In the Americas, settlements at least six thousand years old have yielded the remains of chili peppers along with maize and other subsistence foods. Clearly our ancestors chose to eat local pungent plants long before any trade in exotic goods. And not only pungent plants: such aromatics as coriander and cumin seeds have been found in sites just as old in the Middle East; dill seed, in Swiss lake villages almost as old; both ginger and turmeric, in Indus civilization sites inhabited five thousand years ago.

So these irritating and aromatic materials were eaten from the earliest days of settled life. None is especially nourishing, which suggests that their original value was simply irritation and aromas. They brought sensory stimulation and interest to the daily necessity of eating. Hunter-gatherers ate a varied diet; the rise of cereal and animal agriculture narrowed the range of sensations that food could provide, and intensely flavored herbs and seeds and roots could compensate for its monotonous blandness. Once valued for their distinctive flavors, spices could take on all kinds of religious, medicinal, social, and economic significance.

The progress of trade since Columbus and da Gama has emptied spices of much of that significance. They've become everyday ingredients, notable mainly as flavor landmarks for particular culinary traditions: cumin for the Middle East or Mexico, caraway for northern Europe, ginger and star anise for southern China, complex blends for India. In modern industrial foods, they're often present only by proxy, in the form of synthetic imitations. But to the smell explorer they remain intriguing inventions of Hero Carbon, materials in the making for millions of years through the same design process that made us, more varied and variable and interesting than any imitation.

On, then, to virtual tables of roots and seeds both bland and breathtaking. Again, take advantage of your memory and kitchen and the shopping made possible by modern trade. Try to taste the real things as you read so that these volatile rosters can really inform your experience, as you taste and afterward.

Underground vegetables

Plants depend on extensive underground structures to anchor and support themselves, to absorb water and minerals, and sometimes to produce offspring without having to make flowers and seeds. Most important for us, some plants develop organs to store surplus chemical energy and carbon chains generated by the leaves during photosynthesis. These organs include enlarged roots or lower stems, both sometimes called *tubers*, as well as bulbs, which are lower stems with stubby leaves surrounding a new bud. Many are valuable nourishment for us—and for soil-dwelling animals and microbes. So as they do for their leaves, plants often defend their underground organs with chemical weapons. Some carry toxins or bitter or astringent molecules that aren't volatile and don't smell, but most rely on the same volatile defenses that aboveground parts do. We usually eat many of these vegetables cooked, and of course cooking changes their flavors. We'll sniff at those transformations in chapter 18. In this section of the market, we'll sample some of our most common roots and tubers and bulbs as they smell fresh from the ground and served raw.

The first stand has a few of the starchiest and blandest: three unrelated tubers from the New World. The **potato** is far more familiar cooked than raw but has a characteristic smell, especially when shredded, from notes generated by its strongly active lipid-oxidizing enzymes, which break long-chain molecules from cell membranes into short-chain reactive aldehydes that inhibit microbes. It also has earthy, bell-pepper, and green-pea notes from nitrogen-containing pyrazines that it shares with those vegetables. The **sweet potato** is also rich in simple aldehydes, but it includes a couple of benzenoids with nutty and honey notes. And the knobby **sunchoke or Jerusalem artichoke** owes its woody, herbaceous smell to terpenoids, sweetened with the benzenoid phenylacetaldehyde.

SOME ROOT VEGETABLES FROM THE AMERICAS

Tuber	Component smells	Molecules
potato (*Solanum tuberosum*)	fresh, fatty, mushroomy, whisky, earthy	hexanal, heptanal, octenal, decadienal & other aldehydes, methylbutanol, isobutyl & isopropyl methoxypyrazines
sweet potato (*Ipomoea batatas*)	green, mushroomy, fatty, cocoa, almond, honey	C6–C10 aldehydes, C6–C8 alcohols, methylbutanal, benzaldehyde, phenylacetaldehyde
sunchoke (*Helianthus tuberosus*)	woody, camphor, minty, honey	bisabolene, bornyl acetate, verbenone, menthadienol, phenylacetaldehyde

Now comes a larger stand of similarly mild Old World roots, all members of the celery family. **Celery root or celeriac**, really a swollen stem, shares the distinctive benzofuran phthalides with celery stalks and leaves (see page 257), but muffles them with citrusy and piney terpenoids. **Carrots**, mostly bright orange but sometimes red or purple or pale, stock up mainly on woody, piney terpenoids, with a touch of the nutmeg-like benzenoid myristicin. **Parsley root** has a similar set of volatiles, along with a weakly parsley-like benzenoid called apiole, but none of the terpenoids that help give parsley leaves their distinctive aroma. **Cilantro root,** a frequent ingredient in Thai aromatic pastes, is dominated by the same simple aldehydes as its leaves, but their green and soapy qualities are balanced by a woody terpenoid, thujene. **Parsnips** are outliers in the celery family: they carry unusual esters and lactones that give them fruity and creamy qualities.

SOME EURASIAN ROOT VEGETABLES IN THE CELERY FAMILY

Root	Component smells	Molecules
celery root (*Apium graveolens* var. *rapaceum*)	fresh, citrus, pine, celery	limonene, pinene, terpinene, phthalides (benzofuranones)
carrot (*Daucus carota*)	woody, citrus, pine, turpentine, nutmeg	bisabolene, sabinene, caryophyllene, pinene, terpinolene, myristicin
parsley root (*Petroselinum crispum* var. *tuberosum*)	nutmeg, pine, woody	apiole, myristicin, terpinolene, phellandrene
cilantro root (*Coriandrum sativum*)	green, soapy, cilantro, woody	dodecenal, decanal, decenal, thujene
parsnip (*Pastinaca sativa*)	waxy, fruity, creamy, green, earthy	octyl butyrate & acetate, hexyl butyrate, ocimene, g-octadecalactone

The next stand of three other Eurasian roots is by far the most colorful in this section of the market, with deep reds, purples, and golds alongside both black and white. The **beet,** here in red, yellow, white, and red-white ringed varieties, is a member of the amaranth family, related to spinach and the herb epazote. Beets are aromatized with a mixture of benzenoids and terpenoids. The most important and unusual is the terpenoid geosmin, which is one of the major volatiles in moist soil; we therefore experience it as earthy and musty—not an appealing quality for many people! (Much more about geosmin in chapter 15.) Biologists long assumed that beets get their geosmin from the soil they grow in—until experiments demonstrated in 2003 that they can make their own. More recent studies have found that some varieties produce a lot, others very little, with the highest levels in the roots' outer couple of millimeters. Sample the lighter colors first: the dark red varieties usually smell earthiest.

BEETS, TURNIPS, RADISHES

Root	Component smells	Molecules
beet (*Beta vulgaris*)	almond, honey, pine, eucalyptus, earthy	benzaldehyde, acetophenone, pinene, eucalyptol, nonanal, geosmin
turnip (*Brassica rapa*)	fresh, grassy, fruity, floral, sulfurous, pungent	limonene, hexenol, hexenyl acetate, geranyl acetone, dimethyl disulfide, isothiocyanates
radish (*Raphanus sativus*)	pungent, sulfurous, garlicky, meaty, cooked onion	isothiocyanates, dimethyl sulfide, dimethyl tri- & tetra-sulfides

Alongside the beets are two vegetable roots that begin to hint in the direction of spiciness. **Turnips** are white flattened spheres with a red or purple blush, and this selection of **radishes** includes small salad types with red or white or purple skins, long white Japanese daikons, and large black-skinned "Spanish" types. All are members of the cabbage family, a clan that defends damaged tissues with enzyme-generated volatiles that contain both sulfur and nitrogen, the isothiocyanates (see page 182). These molecules are sulfurous-smelling and also variably pungent, depending on their particular structures: some sting both the nose and the mouth, giving mustard and horseradish their power. As does cabbage itself, turnips stay well on the vegetable side, with mild grassy and fruity qualities and only a touch of pungency. But radish pungency can range from a touch to a punch.

Pungent roots and bulbs: the cabbage and garlic families

On the neighboring stand are two of the cabbage family's most pugilistic roots, so powerful that they're used only sparingly as condiments—like the seeds of their fellow family members the mustards (see page 287). **Horseradish** and **wasabi** emit more isothiocyanates than any other volatiles, and some of the most irritating.

Wasabi, a small greenish root native to Japan and a standard condiment for raw fish served as sushi and sashimi, has a more diverse aroma mix, with celery and sweet-creamy notes, than the larger off-white Central Asian horseradish. But it's harder to cultivate, expensive, and so is usually imitated with green-tinted horseradish. Both roots are at their most complex and interesting when freshly grated, but they quickly change as their components react with each other and escape into the air. Their isothiocyanates are so irritating and abundant and volatile that simply breathing in through the mouth while eating them can sting our airways and cause coughing and choking. The remedy: inhale fresh air through your nose, then close off the back of your throat and exhale the irritants through your mouth.

One stand over, rivaling horseradish and wasabi for potent and distinctive chemical weapons, are piles of our most familiar and strongest-smelling bulbs, also sulfur specialists, so weaponized that we either use them in small doses or defuse them by cooking. **Onions** and **garlic**, sister species in the genus *Allium*, develop bulbs that emit complex mixtures of sulfur volatiles when they're cut or crushed (see page 183). Some of these volatiles are intensely irritating as well as sulfurous, causing the eyes and mouth to burn. Because they're actively generated by enzymes in damaged tissues, and the enzymes are inhibited by acids and inactivated by high heat, cooking and pickling tone down the pungency and sulfurousness.

SOME BULBS AND ROOTS IN THE CABBAGE AND GARLIC FAMILIES

Root, bulb	Component smells	Molecules
horseradish (*Armoracia rusticana*)	pungent, green, horseradish, watercress	allyl isothiocyanate, phenethyl isothiocyanate
wasabi (*Wasabia japonica*)	pungent, green, metallic, sulfurous, celery-like, creamy	allyl, pentenyl, & hexenyl isothiocyanates, octadienone, methyl butenethiol, methyl decalactone, vanillin
onion (*Allium cepa*)	sulfurous, pungent, oniony, cooked onion, cooked vegetable, meaty	thiosulfinates, dimethyl di- & tri-sulfides, thiols
garlic (*Allium sativum*)	pungent, sulfurous, garlicky, cooked vegetable, cooked onion, meaty	allicin (thiosulfinate), dimethyl sulfide, dimethyl di- & tri-sulfides

Aromatic roots: the ginger family, licorice, valerian

With the next stand, we shift shapes and smells pretty dramatically to irregular and terpenoidal roots. The first set are members of the ginger family, ginger itself and its cousins turmeric and galangal, all originally from East Asia. They grow in ramifying masses that are likened to hands and fingers. Ginger and turmeric are among the earliest spices known to have been used in the prehistoric Indus River civilization; later the dry roots were important items of trade with the West. All three are somewhat pungent, but the responsible molecules aren't volatile, so they irritate only the mouth, not the eyes and airways. Their volatile defenses are mainly terpenoids, some unusual and some shared with many other plants.

Versatile **ginger** emits a distinctive, fresh-smelling terpenoid dubbed zingiberene along with fresh-woody pinene and eucalyptol, as well as lemony neral and geranial, floral linalool, and a woody note from terpinolene. **Turmeric**, a mainstay in Indian and Thai cooking, comes in both yellow and white species, the white sometimes called zedoary. Both carry several unusual spicy-woody sesquiterpenoids, including several named after them, the turmerones. Eucalyptol enlivens fresh turmeric root but is absent in the dried version. **Galangal**, plumper and more fibrous than its relatives, is an essential component of many Thai curry pastes. Its prominent eucalyptus, turpentine, and camphor terpenoids give it a fresh, bracing, medicinal quality, with sweet and woody notes from a terpenoid ester and the unusual sesquiterpenoid guaiol (not to be confused with the smoky phenolic guaiacol).

SOME AROMATIC ROOTS IN THE GINGER FAMILY

Root	Component smells	Molecules
ginger (*Zingiber officinale*; Asia)	lemony, woody, ginger	neral, geranial, pinene, eucalyptol, linalool, terpinolene, zingiberene
turmeric, zedoary (*Curcuma longa, zedoaria*; Asia)	woody, warm, spicy, nutty, fresh	turmerones, zingiberene, bergamotene, vinylguaiacol, eucalyptol
galangal (*Alpinia galanga*; Asia)	eucalyptus, turpentine, camphor, fresh, woody	eucalyptol, pinene, terpineol, camphor, fenchyl acetate, guaiol

The next stand is sparsely laden with two unrelated, odd, fibrous aromatic roots native to Eurasia. **Licorice** names the thin, twisted, woody roots of a plant in the bean family. It's been cultivated for thousands of years for the intensely sweet liquid that can be extracted from it (the name comes from the Greek for "sweet root"), and for its purported medicinal value (it does raise blood pressure). Raw licorice emits an eclectic blend of terpenoids and benzenoids with herbal (thyme, oregano) and spicy (anise, clove) as well as floral and beany qualities. When boiled to make the syrup for licorice confections, these are largely replaced by sweet caramel, buttery, and vanilla aromas. Most commercial licorice candies and confections enliven this undistinctive flavor with added doses of either the anise benzenoid anethole, or—mainly in Scandinavia—with ammoniacal ammonia derivatives: very much an acquired taste!

LICORICE AND VALERIAN ROOTS

Root	Component smells	Molecules
licorice root (*Glycyrrhiza glabra*); licorice extract	green, thyme, floral, beany, anise, clove; sweet fenugreek, caramel, butter, smoky, vanilla	nonadienal, carvacrol, thymol, linalool, methoxypyrazines, anethole, estragole, eugenol; sotolon, furaneol, diacetyl, guaiacol, vanillin
valerian (*Valeriana officinalis*)	cheesy, sweaty, rancid, camphor	valeric & isovaleric acids, bornyl acetate, acetoxyvaleranone, valerenol

Valerian root isn't really an edible root, but the smell explorer should know it: it's the natural material in which five-carbon valeric acid was first discovered and from which it got its name. Valeric and isovaleric acids (*isovaleric* is a synonym for *methylbutyric*) both smell cheesy, sweaty, and rancid. The roots have been used in herbal medicines, and as food savers rather than foods, dried and strategically placed to repel insects, rodents, and even skunks. Not as pleasant a pest repellent for us as citronella and many other aromatics, but apparently effective!

Nourishing seeds: grains

We now cross a wide aisle in the market as our exploration of plant edibles turns from roots to seeds, from underground anchors to unanchored offspring. Seeds are lifeboat-like structures that transport embryos of the next generation out into the wide world, along with the compact food supply they need to get growing. Of course food for embryos is also food for microbes and animals, so plants have various strategies for defending their seeds against disease and predators. Chemical weapons and physical armor, hard or prickly outer layers, are two. A third and very different tack is to flaunt fecundity: to launch large numbers of small seeds from every plant and rely on the likelihood that at least a few will survive and thrive. This is the strategy that most members of the grass family use, and it's been highly successful. It's also why the grasses we call wheat, rice, and corn (maize) are such important crops throughout the world. Early humans learned to gather and store the bounty of dry, nourishing seeds from natural grasslands, then eventually cultivated them in numbers large enough to support settlements, then cities, then civilizations. Today the annual world production of the major grains is many thousands of trillions of seeds, enough to carpet Earth's entire surface by about an inch.

The first seed stand in our market is decorated with corncobs, the most familiar evidence of the grasses' strength-in-numbers strategy. The cobs encircle nine zip-top bags of dry seeds and flours representing some of our favorite grains. Some of the seeds and flours are brownish from the tough protective coat or bran that surrounds grass seeds; some are off-white "refined" samples with the bran removed. We don't normally experience grains in the raw, but they do have distinctive smells that simple boiling amplifies and rounds out, while more elaborate cooking and fermentation methods transform them, as we'll see in chapters 18 and 19. Here, stick your nose into the bags that grains often come in, which help trap the volatiles, and then crunch a few seeds and lick finger-dustings of flour.

You'll find that the smells are generally mild and faint. All the samples emit aldehyde chains six to ten carbon atoms long, fragments broken from longer lipid chains that generally have green, fatty, and metallic smells. We perceive low levels of these mixed aldehydes as nondescriptly pleasant. But as grains sit in storage for weeks and months, more of the long chains are broken apart by exposure to oxygen and light, and their fragments accumulate. High levels of short-chain aldehydes smell cardboard-like, and define the quality we call *staleness* (see page 536).

Like other common grains, **wheat** carries traces of the sweet-smelling furanone sotolon, echoing caramel and the spice fenugreek (coming up later in this chapter).

The bran of whole grains and their flours is often rich in phenolic ring molecules that help to strengthen seed coats as well as provide a first layer of chemical defense. **Whole grains and flours** therefore tend to have a distinctive aroma from traces of such carbon-ring volatiles as vanillin, honey-like phenylacetaldehyde, and spicy-smoky guaiacol and vinylguaiacol. **Rye** is unusual for its prominent mushroomy and cooked-potato volatiles. The signature aroma of **buckwheat** includes an unusual wintergreen-like benzenoid and spicy-smoky vinylguaiacol, while the oatiness of **oats** appears to come from one particular and unusual nine-carbon aldehyde with three double bonds. The essence of **barley** arises from short branched-chain aldehydes that are especially prominent in malt syrups made from the germinated grain—and in cocoa (see pages 576, 530).

And the corniness of corn? Surprisingly, I haven't found any published studies of raw corn kernels or grits or flour.

SOME GRAINS

Grain	Component smells	Molecules
wheat, whole (*Triticum aestivum*)	vanilla, fatty, caramel, fenugreek, sweaty, potato, honey	vanillin, C9 & C10 aldehydes, sotolon, methylbutyric acid, methional, phenylacetic acid
wheat, refined	vanilla, caramel, fenugreek, fatty, honey	vanillin, sotolon, C9 & C10 aldehydes, phenylacetic acid
rye, semi-refined (*Secale cereale*)	cooked potato, mushroomy, caramel, fenugreek, fatty	methional, octenone, sotolon, C9 & C10 aldehydes
buckwheat (*Fagopyrum esculentum*)	medicinal, wintergreen, caramel, fried, honey, clove, smoky	salicylaldehyde, furaneol, decadienal, phenylacetaldehyde, vinylguaiacol
oats (*Avena sativa*)	sweet oat, cooked apple, cheesy, vanilla	nonatrienal, damascenone, butyric acid, vanillin
barley (*Hordeum vulgare*)	malty, cocoa, fatty	methylbutanals, decanal
rice, brown (*Oryza sativa*)	fruity, vanilla, clove, smoky, caramel, fenugreek	aminoacetophenone, vanillin, vinylguaiacol, sotolon

Grain	Component smells	Molecules
rice, fragrant (basmati, jasmine)	pandan, popcorn	acetyl pyrroline
rice, black	pandan, popcorn, smoky, rose, citrus peel	acetyl pyrroline, guaiacol, nonanal, decanal

Rice is remarkable for offering varieties that are prized specifically for aromas that you can smell the moment you put your nose into the bag. These basmati (from the Hindi for "fragrant") and jasmine (Thai, for the fine white color) and popcorn (American, of course) varieties emit the distinctive nitrogen-containing molecule acetyl pyrroline, which also happens to be formed by heat when popcorn is popped and is the main volatile in pandan leaves, the Asian herb sometimes cooked with rice to intensify its scent (see page 263). It's also in plain white rice, but in barely detectable traces, masked by the generic grain aldehydes. By itself, the pyrroline-ring portion of acetyl pyrroline has an animal, spermous smell, and this facet sometimes appears in basmati and pandan. Once you've bought basmati or jasmine rice, keep it cold to slow the inevitable loss of its special fragrance. Black or "forbidden" rice is sometimes well endowed with acetyl pyrroline but mixes it with smoky guaiacol and floral, fruity chains into an aroma all its own.

Some beans and the peanut

Over to the next stand, also covered with piles of seeds, many of them bigger and plumper than the grains and more eye-catching, with reds and blacks and many shades of white and brown, sometimes dappled with more than one color. These are seeds from members of the bean family, the legumes. Though not grown in nearly the Earth-carpeting numbers as the grasses, they too provide substantial nourishment throughout much of the world. They're significantly richer in protein than the grains and are often eaten along with them. Soybeans are also a good source of oil. Beans of many sorts, as well as peas and lentils, are usually dried and then cooked, though some are eaten when mature but still moist. Most share a common family quality that's often described as "beany": it arises from a particular mix of short-chain aldehydes generated by the seeds' own active enzymes.

There are also nitrogen-containing methoxypyrazines that are characteristic of fresh green beans and peas and so evoke them, but they're usually more prominent in the fresh pods than in seeds that have matured and dried out. The distinctive aldehydic signature has been a challenge for manufacturers to erase from pea protein extracts, which can replace animal proteins in vegetarian imitations of meat and eggs, but whose smell gives their beany-peasy origins away.

One lesser-known bean has its very own flamboyant signature. The **petai or stinky bean**, borne on a large tree related to the mimosas and a popular green vegetable in Malaysia and Indonesia, is infamous for its strong sulfurousness, which persists long after eating. It's a true virtuoso of sulfur volatiles, forming rings or chains with three, four, and five sulfur atoms. I've found petai beans in the frozen section of Asian groceries.

Last in this selection of beans is the odd one out in the family. The South American **peanut or groundnut** develops underground, where the bushy parent plant buries its pea-pod-like fruits. Peanuts resemble tree nuts in their high oil content and their tenderness compared to most other dry beans, but their raw smell is strongly reminiscent of green peas and beans from shared methoxypyrazines.

SOME BEANS AND THE PEANUT

Legume	Component smells	Molecules
red bean (*Phaseolus vulgaris*)	green, mushroomy, earthy	C5–C10 aldehydes & alcohols, octadienone
pea (*Pisum sativum*)	pea	isopropyl methoxypyrazine
petai bean (*Parkia speciosa*)	grassy, mushroomy, sulfidic, garlicky, rotting	hexanal, dimethyl trisulfide, trithiolane, hydrogen sulfide, methanethiol
peanut (*Arachis hypogaea*)	pea, earthy, bell pepper, fatty, metallic	isopropyl & isobutyl methoxypyrazines, decenal

Tree nuts

On the next table is another arrangement of dry seeds, again dull in color, this time ranging in size from fingernails to cupped hands. These are nuts: in common rather than strict botanical parlance, edible seeds enclosed in hard shells. They tend to store energy for the embryo in oil rather than starch, so they're a more concentrated source of calories than are grains and legumes. The long carbon chains of oil molecules are vulnerable to attack by oxygen and light, which over time generate the mix of aldehyde fragments that we perceive as rancidity (see page 536), so nuts are best stored undisturbed in their shells.

Pine trees grow across the Northern Hemisphere, and a number of species bear cones with edible seeds that are large enough to bother extracting. Diminutive **pine nuts** bear the terpenoid hallmarks of their conifer heritage, with pine, fir, and citrus aromas. The pistachio tree isn't a conifer, but it is a close western Asian relative of the terebinth tree, which provided ancient peoples with a useful aromatic resin and thereby lent its name to both turpentine and terpenoid molecules (see page 417). **Pistachio nuts**, often sold in their gaping shells, express that family relationship as well, with pine, fir, turpentine, and citrus volatiles like those in pine nuts, plus lactones and an ester that can shade the aroma surprisingly close to that of terpenoid-rich mango fruits (see page 328). I was struck by that fruity quality when I tasted pistachios grown for the Turkish baklava industry, harvested before full maturity and therefore small and intensely green and flavorful. I brought some to a chemist friend then at the Davis campus of the University of California, Arielle Johnson—every smell explorer should have such a friend!—and she determined that they had two to ten times the levels of major volatiles compared to everyday California pistachios. Fresh young pistachios are worth seeking out.

Almonds come from Eurasian trees that are close relatives of the stone fruits, peaches and apricots and cherries, and they share in their pits notable levels of benzaldehyde, the aroma-defining volatile in almond extract. Ordinary "sweet" almonds are fairly bland, and most carry only a hint of benzaldehyde compared to the fruit pits and "bitter" almonds. Benzaldehyde is part of a seed defense system that also generates potentially deadly cyanide, so only low-benzaldehyde nuts are safe to eat in any quantity. Bitter almonds and stone-fruit pits are sometimes used much like a spice, in small quantities to contribute their aroma to marzipan and other nut pastes.

Various species of the bushy hazel tree grow throughout the Northern Hemisphere, and their nuts are variously called **hazelnuts**, **cobnuts**, or **filberts**. The cultivated Eurasian varieties are flavored with the unusual and char-acterizing molecule aptly known as **filbertone**, a seven-carbon chain with a one-carbon branch. Toasting the nuts boosts filbertone levels and their aromatic distinctiveness.

The walnut family has representatives native to Eurasia and the Americas. Their various nuts share a peculiar puzzle-like interior that locks the irregular nut meat into its shell, but the flavors are very different. The common **walnut** comes from a Eurasian species sometimes called English or Persian; its mild aroma is the product of several simple-chain aldehydes and alcohols. **Black walnuts** come from a notably hard-shelled walnut species native to the American East and Midwest, and have an earthy, woody, fruity character arising from a number of unusual esters and a sweet, musty furanone. Other American trees in the walnut family are the source of pecans and hickory nuts. **Pecans** have a distinctive fatty, spicy, sweet, creamy smell that may come in part from lactones, volatiles that we most commonly encounter in dairy products, in peaches, and in the last and largest nuts on the stand: **coconuts**, natives of tropical Asia. Their milky-white meat is rich in a particular group of lactones that generate their landmark smell, often simulated in sunscreens and lotions. Think of the pecan and coconut as offering temperate and tropical studies in lactones.

SOME TREE NUTS

Nut	Component smells	Molecules
pine nut (*Pinus* species)	pine, fir needle, camphor, citrus	pinene, limonene, hexanal, camphene, carene
pistachio (*Pistacia vera*)	pine, turpentine, fresh citrus, creamy, fruity	pinene, terpinolene, limonene, g-butyrolactone & hexalactone, ethyl acetate
almond (*Prunus dulcis*)	almond extract, floral, solvently, green	benzaldehyde, benzyl alcohol, phenylethanol, toluene (methylbenzene), hexanal
hazelnut, filbert (*Corylus* species)	hazelnut, sweet, hay	filbertone (methyl heptenone), dimethoxybenzene
walnut (*Juglans regia*)	green, fruity, whiskey	hexanal, pentanal, hexanol, pentanol, ethyl toluene

Nut	Component smells	Molecules
black walnut (*Juglans nigra*)	earthy, woody, fruity, caramel, musty	methyl isovalerate, methyl hexenoate, methylethyl hexanoate, mesifuran
pecan (*Carya illinoinensis*)	grassy, fatty, sweet, creamy, coconut, buttery	hexanal, g-decalactone & massoia lactone, diacetyl
coconut (*Cocos nucifera*)	creamy, coconut, nutty, fruity, fatty	d-octalactone & decalactone, ethyl octanoate & decanoate

Spice seeds and a resin: the celery family

Another broad aisle marks our passage from the mildness of nourishing seeds to the aromatic intensity of spice seeds: some of the durable materials that cooks use to lend flavor to nourishment. Until global trade turned tropical spices into affordable commodities, most people in the Western world cooked with aromatics from their own region. The ten piles of small seeds on the first stand come from related plants indigenous to Europe and western Asia. They're members of the same celery family that provides many aromatic herbs (see page 256), and some of these seeds are produced by those herbs. Celery-family "seeds" are actually dry fruits just a couple of millimeters in size, with a woody outer layer instead of a fleshy one, and numerous volatile-filled canals surrounding and protecting the true tiny seed within. Crushing the whole package releases its volatiles.

Celery, dill, and caraway seeds all share the fresh, citrusy quality of the terpenoid limonene. **Celery seeds** owe their specific scent to the same unusual volatiles that dominate fresh celery, the benzofuranone phthalides, along with an herbal terpenoid, selinene. **Caraway** is defined by its distinctive terpenoid carvone (not the mirror-image mint carvone), and vice versa: the molecule's smell is usually named by the spice. **Dill seeds** also contain the caraway carvone, but it's less prominent, evenly blended with other terpenoids.

Then there's the anise subfamily, with two different benzenoid volatiles that have an anise quality: anethole, which also has sweet and medicinal facets, and estragole, an important volatile in some basils, which is also green and minty. **Anise seed** combines anethole and estragole, and includes a hint of cinnamon and clove from methyl eugenol. **Fennel seed** is also distinctly anise-like, but comes in two types. In the first, including "sweet" fennel, sweet-medicinal anethole accounts for as much as 90 percent of the volatiles; in the second, including "bitter" varieties, the proportion is more like 65 percent, with minty, camphor-cedar-like fenchone contributing up to 20 percent. Anethole is sweet to the taste as well as aromatic, and both anise and fennel seeds are prominent in the Indian breath-freshening nibbles called *mukhwas*.

SOME AROMATIC SEEDS AND A RESIN IN THE CELERY FAMILY

Spice	Component smells	Molecules
celery (*Apium graveolens*)	citrus, green, celery	limonene, selinene, phthalides (benzofuranones)
caraway (*Carum carvi*)	caraway, fresh, shiso	caraway carvone, limonene, perilla aldehyde
dill (*Anethum graveolens*)	citrus, caraway, spearmint	limonene, caraway carvone, carveol
anise (*Pimpinella anisum*)	anise, sweet, medicinal, cinnamon	anethole, estragole, methyl eugenol
fennel (*Foeniculum vulgare*)	anise, sweet, minty, camphor, fresh	anethole, estragole, fenchone, limonene
coriander (*Coriandrum sativum*)	floral, fruity, turpentine, camphor	linalool, geranyl acetate, terpinene, camphor, pinene
cumin (*Cuminum cyminum*)	cumin, sweaty, fatty, woody	cuminaldehyde, terpinenal, terpinene
black cumin (*Bunium bulbocastanum*)	cumin, sweaty, pine, woody	cuminaldehyde, terpinenal, terpinene, cymene
ajwain (*Trachyspermum ammi*)	medicinal, thyme, woody, pine	thymol, cymene, terpinene, pinene
asafetida (*Ferula assa-foetida*)	cooked onion, garlic, sulfurous	propenyl disulfides, di-, tri-, & tetra-sulfides, pinene

Two other celery-family seeds are foundational ingredients in Indian spice mixes. **Coriander seeds** are borne by the herb also known as cilantro, but have nothing like the green soapy smell of its leaves. The seeds are startlingly perfumed, flowery and citrusy thanks to the terpenoid linalool and a terpenoid ester, with turpentine and camphor in the background. They're also unusual in having a thick round fruit coat that's usually ground up with the two inner seeds and that has its own aroma, with less than a third of the seeds' linalool levels and proportionally more camphor, pinene, and woody caryophyllene. **Cumin**, another essential Indian spice and prominent in Mexican cooking from colonial times, isn't at all reminiscent of flowers or fruits. Its defining volatile, the unusual terpenoid cuminaldehyde, is aggressive and has an animal, sweaty quality. **Black cumin**, as its name suggests, has a very similar smell and is the seed of a sister species grown mostly in Central Asia. **Ajwain or ajowan** comes from that same region and is less known in the West than its relatives. Its smell is distinct: it's a seed version of the herb thyme, dominated by thymol.

After all the little seeds, there's a jumble of brownish chunks of a hard, shiny material, too hard to scratch into with a fingernail. Find the smallest piece, take a bite, and when you need relief, reach for the sweet fennel and anise. This is **asafetida**, the resin of a celery family member native to Iran and Central Asia, whose various names reflect its very particular and not especially appealing smell. The English is a hybrid of Persian and Latin meaning "stinking resin"; the French *merde du diable* and German *Teufelsdreck* both mean "devil's dung." All comprehensible: asafetida carries a remarkable range of sulfur volatiles, from simple and common sulfides to oniony and garlicky sulfide chains. The tall weedy parent plant exudes the protective resin when its lower stem is wounded; the resin is collected and dried and often sold in pulverized form, diluted with starch or flour. Asafetida is used in Central Asian and Indian cooking (its Hindi name is *hing*) much as onions and garlic are, to provide savory, meaty notes.

Unallied Eurasian spices:
fenugreek, mahleb, saffron

At the next stand, we tarry at the geographic border between West and East for four more aromatics from that region, three unrelated seeds and a dried flower. To begin, a relatively obscure spice and a major olfactory landmark.

Fenugreek is the small, oblong, tan seed of a plant in the bean family, often included in Central Asian and Indian spice mixes. It's scented in part by familial green-bean, green-pea notes from nitrogen-ring pyrazines, fatty lactones, and sweaty acids. But its most distinctive, sweet quality arises from the furanone sotolon (see page 172). Traces of sotolon are present in grains like wheat and rye, but apart from fenugreek, it's prominent only in certain cooked foods, so its smell is usually described with reference to caramel, brown sugar, and maple syrup. Millions of New Yorkers encountered sotolon in 2005 and 2006 as a citywide maple-syrup smell that wafted over from a New Jersey flavor factory processing fenugreek seeds; we'll encounter it often in chapters 18 and 19. There, to give plants credit for priority in creating it, I'll describe it as fenugreek-like! Unlike most spices, which lose some of their raw aroma and distinctiveness when heated, fenugreek's furanone levels are significantly boosted by heat, especially in acidic solutions (some sauces). We can smell like fenugreek for a while after eating it; either soto-lon itself or a similar-smelling derivative ends up in our sweat and urine.

EAST MEDITERRANEAN AND WEST ASIAN SPICES:
FENUGREEK, NIGELLA, MAHLEB, SAFFRON

Spice	Component smells	Molecules
fenugreek (*Trigonella foenum-graecum*)	maple syrup, sweaty, green bean, coconut	sotolon, butyric & methylbutyric acids, methoxypyrazines, g-nonalactones & hexalactones
nigella, kalonji, onion seed (*Nigella sativa*)	woody, medicinal, spearmint, anise	cymene, thymoquinone, carvacrol, anethole
mahleb (*Prunus mahaleb*)	hay, almond extract, resinous	coumarin, benzaldehyde, benzyl alcohol
saffron (*Crocus sativus*)	fresh, medicinal, woody, sweet, green, tea, tobacco	safranal, isophorones

Alongside the brown fenugreek is a heap of much smaller, black, angular seeds. **Nigella** is in the buttercup family and related to a number of garden flowers; its seeds are often sprinkled on breads. They're relatively gentle in flavor, woody, with a medicinal accent from thyme-oregano terpenoids, and sometimes hints of spear-mint or anise.

Next to the nigella are round, cream-colored seeds a couple of millimeters across. **Mahleb** is the seed of a tree, a sister species to the almond and stone fruits, whose small fruit is often called a type of cherry. Like the bitter almond, the small mahleb seed is removed from the hard pit and used as a flavoring. It also contains almond-extract benzaldehyde, but mixes it with the sweet, haylike benzenoid coumarin, the main volatile in tonka bean (see page 291).

Last on this table: not seeds, but instead a very small pile of bright red threads, a half inch to an inch long. These are the dried stigmas (pollen-accepting parts) of the **saffron** flower, a species of crocus domesticated in Greece three thousand years ago for use both as a spice and as a colorant: its threads are rich in terpenoid pigments that dye both foods and fabrics a vibrant yellow-orange. The living saffron flower has a pleasant but unremarkable scent (see page 223), but drying causes a number of pigment and related terpenoids to be fragmented into unusual many-faceted volatiles with medicinal, woody, camphor-like, hay, musty, tea, and tobacco qualities. A unique spice and an expensive one: the threads are picked by hand on the day that each flower opens, and it takes hundreds of flowers to yield a gram of saffron.

Pungent spices worldwide: mustards and peppers

Another wide aisle and wide shift in focus: this time to the *spicy* spices, the seeds and fruits that sting as well as smell. The term "spicy" is often used to mean "hot"—pungent and pleasantly painful—and we use peppers and mustard to "spice up" everything from breakfast eggs to sandwiches to cookies. These materials are beloved for their pungency or tingling effect, but their smells are also worth noticing and contrasting, so this stand brings together samples from all over the world.

First on the stand are several piles of small round seeds, some yellow, others various shades of brown. **Mustard seeds** come from plants in the cabbage family that grow widely throughout Europe and Asia. Like that family's vegetables and herbs, they're defended with nitrogen- and sulfur-containing isothiocyanates that irritate the airways as well as the mouth and have a generically sulfurous smell. The seeds store them in a nonvolatile combination with other molecules and emit them when physical damage releases a volatile-liberating enzyme to do its work. The enzyme is inactivated by high heat, so toasting the dry seeds drastically reduces their pungency. The enzyme also requires moisture to function, so prepared mustards are made by grinding the seeds and wetting the powder to form a paste. Mustard pungency tends to fade with time unless the isothiocyanates are stabilized by lowering the pH, which is one function of vinegar in prepared mustards.

By contrast, the spices commonly called **peppers**—black, white, chili, and Sichuan, all piled here alongside the mustard—store pungent or tingling molecules in their active form, ready to attack even when dry. Those molecules—piperine in black and white pepper, capsaicin in chilis, sanshools in Sichuan pepper, all nitrogen-containing chemical relatives of one another—are much larger and less volatile than the isothiocyanates, with little or no smell of their own. Unless they're launched into the air by the high heat of cooking or by splashing from dishes as they're washed, they irritate only the mouth. All of these peppers start out as small, seed-containing fruits that dry easily for long keeping. The English word *pepper* comes from *pippali*, the ancient Sanskrit name for what we now call long pepper. Both long and black pepper are species in the large genus *Piper*, which belongs to one of the most ancient lineages of flowering plants. The other, unrelated "peppers" are so called for their resemblance to these models.

The first pepper pile collects dark pencil-thick masses a half inch to an inch long. This is **long pepper**, the fruit of a woody vine native to India and Southeast Asia that takes the form of many small berries fused together. It was apparently the first pepper in the ancient spice trades, later superseded by black pepper. Long pepper is relatively rare today but worth seeking out, for its historical significance and for the pleasant aroma that accompanies its pungency, woody and turpentiney, with a hint of clove.

Alongside the long pepper are the familiar fruits of its far more popular sister species **black pepper**. It's from the same homeland, also grows as a vine, and carries mainly woody, piney terpenoids, including the distinctly peppery sesquiterpenoid rotundone. The black outer coating is the dried remains of the unripe fleshy fruit that surrounded the seed when it was harvested. **White pepper** is made from the same fruits as black pepper, but by letting the pepper berries ripen and then removing the softened fruit coat before drying. The traditional practice is to wet the ripe fruits and let ambient microbes attack and digest the fruit layer so that it can be washed away. In the tropical heat it's easy for this controlled decomposition to get out of control, and this is why some batches of white pepper end up with barnyardy, fecal off-aromas. White peppercorns end up with pretty much the same volatiles and nonvolatile pungency as black pepper, but they don't speckle food with dark particles.

After the two true peppers are a couple of heaps of small mottled-looking spheres, one green and one red. **Sichuan pepper**, *hua jiao* in Chinese, and the Japanese equivalent, **sansho**, are the small dried fruits of Asian trees in the citrus family, though not citrus species themselves. Their main nonvolatile defenses, the sanshools, trigger very strange sensations in the mouth, not really pungent, but metallic and tingling and numbing. Both the nonvolatile and the volatile defenses are concentrated in the thin fruit layers surrounding the hard seeds, the mottling revealing storage glands that resemble the glands in the rinds of true citrus fruits. The volatiles reflect the parent trees' botanical affiliation, with floral-citrus notes on a woody and piney base, and noticeable differences among green and red, Chinese and Japanese varieties. (A number of other species are used in Korea, India, Indonesia, and nearby countries.)

PUNGENT SPICES: MUSTARDS AND PEPPERS

Spice	Component smells	Molecules
mustard seeds (*Brassica nigra* & *juncea, Sinapis alba*)	sulfurous	isothiocyanates, thiocyanates
pepper, long (*Piper longum*)	woody, terpenic, clove, citrus, fresh	caryophyllene, carene, eugenol, limonene, zingiberene
pepper, black (*Piper nigrum*; Asia)	peppery, floral, fresh, pine, woody, resinous	rotundone, pinene, caryophyllene, limonene, myrcene, linalool
pepper, white (*Piper nigrum*)	peppery, citrus, floral, pine, eucalyptus	rotundone, limonene, linalool, pinene, eucalyptol
carelessly processed	fecal, barnyardy	skatole, cresol
pepper, Sichuan green & red (*Zanthoxylum simulans* & *Z. bungeanum*)	floral, sweet, pine floral, fresh, pine, rose	linalool, terpineol linalool, eucalyptol, phenylethanol, limonene, myrcene
sansho (*Zanthoxylum piperitum*)	fresh unripe: floral, green, citrus, pine; ripe, dried: floral, green, fruity	geraniol, linalool, citronellal, myrcene, limonene, phellandrene; geraniol, linalool, citronellal, geranyl acetate, methyl cinnamate

continued

Spice	Component smells	Molecules
pepper, chili (*Capsicum*)	earthy, fruity, waxy, cedar, green, violet	octenol, damascenone, hexyl methylbutyrate, himachalene, safral, ocimene, toluene, methyl heptenone, ionone
pepper, pink (*Schinus molle* & *S. terebinthifolius*)	turpentine, fresh, citrusy, woody, herbal	pinene, limonene, sabinene, germacrene; phellandrene, carene, pinene, germacrene

After these fairly uniform piles of long and black and Sichuan peppers comes a bunch of elongated and variously reddish pods ranging in size from fingernail to whole hand. These are **capsicum or chili peppers**, from small annual plants native to the Americas, whose fruit layers are much larger than those of the other peppers. Their seeds and the seed-bearing inner ridge of tissue carry most of the pungent capsaicin, while the volatiles of the dried spice come mainly from fragments of the outer fruit, which is rich in red and orange carotenoid pigments. Strangely, the volatiles of dried chilis haven't been much studied, but it looks as though the main ones come from the terpenoid metabolic highway that leads to the pigments and from fragments of the pigments themselves, woody and floral and fruity.

Closing out the pepper stand is a heap of pink peppercorn-size spheres: **pink peppercorns**, of course! Like the chilis, they're from the New World, produced by a couple of different trees native to South America. They don't actually have any pungent defenses, but got their name from being about the same size as black pepper seeds and having a somewhat similar aroma derived mainly from fresh and woody terpenoids—though not black pepper's quintessential rotundone.

Asian spices: cinnamon, clove, cardamom

We move now to a stand with a half-dozen familiar aromatics from Asia, some seeds and some not, including a couple that are also somewhat pungent. First, reddish-brown cylinders a couple of inches long, some a single thick piece curled onto itself, some a roll of many papery layers. These are versions of **cinnamon**, the bark of trees in the laurel family native to Southeast Asia, whose smell is familiar from breakfast pastries and apple pies. We've already taken a whiff of them in the arboreal section of our imaginary garden (see page 197). The main and defining cinnamon volatile, the benzenoid cinnamaldehyde, also happens to be warming and pungent in high concentrations, and it's the key ingredient in the candies known as Red Hots. Different varieties of cinnamon have different levels of cinnamaldehyde, some smelling much like the one-note candy, others with more balanced, complex aromas, honey and hay and floral notes from related benzenoids and the terpenoid linalool.

Alongside the cinnamon quills is a pile of much smaller brownish-black oblongs with one enlarged bulbous end. These are **cloves**, the sun-dried unopened flower buds of a Southeast Asian tree in the myrtle family; its cousins include allspice as well as eucalyptus and guava, all strongly aromatic. Cloves get their defining and landmark smell from the benzenoid eugenol, whose nerve-numbing properties have been taken advantage of for thousands of years to treat toothache. Its continuing use in dentistry and mouthwashes can give the smell of cloves an unfortunate medicinal association. Cloves represent a remarkable investment of their trees in chemical defense; the dried flower buds are as much as 20 percent volatiles by weight, and up to 90 percent of that is eugenol. One or two cloves go a long way in cooking.

SOME ASIAN SPICES

Spice	Component smells	Molecules
cinnamon (*Cinnamomum verum*)	cinnamon, honey, flowery, woody, clove, hay	cinnamaldehyde 5–15%, cinnamyl acetate, linalool, caryophyllene, eugenol, coumarin
cassia, Indonesian cinnamon, Saigon cinnamon (*Cinnamomum cassia, burmannii, loureiroi*)	cinnamon, pungent; sweet, hay	cinnamaldehyde 15–20%, 20–50%, 55–70%; cinnamic acid, coumarin
clove (*Syzygium aromaticum*)	clove, sweet, floral, woody	eugenol, eugenyl acetate, caryophyllene
nutmeg, mace (*Myristica fragrans*)	balsamic, sweet, woody, spicy, sassafras, pine	myristicin, safrole, sabinene, pinene, terpinenol

continued

Spice	Component smells	Molecules
star anise (*Illicium verum*)	anise, medicinal, sweet, floral, green	anethole, anisaldehyde, foeniculin
cardamom (*Elettaria cardamomum*)	citrus, fresh, eucalyptus, lavender, floral	terpinyl acetate, eucalyptol, linalyl acetate, linalool

After the cloves come a few nonpungent Asian spices. There's a handful of inch-long brown seeds, each contained in a coarse, bright red netting. The seed is **nutmeg**, and **mace** is the garment that becomes visible when the surrounding fruit is ripe and splits open—it catches the eye of birds that take the fruit and drop the seed. The nutmeg tree is native to Indonesia and a member of the ancient magnolia lineage. Nutmeg and mace volatiles include common terpenoids but are dominated by two unusual benzenoids. Resinous-sweet myristicin is also a minor volatile in carrot and parsley, while safrole is the main volatile in sassafras root (see page 198).

Next to the nutmegs are smallish, brownish, rough-looking stars, with each of their eight rays partly split and showing a seed inside. This is **star anise**, the woody dry fruit of a small Chinese tree that, as its name indicates, is rich in the same benzenoid volatiles as anise seed, and in fact is now the primary commercial source of anise extract. It's the stars, not the seeds, that carry the volatiles. Star anise is a prominent spice in many Asian cuisines, particularly Chinese and Vietnamese.

The last spice on the Asian stand is a pile of smallish oblong flat-sided pods, some tan, some green. These are bleached and unbleached seed pods of **cardamom**, a sword-leaved, shrubby relative of the ginger plant native to India and Southeast Asia. It bears flowers on long stalks, and these flop onto the ground where its fibrous, flavorless green seed pods mature. Inside are seeds defended with a distinctive mix of terpenoids and terpenoid esters, surprisingly flowery and fruity given their intimacy with the soil, and with a fresh eucalyptus-camphor edge. (A number of other ginger relatives produce larger and darker pods that are also called cardamom; they have seeds that are more medicinal and less widely used.)

Spices from the Americas: allspice, achiote, tonka, vanilla

We come now to the last two stands of edible aromatics, the New World's few but outstanding contributions to the world spice rack. There are three piles of seeds on the first stand: in the first, the seeds are round, medium-size, and brown; in the second, they're small and orange-red; and in the third, large and beanlike. Round **allspice** got its name from the fact that it offers the benzenoid aromas of two major Asian spices, clove and cinnamon, in one compact dried fruit. It's borne on a tree in the clove family that's native to the West Indies. The cinnamon note comes not from cinnamon's primary cinnamaldehyde, but from one facet of clove-like, cinnamon-like methyl eugenol, so it's relatively subtle.

Orange-red **achiote** is the spice obtained from the **annatto** tree, a shrubby evergreen that bears prickly seed pods. Inside the pods are small seeds covered with an intensely orange-red, aromatic pulp. The dried pulp layer is used to color both foods and textiles, and to foods it brings a mild forest aroma from several terpenoids, including an unusual earthy sesquiterpenoid, **spathulenol**.

Tonka beans come from trees that are in fact in the bean family. They're rich in coumarin, the benzenoid that's found in clovers and has the sweet smell of cut clover and drying hay (see page 206; it's also in mahleb seeds and cinnamon). Tonka beans and extracts had long been used to flavor foods and tobacco—and replace costlier vanilla—until coumarin came under suspicion as potentially toxic. It's now illegal to sell tonka beans in the United States, though they're not hard to find. European countries allow them as an ingredient in foods and drinks, but set upper limits on total coumarin levels in foods (cinnamon is sometimes a significant source).

SOME SPICES FROM THE AMERICAS

Spice	Component smells	Molecules
allspice, Jamaica pepper (*Pimenta dioica*)	fresh, cinnamon, clove, pine, woody	methyl eugenol, eugenol, pinene, caryophyllene
achiote, annatto (*Bixa orellana*)	woody, resinous, earthy, fruity	humulene, pinene, spathulenol, methyl heptenone
tonka bean (*Dipteryx odorata*)	sweet, warm, hay, caramel	coumarin, hydroxymethyl furfural, methyl hydroxyphenyl propionate

Finally, the spices' last stand: three groups of long brown-black pods emitting an immediately recognizable aroma. These are versions of **vanilla**, whose flavor is so beloved and ubiquitous in the West that "plain vanilla" means something ordinary, not special. The true spice is in short supply and expensive; its flavor is everywhere only because chemists are able to approximate it with a synthetic mixture of its major volatiles, and sometimes simply with the one volatile that gives it its essential sweet character: the landmark benzenoid vanillin. Vanillin is rare elsewhere in the plant world; it lends touches of its sweet warmth to benzoin resins and to the wood of some oak species, where its levels can be raised by the heat treatments used to make wine and whiskey barrels.

Vanilla pods are the heat- and time-cured fruits of a vine in the orchid family that's native to the American tropics. They're long and thin, with a leathery skin enclosing a sticky mass of tiny black seeds. The key to their appeal is a rich set of benzenoids led by distinctive and sweet vanillin, creamy and milky vanillic acid, a couple of fruity and floral volatiles, and then several that are less intrinsically pleasant but whose smoky, medicinal, barnyardy notes provide depth and complexity. The best way to detect and appreciate those secondary smells is to sniff a bottle of artificial vanilla flavoring that's only or mostly vanillin, then some true vanilla extract or a freshly scraped vanilla pod. The deeper, edgier qualities of the real thing can be startling.

It's also startling to sample rare Tahitian vanilla alongside "plain" vanilla: its main vanilla quality is much weaker, and it has flowery, fruity, almond, and anise notes, all from its own set of benzenoids. Tahitian vanilla was found on its namesake Pacific island far from the American tropics, and its origins remain something of a mystery.

The current guess is that it's a hybrid of two or more different vanilla species that were brought from the Americas to the Pacific for cultivation. One of the possible parents, pompona vanilla, is also beginning to find its way into the food and perfume worlds; it carries some of the same benzenoids that dominate Tahitian vanilla.

SOME VARIETIES OF VANILLA

Vanilla species	Component smells	Molecules
common (*Vanilla planifolia*; tropical Americas)	vanilla, sweet, creamy, almond, smoky, tarry, barnyardy, floral	vanillin, vanillic acid, hydroxybenzoic acid & alcohol & aldehyde, guaiacol, phenol, cresol, anisyl alcohol
Tahitian (*Vanilla* x *tahitensis*; Pacific islands)	floral, fruity, almond, caramel, anise, vanilla	anisyl aldehyde & alcohol & acetate, methyl anisate, vanillin
pompona (*Vanilla pompona*; tropical Americas)	floral, sweet, spicy, resinous, anise, smoky, tarry, vanilla	octadienone, methyl cinnamate, anisyl aldehyde & alcohol & formate & acetate, guaiacol, phenol, vanillin

Among all the spices, vanilla is unusual for the elaborate process it's put through in order to intensify its flavor. Rather than simply allowing the pods to ripen and dry out, vanilla producers harvest them while they're still immature, expose them to temperatures of 140°F (60°C) or more to arrest their development, and then reheat them regularly over the course of several weeks before drying and resting them. Most of the benzenoid volatiles are stored in the inner fruit tissues bound to nonvolatile sugar molecules, and they become perceptible only when the fruit is damaged and enzymes separate them from the sugars. The curing process stops the pod's own ripening without completely inactivating the volatile-liberating enzymes, and also allows for heat-tolerant bacteria and yeasts to grow and contribute enzymes of their own. The result is a semidry fruit that resists spoilage and is full of free volatiles that would otherwise still be locked up in their unsmellable stored form.

Vanilla offers a sweet endpoint for this chapter's aromatic gamut from the blandest grains to the boldest spices. Of the more than three dozen spices we've sniffed, fewer than ten highlight volatiles that aren't also common in fresh greens and herbs—notably cumin, fenugreek, ginger, cinnamon, and vanilla. The others work with more common terpenoids and benzenoids to make their own proprietary blends. Like vanilla, though, they all emit their volatiles to discourage other life forms from feeding on them—at least those life forms less solicitous than we are of their long-term welfare. The next chapter brings us to the remarkable plant structures and volatiles that actually invite animals like us to eat and enjoy them—including, strangely enough, chili peppers and vanilla! On to fruits.

Chapter 13 · FRUITS

....

I . . . landed safely on the banks of a fine meadow, which lay on the opposite shore,
where I . . . turned out my steed to graze, and then advanced into the strawberry
plains to regale on the fragrant, delicious fruit, welcomed by communities of the
splendid meleagris, the capricious roe-buck, and all the free and happy tribes, which
possess and inhabit those prolific fields, who appeared to invite, and joined with me
in the participation of the bountiful repast presented to us from the lap of nature.
> · William Bartram, *Travels through North
> and South Carolina*, 1791

This pulp is the eatable part, and its consistence and flavour are indescribable. A
rich butter-like custard highly flavoured with almonds gives the best general idea of
it, but intermingled with it come wafts of flavour that call to mind cream-cheese,
onion-sauce, brown sherry, and other incongruities. . . . The more you eat of it the
less you feel inclined to stop. In fact to eat Durians is a new sensation, worth a
voyage to the East to experience.
> · Alfred Russel Wallace, *The Malay Archipelago*, 1869

There is in the Apple a vast range of flavours and textures, and for those who
adventure in the realm of taste, a field for much hopeful voyaging.
> · Edward A. Bunyard, *The Anatomy of Dessert*, 1929

So far in our exploration of edible plants we've smelled dozens of pleasant
leaves and roots and seeds, and we've seen that our pleasure has nothing to
do with why the plants produce the smells they do. We can savor herbs and
spices because we're bystanders to the network of life-and-death relationships that
make them aromatic. We adjust how much of them we ingest and enjoy them as
the source of enlivening sensations.

With fruits we're bystanders no longer. We finally encounter plant smells that involve us as the animals that we are, an affiliation delightfully embraced by the naturalist William Bartram as he joined wild turkeys and deer to feast on their strawberry fields. Like all fleshy fruits, the berries that fed bird and quadruped and primate are made by rooted mother plants to hail rides for their offspring seeds. Fruit flesh is fuel for mobile animals, and the smells that advertise it are bound up with the most involving and animalistic sensations we feel: hunger and its satisfaction.

Fruits are structures built by a plant from the tissues of its pollinated flowers, and initially they protect the developing seeds with various unpalatable defenses. When the seeds are mature, fleshy fruits *ripen*: they soften, change color and flavor, and become aromatic. Most fleshy fruits—hereafter I'll just call them fruits—have evolved to catch the attention of mammals and birds, and birds identify ripe ones mainly by their bright colors, not by smell. Many fruits probably smell the way they do thanks to the immemorial mutual influence of plants and mammals small and large: our ancestral primates, but also rodents and bats, elephants and bears.

Fruits are not our primary foods, but they are *model* foods: models of deliciousness. Ripe fruits at their best are the supreme example of how stimulating and satisfying foods can be to our senses: how visually striking, how succulent, how balanced in taste, how intense and full in aroma. The English essayist Walter Pater wrote that "all art aspires to the condition of music," to music's emotional immediacy. I've long thought that all cooking aspires to the condition of ripe fruit. Their sensory richness is why home cooks and accomplished chefs alike can find it appropriate to end a meal with a plate, a knife, and a peach or couple of figs at their peak. Fine fruits have already been prepared to a kind of perfection. And beyond their deliciousness and beauty, fruits are vessels for meaning and emotion. They're emblems of relationship, symbiosis, cooperation, generosity: the plant kingdom transmuting its bounty of sunlight and air into animal nourishment and pleasure.

Fruits are also wonderfully diverse. Most of us are familiar with a few kinds of apples and pears and citrus, and maybe a handful of other fruits besides. But there's a universe of deliciousness beyond these, a banquet of flavors for which Hero Carbon did eons of prep work, and which fruit fanciers have elicited through centuries of plant breeding and connoisseurship. And there's more to seek out than deliciousness. Not all fruit fanciers would agree with the British explorer Alfred Russel Wallace that a taste of durian is worth a long sailing voyage from

England to the Far East. Rather than almonds and onion sauce and sherry, the eminent botanist E. J. H. Corner later wrote that it "smells of a mixture of onions, drains, and coal gas." But to see and smell and taste a spiky, basketball-size durian is an unforgettable experience, and an occasion for reflecting on what fruits and fruity pleasures are all about. These days many fruits travel by air to us. It's never been easier for a smell explorer to track down unusual fruits of all kinds, some intoxicatingly delicious and some provocatively not.

In this chapter, we'll sample several dozen of the fruits that are commonly available in the temperate West and often eaten raw, including a few that we tend to think of as vegetables (we'll sample cooked aromas in chapter 18). Their geographical origins and family relations are fascinating, so that's how I've grouped them, a bit roughly. After a virtual visit to the wild fruit nursery of the Tian Shan mountains, we'll pick up again at our imaginary market stalls, each dedicated to a small group related by lineage or homeland. Head right for your favorites to clarify what it is you love about them, or walk straight through from apples to durians, or hunt for the unfamiliar, or just browse. Best of all, assemble thematic platters of real fruits, taste them with curious friends, and help each other articulate what it is that makes them delicious. Take field trips to farmers markets and farms to find unusual varieties picked at the peak of ripeness. Today supermarket fruits are commonplace commodities, often watered-down versions of what they can be when grown for flavor rather than productivity and bruise resistance and looks. At their best, they can be what Thoreau said of wild apples: Earth's true nectar and ambrosia.

Fruit volatiles:
essential esters, sulfur accents

Many fruit volatiles and their smells will already be familiar from our tour of vegetables and herbs and spices. Grassy notes from green-leaf volatiles are prominent in Granny Smith apples and in kiwi fruits. Piney-floral terpenoids scent citrus fruits, and almond-vanilla benzenoids do the same for stone fruits and cranberries. Evidently such molecules do double duty in ripe fruits: they deter microbes and insects and at the same time beckon larger creatures sniffing for a meal to go. But two other groups of volatiles are especially important in fruits. One helps define the general quality of "fruitiness," and the other provides special nose-catching accents. Here's a brief introduction to them.

The most important fruity group is the **esters**. The aroma of nearly every fruit we'll sniff comes in part from one or more of these molecules. As we saw when we explored the families of plant volatiles, the common esters are formed by the union of two different kinds of short chains, an alcohol and an acid. The progeny molecules are generally far more pleasant to our noses than the parents: less reactive, less irritating, products of molecule building rather than metabolic breakdown. Ester names are double-barreled, with the name of the alcohol first, then the acid's plus the suffix *-ate*. Take the example of ordinary *ethyl* alcohol and *acetic* acid, which respectively smell penetrating and vinegary. Their ester, *ethyl acetate*, is a molecule that we can find and smell pretty much unmixed; it's the solvent in some nail polish removers and plastic glues. It smells sweet and intoxicating and generically fruity, and it's a component in many fruit aromas.

Plants routinely make dozens of different esters from short alcohol and acid chains. The table on pages 158–59 lists the most common of them along with their individual smell qualities. Nearly all the descriptions name more than one fruit because nearly all esters are found in more than one fruit. Similarly, any given fruit usually emits a number of esters, which in the aggregate often provide a kind of background hum of fruitiness. In the table below, I list some of the esters that are especially prominent in familiar fruits, rising above the hum to contribute their distinctive aromatic identities to apple, banana, pear, and so on. Atypical and worth singling out is **methyl anthranilate**, whose acid portion is a nitrogen-decorated carbon ring. Its floral-earthy quality helps define the smells of Concord grapes (and artificial grape flavoring) and of "wild" Alpine strawberries. In strawberries it inhibits both fungal and bacterial disease and also the germination of its own seeds, thus encouraging them to grow only when separated from the fruit. And in concentrated form it's an effective bird repellent!

SOME COMMON ESTERS IN FRUITS

Ester	Important in
ethyl acetate	many fruits
butyl acetate	apple

Ester	Important in
methylbutyl (isoamyl) acetate	banana
methyl butyrate	strawberry
ethyl butyrate	strawberry, citrus juice, mango
hexyl butyrate	passion fruit
ethyl methylbutyrate	strawberry, citrus juice, mango, pineapple
ethyl hexanoate	table grape, blackberry
hexyl hexanoate	passion fruit
methyl, ethyl decadienoate	pear
methyl, benzyl benzoate	papaya
methyl anthranilate	N. American grapes, Alpine strawberries

In addition to the common alcohol-acid esters, there are two other subgroups that are formed differently, include carbon-oxygen rings in their structures, and play important roles in specific fruits. The **lactones** (see page 160) lend qualities reminiscent of the few materials we encounter them in, coconut and peach fruit, but also dairy fats like cream and butter. And a small handful of ring molecules called **furanones**, most prominently **furaneol**, lend a sweet, caramel quality along with other nuances to the smells of strawberries, muskmelons, pineapples, and tomatoes. For all their likely primal appeal to us (furaneol and others have been found in breast milk), the furanones and lactones have defensive roles as well; they resemble molecules that microbes use to signal each other, and so can interfere with their invasion. Particularly noteworthy is the crossover ten-carbon, two-ring furanone known as the **wine lactone**, which contributes a coconutty sweetness to fresh fruits as well as wines, but was first discovered in the urine of the koala: it's also a metabolic by-product of dining exclusively on pinene-rich eucalyptus leaves!

SOME FRUIT LACTONES

Lactone	Smell qualities	Important in
g-hexalactone	coconut, hay	papaya
d-octalactone	coconut, dairy	mango, pineapple

continued

Lactone	Smell qualities	Important in
g-octalactone	coconut, fruity, green	plum, papaya
d-decalactone	peach, dairy, coconut	peach, pineapple
g-decalactone	fresh, peach, creamy	apricot, peach, plum, mango
g-dodecalactone	dairy, fruity	apricot, plum
wine lactone (two-ring g-lactone)	sweet, coconut	apple (Cox), mandarin, grapefruit, yuzu

SOME FRUIT FURANONES

Furanone	Smell qualities	Contributes to
furaneol, strawberry furanone (a hydroxydimethyl furanone)	caramel, fruity	garden strawberry, pineapple, mango, tomato, cantaloupe
sotolon, fenugreek furanone (a hydroxydimethyl furanone)	fenugreek, caramel, maple syrup	pineapple
mesifuran, berry furanone (methoxydimethyl furanone)	caramel, musty, smoky	Alpine, musk strawberries
maple furanone (ethyl sotolon)	caramel, fruity	blackberry, raspberry
wine lactone (dimethyl benzofuranone)	sweet, coconut	apple (Cox), mandarin, grapefruit, yuzu

Now for what I call the nose-catching group of fruit volatiles. They come in many different chain and ring structures but share one common feature: the presence of sulfur atoms. Their smells run the gamut from haunting to defining to disgusting. Do you remember the sulfidic, stinky, rotting-vegetable sulfides and thiols produced by microbes in oxygen-poor habitats like our mouth and innards? And the sulfur-containing chains that perfume tomcat urine and human sweat? Some fruits produce these very same and similar molecules—but usually in trace amounts that keep them subordinate to the estery hum while contributing another dimension to the overall smell. That quality is often described as "exotic" or "tropical" because it's more characteristic of Asian and South American fruits, where animalic muskiness may have helped broaden their appeal to the local fauna. As we'll see, frankly sulfurous durian has many jungle admirers.

The names of sulfur volatiles are many and confusing. **Sulfides** are familiar from the mineral and animal worlds. The terms *sulfanyl*, *thiol*, and *thia-* all indicate the added presence of sulfur in starter-set carbon chains, in esters, and even in terpenoids. Sulfanyl alcohols, esters, and ketones are essential to the identity of grapefruit, guava, and passion fruit, and important to a number of others, from tropical mango to strawberries and European wine grapes. Methyl sulfanyl hexanol, found in passion fruit, is the meaty, fruity sulfur component of our underarm sweat, and methyl sulfanyl pentanone, in several fruits, is the cat-pee ketone!

SOME IMPORTANT SULFUR VOLATILES IN FRUITS

Sulfur molecule	Smell qualities	Important in
dimethyl sulfide, dimethyl di- & tri-sulfides	sulfurous, cooked vegetable	lychee
methanethiol, ethanethiol, propanethiol	sulfurous, rotting cabbage	durian, lychee
sulfanyl hexanol	citrus, tropical, grapefruit	grapefruit, guava, passion fruit, wine grapes (Sauvignon Blanc, others)
methyl sulfanyl hexanol	meaty, fruity, sulfur	passion fruit
sulfanyl hexyl acetate	blackcurrant	guava, passion fruit, strawberry
methyl sulfanyl acetate & butyrate	cheesy, garlic, cabbage	strawberry
methyl sulfanyl pentanone	tropical, cat urine	grapefruit, yuzu, mango; blackcurrant?
methoxymethyl butylthiol	blackcurrant, sulfur	blackcurrant
methyl butenethiol	sulfurous	mango
menthenethiol	grapefruit, juicy	grapefruit
ethyl sulfanyl ethanethiol	roasted onion	durian
dimethyl trithiolane	sulfurous, onion	durian

So these are some of the special volatile ingredients of fruit aromas: fruitiness-defining common esters, creamy-coconut lactones, sweet caramel furanones, and exotic, animalic sulfur molecules. On now to the fruits themselves, after the usual disclaimer. Like the flowers they emerge from, fruits emit many, many volatiles, and here I'm able to name only a handful of the most prominent—and often for only one of many varieties of a given fruit. What follow are sketches, not portraits. But sketches can clarify what features make particular fruits recognizable as themselves, help us notice those features, and help us appreciate the qualities that round out their appeal.

Temperate fruit forests of the Tian Shan

Before we return to our orderly sequence of market stalls, let's take a quick virtual excursion to remnants of the wilderness that supplied them. Imagine flying as John Milton's Raphael did through Eden's spicy forests, but instead to real forests in the Central Asian mountain range system known in China as Tian Shan, or "heavenly mountain." It lies north of the Indian peninsula and the Himalayas and spans elevations from below sea level to more than twenty-four thousand feet, or four and a half miles (seven thousand meters), high. The region encompasses countless different combinations of soil type, exposure, and climate, and some of these ecological patches harbor the ancestors or close relatives of a remarkable number of our favorite fruits.

At medium elevations in Kazakhstan and Tajikistan, travelers over the centuries have reported finding extensive stands and even forests of wild apple trees, including one of the primary parent species of our cultivated apple. Not far away is the former Kazakh capital city of Almaty, whose name derives from the local name for that fruit, and where in the 1990s journalist Frank Browning observed apple trees "growing everywhere, along fences, between cracks in the sidewalk." Some of the mountain trees bear apples as large and delicious as modern "improved" varieties. And apples are only the beginning. Alongside them in the fruit forests of the Tian Shan region grow wild species of pears, apricots, cherries, plums, and mulberries. And figs and pomegranates, grapes and currants and gooseberries. And strawberries, and raspberries and blackberries, cranberries and elderberries. (*And* walnuts, almonds, hazelnuts, and pistachios.)

Most of these fruits are only barely suggestive of the deliciousness that millennia of human selection and breeding would elicit from them, but the Tian Shan forests are something close to a paradise of raw materials and promise for the fruit lover. And their wild yet good-size, sweet apples are evidence of the power of animal need to direct plant creativity. Plant biologists Barrie Juniper and David Mabberley conclude that trees producing large sweet fruit have received preferential attention and dispersal from the Asian brown bear, which gorges on foods of all kinds in the autumn to prepare for hibernation. As they see it, bears were the original Johnny Appleseed.

Many of the Tian Shan tree fruits and berries are members of the rose family; more than twenty species of rose itself grow there as well. The rosaceous tree fruits fall into two different groups: *pome* fruits, apples and pears, with their cluster of small seeds in a tough core running through the center, and then the *stone* fruits, cherry and plum and apricot and peach, with a single seed enclosed in a large, central, hard pit. The family resemblance may not be obvious from their shapes or flavors, but you can get a hint of it from the common smell when you pick out a seed or two from the tree fruits and chew attentively. They're all bitter, evidently a defense to discourage animals from chewing, and they all give off the smell of almond extract. We recognize that smell because almonds are seeds of a sister species to the stone fruits, and when they're crushed, aromatic varieties respond by emitting the volatile benzaldehyde as a warning molecule, along with the bitter and potentially deadly toxin hydrogen cyanide (see page 279).

The rose family has been quite successful at colonizing much of the Northern Hemisphere, often thanks to seed-carrying birds. Many fruits of the Tian Shan region came originally from elsewhere and have relatives in Eurasia and North America, and the fruits we eat today often have complex multicontinental ancestries. As we return to our virtual market, I'll start with the fruits that we've glimpsed in the mountains, and loosely group them and others by family and homeland. Fruits are familiar enough sights in everyday life, so I'll mostly dispense with non-olfactory descriptions.

Pome fruits: apples, pears, quince

First up: a stand devoted to the apple and fellow tree-borne pome fruits, which are model ester-emitters. Setting aside wine grapes and the starchy staple banana and plantain, apples are the world's most popular fruit: annual production is some twenty pounds (ten kilograms) for every person on the planet, and there are thousands of named varieties. **Apple** esters are dominated by the pairings of acetic acid with alcohols two, four, and six carbons long, and green-leaf volatiles also contribute a fresh quality. Among widely grown apple varieties, green-skinned Granny Smith is remarkable for its relatively low level of esters and an aroma dominated instead by the green-leaf aldehydes. Red-green Macintosh is similar, while more uniformly red and yellow varieties such as Fuji, Delicious, Gala, and Pink Lady are more intensely estery. Overripe fruit has a solvent-like quality from the dominance of ethyl acetate. Generally, varieties that ripen early in the season tend to be fragrant and fragile, while late-season varieties keep better and reveal their aroma only when chewed.

In 2007, along with my friend and colleague Dave Arnold, I had the opportunity to sample several hundred apple varieties at the USDA apple germplasm collection in Geneva, New York. Among them we caught notes of anise, pear, banana, flowers, tea, citrus, and coconut, thanks to traces of the benzenoids, terpenoids, and unusual esters that characterize those materials. Researchers there told us that they had come across fruits reminiscent of tomatoes, butter, cabbage, wax crayons, and cat urine. Apple connoisseurship reached a high point with the early twentieth-century English writer Edward Bunyard, who described the flavors and ideal cellar-aging periods—sometimes months—for more than fifty varieties. The Gravenstein, my favorite small-farm specialty in Northern California, he described as "scented with the very attar of apple," a perfume that "comes out on the oily skin and remains on the fingers." Wonderfully true! And there's clearly more to his favorite fruit than even he was able to taste or dream of.

APPLES, PEARS, QUINCE

Fruit	Component smells	Molecules
apple (*Malus domestica* or *pumila*)	sweet, ripe, apple, solvent, green, fresh	esters (ethyl, butyl, methylbutyl, hexyl acetates), butanol, hexenal & hexanal
apple, unusual accents	anise, spicy, rose, orange, strawberry, pineapple, citrus, coconut	estragole, anisole, rose oxide, damascenone, wine lactone

Fruit	Component smells	Molecules
pear, Eurasian (*Pyrus communis*)	sweet, ripe, apple, pear, floral	esters (butyl & hexyl acetates, methyl & ethyl decadienoates, phenylethyl acetate)
pear, Asian (*P. pyrifolia, serotina, ussuriensis*)	sweet, ripe, apple, solvent, floral	esters (ethyl acetate & butyrate & hexanoate, hexyl acetate), farnesene, toluene, phenylethanol
quince (*Cydonia oblonga*)	green, woody, fruity, sweet, winey, pear, floral	farnesene, esters (ethyl hexanoate, octanoate, decadienoate), vitispirane, nerolidol

Bunyard called the apple the King of Fruits and the pear the Queen. The **European** species of **pear**, also widely grown in the Americas, shares some of the same short esters found in apples, and this is why hexyl and butyl acetates can be reminiscent of both apple and pear. The highly distinctive pear esters are built on long kinked ten-carbon decadienoic acid. Their most intense source is the variety known as Bartlett or William, but Comice produces a greater quantity of volatiles overall, including strawberry-like ethyl methylbutyrate and a floral benzenoid ester, phenylethyl acetate. Other common varieties such as Anjou and russeted Bosc are weaker volatile emitters and have relatively mild aromas. **Asian pears**, increasingly available in Western markets, come from several other species, usually less pear-shaped than the Europeans, rounder and russet-brown, with a crisper texture and a very different aroma. They emit little or none of the common pear esters and have a more general fruitiness, sometimes candy-like, sometimes floral, sometimes with a prominent solvent note from toluene, which we otherwise encounter in glues, plastics, and printing inks. The **quince** is also native to Central Asia and looks like a fuzzy-skinned, more or less oblong yellow apple. Its flesh is denser and harder than an apple's, so it's usually cooked. Even a raw quince is wonderfully aromatic when ripe, richly fruity from several longer-chain esters than the apple carries, and augmented by several terpenoids that give it hay, winey, and floral qualities.

Stone fruits, and fig, pomegranate, persimmon

Next stand: stone fruits from rose-family trees, two other unrelated tree fruits also found in Central Asian forests, and one more from farther east. The stone fruits are sister species of the almond in the genus *Prunus* and share with it a number of benzenoid volatiles. The kernels are protected by the volatile almond-essence warning signal benzaldehyde and are sometimes used to flavor alcohols and syrups. The flesh of **sour cherries**, favorites in pie making, has the strongest almond smell, with mainly benzenoid supporting notes of flowers, honey, clove, and vanilla. The **sweet cherry**, not just another variety but a different species, shares the floral and honey notes but downplays the other benzenoids in favor of fruity esters and green-leaf volatiles.

Plums are very diverse: there are a dozen species and a couple of thousand named varieties, many of these hybrids, as well as hybrids between plums and apricots (plumcots, pluots, apriums). Even within the main European species there are purple, green, and yellow varieties—prune plums, greengages, damsons, mirabelles—with distinctive flavors, and commercial plums in North America are generally American-Asian hybrids. Most plums seem to share benzaldehyde, floral linalool, a number of esters, and sweet coconut and peach notes from lactones, with European types having more diverse alcohols and esters, American-Asian hybrids more diverse lactones.

Compared to the plums, **peaches and nectarines** are much easier to characterize. They share the family benzaldehyde and linalool and esters, but these are overshadowed by the ten-carbon lactones, so much so that the quality of those pure molecules is described as peachy. Because we encounter the same or very similar lactones in milk products and tropical coconuts, mangoes, and pineapples, peach aromas can suggest those nuances as well. Sweet lactone notes are also prominent in **apricots**, mixed with an important violet quality from ionone and green freshness from eight- and nine-carbon chains more typical of leaves and cucumbers.

SOME STONE FRUITS

Fruit	Component smells	Molecules
sour cherry (*Prunus cerasus*)	almond, floral, honey, clove, vanilla	benzaldehyde, benzyl alcohol, phenylacetaldehyde, eugenol, vanillin
sweet cherry (*P. avium*)	green, floral, almond, fruity	hexenal, hexenol, hexanal, ionone, benzyl alcohol & aldehyde, esters (ethyl acetate & hexanoate)
plum, European (*P. domestica*)	floral, fruity, almond, alcohol, green, floral, peach	linalool, esters (ethyl butyrate, hexyl acetate, many others), benzaldehyde, methyl butanol, hexanol, nonanol, g-decalactone
plum, Asian, American (*P. serotina, americana*, others & hybrids)	floral, almond, fruity, coconut, peach, creamy, fresh	linalool, benzaldehyde, esters (butyl acetate & butyrate), g-octa-, g-deca-, g-dodeca-lactones, nonanal, hexenal
peach, nectarine (*P. persica*)	peach, coconut, creamy, fresh, almond, fruity, floral	g- & d-decalactones, hexenal & hexanal, benzaldehyde, esters (butyl, hexyl, hexenyl acetates), linalool
apricot (*P. armeniaca*)	floral, peach, creamy, geranium leaf, cucumber	ionone, g-deca- & g-dodeca-lactones, linalool, octadienone, nonadienal

Alongside the stone fruits are two non-rosaceous tree fruits that spread to Central Asia from original homes in the Mediterranean. The **fig** starts out as a bulbous hollow flower that admits tiny pollinating wasps through a small pore, then fills out and ripens into a sweet soft mass with small crunchy seeds or pseudoseeds inside. Its volatiles include acetate esters but also a four-carbon alcohol and ketone that have creamy, buttery qualities, and almond and floral notes; the overall effect can suggest honey. The **pomegranate** has a thin, leathery skin enclosing dozens of individual, juicy, typically dark red fruitlets with a hard seed at their center. The fruitlets have a mild but unique aroma, with green and piney-woody accents to its estery core.

The last fruit on this stand is the **persimmon**, a fist-size orange-fleshed fruit of a tree native to China and now grown in Japan, Israel, Italy, and California.

Its American sister species, smaller and seedier, is less widely appreciated and harder to find. The Asian persimmon is relatively mild, with honey, caramel, and balsamic sweetness and a hint of potato from the sulfur-containing aldehyde methional.

SOME NON-ROSACEOUS TREE FRUITS

Fruit	Component smells	Molecules
fig (*Ficus carica*)	fruity, creamy-buttery, woody, almond, floral	esters (butyl, isoamyl, hexyl acetates), acetoin (hydroxybutanone), germacrene, benzaldehyde, linalool
pomegranate (*Punica granatum*)	green, fruity, pine, woody, musty, mushroomy, floral	hexenal, ethyl methylbutyrate, pinene, myrcene, caryophyllene, heptenal, ethyl hexanol
persimmon (*Diospyros kaki*)	green, potato, honey, caramel, balsamic	hexenal, methional, phenylacetaldehyde, furaneol, methyl cinnamate

Strawberries

We shift our view now from sizable fruits borne on tree branches to the small fruits borne lower down, on vines, canes, small shrubs, and ground-hugging plants, the ones often loosely called *berries* (the botanical definition includes bananas and excludes raspberries; we'll ignore it). The rose family evolved a number of excellent wild berries in Central Asia and elsewhere, all of them cultivated and improved by Hero Carbon's fruit fanciers, and none more remarkably than the strawberry. The saga of the common garden strawberry has filled books. Here's the short, volatile version, history you can smell behind the several piles of different-size, different-smelling fruits on the stand.

The strawberry's botanical name, *Fragaria*, comes from its name in ancient Latin, *fraga*, which meant "fragrant berry." Of the twenty species in the genus *Fragaria*, the **Alpine or "wild" strawberry**, French ***fraise des bois***, is the most common and widely distributed in the northern reaches of the Northern Hemisphere, including the Tian Shan region. Both white and red types were taken from the wild into the European garden many centuries ago. The small, spongy fruits are still grown as a specialty crop and have an aroma very different from the modern strawberry, with resinous terpenoids and the fruity-floral nitrogen volatile methyl anthranilate, which it shares with Concord and related American grapes (see page 313).

A second indigenous north Eurasian strawberry is the **musk strawberry**, the hybrid offspring of the Alpine and another wild European species, *F. viridis*. Its fruits are somewhat larger, unevenly red outside and white within, and as its name suggests, it has a richer aroma than the Alpine: with a variety of esters in place of terpenoids, including an exotic one that includes sulfur, some methyl anthranilate, and two furanones—mesifuran, which contributes both desirable caramel and less desirable mustiness, and caramel-fruity furaneol. It can be delicious, but the plant bears fruit for only a couple of weeks every year.

And that was pretty much it for the European strawberry—methyl anthranilate plus terpenoids, or esters plus furanones—until the discovery of the New World. Then two American strawberry species came into the picture, both hybrid descendants of the Alpine species. A French explorer with the impossibly apt name of Frézier collected specimens from the Pacific coast of Chile of a type that had been cultivated for centuries by the Mapuche and Huilliche peoples. This **beach strawberry** was white, large, and richly aromatic, with fruity esters, a peachy lactone, and the finer furanone, furaneol. English colonists found another species in distant eastern North America: the **Virginia strawberry**, which William Bartram described growing in such fragrant abundance, red and productive, similarly endowed with esters, furanones, and a lactone. The two American species ended up in a collector's garden in France, and sometime early in the eighteenth century, the pollen of the Virginia fertilized the ovaries of the beach.

The result was the birth of the modern hybrid **garden strawberry** *Fragaria* x *ananassa*, so named at the time, as Antoine Duchesne explained (see page 160), because its smell suggested the pineapple (*anana*) in both intensity and quality. Sure enough, the same volatiles take leading roles in both fruits: sweet furaneol and butyrate esters. And unlike the Alpine and musk berries, neither the American parents nor their hybrid progeny emit floral-fruity methyl anthranilate. To European noses, the new hybrid would have smelled more like exotic pineapple than like the strawberries they knew. (The only known exception: a German heirloom hybrid, Mieze Schindler, which emits some methyl anthranilate.)

SOME STRAWBERRY VARIETIES

Strawberry variety	Component smells	Molecules
Alpine, "wild," *fraise des bois* (*Fragaria vesca*)	resinous, woody, wild-strawberry	pinene, myrcene, terpinenol, phellandrene, myrtenyl acetate, methyl anthranilate
musk (*F. moschata*)	fruity, sweaty, musky, wild-strawberry, musty, caramel, tropical	esters (hexyl, octyl, myrtenyl acetates; methyl butyrate), methylbutyric acid, methyl anthranilate, mesifuran, furaneol, sulfur volatiles (sulfanyl hexyl acetate)
beach (*F. chiloensis*)	fruity, sweet, caramel, sweaty, peach	esters (butyl & ethyl acetate, ethyl butyrate & hexanoate), mesifuran, furaneol, methylbutyric acid, decalactone
Virginia (*F. virginiana*)	fruity, sweet, caramel, resinous, cheesy, coconut	esters (methyl & ethyl butyrate & methylbutyrate), mesifuran, furaneol, terpinenol, butyric acid, octalactone
common, garden, Virginia x beach (*Fragaria* x *ananassa*)	sweet, caramel, green, fruity, buttery, cheesy, sweaty, tropical	furaneol, hexenal, esters (methyl & ethyl butyrate & methylbutyrate), diacetyl, butyric acid, sulfur volatiles (methanethiol, methyl thioacetate)

The now common hybrid strawberry also has buttery and sulfurous notes, and a cheesy one from butyric acid, an ingredient in the butyrate esters. Once you realize these not-so-fruity notes are there in the strawberry, you can sometimes catch them and appreciate the depth they give to the flavor. To reveal them more clearly, put a basket of fresh strawberries in a plastic bag and tie it off, leaving it inflated with plenty of air. After a few hours, carefully open the bag and take a whiff. It'll be surprisingly stinky: perhaps because the trace emissions of butyric acid and/or sulfur molecules accumulate to levels high enough to compete with the esters for our perception.

For many generations, fruit breeders were primarily concerned with developing larger, juicier, more productive strawberry varieties, and the aroma of the anthranilate-deprived common strawberry didn't change much, or only faded in intensity and complexity. That changed in the 1990s with the introduction of the French **Mara des Bois** variety, which was specifically bred to bring the Alpine or *fraise des bois* anthranilate character into the common strawberry. It was well received and bodes well for an increasing diversity of flavors in good-size, succulent fruit. In the meantime, keep on the lookout for the diminutive Alpines and musks. If you have access to a sunny patch, plant a strawberry sampler.

Berries from cane, bush, and vine

On now to a stand with a half-dozen piles of small berries, a few brown and fuzzy oblongs, and then several different clusters of grapes. First, two sister species in the generous rose family. Blackberries and raspberries are called caneberries because they're borne on long, thin, stiff, canelike stems, usually thorny. They're members of the prolific and confused genus *Rubus*, which includes hundreds of different species and hybrids, among them dewberries, salmonberries, cloudberries, boysenberries, and regional versions of the dark-purple blackberry and red raspberry. Native caneberry species are found on most continents, due at least in part to the bird-friendly small, brightly colored fruits. Of the most common varieties in markets and gardens, **blackberries** are characterized by prominent caramel-sweet furanones, floral linalool and violet-floral ionones, and a coconut-cheesy nuance from a seven-carbon ketone, heptanone. **Raspberries** are dominated by floral volatiles and emit an unusual molecule specific enough to them that it's been named raspberry ketone; it's a benzenoid ring decorated with a four-carbon chain, and has berry, floral, and "jammy" cooked-fruit qualities.

Now we leave the rose family behind for a couple of unrelated but prolific berry makers. The genus *Vaccinium* includes several hundred mostly northerly shrub species that thrive in wet and woodland settings with acidic soils; it gives us blueberries, cranberries, huckleberries, bilberries, lingonberries, and many variations thereon. **Blueberries** come from several different species with broadly different volatile mixes. Lowbush varieties tend to have stronger and more estery aromas, while the more common highbush types have fresh, floral, and tea notes from aldehydes and terpenoids. **Cranberries** are rich in almond, resinous, and sweet-balsamic benzenoids, with copious benzoic acid also accounting for their tartness and resistance to spoilage. The Euro-American species of cranberry (*V. oxycoccos*) is said to have higher volatile emissions and a more intense aroma than the commercial American species.

SOME COMMON BERRIES

Fruit	Component smells	Molecules
blackberry (*Rubus fruticosus* & others)	caramel, fruity, violet, floral, coconut, cheesy	furaneol, maple & other furanones, ethyl hexanoate, ionones, linalool, heptanone, heptanol, hexanal
raspberry (*Rubus idaeus* & others)	raspberry, violet, floral, pine, fruity	raspberry ketone, ionones, damascenone, pinene, esters (ethyl acetate & heptanoate)
blueberry (*Vaccinium corymbosum* & others)	fresh, floral, fatty, fruity, tea	hexenal, linalool, nonadienal, methyl methylbutyrate, geranyl acetone, damascenone
cranberry (*Vaccinium macrocarpon, oxycoccos*)	almond, floral, resinous	benzaldehyde, benzyl alcohol, benzyl benzoate, terpineol, ethyl methylbutyrate, ionone, heptenal
blackcurrant (*Ribes nigrum*)	fruity, fresh, pine, musky, sulfurous	esters (ethyl butyrate, hexyl acetate, methyl benzoate), hexenal, pinene, terpinenol, methoxymethyl butylthiol; cat ketone?
gooseberry (*Ribes uva crispa*)	grassy, mushroom, pineapple, apple, floral	hexenal, octenol, ethyl & methyl butyrate, hexenol, acetophenone
Chinese gooseberry, kiwi (*Actinidia chinensis* var. *deliciosa*)	green, solvent, fruity	hexanal, hexenal, pentanone, esters (methyl & ethyl acetate & butyrate), pinene
gold kiwi (*A. chinensis* var. *chinensis*)	fruity, fresh, fatty, tropical	esters (ethyl & butyl butyrate), hexenal, heptanal, octanal, eucalyptol, dimethyl sulfide

The shrubby genus *Ribes* includes around 250 species native to the temperate Northern Hemisphere, and gives us several relatively minor berries, the currants and gooseberries. **Blackcurrants** in particular have a distinctive musky undertone from the presence of several sulfur-containing volatiles, possibly including the "cat ketone" (see page 81); studies disagree. **Gooseberries** are larger, tarter relatives of blackcurrants; they're dominated by green and fruity volatiles, with a flowery note from acetophenone.

The yet larger fuzzy brown oblongs alongside the gooseberries are known as **Chinese gooseberries** in New Zealand, where growers experienced a boom in the 1960s when their crop was promoted in the United States as **kiwifruit**. They're borne on a woody vine native to China and are more remarkable for their bright green flesh and halo of small black seeds than for their aroma, which does resemble the unrelated European gooseberry's. Fittingly for its color, green-leaf aldehydes dominate, with esters and a solventy ketone moving the grassiness into the fruity domain. A less common variety, the **gold kiwi**, ripens to a yellowish flesh and overlays the green notes with more abundant esters, eucalyptol, and a musky-tropical sulfide.

On now to the familiar-looking clusters. They too have been harvested from vines, long-stemmed plants that support themselves by clinging to nearby objects or by trailing along the ground. The word *vine* comes from the Latin for "wine," and the wine grape is the prototypical vine. Its wild ancestors climbed trees in forests throughout central and western Asia, and sister species did the same in North America.

Most modern **table grapes** are varieties of the European wine grape, and most have a fairly simple aroma arising from a couple of fruity esters, fresh aldehydes, and a floral alcohol, with no prominent terpenoids. **Muscat grapes** are one exception: they're an ancient European variety, possibly the first to be recognized as a distinctive type, the name (and synonym **muscadine**) possibly related to similarly prominent musk-deer scent (see page 466). They're rich in terpenoids and their acetate esters, and they have a strong sweet aroma with rose and citrus aspects. (For wine grapes and their volatiles, see page 571.)

Two other exceptional grapes are American species with the assertive aromas once known as "foxy." That quality, which comes from the rare nitrogen-containing benzenoid aminoacetophenone, is important in some tropical flowers, strong-smelling chestnut honey, and corn tortillas. Aminoacetophenone hasn't been found on fox bodies, but it does turn out to be an important volatile for some New World species of bats (Asian fruit bats are called flying foxes), whose caves can reek of it. **Muscadine grapes** and their products, hard to find outside their native range in the Southeast, mix this muskiness with caramel, fruity, and floral volatiles as well as a sweaty branched-chain acid. **Concord grapes**, familiar from common American grape jellies, add the nitrogenous candy-like, fruity-flowery ester methyl anthranilate.

SOME EURASIAN AND AMERICAN GRAPES

Grape	Component smells	Molecules
table, non-aromatic (*Vitis vinifera*)	green, fresh, waxy, fruity, floral	esters (ethyl acetate & hexanoate), hexanal, octanal, nonanal, decanal, phenylethanol
muscat (*V. vinifera* var. Muscat)	sweet, floral, rose, citrus, fruity, lavender	linalool, geraniol, citronellol, esters (linalyl, geranyl, citronellyl acetates)
muscadine (*V. rotundifolia*)	musky, caramel, floral, fruity, sweaty	aminoacetophenone, mesifuran, furaneol, phenylethanol, esters (ethyl methylbutyrate & butyrate), methylbutyric acid
Concord (*V. labrusca*)	fruity, floral, musky, strawberry, caramel	methyl anthranilate, aminoacetophenone, esters (ethyl & methyl hydroxybutyrate, ethyl decadienoate), damascenone, furaneol, mesifuran

Fruits of the gourd family: cucumber and melons

We move on now to several market stalls devoted to individual families of fruits. The first displays cucumbers and melons, all members of the cucurbit or gourd family, whose fruits are borne on nonwoody vines that die back every year. The same family gives us edible squashes and gourds, which usually need cooking to make them palatable (see page 517). The cucurbits appear to have arisen in Asia and then spread across the globe, sometimes floating across oceans; the Americas and Africa are the respective homelands of the vegetable squashes and the watermelon. Their family chemistry is dominated by several nine-carbon aldehydes that have one or two double bonds along the chain (nonenal, nonadienal) and give them their distinctive cucumber or melon aromas. As we'll see, some shellfish and fish emit the same molecules and are therefore said to have a fresh, cucumber-melon aroma. However, it's likely that they were the first to deploy these aldehydes, so the all-smelling Chef of the cosmos would say that cucumbers and melons have fresh aquatic notes!

The **cucumber** is an unusual fruit. It's not sweet and it doesn't emit any generically fruity esters, so we don't treat it as we do other fruits. It's at its best as a refreshingly moist, crunchy vegetable, a natural in raw salads. Cucumber aroma comes mainly from the nine-carbon aldehydes, along with fresh, green-smelling six- and eight-carbon chains.

The very sweet true **melons** are deployed by a sister species to the cucumber and were developed from moist but bland precursors in Persia around a thousand years ago. Our name for them derives from the Greeks, who called the fruit *melopepon*, "apple-gourd." The many varieties of melon fall into two broad groups that are defined in part by their typical smells. One group tends to emit few or no volatiles from the intact fruits, which are usually smooth-skinned, ripen slowly, and remain edible for some days or weeks after ripening. The second group tends to emit copious volatiles from their rough, netted skins; these melons ripen quickly and quickly deteriorate. The two groups emit very different volatile cocktails from their flesh, but modern melon breeders have crossed their members to create new varieties with intermediate qualities, so the distinctions aren't as clear as they once were.

SOME CUCURBIT FRUITS

Fruit	Component smells	Molecules
cucumber (*Cucumis sativus*)	cucumber, melon, fatty, green, geranium-metallic	nonadienal, nonenal, hexanal, octadienone
honeydew melon (*C. melo inodorus*)	cucumber, fresh, floral, sweet, creamy, melon, medicinal	nonadienal & nonadienol, phenylethanol, phenylacetaldehyde, nonenal, guaiacol
cantaloupe melon (*C. melo reticulata*)	sweet, fruity, floral, fresh, fatty, cooked vegetable, sulfurous, caramel	esters (ethyl acetate & butyrate & hexanoate), octenal, octenol, benzyl acetate & alcohol, sulfur volatiles (dimethyl di- & tri-sulfides, sulfanyl esters), furaneol

continued

Fruit	Component smells	Molecules
fragrant, pocket melon, dudaim (C. *melo dudaim*)	sweet, fruity, floral, clove, medicinal, almond, peach, coconut	esters (ethyl butyrate & hexanoate), eugenol, chavicol, benzaldehyde, sulfur volatiles (sulfanyl esters), hexa-, octa-, deca-lactones
watermelon (*Citrullus lanatus*)	fresh, waxy, cucumber, melon, fruity, apricot, floral, violet	hexanal, nonanol, nonanal, nonenol, nonenal, nonadienal, methyl heptenone, ethyl methylbutyrate, dihydroactinidiolide, geranyl acetone, ionone

The **non-aromatic melons**, typified by the green-fleshed **honeydew**, share the cucumber aldehydes but augment them with floral and honey-like benzenoids. The **aromatic melons**, often called **cantaloupes or muskmelons**, eschew cucumbery aldehydes for full-bore fruitiness, with copious quantities of several apple-pineapple esters, floral benzenoids, a number of sulfur-containing molecules that provide the musky aspect, and caramel-fruity furaneol.

In a class of its own is the small **dudaim or pocket melon**, which is a couple of inches in diameter and has only a thin layer of bland flesh. It's grown in the eastern Mediterranean and western Asia just for its rich scent, which is emitted mainly by the skin and can perfume a room. The scent augments cantaloupe volatiles with the clove note of eugenol, medicinal chavicol, and sweet hay and coconut lactones.

The **watermelon** is a cousin rather than a sister species to the cucumber and true melons, from a genus that evolved in Africa rather than Asia. It may have been cultivated first as a source of water, which it can scrounge and store up in arid conditions. A cucumber-like watermelon appears to have been domesticated in northeastern Africa about five thousand years ago, sweet varieties in the eastern Mediterranean two thousand years ago, and the familiar deep red flesh had appeared in Italy by the fourteenth century. The aromas of modern varieties bear some resemblance to the cucumber and smooth-skinned melons, but to the nine-carbon aldehydes they add fresh, floral alcohols and fruity esters. Red-fleshed watermelons also emit apricot- and violet-like fragments from the red carotenoid pigment lycopene. Yellow and orange varieties are colored by related but different carotenoids and tend to lack those notes.

Fruits of the nightshade family: tomatoes and capsicum peppers

The next market stand is covered with an array of tomatoes, capsicum peppers, and a couple of their relatives in the nightshade family, which also includes the eggplant and potato—and tobacco. Most of the edible nightshades come from the New World, and we use most of these as vegetables rather than fruits, in part because they don't sweeten much when ripe, and in part because their smells aren't generically fruity; esters are not their major volatiles. The **tomato** is marked by fresh, green-vegetal, mushroom, floral, cooked apple, and malty notes, and the sweet caramel of furaneol. It's a unique and complex, non-estery mix, one that leans toward the savory more than the fruity-sweet.

Alongside the tomatoes are handfuls of what look like small clusters of dead leaves. Inside the papery husks are the fruits of two tomato relatives. *Tomatillo* means "little tomato," and these fruits from a different genus in the family do resemble green or purple versions of a small tomato. Their smell is also diminutive by comparison; the range and quantities of its volatiles are smaller. The tomatillo's sister species, the **ground cherry or cape gooseberry**, breaks the tomato mold: its berry-size orange fruits are far more typically fruity, with lactones and an ester reminiscent of stone fruits and coconut.

Next to the diminutive dull husks is a visual riot of colors and sizes and shapes: all hollow-fruited members of the New World *Capsicum* genus, commonly known as "peppers" because some carry a pungency that reminded early European explorers of black pepper. Many of the pungent **chilis**, to use a version of their original Nahuatl name, are used in such small quantities that their aroma isn't noticeable. Nonpungent varieties are familiar as vegetables, and green and ripe versions of them have very different smells. Even prominent chefs with very catholic tastes have professed their hatred for **green bell peppers**. Their smell is dominated by an unusual nitrogen-containing carbon ring, a pyrazine with a strong green-vegetal quality—it's also found in green peas and beans, in lettuce and spinach. Add to that the sulfurous, cooked-vegetable notes of a sulfide and thiol, and the result is vegetality squared, with little subtlety. Let that same bell pepper ripen into a **ripe bell pepper**, however, and the color changes to more fruity red or orange or yellow, the sugars and sweetness rise, and the smell is transformed. Whereas in most fruits ripening brings a more complex and stronger smell, in capsicums the levels of nearly all the volatiles decrease, and an ester and lighter green-leaf volatiles become more prominent than the pyrazine.

There are a couple of dozen species of *Capsicum*, and some of the less common are prized for both pungency and aroma. One example is the **habanero chili**, which when ripe has an estery fruitiness reminiscent of banana and pineapple. There are also nonpungent varieties of habanero whose aroma can be enjoyed painlessly.

TOMATOES, PEPPERS, AND RELATIVES

Fruit	Component smells	Molecules
tomato (*Lycopersicon*)	green, cooked apple, metallic, coconut, mushroom, floral, malty, caramel	hexenal & hexanal, damascenone, epoxy decenal, wine lactone, octenone, linalool, methylbutanal, furaneol
tomatillo (*Physalis ixocarpa*)	green, wintergreen, tomato-vine, malty, honey	hexanol, hexenal & hexenol, methyl salicylate, phenylacetaldehyde, isobutyl thiazole, methylbutanal
ground cherry, cape gooseberry (*Physalis peruviana*)	peach-apricot, sweet, coconut, fruity-pear, waxy-citrus	g-octalactone, g-hexalactone, ethyl octanoate, heptanone, nonanal
chili, bell pepper (*Capsicum annuum*)	unripe: vegetal-earthy, cucumber, sulfurous; ripe: green, fruity, sweet, woody, vegetal	unripe: methoxypyrazines, hexanal & hexenal, nonanal, ocimene, sulfur volatiles (dimethyl disulfide, heptanethiol); ripe: hexanol, hexenal, methylheptyl propenoate, ocimene, methyoxypyrazines
habanero pepper (*C. chinense*)	fruity, floral	esters (hexyl butyrate & pentanoate), cyclohexanol, ionones

Citrus fruits: peel and juice

Now to a market stand crowded with lemons, limes, oranges, grapefruit, and a handful of similar-looking spheres. Among all of our favorite fruits, the citrus family stands out as a distinctive group: its members share thick, finely dimpled peels, an inner division into segments filled with small juice-filled sacs, and fresh, sharp smells more similar to conifer trees than to sweet-fruity apples and strawberries. In fact, the family gets its name from that similarity to aromatic trees. The word *citrus* comes from the Latin *citrum* for the timber of a kind of Mediterranean cypress (*Tetraclinis*) that was prized for scenting rooms and clothes; and *citrum* in turn may come from *cedrus*, Greek *kedros*, the name for the aromatic juniper. Citrus fruits are natives of subtropical and tropical Asia and were unknown in the Mediterranean until the conquests of Alexander the Great around 330 BCE. The first example to arrive contained a small portion of inedibly sour juice, but the massive peel could perfume whole rooms, its terpenoid volatiles suggesting both conifers and herbs that we now describe as lemony. This foreign fruit was given the name *citron*. After Muslims brought several sister species to the region a millennium later—the sour orange first, then the lemon and pomelo—the citron provided the family with its name.

Citrus fruits have split personalities: the peel and interior have very different structures, flavors, and culinary uses. The relatively dry, spongy **citrus peel** protects the inner seeds and juice sacs and stores its chemical defenses in the spherical oil glands that are visible as tiny contrasting dots in the surface. Pressing or cutting into the fruit releases the gland contents in a defensive spray. This anatomy is what makes it possible for bartenders to aromatize a cocktail by twisting a piece of orange peel over it, or squeezing out and igniting the oils in a showy flash.

The peel volatiles are dominated not by fruity esters but by terpenoids, with their references to pine trees, resins, aromatic herbs, and flowers; they're usually joined by aldehydes eight and ten carbon atoms long, longer than the six-carbon green-leaf volatiles. A few terpenoids are so dominant in citrus fruits that they define each other's smells. This is true for **neral and geranial**, co-occurring mirror-image terpenoids that define the smell of lemon. Another terpenoid, **limonene**, is the most abundant volatile in most citrus fruits, with a more generic, "citrusy" quality, fresh and slightly minty. The particular smells of different citrus peels appear to be due largely to different proportions of shared terpenoids, with longer aldehydes and some esters also helping to distinguish one species from another. Two, though, owe their distinctive smells to specific terpenoids. Lemon peel is defined by neral and geranial, and grapefruit by the woody-grapefruity sesquiterpenoid nootkatone.

Beneath the peel's terpenoid shield are the **citrus segments**, the portion of the fruit that attracts animal seed dispersers with valuable water and sugars. The many individual little sacs contained in each segment all produce their own volatiles, which include a mixture of terpenoids and more typically fruity esters and aldehydes. Because the largely hydrocarbon terpenoids (limonene, terpinene, pinene, myrcene) don't mix well with water, they're less prominent in citrus fruit segments compared to the esters, aldehydes, ketones, and alcohols. They tend to bind to the pulpy segment and sac debris when the fruits are juiced. This is why juices with pulp are more flavorful. And machine juicers that crush the peel along with the pulp fortify the juice with peel terpenoids, something that gentle hand juicing does not.

The citrus genus is promiscuous: different species readily mate with each other to produce new hybrid species. Genetic studies indicate that most of our common citrus fruits are descendants of three parental species, the citron, mandarin, and pomelo. Crossings involving the citron gave the small "Mexican" lime and lemons. Crossings between the mandarin and pomelo gave sour orange and sweet orange, and clementine and satsuma. And crossings among parents and offspring gave grapefruit (pomelo x sweet orange), larger "Persian" lime (Mexican lime x lemon), and bergamot (lemon x sour orange).

Sometimes lineage seems to be reflected in aromas, sometimes not. Here on the market stand are a dozen or so of the more familiar citrus fruits, loosely grouped by their family relations. Again, remember that there are many varieties of each, volatiles can vary among fruits from the same tree, and analyses of the same fruit often vary as well. The tables below are only a rough guide.

Citrus fruits: citron to yuzu

First, the sour and aromatic citrus fruits. The citron and its immediate offspring, the lemon and Mexican lime, have either sparse or very sour juices that are not strongly aromatic; their chief interest comes from their peels. In addition to copious but generic limonene, the citron endowed lemons and limes with a variable mix of piney-turpentiney, distinctively lemony, and floral terpenoids. It seems that the relative proportions of these qualities, especially the balance between conifer and lemon, are what define their different flavors, so I've included some illustrative numbers in the table (for sources, see pages 618–19). The **citron** itself emits the most of both conifer and lemon volatiles, the high terpinene level consistent with the ancient Greek association to cedar and juniper woods. **Lemons** reduce both the conifer terpenoids and lemony neral + geranial, but give the lemony more relative weight. The small **Mexican or key lime** does the reverse: it lowers the lemony and raises the conifer by multiplying the number of woody, resinous terpenoids. The larger, thicker-skinned **Persian lime**, the most commonly available version of lime in temperate countries, is a cross between the Mexican lime and the lemon, and its volatile profile seems to reflect that: it's rich in both conifer and lemony terpenoids, and adds to them a couple of fruity-floral terpenoid esters. The **Meyer lemon**, whose complex parentage involves both sweet and sour oranges, carries only a trace of lemony neral + geranial, includes the resinous myrcene found in the limes, and adds a quirky touch to the mix: thymol, the main medicinal smell of thyme, which it shares with yuzu (see below). The overall effect is intriguingly different from the more straightforward lemon varieties, and along with its less acidic juice has made the Meyer an appealing alternative to standard lemons.

CITRON, LEMONS, AND LIMES

Fruit & part	Component smells	Molecules (percentage of total volatiles)
citron peel (*Citrus medica*)	citrus, turpentine, pine, lemon, floral	limonene (52%), terpinene (27%), pinene (4%), ocimene, neral + geranial (4%), linalool
lemon peel (*C. limon*, citron x sour orange)	citrus, pine, turpentine, lemon, floral	limonene (65%), pinene (12%), terpinene (6%), sabinene, neral + geranial (2%), linalool
lime peel, Mexican, key (*C. aurantifolia*, citron x papeda)	citrus, turpentine, pine, resinous, woody, lemon, floral	limonene (65%), terpinene (7%), pinene, sabinene, myrcene, terpinenol, terpineol, neral + geranial (1%), linalool
lime peel, Persian, Tahiti, Bearss (*C. latifolia*, Mexican lime x lemon)	citrus, turpentine, pine, resin, lemon, herbal, floral	limonene (40%), terpinene (20%), pinene, sabinene, myrcene, neral + geranial (4%), bisabolene, esters (terpinyl & neryl acetate)
Meyer lemon peel (*C. meyeri*, citron x oranges)	citrus, turpentine, pine, thyme, floral, lemon	limonene (80%), terpinene (7%), myrcene, cymene, pinene (2%), thymol, linalool, neral + geranial (0.1%)

Next, the sweeter fruits of the mandarin clan. Its worldwide megastars are the sweet oranges, but that may have more to do with their size and physical robustness than with their flavor. The parental **mandarins** are small and thin-skinned but intensely flavorful, and they passed along both flavor and orange color to their progeny. Their name comes via Portuguese from Malay and Hindi words meaning "counselor"; the Portuguese introduced them to the Mediterranean from China sometime after they brought the sweet orange in the sixteenth century. *Tangerine* comes from the name given to particular mandarins that came from Tangiers. Clementine and satsuma are two of many commercial varieties. All of them have peels that emit citrusy, woody, resinous, floral terpenoids and a couple of waxy, fatty aldehydes; their pleasantly sweet-tart juices combine some of these with fruity esters.

The first oranges to reach the West were **sour or bitter or Seville oranges** brought by the Muslims to Spain in the tenth century. Though we now associate the fruit's name with its color, *orange* is thought to come via the Sanskrit *narang* from a South Indian word related to fragrance. (In the West, *orange* didn't name the color until the sixteenth century.) As its attached adjectives suggest, the sour or bitter orange isn't edible as is, but it became a standard ingredient in marmalades. Its peel and juice both carry fresh, conifer, and floral terpenoids.

Sweet oranges are the pleasingly intermediate product of the small, delicate, intense mandarin and the large, robust, relatively neutral pomelo. The sweet orange peel has the familiar conifer-lemon-floral constellation, but adds several longer and fairly characteristic aldehydes of eight to ten carbons that contribute smells usually described as waxy and citrus-peel-like. And the sweet-tart juice is more conventionally fruity, with a number of common esters dominant. **Blood oranges** are varieties of sweet orange that augment their orange carotenoid pigments with red anthocyanins, and seem to have a more intense flavor, more like raspberries and strawberries than oranges. Their color helps suggest this, but compared to other oranges, blood oranges do have larger proportions of particular volatiles that help characterize berries: three to six times the butyrate esters, which are especially prominent in strawberry and pineapple, the same multiple for woody myrcene, and as much as ten times the floral linalool, also important in strawberry and raspberry.

Many orange producers coat their fruit with wax to reduce moisture loss, slow metabolism, and prolong shelf life. But because the coating limits the living fruit's oxygen, its inner tissues switch to anaerobic metabolism and produce ethyl alcohol. The alcohol can accumulate and also increase the production of fruity ethyl esters, so older overripe fruit has a noticeable alcohol and solvent smell, and unusually strong fruitiness.

Modern breeders have extended the mandarin family with crosses that amplify the mandarin flavor in orange-size fruits; these include the Temple orange or tangor (mandarin x sweet orange) and tangelo (mandarin x grapefruit).

MANDARINS AND ORANGES

Fruit & part	Component smells	Molecules
mandarin peel (*Citrus reticulata, deliciosa, clementina, unshiu . . .*)	floral, fatty, sweet, pine, metallic, citrus	linalool, decadienal, wine lactone, pinene, myrcene, octanal, limonene
mandarin juice	fresh, fruity, resinous, floral	hexanal, ethyl methylbutyrate, myrcene, terpinene, linalool
sour orange peel (*C. aurantium*, citron x mandarin x pomelo)	citrus, resinous, floral, pine	limonene, myrcene, linalool, pinene
sour orange juice	citrus, pine, floral, turpentine	limonene, pinene, linalool, heptanone, terpineol, cymene
sweet orange peel (*C. sinensis*, mandarin x pomelo)	citrus, geranium, pine, floral, green	limonene, myrcene, pinene, terpinene, linalool, octanal, nonanal, decanal, decenal
sweet orange juice	fruity, citrus, sweet, green, floral	esters (ethyl butyrate & methylbutyrate), limonene, hexenal, linalool, decenal
blood orange juice	fruity, citrus, floral	esters (methyl & ethyl butyrate, ethyl methylbutyrate), limonene, linalool, myrcene, nootkatone

The **pomelo** is, as its name in botanical Latin indicates, the big citrus; the odd common name is a complicated mix of French, Dutch, and Portuguese, also rendered as pummelo. It's usually six to ten inches (fifteen to twenty-five centimeters) across, though surprisingly light for its size because much of its volume consists of spongy peel tissue. Its peel and juice are relatively mild in flavor, with modest amounts of the standard citrus terpenoids supplemented by an unusual woody sesquiterpenoid, nootkatone, named for the cedar tree in which it was first found.

Unremarkable as its own flavor is, the pomelo is a reticent parent of sweet and sour oranges, and the more forthright parent of the **grapefruit**, which arose in a cross with the sweet orange on the island of Barbados in the early nineteenth century. The grapefruit peel is relatively poor in terpenoids except for nootkatone, dominated instead by waxy, citrusy straight-chain aldehydes. Grapefruit juice carries some esters and fresh and citrusy aldehydes along with nootkatone, but its most notable components are a couple of sulfur-containing molecules, a terpenoid and a five-carbon chain. These contribute a nonfruity, nonconifer, noncitrusy, somewhat vegetal, somewhat animal quality that's often described as exotic or tropical, because similar molecules and notes are found in guavas and passion fruit (see page 329). The juice of a red variety of grapefruit, which accumulates the red terpenoid pigment lycopene, has been found to contain woody caryophyllene and sweet-fresh valencene, both sesquiterpenoids, and floral linalool oxide.

POMELO AND GRAPEFRUIT

Fruit & part	Component smells	Molecules
pomelo peel (*Citrus maxima*)	citrus, floral, pine	limonene, geraniol, nerol, linalool, terpineol
pomelo juice	pine, fatty, lemon, rose, waxy, floral, woody-grapefruit	pinene, decadienal, citronellal, nonanal, linalool, limonene, nootkatone
grapefruit peel (*C. paradisi*, pomelo x sweet orange)	fresh, green, floral, eucalyptus, woody-grapefruit	octanal, decenal, dodecanal, eucalyptol, nootkatone
grapefruit juice	fruity, coconut, grassy, woody-grapefruit, sulfurous, catty	esters (ethyl methylpropanoate, butyrate, methylbutyrate), wine lactone, hexenal, decenal, nootkatone, sulfur volatiles (menthenethiol, methyl sulfanyl pentanone)

To round out this very mixed salad of citrus fruits, one that's eaten whole and three that are rarely seen in Western markets. **Kumquats** are bite-size morsels that are mostly peel; they lack the nonvolatile terpenoids and phenolics that give other citrus peels a bitter taste. Their volatiles and aroma resemble the peel of sweet oranges. **Yuzu**, small and sour, yellow-orange when ripe, doesn't belong to any of the three major clans and seems richer and sweeter-smelling than most citrus, possibly due to its unusually complex mix of terpenoids, thymol, coconut-sweet wine lactone, and a sulfur volatile. It's especially enjoyed in Korea and Japan, where the whole fruit is used to perfume bathwater and made into a marmalade for mixing with hot water into a tea. In the Japanese kitchen, yuzu peel is ground with salt and chilies to make the condiment yuzu koshu, and its juice is mixed with soy sauce to make the dipping sauce ponzu. The small **makrut lime** is also one of the family's odd species out, used in Southeast Asia to lend its piney-lemony qualities to various foods (though less often than the makrut tree's leaves, which are dominated by lemongrassy citronellal).

UNUSUAL CITRUS FRUITS: KUMQUAT, YUZU, MAKRUT, BERGAMOT

Fruit & part	Component smells	Molecules
kumquat (*Fortunella/Citrus* species)	citrus, resinous, pine, waxy, floral	limonene, myrcene, pinene, octyl & geranyl acetates
yuzu peel (*Citrus junos*)	floral, green, balsamic, sulfurous, thyme, fatty	linalool, undecatrienone, sulfur volatile (methyl sulfanyl pentanone), thymol, decadienal
yuzu juice	floral, coconut-woody, green, thyme, fatty	linalool, wine lactone, undecatrienone, methyl epijasmonate, thymol, hexenal, decadienal
makrut lime peel (*Citrus hystrix*)	pine, woody, lemony	pinene, sabinene, citronellal, terpinenol
bergamot peel (*Citrus bergamia*, lemon x sour orange)	floral, pine, lemon, lavender, citrus	geraniol, pinene, linalool, neral + geranial, linalyl acetate, limonene oxide

Then there's orange-size **bergamot**, probably most recognizable as the distinctive scent of Earl Grey tea: not especially citrusy, more reminiscent of lavender flowers. Its peel volatiles suggest exactly that, with predominantly floral terpenoids and linalyl acetate, one key to the flowers' rich smell (see page 227). Where did the lavender-like volatiles come from? Bergamot is a cross between lemon and sour orange, neither of whose fruits hints of lavender—but the flowers and leaves of the sour orange do. Both are prized in perfumery as the sources of neroli and petit-grain oils (see pages 458, 460), and bergamot oil was one of the ingredients in the original Eau de Cologne (see page 478). The fruit's unusual name apparently comes not from the Italian city Bergamo, but from the resemblance of its slightly elongated shape to that of another highly regarded fruit, the bergamot pear, whose name came from the Turkish *beg-ármûdi*, "the prince's pear." It may have arisen somewhere in the Mediterranean and is now produced mainly in the southern Italian region of Calabria. The whole fruit is sometimes made into a fragrant marmalade.

Other subtropical and tropical fruits: bananas to pineapples

We now take our leave of the prolific citrus family and its terpenoid variations and come to a stand with a handful of familiar and miscellaneous fruits—varied in family background, shape, color, and aromas—that have in common their origins in the Asian and American tropics and subtropics. They run the volatile gamut, from terpenoids to esters and lactones to benzenoids, furanones, sulfur molecules, and even mustardy isothiocyanates. For people in temperate Europe and North America, these fruits were once new and exciting discoveries, and very different, with intense and sometimes strange flavors. To this day professional flavorists describe the qualities of some of their volatiles as "tropical" or "exotic." Why might fruits from warm climates be so distinctive? Perhaps abundant resources help them explore a wider range of volatiles. Perhaps abundant competition obliges them to exploit that wider range and amplify their smells to stand out. Perhaps because bats are common fruit eaters and seed dispersers and respond to the sulfur molecules that characterize many tropical fruits. We don't really know, but it's fascinating to ponder.

Dates are thumb-size fruits borne in large clusters on palm trees native to the arid Middle East, and have been cultivated there and in western Asia for thousands of years. Elsewhere they're most familiar in their sun-dried form, two thirds and more sugar by weight, brown and wrinkled, and with a sweet caramel aroma developed as they dry. At the earlier firm-ripe and soft-ripe stages known as besser and rutab, still plump and tan, dates typically carry a mix of fruity esters and benzenoids suggesting honey, almond extract, hay, and even cinnamon; some varieties also emit citrusy and floral terpenoids.

Bananas are the fruit of large nonwoody plants native to Southeast Asia and the Pacific islands, and the most important tropical fruit in worldwide commerce. They're typically harvested and often sold while still green and bland; as they ripen they turn color, convert their starch into sweet sugars, and develop aroma. Like the temperate apple and pear, the tropical banana owes its volatile identity to esters. Banana volatiles are dominated by various branched carbon chains that arise from the breakdown of amino acids instead of fatty acids. Sweet bananas are dominated by esters formed from the branched-chain alcohol methylbutanol, also known as isoamyl alcohol; these esters and ripe bananas define each other. As bananas ripen, their initial green-banana smell fades to a sweeter fruitiness, with a clove note from eugenol appearing in some varieties. As they pass their prime, they develop fermented and medicinal notes from ethanol, ethyl acetate, and other volatiles that mask the main characteristic esters.

Plantains are banana varieties that convert less of their starch to sugars as they ripen and don't develop the typical banana smell, so they're usually cooked as a starchy vegetable. Plantains emit spicy and woody notes from the benzenoid pathway, along with green-leaf aldehydes. Interestingly, plantains store a number of other volatiles by attaching them to nonvolatile sugars. Mashing the pulp or heating it can liberate some of these stored volatiles and their smells, which include sweaty, vanilla, and floral notes (from methylbutyric acid and alcohol, vanillin, and phenylethanol, respectively).

Mangoes are borne on large trees native to India and Southeast Asia, and in a manner that seems to encourage the attention of the large fruit bats of the region, hanging down from high branches where they're easily located by smell at twilight or night. A few years after Dave Arnold and I tasted our way through the USDA apple field, we visited a private mango collection in Florida that grows dozens of the thousands of different varieties that have been developed in Asia and else-

where. Their flavors spanned quite a range, from green and turpentiney to
pineapple-like, coconut, peach, raspberry, sweaty, floral—all on a sweet, fruity
base, and all recognizably mango-like. Mango analysts at the German Research
Center for Food Chemistry asserted in 2016 that fifteen key volatiles generate "an
orchestral mango aroma" that persists despite the presence or absence of other
smells. Among those key molecules are fruity esters, caramel furaneol, and lactones suggestive of peach and coconut.

SOME TROPICAL AND SUBTROPICAL ASIAN FRUITS

Fruit	Component smells	Molecules
date (*Phoenix dactylifera*)	fruity, sweet, almond-extract, hay, creamy	esters (ethyl & geranyl acetates), phenylacetaldehyde, benzaldehyde, coumarin, butanediol
banana (*Musa* species)	banana, fruity, grassy, clove, buttery	esters (methylbutyl butyrate & acetate, ethyl methylpropionate, hexyl acetate), hexanal, eugenol, diacetyl
plantain (*Musa* species)	grassy, spicy, fruity, woody	hexenal, hexanal, ethylbutyric acid, unusual benzenoids (elemicin, vinylmethoxyphenol)
mango (*Mangifera indica*)	fruity, pineapple, caramel, resin-turpentine, tropical/passion fruit, sulfurous, peach, coconut	esters (ethyl butyrate & methylbutyrate), undecatriene, furaneol, myrcene, ocimene, sulfur volatiles (methyl sulfanyl pentanone, methyl butenethiol), g-deca- & d-octa-lactones
lychee (*Litchi chinensis*)	floral-rose, lemon, violet, green, citrus, sulfurous	rose oxide, citronellal, linalool, phenylethanol, ionone, nonenal, octanol, nonadienal, sulfur volatiles (diethyl disulfide, dimethyl trisulfide, methyl thiazole)

Lychee or litchi is a small tree fruit native to subtropical southern China, and
perhaps the most floral of all fruits: it emits a cluster of terpenoids that give it a
strong rose, citrus quality, one of them named "rose oxide" for its familiar source
and smell. Less immediately prominent but long noted by lychee lovers is a garlicky, sulfurous background that's apparently strong in fruit fresh from the tree,
less evident by the time they get to market.

We shift homelands now from Asia to the Americas, and take up the **avocado**, native to semitropical and tropical regions, whose massive seed and unusual pulp have suggested that it evolved to appeal to large mammals that are now extinct, not to birds or small mammals. The dense pulp surrounding the seed is rich not in sugars but in oil, and its volatiles are mainly nutty, fatty aldehydes that communicate the pulp's wealth of energy. It's usually treated more as a vegetable than as a fruit.

Papayas are native to the Central and South American tropics, and are a tropical fruit second only to lychees in their floral quality. In addition to floral linalool, they carry the common tropical coconut-like lactones, but they stand out for emitting a sulfur- and nitrogen-containing isothiocyanate—a molecule characteristic of pungent members of the cabbage family, especially mustards and radishes! That sulfurous, biting quality is obvious in papaya seeds, where the defensive isothiocyanate is concentrated, while in the flesh it's far less obtrusive but still contributes to the fruit's unique aromatic identity. There are yellow- and red-fleshed papayas, and the red varieties appear to emit more ionone, the violet-flower terpenoid fragment associated with the red lycopene pigment.

Passion fruits are borne on vigorous vines whose large, complexly patterned flowers reminded Europeans of the cross and Christ's suffering, or passion. The most common are yellow and purple varieties of the same species, the yellow type typically having a stronger aroma. Both are set apart from other tropical fruits by their numerous sulfur volatiles, more than fifty, which give them their distinctive smell. It can sometimes verge on the sweaty and animalic; some of the same volatiles are found in the body fluids of our pets and ourselves (see page 301).

Guava and feijoa or pineapple guava are tropical to subtropical American members of the myrtle family, which includes such aromatic members as eucalyptus, clove, and allspice. Varieties of **guava** range from white to yellow to red and are generally characterized by green-leaf aldehydes and fruity esters, with the yellow apparently rich in caramel furaneol and some sulfur alcohols and esters also found in blackcurrants and grapefruit. **Pineapple guava or feijoa** comes from a cousin genus and has a very different smell dominated by benzoate esters, also prominent in cranberries, which have wintergreen, medicinal qualities.

SOME TROPICAL AND SUBTROPICAL FRUITS FROM THE AMERICAS

Fruit	Component smells	Molecules
avocado (*Persea americana*)	grassy, nutty, fatty, sweet, resinous	hexenal, hexanal, pentanal, nonanal, methyl acetate, myrcene
papaya (*Carica papaya*)	floral, radish, wintergreen, resiny, fruity, coconut, violet	linalool, benzyl isothiocyanate, esters (methyl & benzyl benzoate, ethyl butyrate), g-hexa- & g-octa-lactones, ionone
passion fruit (*Passiflora edulis*)	fruity, floral, caramel, blackcurrant, grapefruit, tropical, sulfurous	esters (hexyl hexanoate & butyrate), ionone, jasmine lactone, furaneol, sulfur volatiles (sulfanyl hexanol, methyl sulfanyl hexanol & their acetate, butyrate, & hexanoate esters)
guava (*Psidium guajava*)	grassy, grapefruit, blackcurrant, fruity, caramel, floral	hexenal, sulfur volatiles (sulfanyl hexanol & sulfanyl hexyl acetate), esters (ethyl butyrate, cinnamyl acetate), furaneol
pineapple guava (*Acca sellowiana*)	wintergreen, medicinal, tropical, fruity, green	esters (methyl & ethyl benzoates, ethyl & hexenyl butyrates), hexenal
pineapple (*Ananas comosus*)	caramel, fruity, pineapple, fresh, coconut	furaneol, esters (ethyl methylbutyrate & propionate), undecatriene, d-octa- & d-deca-lactones

Last but not least among familiar tropical fruits is the **pineapple**, a native of South America, where in the indigenous Tupi it's known as *nanas*; Europeans named it after its resemblance to a pine cone. The fruit is formed by the fusion of dozens of individual fruitlets around a common core; it ripens progressively from the bottom, where it's attached to the mother plant. Pineapple is the tropical fruit par excellence, intensely aromatic and sour and sweet, and it made a strong early impression on Europeans. Try imagining what it would have been like for apple and strawberry eaters to experience the rare pineapple, as Charles Lamb famously described in his 1823 "Dissertation upon Roast Pig":

Pineapple is great. She is indeed almost too transcendent—a delight, if not sinful, yet so like to sinning that really a tender conscienced person would do well to pause—too ravishing for mortal taste, she woundeth and excoriateth the lips that approach her—like lovers' kisses, she biteth—she is a pleasure bordering on pain from the fierceness and insanity of her relish—

A century earlier, it was a new hybrid strawberry's aromatic resemblance to the pineapple that suggested its botanical name, *Fragaria* x *annanasa*, and that resemblance was and is real. Compare strawberry volatiles (see page 310) with the pineapple's and you'll see furaneol, the most prominent in pineapple, is missing in the native European Alpine strawberries. Furaneol has several names, including "strawberry furanone" and "pineapple ketone," and it's essential to both fruits. Another of the pineapple volatiles, an eleven-carbon chain with three double bonds, is said to smell like pineapple on its own, and lactones provide nuances of coconut.

Fruits for thought:
ginkgo, vanilla, durian

Now the last fruits in the market, odd but not mere leftovers. A pile of small yellow-orange spheres, a bunch of hand-long, thin, dark brown pods, and a large, salad-bowl-size yellow mass covered with short spikes, one cut end revealing knobby arrays of white flesh within: these three unusual fruits offer slant perspectives on the estery-terpenoid mainstream that our favorites belong to.

The first pile comes from the **ginkgo**, an attractively upward-branching, fan-leaved tree that once grew all over the Northern Hemisphere and still does as a hardy street tree. Individual ginkgo trees are either male or female, and urban foresters generally try to plant only known males to avoid the nuisance of the females' fruits. Or more properly, seeds: the little spheres that drop in the fall are not fruits. They're premonitions of fruits.

The ginkgo is a relic tree that has survived from the Jurassic period, for some two hundred million years. That's tens of millions of years before the rise of the first plants to bear flowers and fruits. The ginkgo is more closely related to cycads and conifers. The fleshy part of its "fruit" is anatomically part of a seed, not the swollen wall of a flower's ovary. But it's clearly playing the role that the fleshy fruit would eventually be invented for, to attract the attention of animal dispersers.

When they mature and fall from the tree, ginkgo seeds generate a head-turning stench, a mix of the smells of sharp cheese, vomit, and excrement. Its sources are simple starter-set molecules, four- and six-carbon butyric and hexanoic acids. Many mainstream fruits produce the same acids during ripening, but they immediately combine them with alcohols to make pleasing butyrate and hexanoate esters. On their own, the acids are usually signs of decomposition.

Ginkgo seeds are a pungent whiff of what plants had to offer animals before they invented true fruits. In China, one country where they're produced for their starchy (and bland) kernels, the freshly fallen seeds are eaten by wild cats, dogs, and civets. What dispersed them back in the Jurassic, long before there were cats and dogs and fruity fruits? Some biologists speculate that ginkgo seeds evolved their smelly flesh to attract small scavenging dinosaurs! So the next time you're unpleasantly surprised by a sidewalk ginkgo fall, take the occasion to appreciate Hero Carbon's inventive drive, how it managed to advance from stinky premonition to pineapple and peach.

The second exceptional fruit, this time a true fruit, is **vanilla**, the spice. We don't normally think of the vanilla pod as a fruit because it lacks an obvious fleshy layer. Instead its ripening walls turn leathery and brown, and split open to expose tiny seeds that are coated in a thin layer of sugary, oily, aromatic fluid, a kind of syrup. And neither the pod walls nor the syrup emit fruity-smelling esters.

What animals in the tropical forests of Mexico found a fruit like this appealing? Not the usual quadrupeds and birds, but bats and bees! Bats sometimes fly away with the whole fruit, and metallic-colored orchid bees, which are mysteriously avid scent collectors, have been observed to pick up the sticky seeds while harvesting the syrup (other insects may well do the same). The pod's sweet, creamy, floral benzenoid volatiles, vanillin and variations on it, attract these dispersers, and they also suppress the growth of spoilage microbes.

So we apparently owe the particular pleasures of one of the world's favorite spices to tropical bats, shiny bees, and troublesome molds. A remarkable set of influences.

SOME UNUSUAL FRUITS AND FRUITLIKE SEEDS

Fruit	Animalic notes	Molecules
ginkgo (*Ginkgo biloba*)	rancid, cheesy, vomit, excrement	butyric & hexanoic acids
vanilla (*Vanilla planifolia* & others)	vanilla, sweet, creamy, medicinal, floral	vanillin, vanillic acid, hydroxybenzaldehyde, anisyl alcohol
durian (*Durio zibethenus*)	fruity, roasted onion, rotting onion, rotting cabbage, sulfidic, caramel	esters (ethyl methylbutyrate & propanoate), many sulfur volatiles (ethanethiol & methanethiol, ethyl sulfanyl ethanethiol, ethane dithiol, dimethyl trithiolane), ethyl furaneol

The third extraordinary fruit is the **durian**, "the king of fruits" in its native Southeast Asia. It's thought to have originated in the tropical jungles of Borneo; Thailand and Malaysia lead in world production. King it may be, and with its large size and spike-armored skin it looks the part more than Bunyard's apple, but like some rulers, it's both loved and hated. In Singapore it's forbidden to carry durians on public transportation or bring them into many buildings, because their smell is intense, pervasive, and offensive. Having seen the peculiar tasting notes from Alfred Wallace and E. J. H. Corner that I quoted at the beginning of this chapter—onions and cream, sewer drains and coal gas—I had to experience this fruit in its homeland. In 2014, I made a pilgrimage to a celebrated durian stall in the outskirts of Singapore and found that most of the half-dozen varieties I tried tasted of strawberries and a mix of fried onions and garlic. I enjoyed them enough to smuggle one into my hotel room, double-bagged, to have the next day. After just an hour or two its royal presence filled the room and became unbearable. I had no choice but regicide, and disposed of the body like contraband drugs, flushing it in pieces down the toilet.

As the consistent onion and putrid impressions suggest, durians are enthusiastic sulfur chemists. They produce a dozen thiols and a brace of sulfides, some of them unusual molecules with two and three sulfur atoms. Whereas many tropical fruits add interesting sulfurous traces to a core estery fruitiness, durians elevate thiols and sulfides to costarring roles. German flavor chemists reported in 2017

that a simple mixture of just two volatiles could reliably suggest durian to people who smelled it: the ester **ethyl methylbutyrate**, a major
component of strawberry and pineapple smells, and **ethyl sulfanyl ethanethiol**, a rare sulfur volatile that smells like roasted onions.

Which animals do durians succeed in recruiting with this mix of fruity and
sulfurous? According to botanist Corner:

> In Malaya the smell of fruiting trees in the forest attracts elephants, which
> congregate for first choice; then come tiger, pigs, deer, tapir, rhinoceros, and
> jungle men. Gibbons, monkeys, bears, and squirrels may eat the fruit in the
> trees; the orang-utan may dominate the repast in Sumatra and Borneo. . . .

More recent observations have added sun bears, giant rats, porcupines, and ma-
caques, with orangutans apparently especially fond of durians—as of course are
some of their human cousins, who cultivate and harvest them by the ton. Most of
these durian dispersers are omnivores, not strict fruit eaters, so the sulfur volatiles
may help suggest animal prey or carrion, richer in fats and protein than most
fruits—as durian in fact is.

If the smell of durian comes down to strawberry and roasted onion, an unusual
but hardly horrific combination, then what accounts for the disgust and prohibi-
tions it inspires? Perhaps its unremitting intensity: valuable in the open air of a
tropical forest teeming with creatures and their competing volatiles, but not so
nice in the enclosed and deodorized spaces of civilization, where minor rotlike
thiols can accumulate to full putridity. I enjoyed Singapore durians on the sidewalk but not in my hotel room. Alfred Wallace experienced the same contrast:

> When brought into a house the smell is often so offensive that some per-
> sons can never bear to taste it. This was my own case when I first tried it in
> Malacca, but in Borneo I found a ripe fruit on the ground, and, eating it out of
> doors, I at once became a confirmed Durian eater.

If you're curious about durian but a journey to the Far East isn't imminent, then
buy one at an Asian market or online and take it outside. If it's even close to ripe

and representative, it will be sensational, one of the most memorable fruits of your life.

With durian we conclude our seven-chapter survey of the kingdom of land plants and its volatile wealth, which scents the countryside air we breathe and the foods we eat. As we've seen, volatile molecules are one indispensable means by which rooted plants protect and renew themselves in a world crawling with creatures hungry for their substance. Most plant volatiles are defensive warnings and weapons, but by controlling our exposure to them, we manage to enjoy the sensations they stimulate—so much so that we volunteer to protect them ourselves. We do the same for flowers and their mixed chemical messages. And in the case of this chapter's fruits, we're beneficiaries of the eons-long negotiation between seed-producing plants and seed-dispersing animals of all kinds. Not only are fruits delicious themselves: fruits school us in the possibilities of deliciousness.

We've come to the plant kingdom's last stand in our imaginary market. Step past it through the exit, and we come to a large pile of discarded flowers and vegetables and herbs and fruits, all wilted, blemished, bruised. Their useful life is finished. But they're about to become starting materials for the creativity of other living kingdoms that helped foster their growth in the first place. We turn now to the posthumous fate of most plants, the kingdom of the fungi and the making of the soil.

Part 4

....

LAND, WATERS, AFTER-LIFE

Chapter 14 · THE LAND: SOIL, FUNGI, STONE

It is certain that the best soil smells like a fine unguent. . . . It is the odor often recognized at sunset . . . at the place where the ends of the rainbow meet the earth, or when rain has soaked the ground after a long drought. Then it is that the earth exhales her own divine breath, received from the sun, and of incomparable sweetness. This is what soil should emit when turned up. . . . Odor is the best way to judge soil.

> • Pliny, *Natural History*

That many natural dry clays and soils evolve a peculiar and characteristic odour when breathed on, or moistened with water, is recognized. . . . It is primarily in arid regions, where the comparative absence of organic matter in the soils and the frequent preponderance of various types of outcropping rocks in the terrain are characteristic features, that this odour is most widely recognized and is frequently associated with the first rains after a period of drought. . . . The diverse nature of the host materials has led us to propose the name "petrichor" for this apparently unique odour which can be regarded as an "ichor" or "tenuous essence" derived from rock or stone.

In petrichor our olfactory senses give us knowledge of one point in a great natural cycle of physical and chemical reactions.

> • Isabel Bear and Richard Thomas, "Nature of Argillaceous Odour," 1964;
> "Genesis of Petrichor," 1966

Smells of the soil, smells of rock and stone: true olfactory land-marks! They come from patches of Earth itself, the clump of cosmic rubble whose star-warmed surface is our home. So they've long been thought to hold clues to the planet's essential workings. And they do, even when we encounter them in flowerpots and sidewalks, those bits of Earth displaced and remade.

For the Roman naturalist Pliny, the smell of fertile soil is the mixed breath of sun and Earth, especially evident when Earth exhales the day's warmth at sunset, and when the heavens touch down with rainbows and rain. A lovely fable. Two millennia later and half the globe away, Australian scientists Isabel Bear and Richard Thomas reached back to Greek mythology to name the smell of long-dry rock and soil when they're freshly moistened. From *petri-*, "rock," and *ichor*, the ethereal blood of the gods, they coined the paradoxical term *petrichor*: the intangible but smellable "tenuous essence" emitted from Earth at its most tangibly solid.

Despite its evocative pedigree, petrichor is an unfortunate term. It reinforces our first impression when we smell freshly wet stone, that the smell belongs to the stone itself. We think we're experiencing its "minerality." But we're mistaken, misdirected by the stone's solid presence. Bear and Thomas found that the smell actually arises from the sparse airborne remnants of Earth's living things. Dry rock happens to collect these remnant volatiles day after day—as do dust particles in the atmosphere—until sudden rain releases them in a smellable burst. So the smell of rock, like the smell of approaching rain, is a tenuous essence of the living planet. A truer hellenized name would be *gaia-ichor*: the exhalations of Gaia, Mother Earth.

Soils contribute to gaiaichor, and their smells are also strongest after fresh rain. The soil is where the land's creatures have lived, died, and then entered an afterlife as Gaia reclaimed their substance and redistributed it to new generations. The smell of the soil is the breath of that cycle up close, just as it begins to diffuse into the open air. It's sweet, neither sterile nor stinking, when it arises from the dynamism of life and after-life feeding each other, actively building the soil and its fertility. It's the smell of a great cycle without which the land would be little more than bare rock.

Soil is the interface of the mineral planet and the living communities that inhabit it. It's where rock, air, water, and life mingle more intimately than anywhere else. It didn't exist until early microbes from the waters first clung to young Earth's rocky shores. Those pioneers and their followers gradually made the land over. They etched and dissolved it with their acidic fluids. When they died, they bequeathed the soft crumbs of their remains to mix with particles of weathered rock. The collage of inorganic and organic bits retained moisture and minerals and air, gave shelter from the sun and weather, and offered oases where scavenging cells could thrive and then die in their turn. Areas of the bare planet floor began to disappear under a carpet of soil increasingly hospitable to new life.

The greatest benefactor of the soil, and its great beneficiary, has been the plant kingdom. Land plants materialize sunlight and air into leaves and stems and trunks and fruits, and over the eons dropped enough of themselves to form a carpet capable of supporting trees three hundred feet (a hundred meters) tall. Our word *soil* comes from an ancient root meaning "sit." The soil is the plant kingdom's low-profile throne, the seat of power from which potentate trees and humble weeds alike bestow food and shelter on the land's other creatures.

In order for the steady rain of plant remains to become soil and feed new generations, the complex, physically tough plant structures must be broken back down into simple usable carbon molecules. Bacteria do their part by invading damaged tissues, metabolizing the free nutrients, and becoming food in turn for small animals. Animals of all kinds, from single cells to earthworms to beavers, help by grinding up plants and their debris, absorbing some, leaving the rest for others to work on, and eventually contributing their own remains.

But the leading decomposers are members of another realm of life altogether: the fungi, the kingdom of molds and mildews, mushrooms and yeasts, whose earliest ancestors may have colonized wet shorelands long before green plants did. Along with some odd bacteria that emulate them, fungi bring their own distinctive sets of volatiles to our everyday lives. They help scent the soil, rotten fruits, and dank closets; they also create delicious foods and drinks, some worth their weight in mineral gold.

Ahead in this chapter, then: autumn leaves, kitchen compost, mushrooms and truffles, puffy bread dough and funky beer, stones wet and dry. A wide range of settings, so we'll shift among them in imagination as needed.

Molds, mushrooms, yeasts, streptomycetes

To get acquainted with the fungi, let's begin by walking in the cool, moist shade of a forest preserve or patch of woodland you're familiar with. We leave the beaten path, and get down on our hands and knees to poke and dig and sniff in a small area an arm's span across. Among the scattered leaves there are bruised fruits, some bubbling with leaking juices, some with velvety green blotches, some with a black fuzz. There's a disintegrating tree branch, one end brown and crumbly, the other white and soft. A half-buried rock with what looks like a splash of orange paint, but intricately lobed. A cluster of grayish mushrooms poking into the air, one of them showing rootlike filaments at its base. A patch of ground that's pale green and crusty. The smells: variously winey, sharp, rotten, musty, mushroomy. Earthy.

Each of these inconspicuous features is an outpost of the kingdom of the fungi, each a particular fungus doing the dirty and invaluable work of decomposition. The bubbling juices: yeasts. The green velvet and black fuzz: molds. The brown and white branch: less and more effective wood rots. The paint splash: a lichen. The mushrooms: temporary pop-ups from an underground network. The green crust: an integrated community of fungi, lichens, bacteria, algae, and mosses, knitting the surface soil together. Along with plants and animals, the fungi make up the third kingdom of complex life forms that dominate our everyday experience. And their volatiles and smells are very much their own.

Fungus first meant "sponge" in Latin and was later applied to mushrooms, which are spongy and release an impressive amount of water when squeezed by heat. Scientists adopted the word to denote the entire evolutionary clan that includes the mushrooms, an estimated million different species that have colonized both the waters and the land—and jet fuel and paint!—with structures that range from infinitesimal cells to basketball-size truffles.

Though mushrooms seem plantlike in their immobility and habitat, they and their clan are more closely related to animals, both genetically and in their basic life strategy: they're unable to photosynthesize, so they need to get their nourishment from other living things or their remains. They start life as a microscopic seedlike spore that germinates into an actively metabolizing cell. Yeast cells remain solitary, living much like single-cell bacteria and budding off independent progeny, but most fungi form long, branched, many-cell filaments called *hyphae* (from the Greek for "web"). Their fine white haze is the first visible sign of mold growth on fruits and bread. The hyphae of a soil fungus are often bundled together in a mass of interwoven white threads collectively known as a *mycelium* (from the Greek for "mushroom"). The growing tips of these mycelia can weave their way through miles of soil and penetrate solid plant and animal tissues, dead and living, to gather nutrients.

Some fungi attack living plants or animals and cause diseases, while others, including a number of mushroom and truffle species, develop mutually beneficial relationships with specific trees and their roots, supplying soil minerals in exchange for sugars and other carbon-chain nutrients. Then there are the many soil dwellers that gather nutrients from the tissues of the fallen dead. Unlike animals, which take food into their bodies, digest it there, and expel the residues, nonsymbiotic fungi export powerful digestive enzymes into their surroundings, and then absorb the building-block molecules that the enzymes release from the local detritus. Whatever other creatures are nearby can also help themselves to the feast. These soil fungi recycle fallen fruits, leaf litter, animal dung, and the complex matter of soil itself, unlocking stored chemical energy and building blocks and nourishing the neighborhood.

The most massive land plants are trees, and their wood is especially dense and difficult to unlock. A number of bacteria as well as the brown-rot fungi can digest cellulose and hemicellulose, two of the main components of wood. But the third, lignin, is useful to living trees precisely because it resists microbial attack. White-rot fungi are among the planet's few specialists in digesting the more complex lignins of hardwood trees; shiitake and oyster mushrooms are their most delicious representatives.

Mushrooms are special structures that some fungi form to make and disperse their spores, soft showerheads that rise from rotting wood or the soil to rain spores by the billions into the passing air currents. Other species, *filamentous* fungi like those that rot our foods with visible surface patches, raise tiny spore launchers on short filaments that give them a velvety or furry or fuzzy look. Still others, including single-cell yeasts, release their spores invisibly.

SOME FAMILIAR FUNGI: MUSHROOMS, MOLDS, YEASTS

Fungus	Common habitats	Uses
common white/brown mushroom, *Agaricus bisporus*	plant litter, dung	food
shiitake, *Lentinula edodes*	deadwood	food
porcini, cèpes, boletes, *Boletus* species	symbiosis with tree roots	food

continued

Fungus	Common habitats	Uses
truffles, *Tuber* species	symbiosis with tree roots	food
Aspergillus species	moldy vegetables, grains	making soy sauce, miso, rice alcohols
Penicillium species	moldy citrus fruits	making cheeses, sausages, antibiotics
Rhizopus species	moldy bread, fruits	making tempeh
brewing, baking yeasts, *Saccharomyces cerevisiae*	spoiling fruit, damaged plants	making wine, beer, bread, soy sauce, yeast extract
Brettanomyces species	wineries, breweries	making beer

Quite apart from their role as soil makers, fungi are virtuosic chemists whose creations variously please, intoxicate, and cure us. We value mushrooms and truffles more for their distinctive smells and tastes than their nourishment. With molds we turn bland milk and soybeans into aromatic cheeses and savory soy sauce; with yeasts, fruits and grains become wine and beer. Many fungi try to limit the freeloading of opportunistic neighbors with chemical weapons to repel or suppress or kill them. While some of these molecules are toxic for us—and make eating wild mushrooms a risky business—others have turned out to be useful. Alcohol and LSD and penicillin are among them.

One last introduction to some major soil scenters: microbes that are not true fungi, but such good fungus impersonators that they were initially misidentified and named for them. The **streptomycetes** (from the Greek for "twisted" and "mushroom") are bacteria whose family arose around the time when plants were taking hold on land, and apparently evolved fungus-like characters to adapt to life in soil. Like most fungi, streptomycetes secrete digestive enzymes into their surroundings and form spores and hyphae-like chains of cells. They also generate powerful antibiotic molecules to defend themselves: streptomycin, tetracycline, and a number of other drugs derive from streptomycetes.

Soil's beginnings: fallen leaves, kitchen compost

Soil creation begins with the delivery to the ground of its raw ingredients: primarily dead or dying plant tissues. It's estimated that some ten billion tons of tree needles and leaves are moldering away worldwide at any given time. Though most of the volatiles that they release are absorbed into the soil, the remainder that make it into the air are enough to be a major contributor to organic volatile emissions from the land, its tides of scent and gaiaichor precursors.

Leaves and other plant parts drop with their resident microbes, and these start the process of converting plant litter into soil by metabolizing the ready sources of energy, sugars and starch and some cellulose, membrane lipids and proteins. Molds soon join in. In temperate forests, it takes a few weeks for various microbes and fungi to consume half of the carbohydrates in leaf litter. When what remains is mostly structure-reinforcing lignin, the lignin-specialist fungi dominate, along with streptomycetes. It can take six months for half the lignin in a leaf to be broken down.

You don't have to go to the forest to smell these initial stages of soil making: **autumn leaves** in the yard or gutter undergo the same process. Most decomposing vegetation tends to release many of the same simple starter-set carbon chains, ethereal and solventy methyl and ethyl alcohols, acetaldehyde and acetone, all products of the partial breakdown of complex carbon chains. Fallen needles in conifer forests release large amounts of their typical terpenoids, pinene and carene and limonene, while broadleaf foliage instead emits the nearly odorless terpenoid relative isoprene (see page 207). Many leaves are coated with a protective layer of wax, and soil fungi have been found to exploit the wax on pine needles and generate large ring molecules with twelve to sixteen carbons that are close relatives to the "musk" molecules highly prized in perfumery! (See page 465.)

Probably the most distinctive volatiles of leaf litter are sweet-smelling, tobacco-like **furfural** and hydroxymethyl furfural, five-cornered rings with one oxygen that are generated by microbes and fungi as they break down cellulose. Fungal metabolism of the more resistant lignin releases traces of toluene and xylene. We know these molecules best as solvents manufactured from petroleum, so to us they have a solventy "chemical" smell (see page 420), but they contribute quite naturally to the heady scent of the forest floor.

SOME SMELLS OF LEAF LITTER

Smells	Molecules
alcohol	methyl, ethyl alcohols
green, fresh	acetaldehyde
solvent	acetone
sweet, woody, bready	furfural
caramel, tobacco	hydroxymethyl furfural
sweet, plastics glue	toluene (methylbenzene)
plastic	xylene (dimethylbenzene)
warm, sweet, pleasantly animal	C12, C14, C16 lactone rings
sweet, caramel, cake	maltol

There's one remarkable tree whose leaves are appreciated for their autumn smell as much as maple leaves are for their colors. The fallen leaves of the temperate Asian **katsura**, *Kuchenbaum* or "cake tree" in German (*Cercidiphyllum japonicum*), smell like caramel, cotton candy, and baked sweets, thanks to their accumulation of large amounts of the carbon-oxygen ring maltol (see page 490), just before they drop. How might the tree benefit from its carpet of sweet-smelling leaves? Still a mystery.

Garden compost is a backyard, accelerated version of natural decomposition. The piling of garden refuse and kitchen scraps creates a concentrated mass of nutrients that prompts rapid microbial and fungal metabolism and decomposition. In addition to generating chemical energy, these reactions also generate heat, and because the interior of a compost pile is effectively insulated, its temperature can rise high enough to cook a steak beyond well done! The compost microbes that tolerate such temperatures apparently evolved in hot springs, natural vegetation piles, and animal droppings. The smells they generate can be either pleasant or nasty, depending on the compost contents and how they're piled. Fresh "green" refuse retains much of the protein machinery that ran its cells, while dead "brown" matter retains very little. If a compost pile contains more protein machinery from fresh material than there is ready energy from brown matter to build new cells to use it, then the compost crew breaks the machinery down for energy too. Because proteins contain nitrogen and sulfur, the result is a pile that smells of fishy amines and acrid ammonia, and sulfidic, rotting hydrogen and methyl sulfides. If the fresh material is well diluted by dead leaves or wood chips (between twenty-five and fifty parts of brown to one of green), then amine and sulfide production is minimal, and the pile smells like a strong version of the forest floor.

A second factor that determines compost smell is how well aerated the pile is. With plenty of available oxygen in a new pile, the dominant composting microbes are aerobic, and they break their carbon sources all the way down to odorless carbon dioxide and water. As the pile settles and becomes more compacted, low oxygen levels enable the growth of anaerobic microbes. The result is their typical strong mix of sour, vomitous, rancid short-chain acids, fishy amines, and putrid sulfides. These are the smells of compost packed too densely or unable to drain excess water, or of flowerpots soaked by weeks of wet weather. They're also the smells of fresh manure and of natural wetlands (see page 370).

Soil smells: generic life, enigmatic geosmin

Eventually plant litter and compost end up belowground as part of the soil proper, where they contribute to the background of its distinctive smell. Organic remains account for just a few percent of the soil's volume, and most of that portion is *humus*, a mix of persistent molecular remnants that absorb and hold some water and minerals, help keep the soil porous and moist and nourishing, and give it its characteristic brown-black color.

Undisturbed soil teems with life of all kinds, much of it as yet unidentified. It's estimated that there are tens of billions of bacteria in every gram of soil, and tens of thousands of different species of microbes, fungi included—along with minute worms and insects. The bulk of most soils is roughly equal volumes of rock particles and the spaces in between them, either empty or filled with water. The small wet pores are usually occupied by bacteria, and fungal hyphae thread their way through the larger ones to find rare oases of plant or animal remains, or living plant roots.

With such a diverse cast of creatures in the soil, it's no surprise that its volatiles are also legion. They're dominated by several odorless gases: methane, carbon dioxide, and nitrogen oxides, the first the product of microbial metabolism in the oxygen-poor regions of the soil, the others from aerobic breakdown of plant and animal remains. The remaining emissions, a tenth to a third of the total, are largely the usual starter-set volatiles (see pages 47–49), along with terpenoids that originated in needles and leaves. And a small fraction consists of volatiles constructed by the soil denizens as signals, defenses, hormones, and other aids for growing and reproducing.

The distinctive smell of the soil, and its source, were tracked down by chemists searching not for Pliny's divine breath, but for the cause of a dirty taste in river waters and fish. After initial studies by a French chemist and a German bacteriologist around the turn of the twentieth century, English scientists determined in the 1930s that a strong off-smell in drinking water from the river Nile, and an "earthy taint" in Scottish river salmon, were both caused by streptomycete bacteria. In 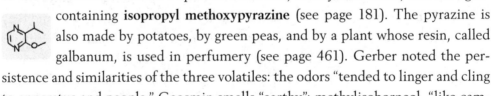 1965, the Rutgers microbiologist Nancy Gerber determined the structure of the earthy-smelling volatile produced by streptomycetes, and named it **geosmin**, from roots meaning "earth" and "smell."

This olfactory landmark is an altered version of a sesquiterpenoid, with ten carbon atoms arranged in two linked rings and decorated with two methyl groups. Today a number of other geosmin producers are known. They include soil- and water-dwelling cyanobacteria and myxobacteria, algae, mosses and liverworts, earthy-tasting beetroots and chard and spinach, cactus flowers, the occasional fungus—though to date not a single mushroom—and millipedes. But by far the most common geosmin emitters are streptomycetes and cyanobacteria.

Nancy Gerber later found two other streptomycete volatiles that contribute to the soil's smell: another terpenoid derivative, methylisoborneol, and nitrogen-containing **isopropyl methoxypyrazine** (see page 181). The pyrazine is also made by potatoes, by green peas, and by a plant whose resin, called galbanum, is used in perfumery (see page 461). Gerber noted the persistence and similarities of the three volatiles: the odors "tended to linger and cling to apparatus and people." Geosmin smells "earthy"; methylisoborneol, "like camphor"; the pyrazine, "musty." All three cause flavor problems in farmed fish (see page 390). Curiously, none of them appears to be a potent chemical defense, especially by comparison to the antibiotics and toxins that streptomycetes make. Instead, they may be a way for these microbes to signal their presence to each other, mark territory, and avoid overcrowding.

SOME DISTINCTIVE SMELLS OF SOIL

Smells	Molecule
earthy	geosmin (modified sesquiterpenoid)
earthy, camphor, musty	methylisoborneol (modified monoterpenoid)
earthy, musty, potato, green pea	isopropyl methoxypyrazine

The prominence of geosmin in the smell of the soil comes partly from its production by ubiquitous microbes, but also from our extreme sensitivity to it: we can detect it in very small amounts. Why should we be so sensitive to molecules that aren't toxic? And why does geosmin smell divine in soil but not in drinking water or salmon? It's appropriate that the Earth should smell earthy, and geosmin is a sign of life, which suggests a hospitable place, moist and fertile—positive things. (In cactus flowers, it may lure animal pollinators as a promise of scarce desert moisture.) But we don't normally ingest earth! In waters and fish, geosmin is earthiness out of place. The same streptomycetes and cyanobacteria that produce geosmin are capable of secreting powerful toxins, as are geosmin-producing penicillium molds that rot fruits and grains. It's good to be sensitive to the presence of these potentially harmful microbes, and wary of earthiness when it doesn't emanate from the Earth or the foods that we pull from it.

So the soil's special smell is a mix of odd terpenoids and a pyrazine overlaid on the common exhaust fumes of living metabolism. Its primary authors are bacteria, not the fungi that are primary authors of the soil itself. We often describe mushrooms and other edible fungi as having an earthy flavor. But they don't smell much at all like the soil they emerge from. Fungi contribute their own set of landmarks to the smellscape.

Fungal volatiles and smells

The leaves of green plants smell "green" when crushed because they emit the six-carbon green-leaf aldehydes and alcohols (see page 154). Most fungi smell "mushroomy" or "moldy" because they emit a characteristic group of *eight*-carbon aldehydes and alcohols. These are called octanal and octanol when they're simple chains, octenals and octenols when two of the chain carbons share a double bond. (There's more than one octenal and octenol because the double bond can be shared by different pairs of chain carbons.) As is true for the six-carbon leaf vola- tiles, the eight-carbon fungal volatiles function as chemical weapons and signals, and fungi boost their production when their tissue is damaged. The most common and characteristically mushroomy molecule, sometimes called *the* **mushroom alcohol**, is an octenol. It's toxic to microbes and repels the slugs that commonly prowl the forest floor. But it and other eight-carbon volatiles also regulate the activities of the fungi themselves. They've been variously found to inhibit the growth of mycelia, the production of spores, and spore germination— perhaps a defensive general systems shutdown.

Why eight-carbon chains for mushrooms, six-carbon chains for plants? Maybe it's been advantageous for the two soil-dwelling kingdoms to maintain separate arsenals and communication channels, to avoid confusion with each other or with the starter-set stews of other soil microbes. It's also a matter of biochemical con- venience. Both sets are made by breaking down the long carbon chains that form membranes in and around cells. The green-smelling six-carbon fragments start with linolenic acid, a copious component of the leaf's photosynthetic machinery with three double bonds. Fungi aren't green, don't photosynthesize, and contain very little linolenic acid; their enzymes work instead with linoleic acid and its two double bonds, and produce eight-carbon fragments, most commonly and abun- dantly the octenols.

The eight-carbon volatiles aren't "earthy" in the way that geosmin is, heavy and clinging. Split a fresh mushroom, hold it to your nose and inhale, and the effect is stinging and ethereal, the abundant eight-carbon alcohols having an effect similar to their two-carbon namesake ethyl alcohol. They're emitted as a blend, their proportions depending on the particular species, its growth stage, and other fac- tors. Isolated and smelled individually, all of the eight-carbon volatiles also tend to suggest the other things in which they make prominent appearances, especially the waxy aspect of citrus peels, and sometimes chicken fat, which is rich in lin- oleic acid and therefore also its fragments. The ketone octenone stands out for its additional metallic, bloodlike qualities, which it suggests because it's the major fragment generated by reactive metals from long carbon chains when we handle keys and coins, wash pots and pans, or bleed iron-rich blood from a bit tongue or a cut (see page 502).

SOME COMMON EIGHT-CARBON VOLATILES IN FUNGI

Molecule	Smells
octanol	waxy, green, orange peel, mushroom
octanal	waxy, citrus
octanone	green, fruity, musty
octenol, "mushroom alcohol" (oct-1-en-3-ol)	mushroom, fatty, earthy, green
octenol (oct-2-enol)	fatty, citrus peel
octenal	fatty, green, citrus peel
octenone	metallic, blood, mushroom, earthy
octadienol (two double bonds)	fatty, chicken broth

We usually notice the generic smells of the fungi when we enter a confined space dank enough to support their growth. Mildew and other molds thrive in bathrooms and basements and storage rooms, even on such minimal nourishment as corner dirt, soap residues in the grout between tiles, and paint. Strong moldy smells may make people feel ill, and the airborne spores that they accompany can cause real illness. However, a few molds apparently improve air quality—by absorbing the volatiles of other fungi and using them for food! Wine cellars and distillers' warehouses have been known for centuries to develop shaggy and sooty deposits on their walls, and these were eventually identified as molds feeding on the vapors of alcohol escaping from the barrels of wine and spirits. One shaggy mold common in the cellars of Germany and Hungary, variously called *Racodium* or *Zasmidium*, can metabolize formaldehyde and other cellar fumes as well, and is sometimes intentionally fed with wine to encourage it to grow and clean the cellar air.

The eight-carbon chains are the most characteristic and recognizable fungal volatiles, but the most popular fungal product by far is a simple two-carbon molecule: the alcohol ethanol. Its solvency smell is pretty nondescript. But yeasts secrete it in copious amounts to suppress their competitors, and billions of us actually enjoy its intoxicating attack on our neurons. There's no other volatile on the planet produced or consumed in greater amounts. We'll take a closer sniff at alcohol and other yeast volatiles in a few pages.

The fungal volatiles that we notice most often belong to the dozen or more common fungi that are delicious enough to eat. These mushrooms and truffles demonstrate how much more their kingdom has to offer than generic mushroom-iness.

Fungal eruptions: mushrooms

For most of us, mushrooms are a kind of vegetable, almost always cooked. To students of the fungal kingdom, mushrooms are "fruiting bodies." Like the seed-carrying fruits of plants, they're specialized structures for getting the next generation out into the world. When the underground parent networks have accumulated enough energy and substance, and the conditions are favorably moist, they organize a mass of hyphae into a structure that can force its way upward. Once it breaks through the soil, it opens an umbrella-like cap that bears small pores or blade-like gills, and these release billions of spores into the passing air currents.

Of course fruiting bodies rising from the soil or rotting wood bear little resemblance to apples and strawberries! The fruits of green plants are tokens of exchange that reward animals for helping to disperse seeds. The primary disperser for mushrooms is the wind, which requires no eye-catching color or nose-catching smell or valuable nutrients. So mushrooms generally don't beckon animals, and may even sicken them with toxins to discourage second helpings. But there are exceptions. The putrid **stinkhorn** emits multiple sulfides and indole, and smells much like the corpse flowers that attract pollinating flies (see page 214). Interestingly, it also emits floral phenylethanol, as do a number of other fungi. Spores are closer in size to pollen than to seeds, so maybe some mushrooms benefit from the attention of insects that visit because they associate floral volatiles with the reward of nectar—which mushrooms don't bother to provide.

Mushroom smells are generally dominated by the eight-carbon chains. Cutting and chopping boost their production. Different species supplement these generic smells with other, often premade volatiles, most commonly benzenoid rings and sulfur molecules, with the occasional fruity esters and piney or floral terpenoids. The result is a spectrum of flavors that share a family resemblance but span quite a range. Mushrooms usually aren't eaten raw, but they're often quite aromatic on the cutting board, and in most cases their native flavors show through even when cooking adds new volatiles to them (see page 516).

Dried mushrooms of all kinds are a wonderful flavor resource. Spongy fungi dry easily, and when they do so, particularly with the help of gentle heat, their aroma intensifies. They retain at least some of their essential mushroomy character, but this is enriched by a host of new volatiles that the cell contents generate as they become concentrated and react with one another. These include meaty-smelling furanthiol and potato-like methional, both sulfur compounds, and a range of roasty, nitrogen-containing pyrazines and nitrogen-sulfur thiazoles. Prized king boletes (cèpes, porcini) are the champions of this transformation, but even ordinary button mushrooms develop great depth of flavor when dried.

Many mushrooms are made by fungi that live independently on leaf litter or decaying wood, and these are the kinds that are easily cultivated and sold inexpensively. The mushrooms of symbiotic fungi, which include boletes and chanterelles, are rarer and pricier because they require a long-standing collaboration with tree roots, typically several years, and that relationship has so far been too complex to simulate by other means. Though some symbiotic fungi are now cultivated in plantations, truly wild mushrooms are gathered from forest lands. This is thought to be a reasonably sustainable practice because the parent body of the fungus remains intact underground.

So let's return to the woodland where we first got our knees and hands dirty, and go hunting for our favorite edible fungi.

Smells of cultivated mushrooms

We'll start at the edge of the woods where there are grass litter and tree stumps, prime territory for the decomposer fungi that are easiest to cultivate. The **common button or white mushroom** and the **brown or cremini or portobello mushroom** are all versions of the same species, the large portobello allowed to mature longest. *Agaricus bisporus* is a humus-loving fungus and grows well on plant residues of all kinds, including horse manure; close relatives are common in suburban lawns and gardens. Its smell is dominated by the typical eight-carbon chains. Mushroomy octenol overtakes green-fruity octanone when the intact mushroom is cut or crushed, then fades when it's dried or cooked and almond-extract benzaldehyde and benzyl alcohol come to the fore. A sister species is called the **almond mushroom** for its prominent amounts of these benzenoids. The **straw mushroom** is the Asian equivalent of the standard button; it's a white-rot fungus that also does well on composted straw, and emits mostly octenol.

There are a handful of different oyster mushrooms, all species of the wood-rotting genus *Pleurotus*, but with their own characters. The standard gray **oyster mushroom** resembles the common white with its eight-carbon chains and almond note, but it also has a couple of floral volatiles that can help it seem more aromatic. The **king oyster, king trumpet, or eryngi** is mostly thick white stalk with a small brown cap; to the usual octenol it adds a couple of sulfur volatiles and their potato and nutlike notes. When cooked, the king develops the caramelly, fenugreeky furanone sotolon. The **abalone** version of *Pleurotus* is remarkable for emitting significant amounts of the nine-carbon floral aldehyde nonanal. And the quirky **pink oyster** carries sweet and medicinal overtones from a furan and phenol. Its pink color comes from a nitrogen-containing indolone pigment, and when it begins to deteriorate, it emits a urinous, ammoniacal smell that cooking can't ameliorate.

The **shiitake** is a widely popular Asian mushroom that grows wild on downed logs of the Japanese chinquapin, or *shii*, a relative of the chestnut. It's now cultivated on wood wastes. The flavor is like no other mushroom's, deriving from a somewhat mysterious set of sulfur volatiles. The aroma of raw and intact shiitake is simply mushroomy. When cut or crushed, it emits the typical mushroomy eight-carbon molecules, but also cabbagey sulfides, and unusual rings with two carbon atoms and three, four, or five sulfurs. The smaller rings are reminiscent of onions, while the five-sulfur molecule largely provides the distinctive shiitake aroma and has been dubbed **lenthionine** after the shiitake's scientific name. The formation of lenthionine in particular seems to involve a combination of mushroom enzyme action and spontaneous chemical reactions: the strongest shiitake aroma results when the fresh mushrooms are dried at moderately high temperatures, around 140°F (60°C), and then rehydrated in hot water. Simply heating them without drying produces a more generic mushroominess.

Shiitake aroma is very specific to that fungus, but there are other mushrooms that echo the sulfurous smells of green plants. As its name indicates, the **garlic**

parachute smells like the most odorous of alliums; it emits familiar sulfides as well as short carbon chains with two and three sulfur atoms.

SOME MUSHROOMS THAT GROW ON LITTER AND DEADWOOD

Mushroom	Component smells	Molecules
white, brown; field, champignon (*Agaricus bisporus* vars. *alba, avellanea; A. campestris*)	mushroomy, almond extract, floral	octenol, octanone, benzyl alcohol & aldehyde & acetate, anisaldehyde, phenylethanol
almond (*Agaricus subrufescens*)	mushroomy, almond extract, anise, sweet	octenol, benzaldehyde, benzyl alcohol
straw (*Volvariella volvacea*)	mushroomy, fresh	octenol, octanol, octadienol, limonene
oyster (*Pleurotus ostreatus*)	mushroomy, almond extract, floral	octenol, octenone, octanal, octanone, nonanal, benzyl alcohol & aldehyde, phenylethanol, linalool
king oyster (*Pleurotus eryngii*)	mushroomy, potato, sweet, nutty; cooked: fenugreek	octenol, methional, pentylfuran, acetyl thiazole; cooked: sotolon
abalone (*Pleurotus abalonus*)	mushroomy-floral, sweet	nonanal, octenol, pentylfuran
pink oyster (*Pleurotus salmoneostramineus*)	mushroomy, sweet, medicinal, fatty; old: urinous	octenol, pentylfuran, phenol, decadienal; old: amines
shiitake (*Lentinula edodes*)	mushroomy; dried: shiitake, sulfurous, onion	octenol, octanone; dried: lenthionine, dimethyl di- & tri-sulfides, trithiolane, tetrathiane
garlic parachute (*Marasmius alliaceus*)	garlicky, fruity	dimethyl di- & tri-sulfides, di- & tri-thiahexanes, benzaldehyde

Smells of symbiotic mushrooms

Now let's head deeper into the woods, where the hyphae of symbiotic fungi grow in close association with tree roots, and their fruiting bodies may betray their presence only with a few fallen leaves elevated above the rest. Among the most prized are various species of *Boletus*, above all the one variously known in English as **cep or king bolete**, in French as ***cèpe***, in the Italian plural as ***porcini***. These are stocky, meaty mushrooms with a cap covering a dense sponge of pores rather than sheetlike gills. They have the usual mushroomy scent when raw, but when dried develop a stronger and much more complex aroma, with meaty, nutty, and roasted notes from a sulfur volatile and a number of pyrazines characteristic of high-temperature cooking (see page 491). Another mushroom whose flavor is transformed by drying is the **candy cap or *Maggi-pilz***, the German name referring to a commercial brand of dry soup mix. The two names reflect the two different qualities attributed to the furanone sotolon (see page 172), which develops as the mushroom dries: maple syrup on one hand, fenugreek and savory Indian spice mixes on the other. Dried candy caps have such a strong maple smell that they're used to make a delicious fungus ice cream!

SOME MUSHROOMS THAT GROW IN SYMBIOSIS WITH TREES

Mushroom	Component smells	Molecules
king bolete, porcini, cèpe (*Boletus edulis*)	fresh: mushroomy, malty dried: mushroomy, meaty, coconut, roasted, potato	fresh: octenol, octenone, methylbutanal & methylbutanol; dried: furanthiol, g-octalactone, octadienol, octanal, methional, methyl & ethyl pyrazines
candy cap, Maggi-pilz (*Lactarius* species)	maple syrup, fenugreek, curry	dried: sotolon, capric acid, methylbutyric acid
chanterelle (*Cantharellus* species)	mushroomy, fruity, green, apricot	octenols, octenyl acetate, methyl octanoate, hexenol, dihydroactinidiolide
black chanterelle, black trumpet (*Craterellus cornucopioides*)	mushroomy, fruity, fresh, almond extract, honey, medicinal	octenols, limonene, benzaldehyde, phenylacetic acid, methyl benzoate
yellowfoot (*Craterellus lutescens, tubaeformis*)	mushroomy, floral, honey, apricot	octenol, phenylethanol, phenylacetaldehyde, nonanal, geranyl acetone, dihydroactinidiolide
maitake, hen of the woods (*Grifola frondosa*)	mushroomy, camphor	octenol, methyl butanone, methyl dihydroxybenzoate

Mushroom	Component smells	Molecules
matsutake, pine (*Tricholoma* species)	metallic, sweet, floral, mushroomy, potato, pine, cinnamon, camphor	octenone, ethyl methylbutyrate, linalool, octenol, methional, terpineol, methyl cinnamate, bornyl acetate
morel (*Morchella* species)	mushroomy, floral, fruity, cocoa, potato	octenol, nonanal, 10 methyl esters, butyl butyrate, methylbutanal, methional

Chanterelles are another favorite among wild mushrooms, in their case for a fruity and spicy aroma from a couple of eight-carbon alcohols incorporated into esters, and a terpenoid fragment that's prominent in apricots. Their close relatives the **black chanterelle** and **yellowfoot** share their distinctive shape, with a flared cap revealing gills that reach low on the stalk, but they have their own interesting smells, fruity, floral, and honeyed. The **maitake**, or **hen of the woods**, which grows as massed clusters of small caps that can weigh 100 pounds (45 kilograms), has a camphor-like quality from an unusual ketone related to the blue-cheese volatiles of *Penicillium* mold, and an odd benzoate ester. The **matsutake**, or **pine mushroom**, much valued in Japan as an autumnal ingredient, is often briefly grilled or simmered in a soup to retain its distinctive mix of pine, medicinal, spicy, and floral volatiles. That mix is at its richest in just mature specimens, then quickly fades to simple mushroominess as they pass their prime.

The **morel** is one of the stranger-looking mushrooms, with a closed hollow cap traversed by a network of ridges; its spores are released from the nooks in between. It's also one of the least studied, even though a number of species are prized throughout the Northern Hemisphere when they emerge, usually in the spring. Some seem to form symbioses with trees, while others grow in decaying plant matter and recent burn sites. The only thorough study so far, of the black morel, found mushroomy octenol accompanied by a cluster of fruity esters and volatiles that provide floral, cocoa, and cooked-potato notes.

Truffles: icons of symbiosis and terroir

There are some fungal fruiting bodies that even the sharpest-eyed hunter can't spot on the forest floor. And their invisibility is the key to their strong, non-mushroomy, and highly prized smells. **Truffles**—a word related to *tuber*, "a swelling"—are spore-dispersing lumps made by soil fungi that live symbiotically with the roots of partner trees. Unlike mushrooms, they never grow above the soil surface. Instead they induce animals—squirrels and other small mammals, wild boars—to dig them up, eat them, and scatter their remains, spores included. Lately they've induced humans to scatter them to distant continents! The inducement for us is their bouquet of volatiles.

Truffles are best known today as a luxury food: they're scarce and expensive, in restaurants sometimes served and billed by the gram. They also happen to be delicious incarnations of symbiosis, and among the truest exemplars of *terroir*, the idea that a food's flavors can reflect the specific place in which it was produced. It takes only a few grams of truffle to lend its flavor to an entire dish, so truffles are a relatively affordable luxury. And they offer a wealth of associations for the smell explorer to savor.

Underground truffles don't bother to make stalks or caps, are less vulnerable to weather and predators, and can take months to develop their dense and nutritious spore-bearing tissues, sometimes several pounds' worth. Because they're hidden, they emit volatile signal molecules that can percolate through the soil to the surface. They need to pump out enough volatiles to be detectable above the general soil smell, and for long enough that there's a good chance that an animal will come across them. Truffle lovers take advantage of this productivity by storing them in a closed container with eggs or butter or rice, which absorb the volatiles as the fungus releases them over time.

Prized truffle species have bouquets like nothing else. Their immature fruiting bodies make some of the typical eight-carbon fungal chains, but when their spores are mature and the truffles ripen, they emit entirely different sets of volatiles to attract animals. Among them are sulfur molecules whose individual smells are vegetal, oniony, garlicky, horseradish-like, meaty, as well as other carbon chains and rings that are cheesy, fruity, leathery, nutty, heady.

Though it's been assumed that truffles make the smells they emit, fungal biologists have recently found that some actually outsource at least part of their volatile production—or *in*source it. Individual truffles turn out to be miniature ecosystems. They develop from an aggregation of mycelia threads that carry soil bacteria and yeasts on their surfaces. Those microbes become enclosed in the developing mass, survive there, and contribute volatiles to it. In 2015, European laboratories reported that the characteristic sulfur volatiles of the small white truffle are synthesized not by the fungus itself, but by its resident bacteria—and somehow just when their host's spores have matured! It seems likely that other truffles also have in-house fragrance specialists.

So truffles from their native regions reflect their home soil more directly than any other food I know. As soil fungi they've helped make the *terre*, the earth of their place. They're fed by and help feed long-standing local trees; they take in and support indigenous microbes, cue them to scent the soil around them, and feed the local animals that spread their spores. Woodlands harbor a mycelium-like web of interdependencies among the many creatures that live there. Truffles are its visible, tangible, smellable nodes.

Savoring truffles, black and white

While cooks usually deepen the flavor of fresh mushrooms by drying or cooking them, truffles are at their distinctive best when raw, shaved onto a dish at the last minute, or cooked only briefly. Serving them is an exercise in compromise. The truest smell of a ripe truffle is most evident when we sniff the intact mass. Cutting or shaving it creates more surface area for that smell to escape into the air and our noses, but the damage also initiates the production of generic eight-carbon mushroominess. So the challenge is to catch truffle flavor before it modulates from animal attraction to damage control. The fine-dining ritual of presenting a box of whole truffles for a look and a sniff is more than just show.

There are about two hundred species in the genus *Tuber*, and they're found across the Northern Hemisphere; some European species are now cultivated as far away from home as Australia. The most prized fall into two broad aromatic groups, the first complex and animalic, and the second predominantly sulfurous, garlicky and oniony.

Black or Périgord truffles are black and warty in appearance and native to southern Europe, usually associated with oaks, beeches, and chestnuts. They've been cultivated in France for many decades. Their basic aroma can be imitated in inexpensive truffle-flavored oils with just two volatiles, vegetal dimethyl sulfide and the malty-cocoa branched chain methylbutanal. But fresh black truffles have a more complex set of volatiles than most other species, with sulfurous, dairy, and fruity notes, as well as unusual animal-leathery carbon rings, ethyl and ethyl-methyl phenols, that resemble *Brettanomyces* yeast volatiles (see page 365). The similar-looking **burgundy truffle** is found throughout Europe and called **summer truffle** when harvested in that season. It's paler inside than the black truffle and less flavorful, with lower volatile levels across the board.

SOME TRUFFLES

Truffle species	Component smells	Molecules
black, Périgord (*Tuber melanosporum*)	sulfur, buttery, green apple, cheese, leather, animal, gasoline, caramel, mushroomy	dimethyl sulfide & disulfide, diacetyl, ethyl butyrate, methyl butanol, ethyl methylphenol, ethylphenol, furaneol, octenol
burgundy, summer (*T. uncinatum, aestivum*)	sulfur, potato, cheese, leather, animal, mushroom, "hazelnut"	dimethyl sulfide & disulfide, methional, methyl butanol, ethylphenol, octenol
white, Alba, tartufo bianco (*T. magnatum*)	garlic, onion, horseradish, cabbage, malty, mushroomy	dithiapentane, dimethyl sulfide & disulfide, methylbutanal, octenol
small white, bianchetto (*T. borchii*)	malty, fresh mushroom, onion/roast meat, cheese, butter, sulfidic	methylbutanal, octenol, octanone, methyl dihydrothiophenes, diacetyl, dimethyl trisulfide
smooth, garlic (*T. macrosporum*)	garlic, potato	thiapentane, thiapentene, butanone, acetone
"Chinese" (*T. indicum*)	mushroomy, potato, malty, whisky, cabbage	octenol, methional, methylbutanal, methylpropanol, dimethyl sulfide

Truffle species	Component smells	Molecules
Oregon white (*T. oregonense*)	sulfurous, mushroomy, cooked potato, floral, fruity	hydrogen sulfide, methanethiol, dimethyl sulfide & trisulfide, octenol & octanol, methional, methylbutanol, ethyl methylpropionate
Oregon black (*Leucangium carthusianum*)	as for white but less sulfurous; + apple, pineapple, sweet	as for white, + methyl methylbutyrate, methylpropyl methylpropionate

Unpublished data for Oregon truffles courtesy of Professor Michael Qian, Oregon State University.

White truffles, light- and smooth-skinned, are the most prized of the sulfurous group. They're associated with Alba and the Piedmont region but are also found in other regions of Italy and in the Balkans; partner trees include oaks, poplars, and willows. Their strong garlicky smell comes from an unusual two-sulfur, three-carbon chain, dithiapentane, which is so distinctive that makers of inexpensive truffle-flavored oils can simply spike the oil with that molecule—no actual truffle needed. **Small white or *bianchetto* truffles** grow throughout Europe and are now cultivated on other continents. Their volatiles overlap to some extent with Alba truffles but include a different set of prominent sulfur volatiles that smell more oniony and meaty than garlicky. The **smooth or garlic truffle** can be found throughout Europe, including the north; it's usually smaller than its cousins, with a smooth surface over a brownish interior. It has an intense garlicky aroma that rivals the white truffle's, derived from one-sulfur, four-carbon thiapentane. Better than two thirds of its emissions can be sulfur molecules, compared to the white truffle's half, and the black truffle's tenth.

Though European truffles have defined what truffles should smell like, a number of Asian and American species are becoming better known and appreciated. **"Chinese" truffles** grow widely in Central Asia but are exported primarily from Yunnan and Sichuan. They have a mild truffle smell, with sulfur volatiles more suggestive of cooked potato and cabbage. **Oregon truffles** (*Tuber oregonense, Leucangium carthusianum*, and others) come in white, black, and brown, and associate with Douglas fir trees in the northwestern United States. I hadn't thought much of the few small Oregon truffles I'd tasted over the years, but in 2017 I had the chance to hunt for them with truffle expert Charles Lefevre and a trained dog, and taste them when large and ripe. They resemble their prestigious European cousins, with the black having especially interesting fruity nuances.

Yeasts: alcohol, fruitiness, flies

Among the microbes that get caught up in the hyphae of developing truffles are other soil fungi—several single-cell yeasts that have been found to produce branched-chain alcohols and sulfur molecules and may well contribute to the overall truffle aroma. One of their cousin yeasts has gotten caught up in the fabric of human culture and now contributes more volatiles to it—billions of gallons every year—than any other living avatar of Hero Carbon.

It's one of the most vivid olfactory memories of my life: returning to my third-grade classroom after lunch on a bitterly cold winter day, I was enveloped simultaneously by the room's warmth and by an intense, perfumy smell emanating from the sunny windowsill, where that morning we'd left cloth-covered baskets containing little balls of bread dough. Exuberantly puffy rolls-to-be were my first experience of the creativity of yeasts.

Yeast comes from an ancient root word meaning "froth," and named the invisible froth-making agent that turns fruits and grains into wine and beer and bread. Today it refers to both the fungus that's primarily responsible for those transformations and the large group of single-cell fungi that it belongs to. Everyday yeast is the species *Saccharomyces cerevisiae*, whose scientific name means something like "sugar-loving, beer-making fungus." It's one of about fifteen hundred yeast species that occupy habitats from ocean sediments to mountaintops to our own skin and insides (see page 113). Common saccharomyces (pronounced SACK-a-roe-MY-sees) is one of a number that exploit sugary plant fluids; it and its close relatives are found on the sap oozing from tree and cactus wounds, in flower nectar and the guts of nectar-feeding insects, and on wounded and overripe fruits.

This most important fungus in our daily lives doesn't smell anything like mushrooms or truffles or molds. It shows an entirely different side to fungal virtuosity, but in similar service to the need to deter enemies and recruit friends. *Saccharomyces* has plenty of potential competitors for those plant juices, including other fungi. It combats them not with the usual mushroomy, musty, moldy eight-carbon chains, but with a volatile that for us and other animals is literally intoxicating: relaxing, freeing, and poisonous.

That double-edged chemical weapon is what we commonly call **alcohol**, the two-carbon starter-set molecule ethyl alcohol, **ethanol** for short. Ethanol is one of the common by-products of anaerobic metabolism, but it interferes with the basic chemical machinery of all cells, so most living things can't afford to accumulate much of it. The special achievement of saccharomyces was to activate its ethanol-producing system even in aerobic conditions, and to develop an unusual tolerance to ethanol. When a saccharomyces cell lands in some plant juice, it converts sugars to ethanol to suppress the growth of other less tolerant microbes. After consuming the remaining juice sugars, it can then use oxygen to metabolize the ethanol to carbon dioxide and extract the energy that it initially passed up for defensive purposes.

Because ethanol's interference with the cells in our nervous system can be pleasant to experience—the heightened feelings that are the positive side of intoxication—saccharomyces fermentations have become hugely important to human cultures all over the globe. Wines made from fruit juices average around 10 percent ethanol; beers from grains, about 5 percent; distilled alcohols, around 40 percent. Current estimates put the total annual consumption of ethanol itself, for each of the billions of adults worldwide, at around 6 quarts (6 liters).

Ethanol is only the start of saccharomyces's volatile creativity. It makes a number of other alcohols, including some from amino acids rather than sugars: winey, fruity branched-chain methylbutanol and floral, rosy phenylethanol. And it combines some of the alcohols with short-chain acids to make a variety of esters, the signature volatiles of ripe fruits (see page 298) and important as well in some flower scents. So the smell of active saccharomyces suggests solvents and fruits and flowers, a pleasant aromatic complexity that it lends to wines and beers and their distilled spirits. (Breads are at their yeastiest while on the rise; baking evaporates most of the yeast volatiles away.)

SOME SMELLS OF *SACCHAROMYCES* YEASTS

Smells	Molecule
alcohol	ethyl, propyl alcohols
nail polish remover, fruity	ethyl acetate
cognac, fruity, banana	methylbutanol (isoamyl alcohol)
floral, rose	phenylethyl alcohol (phenylethanol)
fruity, apple, green	ethyl hexanoate
fruity, peach	ethyl octanoate
sweet, banana, fruity	methylbutyl acetate (isoamyl acetate)
honey, floral	phenylethyl acetate
dairy, cream, butter	acetoin (hydroxy butanone)
boiled egg	hydrogen sulfide
cabbage	dimethyl sulfide

It seems serendipitous that this fungus should scent its defensive ethanol with the volatiles of flowers and fruits instead of the more typical fungal volatiles. In fact yeastiness isn't just a matter of good luck. Those flowery, fruity volatiles serve the yeasts in the same way that they serve plants, by recruiting animal assistants. This is best documented for the small insects commonly called fruit flies or vinegar flies and much studied in biology labs as *Drosophila melanogaster.*

Long research story short: saccharomyces and drosophila flies are a symbiotic couple, and yeasty volatiles are the bond between them. Drosophila are most strongly attracted to fruits not by fruit volatiles, but by the yeast alcohols and esters and other signs of ethanol production. The flies have evolved a tolerance to ethanol, feed on the nourishing yeasts, and lay eggs among them so their larvae can do the same, protected from most other microbes and predators. The flies are repelled by the mushroomy octenol and earthy geosmin produced by other fungi and streptomycetes, which often defend themselves with potent animal toxins. And the yeasts benefit from being fly food when flies carry surviving cells to other yeast-friendly fruits.

So we have insects to thank at least in part for the heightened fruity-floral aromas of wines and beers and products made from them—and quite possibly for a very different set of smells, some less likable! These come from several species of

Brettanomyces, aka *Dekkera*, a yeast that's easier to find in wineries and breweries than in any natural habitats. *Brettanomyces* was named for Britain by a microbiologist at the Carlsberg Brewery who found around 1900 that it was necessary in the secondary fermentation of English beers during conditioning in wooden casks, to give them their "peculiar and remarkably fine flavour."

"Brett" has been described as a survivalist microbe, one of the few that can grow after the main saccharomyces fermentation. In the process it metabolizes some ethanol to vinegary acetic acid and changes the balance of esters. Most important, it generates additional volatiles from protective phenolic ring molecules in fruit skins and grain seed coats. The brett ring volatiles have several qualities: the phenols and catechol mainly disinfectant and animalic, the guaiacols smoky and clove-like (see page 412). Drosophila flies and larvae are attracted by them as well, apparently because they indicate the availability of both nourishing yeast cells and certain plant phenolics, cinnamic acid and its relatives, which for flies are valuable antioxidants.

SOME SMELLS OF *BRETTANOMYCES* YEASTS

Smells	Molecule
vinegar	acetic acid
medicinal	vinylphenol
leather, horse stable	ethylphenol
smoky	vinylguaiacol
clove, spicy	ethylguaiacol
medicinal, horsey	ethyl catechol

Beer and wine drinkers appreciate some brett volatiles but not others. The clove and smoky notes of the guaiacols and even animalic ethylphenol are expected in Belgian-style and sour beers. Although a touch of "sweaty saddle" is a characteristic feature of some traditional wines and is still sometimes appreciated by connoisseurs, many now find that the animalic and medicinal notes tend to mask more delicate aromas and reduce a wine's overall complexity and interest.

We'll take a closer look at the contributions of saccharomyces, brett, and other yeasts to food and drink in chapter 19.

Wet-up: rain-released smells of soil, stone, air

Having zoomed in on the particular smells of Kingdom Fungi and its mushrooms and truffles and yeasts, let's draw back to appreciate the land's ambient smells and the role of rain bringing them to our attention.

Why is the Earth's earthiness so accentuated when rain breaks a drought, as Pliny and surely others noticed long ago? This is one aspect of a phenomenon that soil scientists call "wet-up." They've found that the sudden wetting of dry soil generates a large, hours-long pulse of carbon dioxide. Carbon dioxide itself is odorless, but it helps carry odorous volatiles into the air. Simple physical forces play a role in the wet-up pulse: as the rainwater percolates into the soil, it displaces the gases that had accumulated in its pores and evicts the water-avoiding volatiles that had accumulated on particles of rock and humus. There's also a biological force: sudden rain stimulates long-dormant microbes to resume their metabolic activities and generate both carbon dioxide and smellable volatiles. One study of a California grassland found that the wet-up gas release coincided with the rapid reactivation, within just minutes, of the bacterial group that includes the streptomycetes. The smell of drought-breaking rain can thus be doubly earthy, from the release of slowly accumulated volatiles and a burst of fresh ones.

It's no surprise that the teeming life in soils and spoiling fruit would emit the fumes of metabolism and self-defense. But how can nonliving rock or stone have a smell? They're made of minerals, collections of atoms that are bonded tightly together into some of the hardest of natural materials, which escape into the air only when agitated by very high heat, many hundreds or thousands of degrees. Yet most of us know the very particular smell that concrete and paving stones and bare rock emanate, briefly, when they're suddenly wet by rain or a garden hose. Even the moisture of our breath will liberate it, as Isabel Bear and Richard Thomas noted in their investigations of what they called petrichor.

To identify the petrichor volatiles, Bear and Thomas carefully extracted rock surfaces with a solvent and obtained a yellowish liquid with the characteristic wet-stone smell. When they analyzed it, they found it to be a mix of starter-set carbon chains—hydrocarbons, aldehydes, ketones, lactones, and acids, particularly nonanoic—as well as unusual phenolic rings decorated with nitrogen-containing groups, nitrophenols that they described as having a "peculiar sweet aromatic smell." Sulfur volatiles are surprisingly absent, apparently because they're oxidized on mineral surfaces to odorless elemental sulfur.

So the smell of freshly wetted stone turns out not to come from the stone's own minerals. Petrichor, or gaiaichor, is the veneer of volatiles that had been emitted by microbes, fungi, plants, animals, humans and our technologies, then modified in the atmosphere by sunlight, oxygen, nitrogen, and one another, and accumulated on mineral surfaces. These volatiles are usually too sparse and omnipresent for us to notice them in the air around us. But when rain suddenly drives them in greater amounts from mineral surfaces into the air, the volatiles become perceptible.

The same process is probably what makes rain itself smellable before it even begins to wet rock or pavement. (Ozone generated by nearby lightning can also contribute to storm smells.) The atmosphere is full of dust and other particles that can accumulate gaiaichor when dry, then release it when the air's moisture reaches breathlike levels and begins to condense into water droplets.

It's the sudden volatile release of these wet-ups that reveals an otherwise cryptic phase in what Bear and Thomas called the "great natural cycle" of earthly life, as its small remnant carbon chains and rings fly into the air, then fall again to nourish its next round.

SOME SMELLS OF WET STONE

Smells	Molecules
fresh	acetaldehyde
alcohol, solvent	various alcohols, ketones
waxy, dirty, cheesy	octanoic, nonanoic acids
ammonia, fishy	ammonia, pyridine
sweet, aromatic	nitrophenols

Struck stones and "minerality"

Rain and moist breath aren't the only ways to squeeze volatiles from stones. A good beating can also work, though it doesn't liberate gaiaichor. People have long remarked on the distinctive smell emitted when pieces of the mineral flint are hit sharply with something similarly hard, to generate sparks for fire-making or to discharge an old-fashioned flintlock gun. In the wine world, certain white wines from cool climates have been said to carry this aroma of struck flint or "gunflint." Nearly twenty years ago, French wine chemists connected that aroma to particular sulfur compounds in the grapes and finished wine, smoky benzenemethanethiol and a relative of the cat-pee and grapefruit thiols (methyl thiopentanone; see pages 81, 324). However, the smell of struck flint itself is unlikely to involve such complex organic molecules, and it had never been investigated—until just recently.

Around 2014, scientists in the Geneva laboratories of Firmenich were studying ways to ameliorate the odors of latrines. They noticed that a tank of dilute hydrogen sulfide (common in latrines) had changed smells, from standard egginess to something that they associated with "struck flint" as well as "the cold smell of fireworks or the smell associated with a dentist drilling teeth." The Firmenich group found that the hydrogen sulfide in their tank had been oxidized to form very different and unusual sulfur volatiles, like hydrogen sulfide but with an additional one or two sulfur atoms between the hydrogens, powerfully irritating and unstable. They were able to identify these same **sulfanes** when they struck together two pieces of flint. Sure enough, the Firmenich scientists found sulfanes in Swiss Chasselas wines, and more of them in the wines judged by professional tasters to have a stronger mineral flavor.

Why would rocks emit sulfur volatiles when struck? Flint is a version of quartz, a mineral made up of silicon and oxygen atoms. Silicon oxides have no smell of their own, but flint forms in airless ocean sediments, which carry gaseous hydrogen sulfide and other volatile sulfur compounds from anaerobic microbial metabolism. The sulfur atoms get fixed in flint as trace impurities of solid iron sulfides, the same molecules that color Indian black salt and discolor an overcooked egg (see page 37). When a piece of flint is struck with another hard material, the focused energy of the impact helps oxidize some of the sulfides to sulfanes. So as is also true of gaiaichor, the raw ingredients of struck-flint smell originate in living metabolism. But the backstory of flint sulfanes makes "mineral" an accurate reflection of both their smell quality and their origins: the incorporation of the organic sulfur into iron compounds, the inclusion of mineral particles in sedimentary rocks formed tens or hundreds of millions of years ago, and the instantaneous transformation of sulfides into sulfanes by the simple physical force of mineral striking mineral.

"Minerality" is a quality often touted by wine specialists as somehow derived directly from specific vineyard soils and their underlying limestone or granite or schist. In fact, the volatiles that have been identified to date as actually contributing stony smells to wines are the product of grape-growing and wine-making practices. Wine chemists have found that sulfane structures arise in wines in a number of ways, when traces of copper react with yeast-generated sulfur molecules, and when various wine molecules react with plain elemental sulfur, which is applied to grapevines to control disease (even in "organic" production).

It seems fairly certain that wine minerality doesn't come directly from soil minerals. It comes at least in part from wine molecules that are reminiscent of or identical with molecules that also can be generated from minerals. This is a less simple and picturesque story, less handy as a sales pitch for a wine or a wine critic's expertise, but it's truer to what we actually know—and richer and more thought-provoking. How remarkable, that grapes and yeasts and farmers and vintners can conspire to make the same rare molecules as the combination of ancient seafloor microbes, eons of geological sediment-kneading, and a quick hard whack!

SOME SMELLS OF STRUCK STONE

Smells	Molecules
pungent, sulfurous	sulfanes (hydrogen di- & tri-sulfides)
sulfidic	hydrogen sulfide, CH_3SSH, C_2H_5SSH
cabbage	methane- & ethane-thiols

And there's clearly much still to learn about mineral-associated volatiles. In addition to flint, the Geneva group also tested common conglomerate pebbles from a nearby mountain and found that they emitted a number of other unusual sulfur volatiles along with the sulfanes—smells of shiitake mushroom, skunk, and burnt hair! When I read that detail, I started knocking rocks together in my backyard, on hikes, whenever the thought struck, and sure enough: with a close sniff of the impact zone I often pick up a sharp, singed sulfurousness (though not from nonsedimentary granite and other igneous rocks). There's probably more to "minerality" than flintiness.

Wetlands: swamp gas and salt marsh

To conclude our tour of the land and its smells, let's don a pair of high boots and venture onto the soggier, less firm regions of terra firma. A substantial fraction of the planet's soils are either saturated with water or submerged: constantly in places like the Amazon River basin or the Florida Everglades, seasonally in the subarctic tundra of Canada and Siberia. They're variously called bogs, fens, marshes, mires, mudflats, peatlands, swamps, and wetlands, all of these names deriving from the Germanic of wet northern Europe. In the backyard, long-sodden flowerpots and low spots are small-scale versions of the same conditions, and provide the same olfactory experience when they're disturbed. They stink.

Saturated and submerged soils stink because the air in their pores has been displaced by water. In the absence of oxygen, only anaerobic microbes can thrive. The by-products of their metabolism, which we've already sniffed at on the early Earth and in the residues of the animal digestive tract, include one-carbon methane, familiar as odorless natural gas, and then two-, three-, and four-carbon acids—acetic and propionic and butyric—which smell sour, vinegary, rancid, cheesy, vomitous. Anaerobes also take advantage of versatile sulfur atoms in both sediment minerals and plant remains to pull electrons from carbon chains, and generate a variety of smelly sulfur by-products, including sulfidic hydrogen sulfide, putrid methanethiol, and a variety of sulfides reminiscent of cooked and/or rotten cabbage and onions. Hydrogen sulfide and methanethiol are more evident when storms or human activity disturb wetland sediments; in quiet conditions, they're largely oxidized as they rise in the water and exit as less odorous methyl sulfides.

SOME SMELLS OF WETLANDS

Smells	Molecule
vinegar	acetic acid
cheesy	propionic acid
cheesy, vomit	butyric acid
sulfidic	hydrogen sulfide
rotten	methanethiol
cooked cabbage	dimethyl sulfide

Swamp volatiles are likely also responsible for the many anecdotal reports, going back centuries and from all over the world, of will-o'-the-wisp, elusive flame-like flickering lights seen over various wetlands at night. Flammable methane is one suspect, and another is traces of phosphine and related phosphorus-containing molecules, some of them garlicky, which are more likely than methane to combust in the absence of a spark.

Salt marshes are wetlands that hug the coasts of the continents and receive most of their water from the oceans. Their smell is typically marshy-swampy, but it's dominated by one volatile in particular: dimethyl sulfide, DMS. Some of the DMS comes from the sediments and their anaerobic microbes making use of mineral and plant sulfur, but much of it comes from microbes in the water itself, dealing with the constant biochemical stress of its saltiness. Stir a heaping spoonful of salt into a small glass of water, have a taste, and imagine being constantly exposed to that inside and out! That's one precondition of life in the oceans. It's why salt marshes and ocean air smell the way they do: almost nothing like the land.

We pass now from soil and stone and swamp to Earth's open waters.

Chapter 15 · THE WATERS: PLANKTON, SEAWEEDS, SHELLFISH, FISH

··

The Pacific is my home ocean; I knew it first, grew up on its shore, collected marine animals along the coast. I know its moods, its color, its nature. It was very far inland that I caught the first smell of the Pacific. When one has been long at sea, the smell of land reaches far out to greet one. And the same is true when one has been long inland. I believe I smelled the sea rocks and the kelp and the excitement of churning sea water, the sharpness of iodine and the under odor of washed and ground calcareous shells. Such a far-off and remembered odor comes subtly so that one does not consciously smell it, but rather an electric excitement is released—a kind of boisterous joy.

> • John Steinbeck, *Travels with Charley: In Search of America*, 1962

Intimations of the ages of man, some piercing intuition of the sea and all its weeds and breezes shiver you a split second from that little stimulus on the palate. You are eating the sea, that's it, only the sensation of a gulp of seawater has been wafted out of it by some sorcery, and are on the verge of remembering you don't know what, mermaids or the sudden smell of kelp on the ebb tide or a poem you read once, something connected with the flavor of life itself.

> • Eleanor Clark, *The Oysters of Locmariaquer*, 1964

Electric excitement, boisterous joy, intimations of the ages, the flavor of life itself: sea smells can inspire quite a churn of feeling! And from vagrant weeds and debris and reclusive lumps of flesh. We're at the edge of a bracingly different swath of the osmocosm.

Earth's waters are like a parallel planet. They're a world where oxygen is scarce, sunlight and its heat penetrate only to a limited extent, temperatures and ambient energy levels are low; where water can either counteract the force of gravity or manifest it in crushing pressure. We can visit this world only briefly, and its surface is almost entirely blank, with only scattered signs of the life that thrives throughout its tremendous third dimension. Yet it's where life first took hold, diversified, and fostered the ancestors of all life on land. It harbors unimaginable numbers of living creatures, and it made life on land possible in the first place: it's where photosynthesis was invented, where oxygen first began to seep into the air, and where half of our oxygen still originates.

The smells of the shore and open waters and their creatures can give us a kind of access to that world and its workings that our other senses can't. Remember that Helen Keller delighted in the waves and tides of the land's scents, "filling the wide world with invisible sweetness." The sea's waves and tides also fill the wide world with volatiles, not especially sweet, but contributing just as much of the planet's breath, and revelatory of its hidden life.

I'll leave it to Steinbeck's and Clark's words to prompt the bubbling up of your own oceanic experiences as we make our virtual way through the waters. To make the way more than virtual, head to the beach or cook some seafood!

The planet's waters, salted and sulfured

Like the smells of the land, the smells of the waters come mainly from the living things that manage to thrive there. The carbon chains and rings that they release in life and in death are manifestations of their home's particular challenges, and its particular resources.

Call to mind the iconic image of planet Earth photographed from the moon, with its marbling of deep blue and swirling white. It's a portrait of a world dominated by water. Water's liquid form, which absorbs the red end of the rainbow and reflects the blue, covers more than two thirds of Earth's surface, and its vapor frequently saturates regions of the atmosphere and condenses into clouds of light-scattering droplets. The oceans are nearly as old as the planet. It's still uncertain where all the water came from: maybe the asteroids and other debris that clumped together to form the protoplanet, maybe snowball-like comets that plowed into it later. Whenever the surface oceans formed, they came into direct contact with the planet's underlying mineral body, and from it began to extract the makings of life's fluid home.

Liquid water is a good solvent, capable of dissolving many elements and molecules into itself, and it has long done that with Earth's solid crust and the molten lava and gases propelled through it by the intense heat within. And not only in the ocean basins. When water evaporates from the ocean, it forms clouds of water droplets that absorb volcanic emissions and other gases, grow and merge with one another, and eventually fall back to the surface as rain and snow. When they fall inland, the water gradually flows downward to sea level, dissolving some minerals as it goes, feeding lakes and rivers. But these freshwaters account for less than 10 percent of the Earth's surface waters. And they're relatively poor in minerals, most of which are eventually delivered to the oceans.

Of course the oceans are mostly water, around ninety-six grams out of every hundred. About two grams are chloride, an atom of the element chlorine with an extra electron, and one gram is sodium: the two elements that combined together make sodium chloride, common salt. Sodium comes from Earth's crust, but pretty much all the chloride in the oceans dissolved there early on, when chlorine gas and hydrochloric acid erupted from deep within the planet. The ocean's proportions of sodium and chloride, the equivalent of about a 3 percent salt brine, are unpleasantly salty to the taste—and they're a major key to the smells of ocean life.

Two other aspects of the ocean's mineral chemistry influence its volatile chemistry. Chlorine, the most abundant element in seawater after the hydrogen and oxygen in H_2O itself, is a member of a chemical group known as the *halogens*, or "salt makers": they readily react with metals like sodium to make *salts*, the chemist's term for compounds that, like common salt, will dissolve in water into their halogen and metal portions. Two other halogens found in seawater in significant amounts are iodine and bromine. Along with chlorine, they end up in volatile molecules with distinctive smells.

After chlorine, sodium, and magnesium, the next most abundant element in seawater is sulfur, usually in its oxidized (and negatively charged) form—sulfate, SO_4. As we've seen, sulfur is a versatile element, able to take up extra electrons or give them away, and its presence in volatile molecules gives them distinctive smells, variously sulfidic or cabbagey, oniony or garlicky, rotten or putrid or tropical-fruity. With so much sulfur floating around in the ocean, living cells have made good use of it, and ocean air abounds in it.

Life in the waters

Like the land, the oceans are also home to plantlike primary producers that use sunlight and dissolved carbon dioxide to grow and multiply, and opportunistic animals and other consumers that live off their productivity. Here's a quick introduction to the major players and volatile emitters.

There are staggering numbers of creatures of all kinds in the oceans, apparently far exceeding the number of stars in the known universe, ranging from single cells to whales. The collective term *algae* is often applied to the oceans' photosynthesizers, and they fall into two broad groups defined by their size: tiny **phytoplankton**—from Greek words for "plant" and "drifting"—and sometimes massive **seaweeds**.

Among the phytoplankton, **cyanobacteria**—from the Greek *cyan-* for "blue-green"—are among the most numerous creatures in the seas, and the ones with the longest pedigree: they're the direct descendants of the microbe that invented photosynthesis. Just one of them, *Prochlorococcus*, discovered only in the 1980s, has an estimated worldwide population of a billion billion billion, and is responsible for generating about a fifth of all the new oxygen in the atmosphere. *Spirulina* is a freshwater cyanobacterium that's cultivated and dried into a powdery green nutritional supplement. Various cyanobacteria play a role in the flavor of some fish.

Three other groups of microscopic photosynthesizers are the ones that scent the sea at large. They're all single-cell creatures whose predatory ancestors engulfed a photosynthetic bacterium and made its photosynthetic machinery an integral part of their cells. These microbes are more closely related to the seaweeds than to cyanobacteria, and they go back a mere few hundred million years, not billions. The **diatoms** and **coccolithophores** surround themselves with protective mineralized cages and shells, respectively, the calcium carbonate of the latter giving rise to deposits of what we know as chalk, the mineral of the white cliffs of Dover. The **dinoflagellates** are named for the whiplike flagella that propel them through the water.

Seaweeds are the algae we know, big enough to see. They're large many-cell cousins of the tiny diatoms and dinoflagellates and coccolithophores. Seaweeds fall into three general families named red, brown, and green for their dominant colors, which reflect their different sets of photosynthetic pigments. It was an ancestor in the green clan that gave rise to all the plants that grow on land. Most seaweeds hug the coastlines where they can anchor themselves and take advantage of nutrients washed from the land. Some can survive pounding by wave action and temporary drying and sun exposure in the intertidal zone. Despite their visibility, they're a much smaller fraction of the global aquatic biomass than phytoplankton, and they scent the air only locally.

Of course the animal kingdom also got its start in the waters, and it, too, has evolved a fantastic range of aquatic creatures. These include astronomical numbers of floating zooplankton that feed on the phytoplankton, crustaceans from minuscule copepods to krill to lobsters, mollusks from oysters to octopus, jellyfish and stingrays and tunas and whales. . . . And they do very well compared to land animals. Ecologists have found that land animals, mainly insects, consume something less than 20 percent of the biomass produced by plants every year. By contrast, aquatic animals consume between half and all of coastal seaweed production every year, and the bulk of annual phytoplankton production in the open ocean. Another measure: on land, the biomass of plants outweighs animals by a factor of a thousand; in the oceans, animals outweigh plants by a factor of thirty.

Phytoplankton and seaweeds do deploy chemical weapons to defend themselves against this constant onslaught. Because they and their predators live in water, those defenses are often molecules that persist in the water rather than escaping into the air where we can smell them. This is the case for the toxins produced in large quantities during "red tide" and other blooms of certain cyanobacteria and dinoflagellates, which can contaminate shellfish and make them dangerous for us to eat. Other defenses are volatile and smellable. But by far the single most abundant seaborne volatile arises from an even more fundamental challenge to aquatic life than predatory animals: the oceans' saltiness.

The open ocean: planktonic sulfur

In 1935, a London botanist named Paul Haas set out to identify the nature of the "particularly penetrating, somewhat sickly odor" of a red alga common to the British shorelines, one that forms long branched filaments and is known as "lobster horn." That smell of the intact alga was different from the strongly sulfurous stench of rotting algae, which he attributed to sulfidic hydrogen sulfide. Haas found that the living smell was another sulfide—**dimethyl sulfide**, or **DMS**, still sulfurous-smelling, but with the addition of two carbon groups giving it a different quality. At that time dimethyl sulfide was otherwise known only as a gas often found in crude petroleum. Today it's known as far and away the most abundant molecule emitted into the air by aquatic life, by plankton above all. DMS is at the heart of the sea's smell. And it's there as a product of the system that algae evolved to deal with the saltiness of seawater and the challenge of osmosis.

Osmosis is the process that wrinkles our fingers in the bath and draws moisture to the surface of salted cucumbers or steaks. Living cells are enclosed in thin membranes that allow water molecules to pass through them. When there's an imbalance in the concentration of materials dissolved in the water on the two sides of a membrane, the water naturally flows one way or the other to even out the imbalance. The equalizing flow of water is osmosis (the similar movement of dissolved material is called diffusion). Pure bathwater flows into our part-water skin cells; pure salt pulls water out of low-salt cucumber and meat cells.

The salinity of the open oceans is about three times the concentration of dissolved molecules inside the average living cell. The challenge for aquatic creatures is somehow to protect their intricate inner machinery from the disruptive loss of water and incursion of salt. Microbes and plants and animals manage to thrive in salt water with the help of two biochemical systems. One actively pumps salt's sodium and chloride out of their cells and back out into the water. The other accumulates special dissolved molecules inside cells to help balance the dissolved salt in their surroundings.

The dissolved molecules, or *solutes*, that a cell accumulates for osmotic balancing need to be compatible with the cell's machinery, to leave it unaffected by their presence. There are several kinds of molecules that meet that requirement. Some are the cell's basic building blocks, including the nitrogen-containing amino acids that make proteins, and slight variations on them. And some are built specifically as compatible solutes. They themselves are by definition not very volatile—they prefer the environment of water molecules to the air. But when they're broken down to be used for other purposes, some of the fragments can be volatile and escape from the ocean surface.

By far the most prominent compatible solute in the oceans is a molecule with a jawbreaking name: dimethylsulfoniopropionate, or DMSP for short. DMSP was first discovered by two organic chemists at the University of Leeds as the source of the DMS in Paul Haas's lobster-horn red alga. It also turned out to be the primary compatible solute in dinoflagellates and coccolithophores, as well as some diatoms and green and red seaweeds. These algae are collectively responsible for the emission of tens of millions of tons of DMS into the air every year. Indirectly responsible, though, because most of the DMS is generated when the algae are damaged or eaten, and other creatures—bacteria, single-cell animals, larger zooplankton like krill—break down the DMSP for their own use.

The primary smell of the ocean, then, is the smell of phytoplankton and algae coping with its saltiness, and predators and scavengers dismantling the coping molecules for their own use. Why is DMSP so dominant? One reason may be that ocean sulfur is relatively abundant, whereas the nitrogen needed for amino-acid building blocks is scarce. So ocean dwellers are simply taking advantage of the resources available to them. Another factor is that DMSP is a multipurpose molecule. It and its fragments can act as antioxidants and as reserves for molecule building. And when the DMS molecule is broken off from it, the remainder is a nonvolatile and corrosive carbon chain called acrylate; its presence tends to inhibit zooplankton from further grazing.

The significance of DMS extends far beyond the particular organisms that generate it. It's a general signal for the entire oceanic food web that there's food to be had. Animals at all levels, from krill to filter-feeding whale sharks to seals, are attracted to DMS-emitting communities of plankton. On their thousand-kilometer flights over open waters, petrels, albatrosses, and shearwaters, the tube-nosed seabirds, locate concentrated populations of their prey in part by detecting plumes of DMS. It turns out that plastics tend to accumulate both DMS and microbes that generate it, and this may be one reason why sea creatures ingest ocean trash, with often fatal results. And DMS molecules that make it to the upper atmosphere encourage cloud formation, and by doing so may affect the climate.

DMS is a small and simple molecule, so it's found in many different places in the world, and the quality of its smell depends on the context. In a raw clam or a cooked oyster, it's the smell of shellfish. It's a sulfurous note in the smell of raw truffles, cooked asparagus and cabbage, canned corn, boiled milk. Before all these it was the smell of life and death, and it still is at the shore, omnipresent, subliminal.

The shore: seaweed bromine and iodine

In the distant smell of the Pacific, John Steinbeck picked out the "sharpness of iodine": another sensation that's characteristic of the oceans, especially coastal waters, and nothing like vegetal DMS. Instead of osmotic balance, it reflects the challenges of life at the interface of sea and land.

Iodine is one of the salt-forming halogen elements, which are relatively plentiful in the sea. Their dissolved salts have no smell, but the pure halogens share a distinct family resemblance, a quality that also comes across when they form volatile compounds with carbon. Chlorine is the dominant halogen in seawater, followed by iodine and bromine. Chemists in Switzerland and France discovered all three in the eighteenth and early nineteenth centuries, two in seaweeds, and named one of those specifically for its smell: the Greek root *bromos* means "stench."

Today our most familiar references for halogen smells are household bleach, a solution that releases small amounts of chlorine gas, and the chlorinated water of swimming pools. Various iodine disinfectants were once common and a characteristic smell in clinics and hospitals, and they're still used to purify water and sanitize brewing equipment. To smell iodine, squeeze a few drops of readily available iodine tincture into a small dish of water. Iodine is less soluble in water than alcohol, so the water increases its volatility and presence in the air. It resembles chlorine, but with a less aggressive, rounder, richer quality.

The uses to which we put simple chlorine and iodine molecules, and the health concerns raised by them and by many halogen-containing industrial chemicals— pesticides, flame retardants, solvents, coolants, insulators, plastics—indicate an essential feature of the halogens: they're reactive elements and tend to disrupt the workings of living cells. Bromine compounds and chlorine gas were among the first chemical weapons used in World War I. Seaweeds take advantage of their availability in seawater to deploy halogens against predators and parasites, and also to deal with the chemical stress of exposure at the oceans' edges.

Some of the iodine-bromine-chlorine sharpness that we smell at the seacoast comes directly from the waters. The spray from wave action exposes seawater to the air's oxygen and to the sun's ultraviolet radiation, and that can be enough to transform dissolved halogens into traces of their pure gases. Some dissolved halogens also react with the molecular detritus of ocean life and end up in such volatile one-carbon molecules as **chloroform**, methyl chloride, and **bromoform**, all oddly sweet-smelling, along with the more penetrating, chlorine-like **iodoform**, and the hospital-like mixed halogen **dibromoiodomethane**.

Most of the halogen smell of the seacoast is actively generated by the seaweeds that grow or are washed up there, especially by the large brown seaweeds commonly referred to as *kelp*. That name originally meant any seaweed that could be burned to produce soda ash (sodium carbonate) for glass making. Today it names a group that includes a staple of the Japanese kitchen, kombu, and other "giant" species that form dense underwater forests, rich habitats for all kinds of other creatures that invade, chew on them, colonize their surfaces, and otherwise interfere with their growth and light gathering. In order to limit the damage from these predators and parasites, kelps make their surfaces slimy and regularly slough them off—and they deploy halogen molecules.

SOME SEASHORE HALOGEN MOLECULES

Halogen molecules	Smells
chlorine, Cl_2	bleach, swimming pool
bromine, Br_2	bleach, harsh
iodine, I_2	iodine, rich
methyl chloride, CH_3Cl	sweet
chloroform, $CHCl_3$	sweet
bromoform, $CHBr_3$	sweet
iodoform, CHI_3	iodine, penetrating
dibromoiodomethane, $CHBr_2I$	bleach, iodine
bromophenols, C_6H_4OHBr	iodine, rich

The first line of kelp volatile defense is a steady stream of bromoform and other brominated molecules. In the event of a major breach of its tissue, nearby cells release a blast of strongly oxidizing hydrogen peroxide—another of our household disinfectants—to suppress any potential invaders. But peroxide damages kelp cells as well, so to limit self-inflicted injury, the seaweed releases iodide, the dissolved form of iodine; iodide is a powerful antioxidant that disarms the peroxide.

Iodide is remarkable for being the simplest antioxidant molecule known, the first example of an inorganic antioxidant—the seaweeds don't have to make it; they just accumulate and store it. The kelps are the strongest iodine accumulators and emitters on the planet. Apparently it helps them deal with the special challenges of intertidal life, where they're battered against rocks and sand by the waves and alternately submerged and left high and dry. Even without predation by sea urchins or abalone, direct exposure to the air and sun are damaging enough to trigger the production of peroxides and other forms of oxidative stress, and iodide release limits that stress as well. Once it reacts with peroxide, iodide combines with small carbon chains in the vicinity to make iodoform and other volatiles. When these reach the air, some of them react with oxygen to produce **iodine** gas. This is the backstory to Eleanor Clark's "sudden smell of kelp" at ebb tide. The smell comes from iodine volatiles released by the seaweed as it becomes exposed.

Bromine and iodine volatiles are most prominent at the seashore, but they can also make an impression at home. In Hawaii, red seaweeds known as **limu kohu** are used as a condiment to deliver a strong oceanic accent to various dishes (see page 399). And phytoplankton far from shore generate a different set of halogen compounds that contribute to the full flavor of some wild-caught ocean fish and shellfish. These **bromophenols** are molecules built from six-carbon rings in which the rings are decorated with bromine atoms and oxygen-hydrogen (OH) groups. They accumulate in the bodies of plankton-eating wild shrimp and in krill, through which they work their way up the food chain to larger crustaceans and fish. When not too strong, their iodine-like smell lends a desirable complexity to otherwise mild seafoods. It's thought that bromophenols serve their phytoplankton creators as defensive molecules and antioxidants.

In sum, the smells of the seacoast come from otherwise rare reactive elements that seaweeds mobilize to defend themselves against enemies and physical-chemical stress—and against one of their own defenses.

Fresh seafood: the original vegetality

Open ocean: dimethyl sulfide. Seaweedy shores: bromine and iodine. And now for the fish and shellfish and seaweeds that we take from the waters and experience more intimately and particularly. They may bring along some DMS and halogens, but there's another family of volatiles that contributes to their characteristic flavors, and above all defines the smells of pristinely fresh seafood. It's a very familiar family: plain carbon chains, with no funky sulfur or nitrogen or exotic halogens. We've met many of them before in the starter set of simple chains found in nearly all living things, and some are especially prominent in land plants and fungi. They include molecules that produce the "green" smells of crushed leaves, that scent melons and cucumbers, that create mushroominess. Strange, that they should also be prominent in sea animals, so that oysters can smell like cucumbers, smelts like watermelon. Yet not so strange: oysters and fish were almost certainly emitting these volatiles hundreds of millions of years before cucumbers and melons came on the scene. For the all-smelling cosmic Chef, cucumbers would be evocative of oysters, and watermelons of fish, not the other way around!

Land plants and mushrooms and sea life share these simple carbon-chain volatiles because they have similar systems for making them (see pages 154, 349). The starting molecules are the long carbon chains that make up the bulk of their cell-surrounding and inner membranes. Land plants and fungi construct their membranes from chains that are eighteen carbon atoms long with two or three kinking double bonds. Plant enzyme systems tend to produce fragments that are either six or nine carbon atoms long, with no double bonds or just one. The **six-carbon fragments** are the **green-leaf volatiles,** the **GLVs,** which are especially dominant in crushed leaves. The cucurbit family, which includes cucumbers and melons, has an enzyme system that produces **nine-carbon fragments**, and cucumbers but not melons make a nine-carbon chain with two double bonds. That **nonadienal** is what gives cucumbers their smell identity. Mushrooms in turn specialize in making **eight-carbon chains** (see chapter 14): so those volatiles will make anything that carries them smell mushroomy to us.

Water-dwelling creatures contain plenty of the same membrane-building chains, have similar chain-chopping enzyme systems, and so produce many of the same volatiles that we as land animals associate with green leaves and mushrooms, cucumbers and melons. But they also carry longer, kinkier chains that yield other fragments. Life underwater is generally colder than it is on the sun-warmed land, and at cold temperatures, molecules move more slowly and cell membranes stiffen. To keep their structures and systems optimally limber, aquatic creatures accumulate more irregular and loose-jointed chains, twenty or twenty-two carbon atoms long with four to six kink-making double bonds. (These include the nutritionally valuable ones called omega-3 chains.) Phytoplankton and some seaweeds accumulate these highly kinked chains in large quantities and pass them along to the aquatic animals that eat them. Their enzyme-chopped fragments are more varied than the typical plant and fungal fragments, and often include two or more kinks, and some give a distinctive "marine" quality to the overall smell: notably the seventeen-carbon, three-kink **heptadecatrienal.**

Why do aquatic creatures generate all these volatiles? In algae they're clearly playing a role analogous to the one they have in plants and mushrooms: diatoms and seaweeds release them when their cells and tissues are damaged, and they're known to be toxic to grazing zooplankton. But it's less clear why superbly mobile animals release these vegetal volatiles. One theory is that because aquatic animals also make twenty-carbon chains that regulate their cells' activities as hormones do, the volatile short chains may be those regulators' partly recycled remains. Perhaps they're a prophylactic against microbes and small parasites that attach themselves to animal bodies. Fish tend to release them from their gills and skin, the two organs exposed directly to the water, and especially when they're suffering some kind of stress.

So most water dwellers have fresh smells that are reminiscent of vegetables, but with green-leaf, mushroom, cucumber, and melon qualities often coming from a single creature. On land, oyster plant and borage flowers are among the few that emit a comparable medley. It's the mixture, the volatile vegetal salad, and the occasional distinctive marine molecule, that's the odor signature of life under water.

SOME AQUATIC VOLATILES PRODUCED BY SELECTIVE ENZYMES

Molecules	Smells
pentenol, pentenone	green, pungent
hexanal, hexenal, hexenol	green
octenal	fatty
octenol	mushroom
octenone	mushroom, earthy, metallic
octadienol	mushroom, earthy
octadienone	geranium leaf, metallic
nonenal	fatty
nonenol	green melon
nonadienal	cucumber
nonadienol	watermelon, metallic
heptadecatrienal	seaweedy
ectocarpenes (C11)	tomato leaf
dictyopterenes (C11)	seaweed, moss, green

To decipher the molecule names: the suffixes -enal, -enol, *and* -enone *indicate one double-bond kink;* -dienol *and so on, two kinks;* -trienol *and so on, three kinks.*

One other unusual group of carbon-chain volatiles is worth knowing about despite being restricted to some brown algae and their home waters. Seaweeds mate by releasing male and female reproductive cells, or gametes, into the water, where they have to locate each other. The gametes of some brown algae emit semivolatile chemical signals, constructed from the long polyunsaturated fatty acids, to attract each other. The best-known, named **ectocarpenes** and **dictyopterenes** after their seaweed sources, contain eleven carbon atoms and have fruity, mossy, spicy, and seaweed qualities. One is said to smell like tomato leaves.

Stale seafood: fish-oil fishiness

When most of us think about the smells of fish and shellfish and seaweeds, leafy and mushroomy and cucumbery are not the first qualities that come to mind! Those volatiles are emitted by the living or freshly harvested creatures, their enzymes intact and constantly working. Once the energy supply runs out and the living systems fail, the production of these chains slows and finally stops. Those that escape into the air aren't replaced, and instead, other volatiles develop and accumulate. Oxygen in the air takes over the job of chopping up the long carbon chains, especially at their vulnerable kinks. This nonspecific, nonenzymatic breakage produces a different spectrum of fragments, including **seven- and ten-carbon chains** that are seldom found in living tissues. It's the replacement of enzyme-chopped with oxygen-chopped fragments that turns seafood from mild, fresh, and vegetal to strong and "fishy." It's the odor signature of death and approaching spoilage in aquatic creatures.

The fact that we use the name of the creature to describe the smell of its decay, not its original freshness, testifies to the persistent unpleasantness of one and the fleeting rarity of the other.

SOME AQUATIC VOLATILES PRODUCED BY NONSELECTIVE OXIDATION

Molecules	Smells
hexanal	green
heptanal	green, fatty, fishy
heptenal	boiled fish, burned
heptadienal	mushroomy, painty, rancid
octadienone	geranium, metallic
octatrienone	geranium, fishy
decadienal	fatty, fried
decatrienal	seaweed, fried, painty

This carbon-chain fishiness is easy to experience on its own: just break open a capsule of omega-3 fish-oil supplement. It's blunt, enveloping, persistent. It can't be reduced with a squeeze of lemon, and it hangs around in a kitchen long after the cooking. Though food chemists have looked hard to find a single volatile responsible for fish-oil fishiness—a quality described over the decades as "whale oil," "trainy," "painty," and "minnow bucket"—apparently there really isn't a single culprit. It seems that a small set of six- to ten-carbon aldehydes and ketones is responsible, each not strongly unpleasant on its own, but the mixture a consistent indication that a creature from the waters has been dead for some time. The stronger the smell, the more advanced the deterioration.

Stale seafood: amine fishiness

Fish-oil fishiness is plenty unpleasant, but it's often only half of the experience. Many fish and shellfish from the oceans give it an especially obnoxious boost by way of their salt-balancing compatible solutes.

Whereas algae make their main compatible solute from abundant ocean sulfur, animals make theirs from abundant protein nitrogen in their own bodies. Some animal solutes are tasty amino acids. Shellfish like shrimps and lobsters, clams and oysters and mussels, store sweet glycine and less noticeable alanine, proline, and taurine. Cartilaginous sharks and rays mainly use urea, a storage form for un-needed nitrogen that most animals excrete, along with two purpose-built derivatives of amino acids, **trimethylamine oxide** and **trimethyl glycine** (also known as **glycine betaine**), TMAO and TMG. Bony fish have kidneys, organs with which they actively regulate the contents of their body fluids and tissues, but they also rely to varying degrees on a number of different compatible solutes, particularly on TMAO and TMG, which appear to be especially useful in deep waters.

Neither TMAO nor TMG is volatile; neither has a smell. But when fish die, in the water or on the deck of a boat, these solutes get broken apart, both by the creatures' own failing metabolic machinery and by microbes in their gut and skin. The major product of this breakdown is a volatile and potently smelly molecule indeed: **trimethylamine, or TMA**. The fish richest in TMAO and TMG are the ones that produce the most TMA and smell the strongest. They're why "fishy" is the standard description for the smell of TMA and why TMA defines a primary aspect of fishiness.

TMA is a molecule that we've already encountered several times, always in less than pleasant contexts. It's a common component of animal excrements, including our urine. It's a component of the bouquet of animal death in general. The corpse flower emits it to simulate that bouquet and attract carrion-loving insects. So amine fishiness is a close neighbor to urinous and putrid.

There's a good reason that strongly fishy fish should be off-putting. We're more sensitive to TMA than to its relatives and further breakdown products, dimethyl-amine and ultimately ammonia, by factors of a thousand and ten thousand. It's one of the first signs of deterioration, and as TMA is being produced, so are some unsmellable but toxic amines, notably histamine, which causes sometimes danger-ous allergy-like symptoms. This "scombroid poisoning" is a fairly common form of foodborne illness. (Scombroids are fish in the mackerel family.) TMA fishiness is an early warning signal of possible danger to health.

Fishiness in the kitchen

It's useful to know which fish are most prone to developing fish-oil and amine fishiness, how to minimize their development, and how to cope with them.

In general, freshwater fish—trout, catfish, some kinds of bass—are the least challenging. Because lakes and rivers are usually warmer than the oceans, the oils of freshwater fish tend to be less kinked and less easily oxidized than the oils of ocean fish, and so slower to develop fish-oil fishiness. They also don't make TMAO or TMG and so don't develop amine fishiness—unless their feed includes fish meal with TMAO, as it sometimes does. One special case is freshwater tilapia, whose ancestors lived in the brackish Nile delta; it does make TMAO.

Oceangoing fish develop both forms of fishiness, and more rapidly at a given storage temperature, but they vary in their TMAO content. Sharks and rays rank highest and rely on nitrogenous urea as a compatible solute as well, so they can develop a strong amine-ammonia smell, especially when fermented to make Ice-landic *hákarl* or Korean *hongeo-hoe*. Cod and its relatives, haddock, pollock, and hake, and deep-sea fish rely more on TMAO than most other ocean fish.

Both kinds of fishiness can be minimized by buying the freshest fish possible, ideally with skin and gills that still smell green and vegetal, storing it surrounded by ice, and using it quickly. Much of the undesirable oil fragments and TMA are generated on surfaces by microbes and by oxygen, so thorough rinsing will remove much of them. A few drops of lemon juice can also eliminate amine fishiness. All the amines are chemical bases, and in the presence of an acid will form a bond with the acid's positively charged hydrogen and become positively charged themselves, trapped in the charged network of water molecules and unsmellable. Fish-oil alde-hydes can't be avoided in this way. They're easily scattered throughout the kitchen during cooking, especially frying, and may be lingeringly released from kitchen surfaces for days afterward. To minimize their escape, cook the fish gently in volatile-trapping liquid, or in an envelope of parchment paper or foil or leaf.

Fish: ocean, euryhaline, freshwater

Fish are the lords of the waters, the creatures that move as fluidly as their medium and make the most of its three vast dimensions. Their smells when pristinely fresh and even after cooking are reliable clues to their home waters. Very mild smells indicate a life in the ocean, fresh plantlike smells some time in river estuaries, and earthy smells a career either in freshwater or on the farm, in aquaculture.

Faint, sober, monotonous, empty: these are the words that Japanese researchers applied to the smells of strictly **saltwater** fish in a 1996 review of fish flavor chemistry. **Cod, flounder, hake, halibut, mackerel, snapper, sole, swordfish,** and **tuna** are among the most common saltwater fish, and they're all similarly mild when fresh. Coming from the oceans, they do carry traces of oceanic DMS and fishy TMA, and they emit small amounts of mushroomy, metallic eight-carbon fragments, probably from oxidation of their oils. But they either don't have or don't activate the enzymes that produce vegetal six- and nine-carbon fragments in significant amounts.

SOME SALTWATER FISH

Saltwater fish	Component smells	Molecules
bluefish		
cod		
flounder		
halibut		
mackerel	mushroom	octenol, octadienol, octenone
ocean perch	geranium leaf, metallic	octadienone
ray		
rockfish	ocean	DMS
sablefish	fishy	TMA
sardine		
shark		
skate		
snapper		

Saltwater fish	Component smells	Molecules
sole swordfish tuna turbot	(all above)	(all above)

Watermelon, cucumber, fresh-cut grass: these are the smells that some fish *do* produce in celebrated amounts. The small Japanese **ayu** is known as "sweet fish" or "aromatic fish" and is celebrated for its distinct scent of watermelon. It's a close relative of the smelt family, whose scientific name, Osmeridae, comes from the Greek for "smell." (*Smelt* itself apparently comes from a different root, meaning "soft.") The **smelts** and ayu are fish that can adjust to waters with a range of salinities, from the open ocean to freshwater lakes and rivers. They're usually at their most aromatic in brackish water, often in estuaries where rivers empty into the sea. Another bit of ichthyological Greek: such adaptable fish are called euryhaline, from the words for "broad" and "salt." In addition to the smelts, euryhaline fish include **salmon** and their close relatives **rainbow trout** and **char**, some species of **herring** and **anchovy**, **sea bass** and **striped bass**, **tilapia**, and **eels**. Some, like ayu and salmon, hatch in freshwater, migrate to the ocean to grow and mature, and return to freshwaters, often their river of birth, to spawn. Salmon caught at sea smell very different from salmon caught as they enter the river, with little or none of the vegetal smell that can hang in the air above a salmon stream at spawning time. The mildness of ocean salmon is generally preferred, while ayu are most prized when they're caught in freshwaters on their way upstream.

SOME EURYHALINE FISH

Euryhaline fish	Component smells	Molecules
anchovy ayu	green, grass	hexanal, hexenal, hexenol
barramundi basa, pangasius	mushroom	octenol, octadienol, octenone

continued

Euryhaline fish	Component smells	Molecules
char		
eel	geranium leaf, metallic	octadienone
herring		
salmon	green melon, watermelon	nonenol, nonadienol
sea bass		
smelts	fatty	nonenal
striped bass		
sturgeon	cucumber	nonadienal
tilapia		
trout		

Why do euryhaline fish have more active chain-fragmenting enzymes than saltwater fish? Maybe they're under more pressure from microbes than fish cruising the open ocean, where water nutrient levels and microbial numbers are generally lower.

Then there are strictly **freshwater** fish, the most common being **catfish** and **carp**. Like saltwater fish, their enzyme activities seem to be subdued, and their smell is mild. But that very mildness allows a couple of foreign volatiles to intrude themselves and make the overall impression an unpleasant one. Those interlopers are **geosmin** and **methylisoborneol, or MIB**, which we've sniffed before—in the soil, to which they give its characteristic earthy, musty smell. They find their way into freshwater fish by two routes: runoff or leaching from the soil into the waters, and growth of cyanobacteria that produce them in the waters themselves. Geosmin and MIB get concentrated in fat, primarily in the fish gut and under the skin but also deep in the muscle tissue. We're far more sensitive to them than to the common carbon chains, so we can be offended by relatively tiny amounts.

The muddiness caused by geosmin and MIB can also turn up in **aquacultured fish**, even euryhaline and saltwater species, when they're farmed in earthen ponds, tanks, raceways, or in relatively stagnant offshore pens where fish waste and food debris accumulate and encourage cyanobacteria to grow. Tilapia, rainbow trout, salmon, sea and striped bass, and sturgeon are all examples of cultured fish that can have a muddy off-smell. The only way to get rid of the taint is to hold the live fish in pristine water until the volatiles have largely diffused out again, usually a matter of days, if not weeks. Unfortunately, there's no good kitchen remedy for their presence. Geosmin and MIB are stable to heat and acids, so the only thing to do is trim away all skin and visible fat and overwhelm the flesh with added flavors.

SOME FRESHWATER AND COMMONLY FARMED EURYHALINE FISH

Freshwater & farmed fish	Off-smells	Molecules
big- & small-mouth bass		
carp		
catfish	earthy	geosmin
char	musty	methylisoborneol
eel		
salmon		
sea bass		
striped bass		
sturgeon		
tilapia		
trout		

Fishery scientists are making progress in improving the flavor of farmed fish, which will continue to grow in importance as a sustainable alternative to the depletion of wild stocks. An Australian team has found that they can improve the flavor of freshwater farmed barramundi, also known as Asian sea bass, by finishing them on farmed sea lettuce, which endows them with DMS and a richer, more intense seafood-crab aroma.

Shellfish: oysters

The prime edible representative of the sea is the raw **oyster**. Eleanor Clark has plenty of company in attributing to that bite-size bivalve the power of conjuring the essence of the ocean and the life it sustains. And there are good reasons for that imaginative leap. Oysters are often eaten raw, unaltered by the heat of cooking. They and other bivalve mollusks, mussels and clams, are filter feeders, passing gallons of water every day through their gills, retaining phytoplankton and particulate debris and extracting the nutrients. Their body mass is mainly gill and digestive tissue, suffused with the water to which they adapted with their compatible solutes. So their flavor, both taste and smell, reflects the particulars of their local environment and diet, what has been called "meroir" in analogy to the terroir of grapes and wines (*mer* being French for "sea"). There's a long-standing connoisseurship about oysters that goes back at least to Roman times, celebrating specific waters that reliably produce delicious oysters with the briny, fruity, metallic, floral nuances they can have.

Oysters are also big business, perhaps the prime example of a healthful seafood that can be farmed with minimal and sometimes positive impact on the environment. No surprise, then, that oyster volatiles have been much studied. The scientific apparatus involved is a blunt instrument compared to the palate and experiential database of serious ostreaphiles. It can't begin to account for all the nuances they report, or falsify the more far-fetched. Nevertheless, the results to date can help us begin to understand and appreciate where those nuances come from and what they signify.

Oysters grow attached to rocks or sitting on muddy sediments in the brackish waters of river estuaries, as well as farther out to sea. Today they're farmed in a similar range of environments, often suspended on poles or in bags. They can adapt to a range of water salinities, and their body tissues adjust mainly with sugars and dissolved amino acids, notably taurine, sweet glycine and alanine, bittersweet proline, and bitter arginine. In addition, the "shell liquor" that they retain when harvested begins as a sample of the water they grow in. So the *taste* of oysters is more intense in oysters from high-salinity waters.

Their smells are a different and more complex story. Some belong primarily to the waters and not to the oysters themselves. As they pass large volumes of water through their bodies, oysters pick up volatile molecules that come from other sources—of course the phytoplankton that they ingest, but also from other microbes, from halogen-emitting seaweeds, rain runoff from soils, and releases of industrial or domestic wastes into river waters and estuaries. This same passive absorption is why oysters can concentrate disease microbes and toxins from redtide phytoplankton, all of which can make eating raw oysters a chancy enterprise! Producers sometimes address this problem with *depuration*, or holding harvested oysters in clean or purified water for some time before selling them. Of course this also removes any volatiles characteristic of the home waters—for better or for worse—and can also shift their taste.

The volatile bouquets created by the oysters themselves are more predictable. Taken right from the ocean, they emit mainly the aquatic carbon chains, with little or no DMS, TMA, or halogens. The first products of the oyster's chain-breaking enzymes are fragments that give them the fresh, marine, metallic core of their smell. We associate the smells of some eight- and nine-carbon chains with metals because they also happen to be produced when we handle metals (see page 502). Then as we chew, the enzymes continue to work, oxygen attacks the fragile polyunsaturates, the chains proliferate, and the smell evolves. It becomes stronger and more complex, with new notes that come and go—of mushroom, cucumber, melon, even citrus fruits and flowers.

To savor an oyster straight from the water is thus to get a taste of the water itself *and* the virtuosity of these ancient, "primitive" animals at playing variations on a simple chain of carbon atoms.

SOME SMELLS OF ASIAN AND EUROPEAN OYSTERS

Molecule (carbon chains, C5–C11)	Component smell	Pacific cupped (*Crassostrea gigas*)	European flat (*Ostrea edulis*)
pentenal	green	++	
pentenol	mushroomy	+++	++
hexenol	fresh, green	+	
heptadienal	mushroom, moss	+++	++

continued

Molecule (carbon chains, C5–C11)	Component smell	Pacific cupped (*Crassostrea gigas*)	European flat (*Ostrea edulis*)
octanol	cucumber	+	
octenal	citrus, cucumber	++	++
octenol	cucumber, metallic	+	+
octenone	earthy, metallic	+	+
octadienol	mushroom, moss	+++	++
octatriene	green	+	
nonanol	cucumber	+	
nonadienal	melon, fatty, cucumber	+++	++
nonadienol	fresh, marine, metallic	+	+
decanal	marine	+	
undecanone	fresh, cucumber	+	
cyclohexyl ethanol (C8)	fresh, minty	+	++
lilac aldehyde (C10)	floral	++	++

The carbon-chain bouquet produced by any particular oyster is influenced by various details of the growing environment and by the species. Today the commonest by far is the Pacific cupped oyster, the most vigorous and easiest to farm, but connoisseurs can also find the flat or Belon, originally from northern Europe, the Atlantic oyster, originally from the American coast, the Japanese Kumamoto, the Sydney rock oyster, and others. Connoisseurs, consumers, and experimental taste panels often differ on which oysters tend to be saltier, stronger flavored, finer flavored, and even whether they're distinguishable. This is just another manifestation of their predictable unpredictability, which itself is something to be savored.

Perhaps the most important factor influencing the flavor of any particular oyster is how it has been handled from the moment it was taken from the water, which may be a week or more before it's eaten. Aquatic creatures generally turn up their volatile-generating systems in response to stress. Holding harvested oysters on ice deprives them of water, oxygen, and food, and the extreme cold actually damages their tissues. Holding them in tanks deprives them of food and replaces their home waters with tank water. So many, perhaps most, of the oysters we eat are likely to taste different, less marine and more cucumbery, melony, earthy, than they would if opened at the farm. Eventually the fishy-smelling aldehydes dominate, and traces of DMS from the phytoplankton build up. When it comes to appreciating their native aromas, oysters are best tasted just out of the water.

Other shellfish

Food chemists have paid relatively little attention to the inherent fresh smells of shellfish that aren't commonly eaten raw. Briefly: the oyster's bivalve molluscan cousins get their raw identities from different mixes of short-chain volatiles, **mussels** with winey notes, some **clams** with interesting touches of sweet, almondy benzenoid rings, **scallops** with a hint of chicken fat! In none of them is oceanic, seaweedy DMS especially prominent when they're raw, despite the fact that their digestive organs are often filled with DMSP-storing phytoplankton. But cook them, and all that DMSP generates plenty of DMS: so much so that some marine chemists describe the smell of pure DMS as the smell of cooked clams. Clam chowders, oyster stews, scallop gratins are all powerfully reminiscent of the ocean—the open DMS ocean, not the green or halogenic ocean. Heat tends to dampen specific aromas and replace them with the smell of the ocean at large, intensified.

Squid, **cuttlefish**, and **octopus** are cephalopods, molluscan shellfish turned inside out—the hard exterior now a bonelike stiffener deep within sheaths of muscle—and reshaped into preternaturally agile hunters. Unfortunately, next to nothing has been published about their volatiles, so all I know for sure—from experience as well as the literature—is that they rely on TMAO to adjust to salinity and produce fishy TMA once they're harvested.

SOME SHELLFISH

Shellfish	Component smells	Molecules
mussels	winey, mushroom, fresh, citrus, cucumber	pentanal, pentenol, hexanal, heptanal, octanal, octenol, undecenone, DMS
clams	oily, sweet, mushroom, green, seaweedy, almond, honey, ocean	pentanol, pentenal & pentenol, hexanol & hexenol, heptadienal, benzaldehyde, phenylacetaldehyde, DMS
scallops	fatty, chicken, fermented, green	octadienol, methyl pentanol, hexanal, hexenol, DMS
squid, octopus	mild, marine, fishy	TMA
shrimp	sea air, rock pool, popcorn, seaweed, geranium leaf, metallic, fishy	bromophenols, acetyl pyrroline, DMS, octenol, octadienol, octadienone, TMA
sea urchin, European (*Paracentrotus*)	solvent, fruity, sweet, floral, seaweed, medicinal, marine, meaty/coffee	acetone, methylbenzene, nonanal, decanal, phenol, decanol, benzothiazole

The crustaceans are a different kettle of shellfish. They're mobile; they swim and crawl with their ten appendages, so along with the cephalopods, their bodies are mostly muscle and not digestive organs. Most shrimp flavor comes from the relatively small digestive organs in the "head." Crustaceans are omnivorous, with a diet much broader than the bivalves' phytoplankton. Judging by what's known about **shrimp**, the only crustacean to be studied in detail, they share some simple carbon chains with the bivalves, but stand apart by emitting both DMS and TMA and then two relatively rare kinds of volatiles that make them distinctive. One kind comes from their eclectic diet: the bromophenols sniffed above at the seashore (see page 381), which have an iodine-like smell. The other is an amino acid derivative that contains nitrogen: acetyl pyrroline, which we've encountered in aromatic rices and pandan leaves (see page 181). Its smell is often described as like popcorn, in which it's one of the most prominent volatiles.

Shrimp aroma, then, is a mix of ocean air, iodine, popcorn, green leaves, and depending on how well they're handled, more or less fishiness. Poorly handled shrimp can smell strongly of ammonia. Cooking generally boosts ocean-air DMS but also enhances the popcorn and nutty notes—heat generates more nitrogen-containing carbon chains—and further widens the gap between bivalve and crustacean flavors.

A last example of shellfish flavor is an outlier in many ways: the spine-covered spheres called **sea urchins**, more closely related to sand dollars and sea stars than bivalves or crustaceans, which eat algae along with the occasional animal. Their gonads—masses of sperm and eggs—are rich and delicious raw, at least when they're not overpoweringly animalic. The prized European species has been found to emit a heady mix of solvent-like acetone and methylbenzene, medicinal phenol, and a meaty, coffee-like sulfur volatile, benzothiazole, on top of floral and seaweedy carbon chains. They're unique.

Algae: seaweeds and sea salt

Now for the photosynthesizing creatures that feed us on land as well as their aquatic animal neighbors. Seaweeds have always been important foods along the planet's seacoasts. According to Rachel Laudan's *Food of Paradise*, the Hawaiians once had names for eighty different seaweeds, or *limu*, and still use a number in both fresh and dried-reconstituted forms for their crisp texture and ocean flavors. Seaweeds have become more widely familiar through the influence of Japanese cuisine, in which they're used to flavor soup stocks, to wrap morsels of fish and rice, as a vegetable, and as a garnish; a number are known mainly by their Japanese names. Seaweeds are usually dried immediately after the harvest to preserve them, and this can change their mix of volatiles, but usually leaves their overall character intact.

As we've seen, the seaweeds fall into three general groups named for their dominant colors. The members of each group also tend to resemble each other in their sets and proportions of volatiles.

The brown seaweeds are dominated by the halogens, bromine and iodine and their carbon compounds, and so offer that particular combination of sweetness and disinfectant sharpness. Forests of **giant kelp** decorate the Pacific coast of North and South America and constantly strew shorelines with their impressive stalks and blades and smell. A few browns also produce unusual volatiles, the sesquiter-penoid cubenol with a seaweed-hay smell, and the eleven-carbon molecules called dictyopterenes, some of which smell like the seashore, others like green herbs.

Kombu, the base flavoring for many Japanese soups and stews, is a relative of giant kelp, and happily, it's less chock-full of halogen volatiles. It has enough iodine to be an important contributor to the healthful iodine intake of the Japanese, but it doesn't dominate the smell of soup stock, dashi—unless you make the mistake of boiling it. A gentle simmer keeps the cucumbery and floral volatiles to the fore while it extracts the abundant amino acids that the browns use as compatible sol-utes, most notably the glutamic acid that our taste buds sense as savoriness, or umami. (It was a Japanese scientist investigating the flavor of dashi who discov-ered glutamic acid as a key taste molecule and introduced its purified sodium salt, MSG, as a flavor additive.)

A more delicate brown seaweed, **wakame**, is commonly encountered in seaweed salads—in misleading bright green form, because heating it destroys one of its protein-based photosynthetic pigments and actually makes its chlorophyll partner more visible. It too goes easy on the halogens, instead carrying cubenol, with its seaweed-hay quality. And in Hawaii, the brown alga **limu lipoa**, with narrow fronds a few inches long, grows on offshore reefs. It's gathered after being washed up on the beach and scents the water and air with its herbaceous pheromones. Its flavor is compared to a combination of pepper and sage, and it's served freshly chopped with raw fish.

The red seaweeds are the most diverse group. Some reds are dominated by DMS and other sulfur compounds, others are rich in halogens, and still others seem to emit a little of everything. They include probably the most familiar sea-weed in the West, known in Ireland as **laver** and in Japan as **nori**. These are species of red algae in the genera *Pyropia* and *Porphyra* whose small diaphanously thin blades are pressed together and dried into sheets that are toasted and used to wrap sushi and rice balls. Nori has been cultivated on the seashores of China and Japan for centuries, no doubt partly because it grows on rocks in the tidal zone and sur-vives repeated exposure and drying. The nori algae aren't notable DMSP produc-ers, and instead accumulate sugars, amino acids, taurine, and betaine as compatible solutes. But their aroma is still marked by a number of sulfur volatiles that are enriched by drying and then by toasting. They also have floral notes that can be reminiscent of cured tea and tobacco leaves. The overall effect is appealingly savory and complex.

The other commonly eaten red algae are more dominated by halogen compounds and so are more of an acquired taste. Notable among these are a couple of Irish standbys, **Irish moss** and **dulse**, and in Hawaii, relatively mild **ogo**, whose crunchy filaments are put to many different uses (some species are the source of the thickener agar-agar), and feathery **limu kohu**, which is powerfully halogenic, traditionally dried in the sun and powdered for use as a condiment.

SOME COMMON SEAWEEDS AND SEA SALT

Seaweed	Component smells	Molecules
brown		
giant kelp (*Macrocystis*)	sweet, iodine	chloroform, bromoform, iodoform
kombu (*Laminaria*)	iodine, floral, cucumber	iodo-octane, nonanal, nonenal, nonadienal, octenol, ionone
wakame (*Undaria*)	hay-seaweed, floral, cucumber	cubenol, ionone, nonenal, nonadienal
limu lipoa (*Dictyopteris plagiogramma, australis*)	green, mossy, seaweed, herbal-spicy	dictyopterenes
red		
nori, laver (*Pyropia, Porphyra*)	ocean, sulfurous, floral-tobacco, violet, seaweedy	DMS, hydrogen sulfide, methanethiol, cyclocitral, ionone, heptadecadienal, heptadecatrienal
Irish moss, carrageen (*Chondrus*)	green, fruity, sweet, iodine	hexanal, pentanal, dichloromethane, methyl iodide
limu manauea, ogo (*Gracilaria*)	marine, iodine	bromophenol, dibromophenols
limu kohu, "supreme seaweed" (*Asparagopsis*)	sweet, iodine	bromoform, iodoform, dibromoiodomethane, iodine

continued

Seaweed	Component smells	Molecules
dulse, dillisk, sea parsley (*Palmaria*)	sweet, iodine, fresh, marine	bromoform, chlorobenzene, iodopentane, octanal, nonanal, octatriene
green		
sea lettuce, limu eleele, aonori (*Ulva*, *Enteromorpha*)	oceany, green, cucumber, seaweedy, violet	DMS, hexanal, octenal, nonenal, decadienal, heptadecenal, heptadecatrienal, ionone
aonori (*Monostroma*)	oceany, almond, violet	DMS, benzaldehyde, ionone
sea salt	violet, green, resinous, sweet	ionone epoxide, methyl heptenone, trimethyl cyclohexanone, isophorone

Green seaweeds are found in freshwaters as well as the sea and tend to emit a mix of DMS and carbon-chain aldehydes, including some distinctive long ones such as heptadecatrienal. The types known as **sea lettuce** and **aonori** are delicate and translucent, some of them only one cell layer thick. Their smells are among the most representative of the ocean itself, dominated by oceanic DMS, with the common bouquet of green, mushroomy, cucumbery notes, sometimes a touch of bitter almond, sometimes a floral hint from violet-like ionone.

Finally, the faintly floral scent of phytoplankton emanating from table salt! **Sea salts** are made by allowing seawater to evaporate until its dissolved minerals are so concentrated that they precipitate into solid crystals. The microbes that can survive in such salty brines, bacteria as well as phytoplankton, tend to accumulate orange-red carotenoid pigments to protect their photosynthetic systems, and fragments of these terpenoids and other molecules can persist—along with some dormant cells—on unrefined crystals. Their smells are swamped by whatever food the salt is sprinkled on, but protected in a closed salt cellar and sniffed for as it's opened, they're a wonderfully delicate manifestation of life's robustness.

Invisible volatiles, vital influences

Delicate, delectable sea lettuce became notorious in France in the early years of this century, for its fecundity and for its toxic smell. Over the period from 1989 to 2011, fertilizer-rich runoff from agricultural lands in Brittany regularly caused the offshore population of *Ulva* to explode. Some beaches ended up repeatedly covered with tons of dying, rotting, stinking seaweed. The decomposition of their DMSP-rich tissues and sulfated cell-wall carbohydrates generated enough hydrogen sulfide gas to kill a horse, two dogs, thirty-six wild boars, a jogger, and a seaweed harvester.

The Earth's waters are vast, its creatures' numbers unimaginable, but the cumulative effects of their volatiles are not often this dramatically evident. It's only through modern atmospheric science that we have an inkling of the bigger picture. Like the land's massive emissions of isoprene and terpenoids and green-leaf volatiles, the oceans' release of dimethyl sulfide and halogens affects the chemistry of the atmosphere, the absorption and reflection of the sun's energy, the trapping and release of the Earth's heat, and thus local and global climate. DMS reacts with oxygen to form sulfuric acid and other compounds that initiate the formation of water droplets and marine haze, which can aggregate and develop into cloud cover. Iodine and its oxides do the same, especially locally along seaweedy coasts. Iodine and bromine volatiles also react with and destroy the ozone form of oxygen, which reduces its absorption of damaging ultraviolet light from the sun.

So the smell of the open ocean signals the reinforcement of a cloud-bright shield that helps regulate the Earth's temperature. Coastal pungency at low tide marks the local thinning of the air's shield against solar radiation. Both testify to our planet's dynamism—chemical elements leached into the waters, put to vital use by countless living cells, flung into the air where they modulate the planet's energy balance, then falling back and ending up in the waters to be taken up by life once again.

I've come to love some unlikely tokens of this great natural cycle: not the predictable aquatic edibles, but a handful of Scotch whiskies! Here's the connection. Ocean waves and kelp release bromine and brominated carbon chains into the atmosphere, where they're transported great distances by winds and eventually return to the Earth in rainfall and snow. Most of the bromine compounds that fall on land end up washed into rivers and back to the ocean. But some forms of vegetation have an affinity for bromine and retain it. Among them are the peat bogs formed by sphagnum mosses throughout the Northern Hemisphere. To avoid rotting in the waterlogged bogs, sphagnum mosses fill themselves with protective aggregates of six-carbon phenol rings, which suppress molds and bacteria. When rain carrying bromine falls on peat, the water percolates through, but the bromine stays behind and reacts with the phenols to form bromophenols: the same phytoplankton volatiles that give wild shellfish and fish their intense oceanic flavor.

Peat beds can be thousands of years old and have long been cut, dried, and burned for fuel. Some producers of Scotch whisky dry their malted barley with burning peat, some of that peat can be especially rich in bromophenols, and some of those bromophenols make it through fermentation and distillation and into the bottle. So it is that at home in San Francisco or traveling a thousand miles inland, I can sniff an evening dram of Lagavulin and savor a touch of the ocean, its animals and weeds and long reach.

Chapter 16 · AFTER-LIFE: SMOKE, ASPHALT, INDUSTRY

....

You have the glittering beauty of gold and silver, and the still higher lustre of jewels, like the ruby and diamond; but none of these rival the brilliancy and beauty of flame. . . . The candle alone shines by itself. . . .

The heat that is in the flame of a candle decomposes the vapour of the wax, and sets free the carbon particles—they rise up heated and glowing and then enter into the air. . . . Is it not beautiful to think that such a process is going on, and that such a dirty thing as charcoal can become so incandescent?

> · Michael Faraday, *The Chemical History of a Candle*, 1861

Coal-tar, dark, thickish, neither a liquid nor quite a solid, a useless residue clogging the pipes in the making of illuminating gas, has become . . . a central item in the wealth of nations. . . . The despised by-product is nature's own laboratory, whose magic alembic distills fluids and vapors and scales and crystals for the alleviation of suffering. From coal-tar and allied substances are derived countless synthetics which have replaced the herbs of our forefathers. Among these remedies are the antiseptics, phenol, cresol, resorcin; the local anesthetics, alypin, novocaine, stovaine; . . . saccharin, the permissible sugar of the diabetic. . . .

Parasiticides and perfumes, fuels and photographic supplies, the asphalt of the pavement and the varnish on the roof, are all born in the deep womb of coal-tar.

> · Professor Victor Robinson, "Coal-Tar Contemplations," 1937

We pass now from sea and land to candle flame and coal tar, from great cycles of life and death to life-shattering extremes of heat and gravity, conditions so destructive of complexity that they expose its naked demiurge, Hero Carbon itself, soot-black and incandescent!

The very existence of carbon as a chemical element had only been recognized a few decades before Michael Faraday gave his famous Christmas Lectures on the candle. By the time Faraday spoke, he and others had demonstrated carbon's essential role in *fire*, that rapid release of intense heat energy whose mastery shaped our species, powered the development of civilization, and smoked up everyday life for thousands of years—the previous chapter's peaty whisky included. And these pioneering chemists had discovered coal and tar and petroleum to be carbon's own underground laboratories, dark archives of rings and chains that promised to contribute much more to human life than heat and light.

The coal-tar scholars were among the pioneers of *organic chemistry*, the study of the materials made and used by living organisms during their active existence. Once organisms die, their materials embark on a career of their own: they persist as the remains we call *organic matter*, and because organic matter consists largely of protean carbon, it undergoes its own metamorphoses, generating countless molecules simply because they can be generated, not to serve any organized purpose. The great achievement of the early organic chemists was to discover this posthumous free-form creativity of Hero Carbon. It paved the way for a thorough transformation of material life in industrial countries, the making of the largely synthetic environment that we inhabit today. That environment has its own distinctive smells, which we usually describe as "chemical" or "plastic," loose synonyms for "unnatural." In fact, like the smells of soil and stone, and of smoke as well, they too are traces of the after-life of Earth's organisms. The only unnatural thing about them is their prominence.

Soil and wetted stones get their smells from the shallow burial and quick recycling of Earth's organic matter. Fire and coal tar respectively smell of its cremation and its deep, deep entombment. The smokiness that accompanies fire is a farewell whiff of life's organized structures as they disintegrate into thin air before our eyes. The smells of coal and tar and petroleum arise from those same structures invisibly, airlessly, slowly crushed and simmered into a carbonaceous concentrate. All long predate our primate clan and were olfactory landmarks from our very beginnings.

In the African and Eurasian wild where our clan evolved, smoke would have been an alarming signal of destruction and danger. Once our ancestors learned to control fire, at least half a million years ago, smoke became a constant companion. Its smell continues to pervade the many communities around the world that rely on or tolerate open fires, as my Illinois hometown did when I was growing up: autumn weekends were censed by pyres of raked leaves lining the streets. We know now that combustion fumes are toxic, and societies that can afford to do so limit their release into the air we breathe. Even so, we still enjoy fire's volatile fragments around the fireplace or campfire, on foods, in alcohols, from tobacco. Smokiness remains a mark of human influence, warmth, nourishment, and community.

Surface seeps of tar and petroleum aren't as common as wildfires, but there were plenty in the Middle East for the early hominid migrants out of Africa to encounter. Coal and tar and petroleum got their start millions to billions of years ago, when microbes and plants were buried deep underground, crushed by gravity pulling on the oceans and continents above, and roiled by heat from Earth's core below. Eons of this planetary pressure cooking generated countless variations on carbon chains and rings, including the fragments that give tar and gasoline and paint thinners their familiar smells. Today they also mark human activity and influence, but industrial rather than domestic: the making and maintenance of built structures and roads and machines, the fueling of engines. The abundant chemical energy stored in fossil carbon chains has enabled human activity on an unprecedented scale—and generated enough odorless carbon dioxide to change global climates and ecosystems.

Even though they're mere remnants of life's marvelous self-replicating systems, fossil organics embody another aspect of Hero Carbon's virtuosity: the vast range of molecules that can result from life's dismantling, which rivals the small-molecule inventiveness of life itself! While the fragments generated by fire quickly fly away in smoke and fumes, fossil fragments accumulate in liquids and solids that are far easier to study and play with. That's exactly what the early organic chemists began to do around 1800. Among other discoveries, they figured out how to pluck specific molecules from complex mixtures like tar. They were thus among the first people to smell individual carbon rings and chains on their own. They provided the initial entries for this book's smell tables, the precedents for all the rest.

The rise of organic chemistry marks a turning point in human history and in the history of our corner of the cosmos. No longer was Hero Carbon limited to exploring the possibilities of complexity through the gradual working of natural selection. Now it had active, curious partners in the lab. The result has been two centuries of ever-accelerating change on planet Earth—and the contemporary world's industrial smells, both newly made and eons old.

Fire and life

It would have been a little before the rise of the land plants and their delightful chemical weapons and come-ons, about a billion years ago, that the cosmic Chef would be able to detect the first curlings of the smoke that comes from fire. Not the acrid primeval emissions of volcanoes and lightning strikes, but the sweeter fumes from early land life that stray lava and electrical storms could ignite. But those initial puffs would have been faint and brief. There wasn't much living material to burn, and there wasn't enough oxygen to sustain the burning.

The key to oxygen's role in both life and fire is its hunger for electrons, its aggressive bonding to atoms that have a weaker hold on theirs: namely the carbon and hydrogen atoms out of which all living things are built. In the machinery of living cells, oxygen's electron hunger is actively managed to make it a constructive force: it's channeled into the stepwise disassembly of sugars to provide chemical energy for the cells' workings. Fire results when oxygen goes unmanaged. When goaded by an initial dose of energy from lightning or magma or a struck match, it attacks carbon and hydrogen atoms indiscriminately, ripping them away from each other to form molecules of carbon dioxide and water. Those reactions release enough energy to trigger another round of attacks, and so the reactions and their energy release sustain each other. We sense the energy release in two different ways. There's the heat from invisible infrared radiation and the increased motion of air molecules, and there's the visible light of the flame, both a dim blue glow from the primary reactions, and the orange-yellow glow from superheated particles that haven't yet been fully burned.

While most of early Earth's oxygen came from photosynthetic microbes in the oceans, it was the land-dwelling ancestors of mosses and liverworts and upright plants that helped raise atmospheric levels high enough to sustain burning (15 percent or more of the air's gases). The earliest evidence of persistent fire is charcoal dating to more than 400 million years ago. By 100 million years ago, fire was common enough in seasonally dry regions that it influenced the evolution of plant structures and life strategies, including the thick bark on conifer trees. Many plant species now have seeds whose sprouting is stimulated when they detect chemical signs of recent fire, which often clears the soil of competitors and fertilizes it with their ash. Volatile germination signals called *karrikins* were identified when scientists managed to stimulate seed sprouting with a lab version of the cook's liquid smoke. Fire has also shaped large-scale environments, including the open-woodland savannas of Africa. Grasses dominate there because they can spread by protected underground stems, and in the dry season offer an abundance of thin leaves that burn easily and help eliminate competing plants.

Fire has also had profound effects on our own species. The Harvard anthropologist Richard Wrangham argues that our parent species *Homo erectus* probably controlled fire and began to cook foods far enough back—more than a million years ago—that it fueled the enlargement of the brain, shrank the jaw and chewing muscles, encouraged cooperation and early forms of sociality, and helped turn upright *erectus* into modern, thinking *sapiens*. The earliest clear evidence for human control of fire is much later, around eight hundred thousand years ago, so Wrangham's hypothesis remains controversial. But there's no question that fire and its smells have been a defining part of human life for as long as there have been humans.

The most frequent cause of natural fire is lightning. Across the Earth's surface there are something like a hundred lightning strikes every second. With time, our ancestors discovered that fire could be contained in isolated piles of plant material, maintained as embers to ignite new fires, and started fresh by hitting sparks from iron-rich rocks or heating wood with the friction of rubbing. The smells of fire then became the smells of cooking, keeping warm, gathering in groups, warding off predators, being awake at night, driving game animals, clearing brush and forest to make settlements and plantings, inhabiting caves and marking their walls: the smells, that is, of human activity. For hundreds of thousands of years, most humans would have smelled smoke every day from birth to death.

And what an example fire must have set for sapient animals with a growing capacity to reflect on their experience! A fluid, elusive, painfully hot and bright apparition, ungraspable, covering grasses and bushes and trees and fleeing animals, consuming their substance while sending clouds skyward, then disappearing and leaving behind inert ash and char. It would have suggested many uses but also many ideas: of immaterial presences, tremendous powers, a realm from which fire descends and to which its fumes rise, of sweet-smelling smoke as a means of communicating with the unseen.

Today we can peer down from the sky with Earth-imaging satellites and see hundreds of large fires burning every day across the globe, both wild and set by *Homo sapiens*. Some glow; others, like the periodic underground peat fires of Indonesia, smolder with region-choking smoke. It's the view that the Chef of the cosmos would have had over the last million years as she smelled the volatile inventions of plant life increasingly obscured by the fumes of its incineration.

The life of fire: fumes, flames, embers

Up to this point in our chronicle of the world's smells, the volatiles responsible for them have been formed invisibly and intangibly, mostly by living creatures. Fire makes the creation of smells manifest to all our senses. When we gaze at it and see the flames and smoke, hear the fuel crackle and hiss, feel the embers' glowing heat, we're experiencing the vaporization of life's solid body into a host of volatile molecules.

That flamboyant disappearing act involves two basic processes, one of them mainly generating volatiles and smells, the other mainly destroying them, and each depending on the other for its continuation. The term for the destructive process, *combustion*, is familiar. It comes from the Latin word for "to burn up or consume." Today it usually means the thorough reaction between oxygen and carbon-containing organic materials to yield molecules that can't be spontaneously oxidized any further—namely odorless carbon dioxide and water vapor—together with enough heat to make it self-sustaining. Though combustion proper destroys volatiles, a version of it can actually generate smells. In *incomplete combustion*, there isn't enough oxygen or heat for all the fuel to be broken down all the way to carbon dioxide and water: so some of the fuel carbon survives in larger chains, rings, or in the aggregates we call soot.

Less familiar than combustion is a second process in burning, the more creative one, called *pyrolysis*. That's a term coined by a nineteenth-century chemist, from Greek roots meaning "fire" and "loosen" or "separate" or "release." Pyrolysis names the release of carbon-chain fragments when organic materials are heated to high temperatures *without* oxygen. The heat energy still breaks apart large carbon molecules into smaller ones, but it doesn't quickly oxidize them to odorless gas and vapor. Instead, the fragments escape as volatiles or stick to each other in progressively larger agglomerations of carbon chains and rings. Taken to the extreme, the agglomerating pyrolysis reactions produce forms of solid carbon: charcoal, coal, graphite, diamond.

To understand the roles of combustion and pyrolysis in the creation of smoke, think of the difference between lighting a pure gas flame—on the stovetop or a gas grill or cigarette lighter—and starting a wood fire. Provide a momentary spark to a stream of methane or propane or butane gas, and abundant oxygen readily attacks the abundant carbon-hydrogen molecules, oxidizing them completely to carbon dioxide and water vapor, and producing energy to keep the attack going. Textbook combustion. However, provide that same spark to a piece of wood, and all you get is a tiny scorch mark. Wood doesn't combust as simply as gas does, because it's a solid mass. Only its surface is directly exposed to oxygen and the heat of the spark—and that heat is quickly dissipated by the wood below the surface.

To burn plain wood, we have to heat it for some time, with already burning paper or kindling. As the surface chars, it heats the subsurface wood to several hundred degrees: much lower than the surface, but hot enough to initiate the process of pyrolysis. The subsurface wood begins to break apart and generate volatile fragments that escape as gases to the surface through pores in the wood, hissing as they go. The main pyrolysis gases are one-carbon molecules—the same methane as natural gas, plus methanol and carbon monoxide—along with hydrogen and hydrogen sulfide. When they reach the far hotter surface, oxygen combusts the gases to carbon dioxide, water, and sulfur dioxide, releasing the energy that we see and feel. That energy continues to heat the wood and generate more pyrolysis gases to feed the flames: so pyrolysis and combustion sustain each other.

Along with the mostly odorless main gases, pyrolysis also generates traces of larger fragments that we can smell. The various volatiles emerge unevenly, have different ignition temperatures, and from moment to moment may not have enough nearby oxygen or heat energy to combust all the way to carbon dioxide. The smells of smoke come from the copious free volatiles that escape complete combustion.

The interplay between combustion and pyrolysis shapes the life of a wood fire into several different phases, each marked by a different temperature range and set of chemical transformations, each generating volatile molecules in different proportions, each with its own smells. Here they are in order.

- From room temperature up to 400°F (200°C): evaporation. The release of free moisture and volatiles already present in the wood, for example terpenoids from conifer wood and vanillin and lactones from oak. Heightened woody smells.
- Between 400°F and 500°F (200°C and 250°C): slow pyrolysis. The gradual release of simple breakdown products that aren't combustible, including water, carbon dioxide, formic and acetic acids. Sharp, vinegary smells.
- Between 500°F and 1000°F (250°C and 500°C): fast pyrolysis and ignition. The release of copious combustible volatiles, including hydrogen gas, carbon monoxide, and methane, along with odorous molecules and particles of aggregated volatiles and solids. Glowing surfaces, smoke, and smoky, sweet, spicy smells.
- Between 1000°F and 3000°F (500°C and 1500°C): combustion. Flames fueled by the ongoing pyrolysis and release of volatiles, now including carbon-ring benzenoid and phenolic molecules, and carbon-ring soot makers in low-oxygen or low-temperature regions of the fire. Increasingly harsh, tarry, chemical smokiness.
- Cooling below 1200°F (600°C): smoldering. Dying flames due to declining release of combustible pyrolysis volatiles and declining temperatures; copious carbon monoxide released, some combusting at ember surfaces and heating them enough to glow. Harsh, ashy smell.

So there's a lot going on in that mesmerizing dance of pyrolysis gases and pyro-fog and soot! For the smell explorer, the bottom line is this: flames are dazzling, but smokiness is born in the dark. It's pyrolysis and incomplete combustion that generate the most diverse volatiles in fire fumes and smoke, and that let them survive long enough for us to perceive them.

The smells of fire: sweet, spicy, smoky, tarry

Wood and other solid plant materials burn unevenly. At any given moment, different parts of the material are at different temperatures, have different oxygen supplies, and emit different volatiles. Whenever we smell burning plant materials, we're encountering a mix of all the molecules listed below, their fluctuating proportions determining the qualities we happen to perceive. The major exception to this rule is plain paper, which is mostly cellulose fibers and smells distinctly sweet as it burns.

The smells of burning wood and other plant materials start with the large molecules whose breakdown feeds the flames. Dried plant remains consist mainly of the structures that supported the plants' upright growth and held their living cells together.

The most abundant structural materials are cellulose and hemicellulose. Their building blocks are sugar molecules, small groups of five or six carbon atoms bonded to an equal number of oxygens. Cellulose and hemicellulose have plenty of reactive oxygen built into them, and they disintegrate in the early stages of pyrolysis, around 400°F (200°C) for hemicelluloses and 600°F (300°C) for cellulose. They break apart mostly into starter-set carbon chains and smaller carbon-oxygen rings. Prominent among these are sharp, pungent acids and aldehydes, common indicators of breakdown by microbes as well as heat. More pleasant are the buttery smell of four-carbon, two-oxygen **diacetyl** and bready, nutty, sweet notes from **furans, furanones**, and **furfural**, five-cornered rings with four carbon atoms and one oxygen. We've encountered them occasionally in dried plant materials and in fruits, and will again in many cooked foods, as their smell qualities indicate. Many cooking methods are versions of pyrolysis!

The other primary plant structural material, lignin, consists of hexagonal carbon rings interlocked into large honeycomb-like networks that are strong and stable, with only a smattering of oxygen atoms. Lignin takes more energy to break apart than the sugar-based materials, so it pyrolyzes at higher temperatures, 600°F (300°C) and up. When it does, it generates variations on its basic ring structure.

SOME SMOKE SMELLS FORMED AT LOW PYROLYSIS TEMPERATURES

Component smells	Molecules
ethereal, alcohol	methanol
sharp	formic acid
chemical, suffocating	formaldehyde
pungent, vinegar	acetic acid
pungent, fresh, green	acetaldehyde
pungent, cheesy	propionic acid
acrid, irritating	acrolein (propenaldehyde)
sour milk, cheesy	butyric acid
buttery	diacetyl
solvent, earthy, malty, chocolate	furans
bready, nutty	furfural
sweet, hay, coconut	angelica lactone (furanone)
caramel, sweet, burnt	other furanones

The more characteristic smells of fire and smoke arise when these six-carbon benzene rings break off from lignin and take flight, carrying with them various small chemical groups projecting from the corners. The most representative smell of smoke itself comes from rings that have oxygen-containing decorations. **Syringol**, named roundaboutly from the scientific name for lilac, *Syringa*, is rare outside of woodsmoke. **Guaiacol** is also found in trace amounts in materials that have never been much hotter than a warm day, including oak wood, secretions of the wood-eating beaver (see page 470), and the Caribbean guaiac tree, from which it gets its name. Syringol has a mild, sweet, resinlike smokiness, while guaiacol smokiness has meaty and spicy overtones. Then there are a number of guaiacol-plus molecules, guaiacols with an added carbon-chain decoration, which tend to be less simply smoky (ethyl and **vinyl** name two-carbon decorations; propyl, propenyl, and allyl, three-carbon decorations). One of the guaiacol-plus volatiles is identical to the primary volatile in cloves, eugenol, and a slightly modified version of another is vanillin, the primary volatile in vanilla beans.

Very specific spice smells from burning wood! The clove tree and vanilla orchid use enzymes to construct eugenol and vanillin as chemical defenses, while fire happens to throw them off in the process of deconstructing plant lignin. It almost certainly did so long before cloves and vanilla appeared on planet Earth. When they finally did appear, our cosmic Chef would have described their smells as reminiscent of smoke. For us latecomers to Earth, vanilla and clove notes enrich the smell of woodsmoke with their spicy qualities.

SOME SMOKY, SPICY SMELLS FORMED AT MODERATE PYROLYSIS TEMPERATURES

Component smells	Molecules
sweet, smoky, balsamic, medicinal, woody	syringol (hydroxy-dimethoxy-benzene)
medicinal, smoky, woody, spicy, meaty	guaiacol (hydroxy-methoxy-benzene)
clove, medicinal, curry	4-vinylguaiacol
bacon, clove, smoky	4-ethylguaiacol
vanilla, smoky	3-, 4-methylguaiacol
clove, spicy, sweet	propylguaiacol
clove	eugenol (allylguaiacol)
clove, sweet, woody	isoeugenol (propenylguaiacol)
vanilla	vanillin (hydroxymethoxybenzaldehyde, guaiacol methylaldehyde)

At wood temperatures hotter than those that produce abundant syringol and the guaiacols, the dominant smells become burnt rather than smoky and spicy, suggestive of tar and antiseptics. The higher energies strip the ring volatiles down to just one oxygen-containing decoration. Rings with one hydroxyl (OH) decoration are called **phenols**. **Phenol** itself, named from the Greek for "shining" because it was first found in illuminating gas, has a smell that's sweet but penetrating, reminiscent of disinfectant, plastic bandages, throat sprays, and rubbing creams, because it's found in all of them (see page 426). Add a carbon-hydrogen methyl group, and you get the methyl phenols, alias **cresols**, whose smells depend on which of the ring carbons the methyl group ends up on. One smells medicinal and inky; another medicinal and tarry, but with a leather-like animal note. A third has a strong, pungent smell suggestive of animal excreta in general and horse stables in particular: it's the one most commonly produced by the microbes in animal intestines when they break down amino acids (see pages 59, 67). Phenol and the cresols are sometimes described as "inky" because writing inks were long made from the soot produced by incomplete combustion of various organic materials; *ink* comes from the Greek for "burn."

SOME TARRY SMELLS FORMED AT HIGH PYROLYSIS TEMPERATURES

Component smells	Molecules
sweet, tarry, burnt, disinfectant	phenol (hydroxybenzene)
inky, medicinal	2-methylphenol (o-cresol)
tarry, burnt, leather	3-methylphenol (m-cresol)
stable, fecal	4-methylphenol (p-cresol)
sweet, tarry, burnt	dimethylphenols
medicinal, sweet	vinylphenol

Even more stripped down than the phenol rings are the ring volatiles with no oxygen atoms left, the hydrocarbons **benzene**, **toluene**, **styrene**, and **xylene**, which tend to appear at 1500°F (800°C) and above. All of them have a sweet, ethereal, "chemical" smell, reminiscent of solvents, because that's where we encounter them in their pure form: in paint and polish removers, glues, and marking pens.

SOME HEADY SMELLS FORMED AT HIGH PYROLYSIS TEMPERATURES

Component smells	Molecules
solvent, sweet, plastic	benzene
solvent, sweet, plastic	toluene (methylbenzene)
solvent, sweet, plastic	xylenes (dimethylbenzenes)
solvent, sweet, plastic	**styrene** (vinylbenzene)

So the smells of burning wood are many and various: a mix of sharp and toasted, nutty and bready, smoky and spicy, tarry and medicinal and heady, each quality coming and going as pyrolysis is affected by air currents and shifting logs. Their relative prominence also depends on the fuel. The wood of conifer trees, often called **softwood**, can contain less hemicellulose and more lignin than **hardwood** and is sometimes denser, usually meaning more intense combustion and higher temperatures. Softwood lignin has a higher proportion of guaiacol-producing structures; hardwood lignin, more syringol-producing ones. As a result hardwood smoke is generally milder. Softwoods like pine also tend to burn with more smoke and soot from the incomplete combustion of their terpenoid resins.

Connoisseurs of barbecue and other woodsmoked meats have long known much of this from experience. They choose their woods carefully and manage pyrolysis by various means, particularly by limiting airflow and oxygen to maintain the medium-hot temperatures that favor sweet and spicy and smoky volatiles. Everyday cooks can generate flavorful smoke more simply by wetting wood chips or loosely enclosing them in foil packages to prevent flames and encourage smoldering. In the kitchen, wood can be replaced by a mixture of whole grains and tea, which provide sugar-based carbohydrates for the toasty notes and carbon-ring phenolic acids and tannins for the sweet and spicy and smoky (see page 493).

When we burn wood in the fireplace or smoker, nearly all of it goes up in smoke: only scant ashes of oxidized minerals remain. However, as some of our earliest ancestors discovered, it's possible to capture pyrolysis volatiles and put them to countless uses. That discovery brought a new landmark smell, and a landmark achievement for creatures of Hero Carbon: one that boosted it to a new level in the cosmic game of complexity and anticipated the much later breakthroughs of organic chemistry.

Smoke deconstructed: tar, pitch, turpentine, char

In late 2001, Italian archaeologists discovered the prehistoric remains of an elephant and other animals in the Campitello Quarry, near the river Arno southeast of Florence. The site had apparently been buried by sediments around two hundred thousand years ago. Alongside the bones lay three sharp-edged flakes of flint, apparently stone tools made by our Neanderthal cousins for hunting or butchering. Two of them were covered at one end with a thin layer of hard black material. Those small patches may be some of the earliest evidence of the human family transforming one natural material into an entirely different and useful material. And they mark the introduction of a new family of smells into everyday life. The black patches turned out to be wood tar.

These days *tar* often designates a black, sticky, strong-smelling material deriving from petroleum or coal, but the word comes from the same root as *tree*, and strictly speaking means the black, sticky, strong-smelling stuff made from trees. Its smell is a distinct one, different from petrochemical tar, smoky and "burned." As the roster of smoke volatiles suggests, **wood tar** is rich in cresols, and in fact was the material that led to their discovery, long before they were found in smoke. In 1832, the German chemist Carl Reichenbach extracted from beech tar a colorless liquid that smelled like smoke-preserved meat; he named it *creosote* from the Greek words for "flesh" and "keep safe." Later chemists found a number of individual chemicals in that liquid and named some of them *cresols*. Reichenbach's *creosote* came to mean both a tar extract applied to preserve wooden railroad ties, utility poles, and the like, and also the dangerously flammable deposits left in fireplace chimneys.

The accumulation of creosote in chimneys suggests how early hominids managed to obtain tar from wood and bark. If hot fumes of partial combustion bump into a relatively cool surface before they escape into the open—a chimney wall, for example—then some of their components will condense into a liquid film on that surface. Archaeologists speculate that the Neanderthals noticed sticky black deposits on overhanging firepit stones after birch bark had been burned, or the same stuff oozing from scraps of bark left buried in the ash. They then encouraged the deposits intentionally, scraped them off, and put them to use.

The early tar makers applied it to stone tools to make them easier to grip, and glued stone points onto wooden spears and handles to make the first composite tools. Later on, larger-scale production brought its use to waterproof clothes and containers and shelters; thanks to its oil-like carbon-chain molecules, tar doesn't mix well with water. The solid residue left behind by tar making is what we call charcoal, which burns more cleanly than wood, and is far better at smelting oxidized copper and iron ores into the pure metals. Run the clock forward to late prehistory, and peoples across Eurasia were cutting and pyrolyzing vast areas of forest for charcoal and tar production. The seagoing Minoan civilization of Crete and the Phoenician culture of northern Africa used massive amounts of tar to coat their wooden ships.

By Roman times there were two standard ways to manufacture tarlike materials from wood, and they produced different smells. One was to heat any wood to pyrolysis temperatures in an air-poor kiln, with pipes that condensed and delivered volatiles to the outside. The second way was to start with resinous trees like conifers and terebinth, a relative of the pistachio, slash them and collect the resin that flowed out, and then boil the resin down until it turned thick and black. With no pyrolysis of phenolic-rich lignin to produce cresols, this *pitch* would smell mainly of the resin's terpenoids and their breakdown products.

The Roman naturalist Pliny also described the collection of the most volatile components of resin pitch as it cooked down: "While the pitch is kept on the boil, fleeces are stretched out above the steam rising from it and then wrung out." The wrung-out "pitch-oil" would have been what we now call **turpentine**. Volatiles can make up more than a third of the volume of conifer resin and are usually dominated by pinene, with small amounts of other "terpy"-smelling terpenoids (myrcene, phellandrene, camphene, terpinene). When the resin is heated, the smaller volatile terpenoids evaporate and can be collected as turpentine, while the larger nonvolatile di- and tri-terpenoids concentrate into an ever denser, stickier mass that hardens when cooled. Today this residual *rosin* provides grippiness to violins and ballet shoes and to the hands of gymnasts and baseball players and pole dancers; as a teen I enjoyed its fumes from the molten rosin-core solder as I wired my kit stereo.

SOME SMELLS OF WOOD TAR AND TURPENTINE

Material	Component smells	Molecules
wood tar	smoky, sweet, spicy, bacon; tarry, disinfectant, stable	guaiacols; phenol, cresols, xylenols
turpentine	pine, resinous, solvent	pinene, myrcene, phellandrene, camphene, terpinene

Along with smoke, wood tars and turpentine dominated the smells of daily life in Eurasia for thousands of years. They were so essential to the maintenance of wooden ships that English sailors were known as *tars*, and tar and turpentine as *naval stores*, their production a major industry in the pine-rich Nordic region and the American Southeast. Turpentine was the original all-purpose liquid hydrocarbon, an outdoorsy-smelling cleaning agent for cutting through fats and grease, a base and remover for paints and varnishes and waxes, a lamp fuel, pest repellent, disinfectant, and medicine for both internal and skin disorders—even, thanks to the piney terpenes that it shares with juniper berries, a ready flavoring for cheap gin!

Though largely replaced today by petrochemical hydrocarbons, true turpentine and pine-oil cleaning agents can still be found and smelled. The erstwhile prominence of naval stores survives most noticeably in Nordic countries, where they once coated Viking ships, then wooden churches, then became important export goods. Tar water is still used in Finland in medications, as a scent for saunas, and as a flavoring for many different foods, from meats to ice cream and liqueurs. A few years ago, after a winter dinner at the celebrated restaurant Fäviken in the forested wilds of central Sweden, I sat near a fireplace radiating light and warmth and smokiness and chewed on black tar pastilles, at first doubtfully, then savoring my immersion in the after-lives of trees.

Underground fuels: bitumen, petroleum, coal

While our early relatives in Europe were using fire to transform wood into tar, others in the Middle East discovered that Earth had already done much the same work for them. Around eighty thousand years ago, hominids in present-day Syria had found tarlike stuff oozing from the ground and carried it dozens of miles to use as an adhesive on their stone tools. Some eight thousand years ago, Mesopotamian traders in present-day Kuwait left behind pieces of similar material with impressions of the reed boats and ropes that they had once coated. These ancient relics presage the disappearance of wood tar and turpentine from the modern industrial world, and their replacement by similar and more abundant materials from underground.

Much of the Middle East is arid and tree-poor but richly endowed with surface seeps of **petroleum** and the tarlike sludge known as **bitumen** or **asphalt**; what we know as the Dead Sea was known to early Greek geographers as Thalassa Asphaltites. It was with bitumen that the Mesopotamians and their successors glued together and waterproofed their containers and dwellings and boats, including the legendary arks that Sumerian Utnapishtim and Semitic Noah built to ride out the primeval Flood. Bitumen coated streets and gutters in ancient cities, and petroleum was burned as lamp oil; both repelled pests from date palms and grapevines, preserved mummies (and named them, via the ancient Persian for bitumen, *mumiya*), and were used as medicines. Surface seeps of oil and bitumen aren't rare and were often confused with wood tar and wood pitch (Spanish *brea*); the redundantly named La Brea Tar Pits are an odorous irruption of the underground into the heart of modern Los Angeles. In China more than a thousand years ago, coal was burned in homes and smelting operations to replace increasingly scarce wood; the poet Su Shi called it "black jade" and wrote that thanks to it, "in the southern mountains, chestnut forests can now breathe again."

Petroleum and coal and natural gas are known as *fossil fuels* because their primary use has been as combustible energy sources for stoves and machines and power plants; weight for weight, they provide far more heat than wood and other uncooked organics. However, "fuel" covers only part of their significance. They instigated our understanding of Hero Carbon and continue to give us many of the materials and smells of modern life. A better name for them, and for bitumen, is **fossil organics**.

Fossil organics got their start several billion years ago, when Earth's waters first teemed with life. Masses of cells fell to ocean and lake floors and were buried in the airless sediments of eroded continental rock. The sediments gradually grew to be hundreds and thousands of yards thick, and some were drawn below the continents, which exerted even greater pressure on the organic remains and pushed them closer to the planet's hot core. A parallel process began after plants colonized the land, when early vegetation along shorelines fell into shallow water and formed airless bogs that were then progressively entombed.

Fossil organics are the product of extreme pit cooking. The pits are one to five miles (two to ten kilometers) below us, experience several thousand atmospheres of pressure, and hold temperatures between 140°F and 600°F (60°C and 300°C), the equivalents of a medium-rare steak and a hot home oven. The combination of moderate heat and extreme pressure puts enough stress on organic matter's carbon chains to break them apart, to pyrolyze them without flames. Over the course of millions to billions of years, deep burial transforms the original complex molecules of life into complex *mixtures* of relatively *simple* molecules, made of carbon and hydrogen and just a sprinkling of oxygen and other elements. The relatively shallow and cool pits fill with fluid oil, viscous bitumen, and soft bituminous coal, while deeper and hotter ones accumulate hard anthracite coal and natural gas.

Fossil organics consist of simple carbon chains of varying lengths, assorted rings with occasional double bonds and decorations, benzene-family rings with six carbons double-bonded to each other, and honeycomb-like aggregates with from two to many dozens of rings bonded to each other. The smallest aggregates, like two-ring naphthalene, are volatile, while the largest, the *asphaltics*, are similar to the soot particles in woodsmoke. Petroleum is about three-quarters simple chains and assorted rings, a fifth benzene-family rings, and a few percent solid asphaltics. Solid bituminous **coal** may still contain a quarter small-molecule volatiles by weight; hard coal, a tenth.

These planet-pyrolyzed organics bear some resemblance to fire-pyrolyzed plant smoke and tar. They contain a number of the same volatiles, most notably the tarry cresols. But they lack the smoky-sweet guaiacols. And fossil organics are far more abundantly endowed with simple hydrocarbon chains and diverse rings. We identify these molecules as smelling like gasoline and solvents because that's how we encounter them in everyday life, extracted from crude oil and used as turpentine once was, as **fuels**, **lighting fluids**, **cleaning agents**, and so on.

The shortest hydrocarbon chains, **methane** ("natural gas") and **ethane** and **propane**, are gases at ordinary temperatures and have essentially no smell: that is, we don't seem to have receptors designed to detect them. (They're scented for safety with traces of smelly sulfur compounds.) **Butane**, with four carbons, has a faint gasoline smell, **hexane** a stronger one. Eight-carbon **octane** is familiar as the standard for rating fuel performance in internal-combustion auto engines, and it smells like gasoline. The next several chains, including the mixture called **kerosene**, have oilier, heavier fuel smells. With tetradecane the gaslike quality begins to give way to a faint waxy quality, and chains much longer than that are too large to be volatile.

Crude fossil organics also contain abundant oxygen-free benzenoids that are only trace components of woodsmoke. **Benzene** and **toluene** and **styrene** and **xylene** are common components of solvents, and share a sweet ethereal quality. Less familiar than it used to be is two-ring, ten-carbon **naphthalene**—from *naphtha*, another ancient term for crude oil—once commonly used in mothballs until it was found to be toxic. But a version with three methyl decorations, abbreviated as **TDN** (for **trimethyl-dihydro-naphthalene**), is an important volatile in older Riesling wines, where it contributes what's often described as a kerosene quality: a rare example of a "chemical" smell being valued in food and drink.

SOME VOLATILES FOUND IN PETROLEUM, BITUMEN, AND COAL

Molecules	Smells
propane, butane (C3, C4)	none
pentane, hexane, heptane (C5, C6, C7)	solvent
octane, decane (C8, C10)	gasoline, kerosene
benzene (C6 ring)	solvent, sweet
styrene, toluene, xylene (C6 ring + C1, C2 chains)	solvent, sweet, glue, plastic
naphthalene (C10 double ring)	kerosene, mothballs

So the smells of fossil organics are dominated by solvent, fuel, tar, and medicinal volatiles. Many of these strike us as "chemical" or "industrial" rather than "natural," but they're produced from living things by forces that are just as natural as wildfire. Their extraction from coal and petroleum turned out to be a key to discovering the vast world of volatiles, natural and man-made.

Fossil organics deconstructed: gases, kerosene, coal tar

Coal was the unlikely solid black mass in which a human mind first glimpsed convincing evidence of invisible, intangible, volatile matter filling the air around us. A seventeenth-century Flemish chemist, Jan Baptist van Helmont, somehow managed to burn a lump of coal in an airtight container. He reported that the overall weight of the container didn't change, but when he opened it, he found only a small residue of solid ash. The invisible remainder he described as a *spiritus sylvestris*, or "wild spirit"—wild because it resisted confinement and rushed out when the container was opened, "spirit" because it seemed to be the intangible essence of coal. He named this invisible form of matter *gas*: maybe from Greek *khaos*, the primeval, unorganized state of the material world, or from Flemish *gest* or *geesen*, which could mean "spirit" or gas-making "fermentation" or "yeast."

Once the existence of invisible airborne matter had been demonstrated, chemists and engineers found that there were gases seeping from coal mines, that they were flammable, and that they could be generated at will by pyrolyzing coal. By around 1800, a Scots engineer named William Murdoch had devised a system of

furnaces, pipes, and lamps for illuminating a Manchester cotton factory with
burning coal gas. He reported "the peculiar softness and clearness of this light,
with its almost unvarying intensity . . . free from the inconvenience and danger,
resulting from the sparks and frequent snuffing of candles." In the year 1850 in
Britain alone, "town gas" from six million tons of coal lit streets and buildings
throughout the country. It also left behind tons of pyrolysis by-products: char-
coalized coal, or *coke*—a boon to metal smelters—and coal tar, the "despised by-
product" of this chapter's epigraph from Professor Robinson.

A similar story unfolded for petroleum materials. Fluid varieties of bitumen
known as naphtha were used in the ancient Middle East as lamp oil. By the ninth
century, Arab chemists were heating the naphtha and collecting its vapors, which
condensed when cool into a thinner and cleaner-burning oil. That kind of refined
oil was named *kerosene* by the Canadian geologist Abraham Gesner, who developed a system to produce it in large volumes, and around 1850 it began to replace
turpentine, animal fats, and whale oil in household lamps through North America
and Europe. Around the same time, the Scottish chemist James Young succeeded
in collecting a kerosene-like lamp oil from both petroleum and coal, and found
that the less volatile materials left behind would yield useful oily and waxy fractions.

When petroleum became more available, beginning with the 1859 oil rush in
Pennsylvania, refiners quickly learned how to separate many of its components
from one another, and developed our modern panoply of odorous petrochemical
fuels and solvents and lubricants. After just a couple of decades, an article in the
July 1883 *Century Magazine* nicely enumerated the ten products then—and still—
refined from petroleum, in order from most to least volatile.

1st, rhigolene [mainly pentane], used to produce local anaesthesia [by freez-
ing]; 2d, gasolene, used in artificial gas machines; 3d, 4th, and 5th, three grades
of naphtha, used for mixing paints and varnishes and dissolving resin; 6th, kerosene, the common illuminating oil of commerce; 7th, mineral sperm oil, a
heavier oil for burning in lamps, employed on steamers and railroads; 8th, a
lubricating oil for machinery; 9th, paraffine, from which candles are made; and
10th, paraffine wax. Then there is the residuum, usually called coal tar.

The naphthas are now called mineral spirits or paint thinner, and kerosene is primarily a fuel, notably for jet airplanes. Each of these petroleum products is a selection of carbon chains and rings of particular sizes and structures. The smallest are the most volatile and most flammable, the midsize more viscous and less flammable and easier to handle and control. The largest melt only when warmed, and ignite only at very high temperatures, as at the tip of a candlewick.

Today petrochemical plants can "crack" large fossil organic molecules into smaller ones, generate larger molecules from one-carbon natural gas, and generally produce any mixture to order. The smell of a given petrochemical depends on its population of volatile short chains and small rings. Inexpensive mineral spirits, for example, include significant (and potentially toxic) amounts of benzene-ring hydrocarbons and have a strong solvent smell, while versions tailored for painters and long periods of exposure may be almost odorless. Mineral oils, jellies, and paraffins are deodorized for use in cosmetics and ointments, food coatings and candles.

SOME MATERIALS DERIVED FROM FOSSIL ORGANICS

Fossil organic materials & smells	Typical lengths of carbon chains
methane, natural gas (sulfur smell added)	C1
propane, butane, compressed stove gases (sulfur smell sometimes added)	C3–C4
gasoline	C4–C12
naphtha, white gas	C5–C12
mineral spirits, white spirits, paint thinner, petroleum distillate, charcoal starter	C6–C12
kerosene, jet fuel	C6–C16
diesel	C8–C21
mineral oils (no smell)	C15
petroleum jelly (no smell)	C25
paraffin (no smell)	C20–C40
tar, creosote, asphalt, bitumen	C4–C100

The product that stands apart in this roster of fuels and solvents and lubricants is the "residuum": coal tar, the despised by-product generated by the ton in the early years of gas lighting, for which some uses were found. It supplemented or replaced wood tar in industry, on railroad timbers, in gutters and on rooftops, and in some of the first filled roadways, where it bound crushed rock in an approximation of natural bitumen. Like wood tar before it, coal tar was also touted as a medication for everything from skin to digestive ailments; it's still found in strong-smelling soaps and shampoos meant to help treat various skin disorders.

But coal tar turned out to be far more than a marginally useful residue—hence Professor Robinson's praising it as a treasure chest, the "magic alembic," a source of health and wealth. Human ingenuity did it again. Just as it had worked out how to extract tar and turpentine from wood, illuminating gas from coal, and kerosene and gasoline from petroleum, so it worked out how to make extracts of these extracts and their residues, and teased them apart into *their* components.

This time around, some of the extracted components were no longer complex mixtures: they were individual chemical compounds, separated from each other for the first time. Students of coal tar and illuminating oils were among the first of Hero Carbon's avatars to smell its specific rings and chains. They gave them names by which we still know them, and sparked the generations of research into the volatile world whose results fill this book.

Smelling single volatiles

The first inkling of the very existence of Hero Carbon came late in the eighteenth century, when French chemists demonstrated that three very different materials, charcoal, graphite, and diamond, all behave similarly and simply when burned, and were therefore likely composed of the same basic material. They named this common element *carbone*, from the Latin for "coal" or "charcoal"; in German it was *Kohlenstoff*, literally the "stuff of coal." In the ensuing decades, chemists throughout Europe gradually improved their methods for separating complex materials into their components. In France, Michel Chevreul isolated the starter-set acids from animal fats by making soap from them and became the first to smell their cheesy and goaty qualities unmixed (see page 84). But the most effective general method for extracting volatiles was and remains *distillation*, the basic outline of which goes back to the tar-making Neanderthals: heat the material, collect its volatile vapors, then cool and condense the vapors into a liquid. The early specialists in carbon chemistry controlled the heat and cooling carefully, distilled repeatedly, and began to isolate previously unknown components in woods, resins, tars, turpentine, oils. Often they detected their presence by their distinctive smells.

Take a handful of benzene-ring hydrocarbons that have become mainstays of the modern chemical industry. In 1825, analyzing the residue produced when illuminating gas was made from whale oil and tallow, Michael Faraday isolated crystals "having an odour resembling that of oil gas, and partaking also of that of almonds." The same substance was later extracted in abundance in coal tar and named **benzene** for its chemical relatedness to familiar benzoic acid in the valued tree gum benzoin (see page 195); Faraday's sample likely included traces of almond-extract benzaldehyde. **Toluene**, a benzene ring with one methyl decoration, was first found in pine distillate by a Polish chemist, then in 1841 by a French chemist in tolu balsam, an aromatic exudate from a tree native to South America. **Styrene**, with a two-carbon decoration, was isolated in 1839 from the liquidambar tree, source of the American version of the balsam known as storax (see page 195), and described as having "the same not unpleasant smell of storax." **Xylene** was isolated in 1850 from commercial wood spirits, named from the Greek root for "wood," and described as "aromatic" in smell. All of these names have the ring of synthetic chemicals, but in fact they all derive from natural materials.

These hydrocarbon rings would eventually find many uses, but a couple of rings with hydroxyl (oxygen-hydrogen) decorations were more immediately useful, and responsible for the many disinfectant and medicinal uses to which tars and smoke had long been put. Recall that the smoky-tarry **cresols** were found in **creosote**, the "flesh-preserving" liquid distilled from wood tar in 1832 by the German chemist Karl Reichenbach. Reichenbach used his creosote to concoct the first recorded and taste-tested version of a liquid smoke:

> Its smell is penetrating and unpleasant, but not stinking. From a distance, most people find it similar to [beaver] castoreum. . . . Close up, however . . . I partly perceive the smell of smoked meat. . . . Fresh meat, placed in creosote water for an hour and a half, can be hung without rotting. When I hung some

pieces of beef in the July sun . . . the meat completely dried out within 8 days, became hard, brittle, absorbed a pleasant smell of good smoked meat. . . . People who have made many sea voyages have found it very tasty.

A decade later, popular writers were recommending a creosote-water dip as an alternative to smoking. Soon other chemists identified the individual volatiles in liquid creosote, including the cresols and guaiacol. Like wood tar, creosote was touted as a cure-all; a surviving cough medication called Creomulsion is strong medicine indeed.

The most important hydroxyl-decorated benzene ring is **phenol**. That's the modern name for the molecule isolated in 1834 at a coal-gas works near Berlin by Friedlieb Runge, who called it "coal-oil acid" and "carbolic acid," and described it as having a burned smell and an extraordinary power to deodorize and preserve: carbolic acid "deprives putrefying animal flesh like meat of its stink, as soon as a water solution is poured over it. But it seems not to be identical with the smoky principle, because such meat tastes abominable."

Abominable indeed! The smell of phenol now suggests disinfectants and disinfected spaces, hospitals and public bathrooms: associations arising from developments throughout Europe once plentiful phenol was produced from coal tar. In the 1850s, practical chemists put phenol to use in embalming, water purification, and deodorizing and disinfecting sewage and stables. In the 1860s, physicians recommended its use to dress wounds and incisions, and to wash surgical surfaces and instruments. In 1865, the Glasgow surgeon Joseph Lister used it to prevent the usually fatal infection of a skin-piercing leg fracture in a boy run over by a cart, and through his public advocacy helped inaugurate the era of antiseptic surgery. Nowadays phenol is hard to find in smellable amounts in medications, largely because of its effectiveness: it's strongly reactive and tends to oxidize any carbon chain or ring it touches, including the tissues it's applied to. Other more selective antiseptics have come to replace it.

These two groups of molecules, the benzene-ring hydrocarbons and the hydroxyl-decorated benzene rings, are a small sampling of the volatiles that were first isolated and smelled by the founding generations of organic chemists. There were also the gasoline-like hydrocarbons, like hexane and octane; the analysis of "turpentine oil and the numerous hydrocarbons it contains, which may generally be called **terpenes**," as the German chemist August Kekulé dubbed them in the 1860s; and the key volatiles of many spices and herbs, tonka's coumarin and vanilla's vanillin among the earliest.

Of the various materials that yielded isolated volatiles, coal tar was especially valuable for its revelation that simple heat and pressure could generate useful carbon molecules from organic matter—and therefore that we might be able to do the same in the laboratory and factory. As August Wilhelm von Hofmann, the eminent founding director of the Royal College of Chemistry in London, wrote in his report for 1849:

> It is for this end that the study of the metamorphoses of organic compounds is of so much value. It is not the host of new substances, which we are continually discovering, that interest us so much, but new methods of operation, by which we may imitate, for our special purposes, the formative forces of nature.

Those new methods of operation have been spectacularly successful, so much so that coal-tar molecules and related products of modern chemistry now loiter in the volatile background of nearly every hour that we draw breath.

The volatiles of modern materials: plastics and solvents

Of all the special purposes that organic chemists have pursued, one has been especially far-reaching: to develop new materials for making things. Making is at the heart of being human, and our species has created its astoundingly rich material culture by manipulating the raw stuff that Earth provides. It's a broad set of provisions—minerals and metals, plant fibers and woods, animal bone and skin, beeswax, and silkworm cocoons—but each item comes predefined by very particular properties. What if we could make our own materials, with whatever properties we want? Organic chemistry has done exactly that, starting with the simple rings and chains of fossil organics. The result has been the creation of new forms of Hero Carbon that now pervade everyday life and indirectly scent it.

These new forms are **plastics**, solid materials that can be worked into essentially unlimited shapes and gradations of consistency, from thin stretchy kitchen films to rugged parts for aircraft bodies. Most plastics are made by inducing small carbon rings or chains to bond together into hugely long molecules called *polymers*, from the Greek *poly-*, "many," and *meros*, "part." The polymers in turn bond to each other to form large solid networks.

Plants made plastics before chemists did. The aromatic storax resin of the liquidambar tree is fluid when it emerges from a wound, but then its styrene rings harden it by bonding with each other to form **polystyrene**, the plastic now manufactured by the ton to make takeaway food containers and packing materials and toys. Then there's caoutchouc, or **rubber**, the polymerized wound-repair fluid of a tree native to tropical South America. This milky latex is rich in isoprene, the five-carbon terpenoid relative that trees emit in huge quantities and animals in traces (see pages 207, 104). Other molecules in the latex induce the isoprene to form a polymer network that happens to be elastic rather than rigid, a valuable property in applications ranging from auto tires to surgical gloves.

As early as the 1850s, chemists were making plastic materials using the natural polymer chains of cellulose from plant fibers; celluloid, cellophane, and rayon all belong to this family. The first entirely synthetic plastic emerged a few decades later with the discovery that when carbon-ring phenol and one-carbon formaldehyde are heated together, they form a resinlike solid; the formaldehyde links the rings together into polymers. This phenolic resin became commercially viable around 1907, when a Belgian-American inventor named Leo Baekeland figured out how to manufacture it. His **Bakelite** was soon used to make varnishes, electrical and automotive parts, billiard balls, jewelry, and much else. Early Bakelite objects have become collectables, their genuineness verified in part by the trace smells of phenol and formaldehyde they emit when they're gently heated. Today that peculiar combination is most familiar from electronics circuit boards; I smell it as I type this paragraph when I sniff through the keyboard of my warm laptop computer.

The success of Bakelite inspired the invention of other synthetic resins, and today there are several dozen that weave through our everyday lives, polymers of styrene-like carbon rings (polyurethane, polyester, polycarbonates), isoprene-like short chains (polyethylene, polyvinyl chloride, nylon, acrylics), and hybrids (styrene-butadiene, or SBR, for rubbers; acrylonitrile-butadiene-styrene, or ABS, for plastics). The raw materials for these resins can be manufactured directly and cheaply from fossil organics—no trees needed. There are plastics of one kind or another in infant diapers and bottles and teething rings, in toys, in clothing and shoes, in floors and carpets and walls and furniture, in containers and packaging, in electronics, and in our cars—everywhere! And they all emit volatiles, sometimes undetectably, sometimes obnoxiously.

The polymers themselves are too large and interconnected to be volatile. But the nooks and crannies between the individual polymer molecules are filled with smaller molecules. Some are residual rings and chains that escaped polymerization, like Bakelite's telltale phenol and formaldehyde. Some are fragments broken away from the polymers by exposure to oxygen or light or heat. One polymer often used in screwdriver handles (**cellulose acetate butyrate**) is notorious for breaking down into vinegary and cheesy acetic and butyric acids that accumulate in the toolbox. Other small molecules are residues of the many other petrochemicals used to manufacture the resins and modify them, to give them precisely the properties we want: soft or hard, elastic or stiff, clear or opaque, and on and on. The overall mix of volatile petrochemicals in a given material produces its specific "plastic" or "chemical" smell.

Because they evaporate and exit from the polymer network and become depleted with time, volatile residues are usually most obvious in newly manufactured goods. Flexible products with large surface areas like PVC **shower curtains** and polyester **carpets** and their mixed-plastic backings can be especially smelly when first unfolded. The distinctive rubbery smell of both new and road-warmed **auto tires** comes from a mix of trace petrochemicals and the sulfidic residues of "vulcanizing" the polymer network with sulfur to toughen it. The increasingly popular **3-D printers** form objects by depositing molten plastics, and they emit the noticeable smells of formaldehyde, styrene, and other small petrochemicals. Plastic containers like disposable **polyethylene water bottles** can taint their contents with waxy chemical smells, and some **food wraps**, usually PVC, do the same. I'm still influenced fifty years later by the strong disinfectant-like smell that seasoned every wrapped item in my high school cafeteria; to this day I sniff prepackaged sandwiches before I buy them.

SOME VOLATILES EMITTED BY COMMON PLASTIC ITEMS

Item	Volatiles
polystyrene packing material	styrene
computer & monitor circuit boards, cases	phenol, formaldehyde, toluene, ethyl hexanol
vinyl (PVC) shower curtain	toluene, ethylbenzene, phenol
polyester carpet	chloroform, hexane, toluene, pinene
polyester carpet backings	phenylcyclohexene, vinyl acetate, aldehydes, styrene, benzothiazole
rubber tires	benzothiazole, sulfides, naphthalene, styrene, cresols
polyethylene water bottle	nonenal, octenone, heptenone, toluene, ethylbenzene
rubber bands, balloons	hexane, chloroform, ethyl acetate, toluene, xylenes
pencil eraser	butanol, butyl acetate, octane, ethyl acetate
adhesives & glues: general white glue plastic cement rubber cement	hexane, ethyl acetate, nonanal, chloroform vinyl acetate butanone, toluene heptane, acetone

The most common manufacturing traces in plastics are **solvents**, fluid small-molecule petrochemicals that dissolve the various plastic ingredients so they can react with one another, then mostly evaporate away as the polymer structure forms. There's one particular version of plastics in which solvents can be overwhelmingly prominent to begin with and then stealth volatiles for a long time afterward. **Adhesives** or **glues** are sticky materials that join different materials into more versatile composites. Once made from animal gelatin, milk proteins, and plant starch, they're now largely mixtures of petrochemical solvents and polymers that flow at first, then harden as the solvents evaporate. We smell adhesives directly when we apply glues ourselves or unroll some adhesive tape, but their traces are constantly in the air of almost any indoor space, emanating from plywood, particle board, veneers, and nearly all manufactured construction and finishing materials and furniture.

SOME VOLATILES EMITTED BY MODERN BUILDING MATERIALS

Material	Volatiles
wood particle board, plywood	ethyl acetate, toluene, formaldehyde
plaster, drywall	aldehydes, acids, ketones, xylene, sulfides, thiols
insulation	pentane, styrene, formaldehyde
wood flooring, waxed or varnished	hexanal, pinene, octenone, nonenals, acetophenone
vinyl flooring, wall covering, ceiling tiles, window frames	chloroform, toluene, nonane, decane
floor wax	tetradecane, tridecane, pentadecane, hexane
linoleum	aldehydes, acids, ketones, toluene
water-based paint	toluene, butanol, ethylbenzene, decane, xylenes, acetone
oil-based paint	mineral spirits, C6–C12: hexane, heptane, octane . . .

Volatile solvents have also become indispensable aids for wrangling many other materials. The original man-made solvent was turpentine, whose small-molecule monoterpenoids helped shipbuilders apply waterproof tar to every crevice of ropes and wooden hulls, painters to deposit pigments and varnishes evenly across solid surfaces, and both to clean their sticky stuff up from where it wasn't wanted. Today petrochemical solvents play a host of similar roles in coating and cleaning and polishing. Among the less obvious: sniff the tip of a **marking pen** or the page of a **magazine** or **newspaper** or an **office-printer document**, and you smell solvents that carry and spread their inks: as methanol memorably did on the sniffable purple-print "ditto" sheets common in schools decades ago. **Crayons** carry waxy-smelling medium-length aldehydes, but hydrocarbons as well. And **paraffin candles** emit the same general combustion volatiles as their beeswax look-alikes, but also solvent-derived benzene, toluene, and naphthalene.

SOME SOLVENT VOLATILES EMITTED BY COMMON HOUSEHOLD ITEMS

Item	Volatiles
nail polish & remover, spot removers	butyl acetate, ethyl acetate, acetone
"dry-cleaned" clothing	tetrachloroethylene
shoe polish	naphtha, trimethylbenzene
furniture polish	naphtha, mixed petroleum distillates
marking pens	ethyl & propyl & benzyl alcohols, xylene, phenoxyethanol, butanol, diacetone alcohol
newspaper inks	do-, tri-, tetra-, penta- & hexa-decanes, toluene
magazine inks	ditto + ethyl acetate, hexane, xylenes, nonanal
office printer inks	toluene, styrene, xylenes, ethylbenzene
crayons	hexane, nonane, decane, undecane, decanal, nonanal
paraffin candles (lit)	medium-chain (C11+) hydrocarbons, acids, aldehydes, + benzene, ethylbenzene, toluene, xylene, naphthalene

So our **indoor air**, once dominated by smells of the hearth and oil lamps and tallow-fat candles, stone and wood, wool and leather, turpentine and animal-fat soaps, is now scented instead with a miasma of *fossil* organics from synthetic and composite materials. In newly constructed and furnished spaces, that miasma can contribute to "sick building syndrome," which leaves inhabitants feeling ill. (Another notorious culprit is hydrogen sulfide, which can be released from wall paneling containing the mineral gypsum.) Today some of the most intense and diverse petrochemical lungfuls come from new **automobile interiors**: closed windows and a hot summer day bake the volatiles out of their fabrics, foams, glues, and paints into a small volume of air. Though "new-car smell" may appeal as an indicator of factory freshness, its volatiles can irritate eyes and airways and condense into a foggy film on the windows. Manufacturers have worked to minimize their emission, and many now actively define the smell of their new cars by scenting them with proprietary fragrances.

SOME VOLATILES CONTRIBUTING TO THE SMELL OF AUTO INTERIORS

Source material	Volatiles
polyester (seat covers, floor mats, seat belts, etc.)	styrene, ethylbenzene, hexane, toluene
polypropylene (floor mats)	pentamethylheptane
polyethylene (headrests)	benzaldehyde, nonanal, decanal, benzene, toluene, naphthalene, acetic acid . . .
polyurethane (seats, headrests)	toluene, phenol, nonanal, benzaldehyde
polystyrene	styrene, ethylbenzene
polyvinyl chloride (headrests)	ethyl hexanol
leather (seat covers)	alcohols, methyl pyrrolidine
solvent residues from paints, glues, foaming agents	xylenes, undecane
rubber tires	benzothiazole, sulfides, naphthalene, styrene, cresols
fuel & combustion	benzene, naphthalene, toluene, xylenes, ethylbenzene, formaldehyde, acetaldehyde

In fact, many people in general now try to define the smells of their lives by treating autos and homes with "freshener" sprays and scent diffusers. Just like the residues they're meant to mask, most of the volatiles in these chosen smells are manufactured from fossil organics. Taken together, manufactured petrochemical volatiles—freshener sprays and other personal-care products, solvents, adhesives, inks, cleaning products, pesticides—now pervade our lives to the extent that they're responsible for about half of the harmful air pollution in urban areas.

Tar-barrel intoxications

The pleasures of new-car smell, like my childhood enjoyment of ditto sheets and model airplane glue, and pick-up sticks melted on a light bulb—storax-resin styrene—are casual examples of the sensory appeal that petrochemicals can have.

Professional chemists are more systematic connoisseurs; they have ready access to dozens of solvents and share smelling notes online. Far more numerous are the fans of the *pharmacological* effects of petrochemicals. Many people huff solvents, glues, and fuels (and drink some of them) for the intoxicating high they can induce, akin to the effects of alcohol. When they do, they risk falling victim to SSDS, "sudden sniffing death syndrome." The *toxic* in *intoxicating* comes from the Latin for "poison" and points to the dark side of pyrolyzed organic matter: many of its molecules, the most useful and pleasant included, have turned out to be health hazards. I compiled the tables of everyday materials above in part from the work of toxicologists and from mandatory Safety Data Sheets issued by manufacturers. Long before Victor Robinson eulogized coal tar as the source of new medicines, it was known to be an agent of disease and death. Hero Carbon's virtuosity is far from entirely benign.

Just decades after Shakespeare's death, the Englishman John Evelyn published *Fumifugium*, a pamphlet decrying the effects of inescapable sea-coal smoke on Londoners as "noxious and unwholesome, . . . corrupting the lungs and disordering the entire habits of their bodies" so that "more than half of them who perish in London die of phthisical and pulmonic distempers." That's not a reliable statistic, but many thousands of Londoners were indeed killed by the Great Smog of 1952. And evidence for more insidious effects of pyrolysis volatiles had come as early as 1775. The London surgeon Percivall Pott, physician to Samuel Johnson and Thomas Gainsborough among others, noted that chimney sweeps who cleared out deposits of creosote and soot, typically young boys who worked with little or no clothing and seldom bathed, are "thrust up narrow, and sometimes hot, chimneys where they are bruised, burned and almost suffocated; and when they get to puberty, become peculiarly liable to a most noisome, painful and fatal disease." "Chimney-sweep disease" was cancer of the scrotum.

Following decades of efforts to identify the cause of "tar cancers" prevalent in coal-industry workers, in 1933 a chemist named James D. Cook and his colleagues extracted two tons of coal-tar pitch at the London Gas Light and Coke Company and isolated a few grams of the likely culprit: a faintly aromatic five-ring twenty-carbon molecule called benzo[a]pyrene, or BaP. Around the same time, evidence began to accumulate that the unadorned six-carbon benzene ring, so useful as a lamp fuel and solvent—it was the first agent used in dry cleaning—can cause leukemia and other blood disorders. The benzene rings with one- and two-carbon decorations, toluene, the xylenes, styrene, and ethylbenzene, are less dangerous but can have harmful effects on the nervous system—effects whose early stages can be pleasantly mind-numbing. And all are commonly present in at least trace amounts in modern factories and their products. Toxicologists group benzene and these allies under the acronym BTEX and regularly check their levels in products and environments. Even when they're not high enough to be an immediate danger, they can cause headaches, irritation and breathing problems, and nausea.

Plenty of useful chemical inventions have proved to be toxic. Many are synthetic carbon structures that include chlorine or bromine or heavy metals like mercury and cadmium: among these are the dioxins, PCBs, various pesticides, chemicals for manufacturing electronics, and even medications. Every time we use chlorine bleach and cleaners around the house, we send toxic chlorinated by-products into our indoor air. But coal-tar volatiles, which include the most common volatiles of modernity, are not human inventions. The BTEXes and their ilk were first cooked up either underground from long-dead algae and trees, or by the living trees for which they were named. Solvents like acetone and butanol and isopropanol have been manufactured on an industrial scale by means of bacteria. They're not new or unnatural molecules. What's new is the degree of our exposure to them.

Members of the animal kingdom live by consuming other creatures of all kinds. They've always taken in natural chemicals that their particular chemical machinery can't use productively, which can gum up that machinery if not somehow disposed of—including many pleasing volatiles of herbs and flowers and fruits! As we've seen, animal bodies like our own have evolved systems for disposing of these *xenobiotic* molecules (see page 94), and they handle the low levels found in the natural world fairly well, though not perfectly. At higher levels, the systems get backed up, their drawbacks become more serious, and they can actually make xenobiotics *more* toxic, not less. That's the case with BaP: to excrete it, the human body modifies it in such a way that it can damage critical parts of our DNA.

Our ancestors were certainly exposed to pyrolysis volatiles in wildfire smoke, but only occasionally. Once they brought fire into caves and shelters and breathed its volatiles all the time, their xenobiotic systems would have been challenged. There's genetic evidence that early humans benefited from a random mutation that made their BaP disposal system *less* efficient, so that it would produce less of the DNA-damaging modification. Similarly, humans never inhaled nearly as much benzene and toluene and formaldehyde from domestic fires as we started to do once we broke into Hero Carbon's coal-tar laboratory—nor as much of the citrusy, flowery volatiles that we spray to make our indoor air smell more outdoorsy. Our xenobiotic systems aren't always up to the challenge. Pleasing as the smells of smoke and solvents may be, it's prudent to limit our enjoyment.

Ether and ethereality

It's curious that we should perceive some of these newly abundant xenobiotics as pleasant. As an eight-year-old drawn to molten pick-up sticks and airplane glue, I wasn't huffing for a high; I was just enjoying the smells. One broad description often given to the smells of small petrochemical volatiles is "ethereal." It derives from the Greek *aithēr*, which first denoted the airlike substance breathed by the gods, pure, subtle, light (the root of the Greek word meant "shine"), and later an intangible elemental fluid that permeated the physical universe. Tangible, smellable ether arrived in the eighteenth century when the German chemist August Sigmund Frobenius, working in London, used *Aether* to name the product of a reaction between alcohol and sulfuric acid. He described this remarkable fluid, which boils at our body temperature, as so volatile and flammable that it is "both fire and a very fluid water"; it is, he said, "the purest Fire," burning even when mixed with many times its measure of water, and without a trace of soot or ◠◦◠ ash. Frobenius's *Aether*, which is now known as **diethyl ether**, consists of two two-carbon chains joined together by an oxygen atom, and has found wide use as a solvent—synthesized from petrochemicals—and as one of the first surgical anesthetics.

Ether is the prototypical intoxicating solvent, and "ethereal"—etherlike—suggests the quality of being extremely volatile, diffusive, insinuating, rushing up into the nose and seemingly onward to fill the head, and quickly disappearing. Since ether has quieting effects on neurons in the brain, perhaps it does something similar in the nose, quite apart from triggering particular smell receptors. Solvents are often said to smell "sweet." Ethereality might be a complex sensation indeed, both stimulating and soothing, and perhaps appealing for that very reason.

Whatever the truth might be about their sensory appeal, the commonplace smells of solvents and tar and smoke are also direct experiences of something extraordinary: what the energies of combustion, the weight of oceans and continents, and the fragile, easily altered, immensely capable neural networks of *Homo sapiens* have managed to do with the substance of the planet's living things. For better and for worse.

We turn now to another mixed achievement: the volatile materials that our species has invented in order to scent itself, and its dwellings and automobiles, without and with the help of coal-tar solvents and chemistry.

Part 5

· ·

CHOSEN SMELLS

Chapter 17 · FRAGRANCES

··

In the consideration of the origins of chemistry . . . the incense-burner so prominent
in all temples even today could well have been in ancient times one of the most
important inspirations for those who designed to accomplish wonderful changes in
natural substances by the instrumentation of fire . . . transformations associated with
worship, sacrifice, ascending perfume of sweet savour, fire, combustion,
disintegration, transformation, vision, communication with spiritual beings, and
assurances of immortality.

· Joseph Needham and Lu Gwei-djen,
Science and Civilisation in China, 1974

Essential Oils—are wrung—
The Attar from the Rose
Be not expressed by Suns—Alone—
It is the gift of Screws—
· Emily Dickinson

The smells we've surveyed so far are the world's volatile givens: they're simply there in the air. In this chapter, and in the following two chapters on
cooked and fermented foods, we turn to deliberate gifts: materials and
smells that we coax from the world expressly for our pleasure.

The flowers and herbs and spices that please us seem to do so mostly by happy
accident, the joint boon of the plant kingdom's chemical creativity and the adapt-
ability of our nervous system. But *fragrances* are purposeful products of human
ingenuity. Their makers apply thought and imagination and fire and screws to create olfactory delight beyond what nature has to offer. For incense, smoldering coals

drive trapped wood and resin scents out into the air and into us. For essential oils, devices squeeze scant scent from ephemeral flowers to be bottled for enjoyment anytime. For perfumes, olfactory artists gather aromatics from different continents and kingdoms of life and remix them into entirely new compositions. Fragrances are the materials through which humankind indulges its abiding interest in the world's smells, playing with them and deploying them as we desire, to fill a place of worship or to scent a wrist. Or to spruce up the car, or toilet, or Fido.

Most animals pay attention to smells, and many of our fellow mammals rub themselves with crushed leaves and insects and their own secretions and excreta, to disguise or advertise themselves, or to self-medicate. Our species moved beyond the immediate pragmatics of survival when it discovered that smells could reinforce and even elicit strong feelings, from simple happiness and ease, to physical desire, to awe and the intuition of a world beyond the physical. From early historical times and probably long before, common herbs and flowers and hearth smoke have given everyday pleasure, rare aromatics have helped rulers and priests to intimidate and inspire, and devotees of scent have labored to capture its elusive essence.

Esoteric though incense and perfume materials may seem at first, they're a treasure trove for anyone at all curious about smell. They're a selection of the world's most interesting aromatics, curated by many generations of smell lovers. Some are familiar from the garden and kitchen, but others are like nothing else: wounded Asian trees, for example, and lichens, and strangely appealing stuff from the bodies of beavers and whales. Many offer a hint of the air breathed by ancient and distant cultures. While wearing fragrances isn't to everyone's taste, sniffing them is the chance to explore some of the most complex smells that exist.

Far more familiar are bathroom sprays and dog colognes and the like, the crowd-curated fumigants of modern life. They embody a double dose of ingenuity: not just the capture of pleasant volatiles, but their manufacture without a single flower or tree. A number of molecules invented by plants—pine terpenoids, citrusy limonene, minty menthol, flowery linalool and geraniol—are now produced in factories by the thousands of tons. Most of them go to conjure the illusion of outdoorsy freshness in countless indoor "care" products: care for our skin and hair and mouth, for baby, for pets, for household surfaces, clothes and laundry and cars. We launch these molecules into our surroundings from sprayers and scented candles and diffusers and plug-in essential-oil warmers, and ever more sophisticated nebulizers and vaporizers. Perfumes seep from magazine inserts. Retail stores, hotels, airlines, and automakers scent their spaces with bespoke volatile blends to reinforce their identities with "sensory branding." Enough of these molecules end up escaping their immediate targets to cause measurable damage to city air and water quality.

Even if all these fabricated smells don't signify much beyond mass-market preferences, manufacturing costs, and pollution, the achievement that makes them possible—to produce rose scent without roses, musk smell without musk deer—is genuinely impressive. It's the outcome of a long voyage of discovery into the invisible world of volatile molecules, a generations-spanning effort to which anyone interested in smell is indebted. From their very beginnings, fragrances have depended on the chemical technologies of the day to snatch volatiles from their sources before they disappear into thin air. The rising scent of the incense burner may well have inspired the earliest chemists to study invisible manifestations of matter. Aromatic materials certainly attracted early modern chemists to identify the substances responsible for their smells, and provoked insights into the ways of Hero Carbon just as gunky coal tar did, some of them rewarded with the Nobel Prize. Fragrance chemists continue to advance our understanding of the volatile world to this day. I've gleaned much of the information in this book from their labors.

It's largely thanks to the rise of the fragrance industry and mass market that today a curious amateur can obtain and smell not just fine flowers and woods, but a wide range of their extracts, even refined down to single volatiles. It's now more possible than ever for smell explorers to experience what Bruno Latour described for the perfumer in training: the discovery of a new sensory world through the making of a new nose.

Incense: liberating volatiles

Incense, the "ascending perfume of sweet savour," may well have been the first fragrance. The word comes from an ancient root meaning "shine," and names materials that produce aromatic fumes when they smolder or burn. Today it likely brings to mind either the sweet sandalwood smells of South Asia or the resinous frankincense of churches in the ceremonial branches of Christianity; in ancient

China, the aromatics likely included artemisia, licorice root, magnolia wood, Sichuan pepper. The intentional production of fumes rather than fire might have begun with observations that smoke has the power to mask the smells of human life and death, to kill or repel pests, and to slow the rotting of foods. Distinctive fumes from special fuels found more ceremonial uses, to mark a significant occasion or place, offer tribute to unseen powers above, repel unseen evil powers nearby, or protect the dead on their passage to the next world.

Incense burning had its ups and downs in the Judeo-Christian West and in China, where church reformers and sages denigrated it as a distraction from genuine spirituality. Little such ambivalence afflicted the rest of the Old World. India led the development of manufactured incense blocks and sticks, which are scented with some mixture of aromatic woods, resins, oils, and synthetics. Several billion incense sticks, *agarbatti*, are burned daily in *puja* offerings, as aids to meditation, or simply to sweeten the air. Throughout much of Asia and in the Islamic Middle East, the smell of incense remains a constant presence in public and private life.

What sets incense smoke apart from ordinary woodsmoke is its aromatic richness, the product of the materials themselves and a method of heating that liberates their volatiles rather than consuming them. Raw incense woods and resins are typically "burned" by being placed in small pieces or powders on smoldering charcoal, whose surface temperature runs between 700°F and 1100°F (400°C and 600°C). In that range, most distinctive volatiles evaporate into the air without being modified or destroyed by the heat, while the generic plant materials get broken down to generically smoky-smelling phenolics and aldehydes.

The molecules that define incense materials are often heavy fifteen-carbon sesquiterpenoids and other reticent volatiles that are only faintly evident in the intact wood or resin. After high heat propels them into the air where we can smell them, they settle back on cooler surfaces, linger, and reenter the air slowly as a fainter background aroma. An exception to this pattern is the Native American practice of **smudging** with smoldering bundles of dried herbs; white sage and sweetgrass are often used, and respectively carry pungent, cooling monoterpenoids and sweet, haylike coumarin.

We've met with many of the common incense resins and woods in our virtual forest (see pages 193, 199), so I'll list them here for quick reference. Frankincense and agarwood are especially worth revisiting.

SOME HERBS USED IN SMUDGING

Material	Component smells	Molecules
white sage (*Salvia apiana*)	eucalyptus, camphor, pine	eucalyptol, camphor, pinene
sweetgrass (*Hierochloe odorata*)	sweet hay, baked bread, nutty, almond extract	coumarin, furfural, benzaldehyde

SOME RESINS AND GUMS USED IN INCENSE AND PERFUMES

Incense material	Component smells	Molecules
frankincense, olibanum, luban (*Boswellia sacra* [*carteri*])	incense, old church, woody, piney, peppery	olibanic acids, pinene, myrcene, linalool, cresol, mustakone, rotundone
frankincense, olibanum, luban (*Boswellia papyrifera*)	incense, old church, fresh, green, earthy, mushroomy, floral	olibanic acids, octyl acetate, octanol, limonene, geraniol, linalool
myrrh (*Commiphora myrrha*)	warm, sweet, leathery, mushroomy, undergrowth	furanoeudesmadiene, lindestrene, curzerene (3-ring furanosesquiterpenoids)
gum benzoin (*Styrax tonkinensis* & *benzoin*)	balsamic, sweet, fruity, floral	benzoic acid & benzoate esters, vanillin; cinnamic acid & cinnamate esters, styrene
styrax, storax (*Liquidambar orientalis* & *styraciflua*)	sweet, plastic, piney, woody, floral, fruity	styrene, pinene, caryophyllene, cinnamate esters
copal (*Bursera*, *Hymenaea*, *Protium* species)	woody, spicy, piney, resinous	copaene, germacrene, pinene, sabinene

SOME WOODS USED IN INCENSE AND PERFUMES

Incense material	Component smells	Molecules
sandalwood (*Santalum album*)	woody, creamy, powdery, smoky, urinous	santalols (sesquiterpenoids)
cedarwood (*Cedrus atlantica, libani, deodara*)	resinous, woody, sweet	pinene, himachalenes, himachalol, atlantone
"cedarwood" (*Thuja* & *Juniperus* species)	woody, camphor, pine, green	thujone, sabinene, terpineol, cedrene, cedrol
palo santo (*Bursera graveolens*)	sweet, minty, floral, woody, pine	mint carvone, mint lactones (benzofuranones), pulegone, terpineol, bisabolene
agarwood, aloeswood, oud, *jinkoh, kyara* (*Aquilaria* species)	rich, woody, sweet, floral, vanilla, animal	unique & myriad sesquiterpenoids, benzopyrones, furans

Smoldering woods and resins may do wonders for the imagination and emotions, but like all pyrolyzed matter, they do no favor to the lungs. Incense is known to generate potentially damaging levels of a number of known airborne toxins (nitrogen and sulfur oxides, formaldehyde, carbon monoxide), so it's best for long-term health to make censing an occasional event or use combustionless vaporizers (see page 482).

Incidental incense: tobacco, cannabis, moxa

Tobacco is our modern secular incense: the smoldering stuff whose fumes serve not to propitiate or inspire, but to relieve the craving of addiction to its stimulant drug nicotine. Tobacco smoke did have ceremonial roles among the American peoples who first inhaled it, but it was its introduction to European commerce that culminated in the cigarette, in the words of the historian Robert N. Proctor "the deadliest artifact in the history of human civilization," smoked by the billions daily despite its demonstrated role in causing heart disease and cancer. Toxic though it is, tobacco is a remarkable aromatic material, an instantly recognizable olfactory landmark that can be appreciated with or without smoking it.

Nicotiana tabacum is a member of the plant family that includes both the edible tomato and the toxic belladonna, or deadly nightshade. It has large broad leaves, sticky and bitter, armed with such effective chemical defenses, nicotine included, that they can sicken the fieldworkers who handle them. A century after Arawak emissaries presented a gift of leaves to an unimpressed Columbus, English settlers around Chesapeake Bay were making good money selling tobacco to the European market for chewing or smoking in pipes or leaf-wrapped cigars. To produce a form of tobacco that could survive for weeks in transit at sea without either rotting or crumbling into dust, the colonists developed a curing process that also revealed the leaf's hidden aromatic potential.

That process began with wilting the harvested plants in the field, and slowly drying them over four to eight weeks, sometimes with a smoky fire. The leaves were then piled together when still slightly moist, and allowed to ferment for months, during which microbes and the leaves' own enzymes worked together to make them pliable and resistant to spoilage. After drying and another year of further mellowing, the leaves were ready to be shipped. This strong, dark style of tobacco came to be known as **burley**.

The modern style of tobacco found in most cigarettes is lighter in both color and smell, its smoke easier to inhale often. It started in 1840s North Carolina with the cultivation of pale plants in depleted soils, then developed further with "flue-curing," the smokeless drying of the leaves in just a week before fermentation and aging. The result was mild **bright tobacco**. It became dominant with the invention of the paper-wrapped cigarette, which lent itself to mechanized production and became the mass-market tobacco product after World War I.

There are two chemical keys to the unique smell of smoldering tobacco leaves. Green leaves in general contain far more nitrogen in their active protein machinery than do woods and resins, by a factor of ten or more. Nicotine and other alkaloids also contain nitrogen, and along with protein fragments contribute animalic tobacco notes—nicotine itself becomes volatile and slightly fishy when warmed, pyrolysis-generated pyridines more strongly so. Skatole adds a fecal hint, and reaction with sugars generates earthy-toasty pyrazine rings.

The second key to tobacco smell is the leaves' endowment of large terpenoid molecules, too heavy to be volatile themselves but breakable into fragments that are. The sticky leaf glands contain defensive twenty-carbon diterpenoids that fragment into remarkable molecules also found in labdanum resin and in whale ambergris (see page 472). The orange and yellow carotenoid pigments, evident as the green chlorophyll fades, contain about forty carbon atoms. During curing and aging, the vast majority of carotenoids are broken apart, some to fragments that smell like tobacco and hay, others to sweet, floral, or fruity volatiles.

So the landmark smell of common tobacco is a rich and variable blend of many plant, animal, and cooked notes. Burley-style and cigar tobaccos are stronger in flavor than bright tobacco because they have more nitrogen-containing molecules, and their gradual curing dismantles the leaves' complex molecules more thoroughly. Bright cigarette tobaccos retain more sugars, which pyrolyze during smoking into sweet, caramel aromas and into acids that neutralize the harshness of alkaline nicotine and ammonia. Rarer sun-cured **"Oriental" tobaccos** from the hot and dry eastern Mediterranean have strongly flavored leaves rich in large, persistent "macrocyclic" molecules important in perfumery (see page 465), as well as short branched-chain acids that add buttery, cheesy, sweaty notes. And nearly extinct Louisiana **perique tobacco**, deriving from a native Choctaw process that compressed the leaves during fermentation, develops a distinctive winey, brandy aroma with various short- and branched-chain alcohols and their esters.

The best way to explore tobacco smells is to seek out unblended pipe tobaccos. Manufactured cigarettes, chewing tobacco, and snuff are blends, and "sauced" with additives—most commonly sugar, vanilla, and menthol—to make them less irritating. I don't smoke, but I enjoy sniffing loose pipe tobacco straight from the packet, or heated on a vaporizer (see page 482), or lit in a pipe to smolder like incense.

Like other smokes, tobacco smoke leaves a persistent after-smell. Among the old-smoke remnants are the nitrogen-containing pyridines and pyrazines, the former faintly rotten, musty, and fishy, the latter earthy. These are also the molecules responsible for **smoker's breath**: the mouth offers endless nooks and crannies and sticky molecules to which these volatiles can cling. Cigarette remnants in **ashtrays** also emit various medicinal-plastic-barnyardy phenolic volatiles and fecal skatole.

SOME COMBUSTIBLE LEAVES AND THEIR SMOKES

Leaf or smoke	Component smells	Molecules
cured tobacco leaf, various styles (*Nicotiana tabacum*; Americas)	fruity, floral, violet; warm, animalic; fishy, sweaty, leathery; sweet, toasted, earthy	terpenoid fragments (damascone, damascenone, megastigmatrienones, ionones, dihydroactinidiolide, solanone); pentadecanolide & other large rings, amber naphthofuran; pyridines, methylbutyric & methylvaleric acids; vanillin, benzaldehyde, pyrazines
generic leaf smoke	pungent, solvent, smoky, barnyardy, sweet	formaldehyde, acetaldehyde, acetone, acrolein, furfural, guaiacols, phenol, cresols, benzene, toluene, butadiene, ammonia, pyridines
tobacco smoke	pungent, sharp, toasted, leathery, animalic, fishy, sweet, fruity, floral	generic leaf smoke & cured tobacco volatiles + indole, skatole, dimethyl sulfide, additional pyridines, pyrazines
marijuana, cannabis smoke (*Cannabis sativa* & *indica*; Asia)	earthy, herbal, woody, floral, citrus, pine, sage, diesel, skunk	generic leaf smoke volatiles + myrcene, humulene, naphthalenes, dimethyl trisulfide
moxa smoke (*Artemisia* species; China)	resinous, herbal, cooling	generic leaf smoke volatiles plus pinene, sabinene, thujene, eucalyptol, camphor

Tobacco is one of the largest businesses in the world, and one of the most damaging, so plenty of money has been spent on analyzing its chemistry. **Marijuana**, or **cannabis**, is much less scrutinized despite its smoke having been inhaled for at least as long as tobacco's. A recent comparative survey of tobacco and cannabis shows unsurprisingly that the two smokes share many volatiles and don't share many others. Among the likely defining differences are the resinous terpenoids unique to cannabis, a trisulfide and naphthalenes, and its lack of the fruity-floral terpenoid fragments generated in tobacco curing.

Moxa is the dried and finely divided leaves and stems of species of the terpenoid-rich genus *Artemisia*, commonly known as mugwort, one of which is called wormwood for its use as an anti-parasite medicine (see page 259). Moxa smoke is used in traditional Chinese medicine as a general disinfectant for the air and skin and specific treatment for external ailments, and is said to have some effects on the body when inhaled; its woody volatiles do include cooling camphor. The pyrolysis of moxa is called *moxibustion*.

Perfumery: capturing volatiles

Incense burning liberates the volatiles locked up in aromatic materials so that they're free to scent the air. Perfumery also liberates volatiles from aromatics, but it recaptures them so that we're free to mix them and scent whatever we choose. The word *perfume* comes from the Latin, meaning "through smoke," and aromatic smoke may well have been the first: smoke and incense volatiles cling to clothes and skin and hair. The anthropologist Aïda Kanafani-Zahar vividly describes how Arab women place a smoldering censer under their robes as part of a social scenting ritual, and I can testify from airport encounters to the rich fragrance that emanates from robed Middle Easterners. But incense fumes fly in all directions. Perfumes are precision fragrances. They combine particular aromatics in particular proportions, and they can be sprayed over a broad area or touched discreetly behind an ear.

The subject of perfume is endless: so many ingredients, so many ways to manipulate and mix them, so many cultures, so many centuries! It's a story that runs the gamut from religion to romance to crassness: the anointing oil of Exodus and the blending of sacred and sensuous in the Song of Songs, the Greek Theophrastus noting that scent on the wrist is particularly sweet (*hedys*), the prophet Muhammad's association of scents with purification, and South Asia's with health and the pleasures of cultivated life, perfumed soaps and hair powders and snuff and gloves in the Renaissance, Napoleon's daily bottle of Eau de Cologne and Josephine's love of musk, the geisha's steaming of silken face masks with a rose-filled teakettle, the alliance of fragrance and fashion with Coco Chanel, the recent proliferation of celebrity and novelty perfumes. . . . And on top of all that, it's also a story of Hero Carbon's many guises, and a thousand years of human ingenuity applied to wrangling the best-smelling. We'll follow this lesser-known thread.

In earliest times, perfumes were medicinal, ritual, and cosmetic preparations, compounded with local aromatics and with exotics that trade brought from distant lands. The first extraction methods were to press the juices out of soft materials like flowers, or to soak solid aromatics in liquids. Oils became the extraction medium of choice, an early practical recognition that the essential materials of smell dissolve more fully in oils than water. The Egyptians perfumed their heads with oil-infused wax cones, and Moses anointed his people's altar with olive oil carrying myrrh, cinnamon, and cassia.

A second early insight may have been inspired by a simple observation: when water is heated in a pot, moist vapor escapes from its surface into the air, and if the vapor touches a cooler object held above—a hand, a large leaf, a lid—it condenses back into liquid water. At least five thousand years ago, people in various cultures realized that when aromatic materials are cooked in water, some of their smells escape with the water vapor, and can be condensed and collected along with it as an aromatic extract. Large earthenware pots apparently designed to do just that were found at Tepe Gawra in Iraq, a settlement active around 3000 BCE: the pots have a grooved rim that could collect condensation dripping down along the inner lid surface. The condensation from simmering aromatics is a far more refined extract than pressed juice or infused oil. It's nothing but volatiles and water. Everything else in the flower or herb or spice stays in the cooking water—or, if they're suspended above the water and steamed, in the source materials themselves. Simple and ingenious!

And far-reaching. This method of heating complex materials and collecting their vapors became known as **distillation**, from the Latin for "to drip down." Since particular volatile molecules have characteristic temperatures at which they tend to make the jump from solids or liquids into vapor, increasingly sophisticated versions of distillation became and remain powerful tools for dissecting all kinds of complex volatile materials into their components.

Distillation, essential oils and hydrosols, enfleurage

Early cultures invented a variety of different distillation devices, or *stills*. One of the most influential still designs was already in use three thousand years ago at Pyrgos, a Bronze Age industrial park on the Mediterranean island of Cyprus: a cooking pot whose lid has an opening for a tube that conducts the vapors into a separate collection pot. This two-pot *alembic* (from the Arabic for "cup") was adopted by Greek and Arab alchemists for their investigations of natural materials. One early product was the first refined petrochemical, a lamp oil similar to our kerosene (see page 422). And Arab perfume makers used versions of the alembic to obtain concentrated extracts of flower volatiles.

At the crossroads of East and West, powerful and influential Persia had a deep interest in material culture—including silks, gardens of exotic flowers and fruits, food and drink, and fragrances—and it transmitted that interest to its Arab conquerors. By the year 800, Arab perfumers had distilled an essence of roses that the wealthy used to scent themselves and their food and drink. Seven or eight centuries later, their successors began to separate the distilled essence into its two parts: the large volume of condensed water vapor from the cooking, and the thin oily layer that would slowly form and float on top of it.

The floating stuff of distillation became known as the *attar* or *otto* (from Arabic *itr*, "scent"), and in the West as the **essential oil**: "essential" because it carried the essence of the aromatic material that had been distilled, and "oil" because it behaved like cooking and lamp oils, floating on water rather than mixing with it. The oily layer was the concentrate of the rose volatiles alone, undiluted by water, while the water below carried a share of the rose volatiles that are partly water-soluble. The water solution is now called a **hydrosol**, or simply a floral *water*, and is a far less expensive product than the corresponding essential oil. Rose and orange-flower waters eventually became everyday fragrance materials in the Arab and Arab-influenced world.

A different method for capturing volatiles was prompted by the problem that many aromatic flowers, jasmine and violet and hyacinth among them, would yield little or no scent when subjected to the moisture and heat of distillation. In the practice that came to be known as ***enfleurage***, perfumers confined freshly picked flowers in a container along with oil-rich seeds—sesame in the early Islamic world, melon seeds or almonds in Renaissance Europe. The seed oils would absorb and accumulate volatiles emitted from successive batches of flowers, after which the oils were pressed from the seeds, and the seed solids used as scented paste. French perfumers later replaced the seeds with oil-impregnated cotton sheets, and then with glass plates coated with refined lard or beef tallow.

Ingenious as enfleurage is, it's slow and inefficient. And even when traditional distillation works, it effectively cooks the source material, producing extracts that are often very different from the smells of fresh aromatics. Peter Wilde, a pioneer in the development of modern extracts, famously noted that if you make an essential oil of orange peel, you get the essence not of orange, but of marmalade! More convenient and faithful extracts became possible when perfume makers enlisted the growing expertise of chemists, and the assistance of one particularly prized captive volatile.

Capturing volatiles with volatiles: alcohol

Around the same time that humans began to keep livestock and cultivate crops, they found that they could coax sweet and starchy foodstuffs into becoming intoxicating drinks. Eventually, and perhaps earliest in northern India, peoples around the world developed ways to make those drinks more powerful. A thousand years ago, Arab alchemists discovered that just as distillation can extract the scents of aromatic materials, it could do the same for the intoxicating essence of grape wines. European alchemists later gave the product of this distillation, the mixture of condensed water and essence, prized as both intoxicant and medicine, a number of names, including the Latin *aqua vitae*, "water of life," in French *eau de vie* (plural *eaux*), and *alcohol*, from the Arabic *al-kuhul*, "very fine powder." *Alcohol* eventually lent its name to the intoxicating volatile alone, and then to the chemical family that includes it, so chemists now specify the wine-derived original as *ethyl alcohol* or *ethanol*. Here we'll just stick with the everyday name.

Alcohol is one of our starter-set volatiles, the two-carbon chain H_3C-CH_2OH. The oxygen-hydrogen (OH) pair at one end is the same as two thirds of a water molecule, HOH, and makes alcohol molecules readily mixable with water molecules. But the carbon-chain (C-C) portion of alcohol is more similar to the small chains and rings of other common volatiles (typically four to ten carbons) and to the long chains of fats and oils (sixteen to twenty). It causes the alcohol molecules in a water-alcohol mixture to be less tightly held in the mixture than the water, and so easier to push out of the mixture with heat. Gently heat a wine that's 10 percent alcohol and 90 percent water, and you produce vapors that initially flip that ratio. A simple distillation of wine or beer can produce an eau de vie that's 30 to 40 percent alcohol.

Concentrated alcohol was a boon to perfume makers in several ways. Alcohol's own nature as a carbon-chain volatile helps it dissolve and hold other volatiles more efficiently than water, so eaux de vie do a better job than water or wine of drawing volatiles from aromatics that are immersed in them. As fans of limoncello know, you can produce good extracts of citrus peels simply by steeping them in vodka! By Renaissance times, perfumers were using eaux de vie to make similar extracts of citrus peels and herbs and spices, as well as water-insoluble tree resins and waxy animal materials like musk and ambergris. These extracts were variously called **eaux**, **infusions**, and, when they took up tints as well as volatiles, **tinctures**. They in turn could be redistilled into a more refined extract of an extract, a secondary eau de vie without any pigments or other nonvolatile carbon chains.

The first fragrance to become internationally and lastingly popular was the eighteenth-century eau de vie known as Eau de Cologne, scented with a mixture of citrus peels and herbs, and in its time a sippable medicine as well. Its great appeal isn't just its refreshing smell: it's also the congruent fresh feeling created by the alcohol in its base. Before the age of eaux, most aromatics were extracted and combined in oils. The long carbon chains of oils are not volatile. When applied to the skin, they coat it indefinitely and release dissolved scents slowly; and they're easily fragmented by oxidation into rancid-smelling volatiles of their own (see page 536). But two-carbon alcohol doesn't easily oxidize and is itself more volatile than nearly all scent molecules. When applied to the skin in a scented eau, it evaporates quickly and completely, cooling the skin as it disappears. It takes its pungent but otherwise mild smell with it, leaving only some moisture and the scent molecules.

Alcohol became even more prominent in perfumery after 1800 with the growing availability of distillates that were more than 80 percent alcohol. Their water content is low enough that they leave little or none on the skin: they were no longer eaux but proper alcohols. Concentrated alcohol also mixes reasonably well with oils and melted fats, and therefore can "wash" the volatiles out of fragrance-infused oils and molten enfleurage fats. So it became the fragrance solvent and carrier of choice, the base in which extracts of all kinds can be mixed at will and applied to the body. Today alcohol (often "denatured" with additives to make it undrinkable) accounts for two thirds or more of even the finest bottled perfumes.

Capturing volatiles with volatiles: ether, hexane, CO_2

For all its advantages, alcohol has one important drawback in volatile wrangling: its ability to mix easily with water. This doesn't matter when it's used to extract dry materials like resins and spices. But when used to make infusions of *fresh* herbs and flowers, alcohol extracts watery fluids along with volatiles: and the water and molecules dissolved in it react with the volatiles and change the extract's smell. The ideal fragrance solvent would be water-repellent like fats and oils, so that it extracts only carbon-chain molecules, but also very volatile like alcohol, so that it could be evaporated away once it had done its extracting.

Once again chemists came up with a better way to corral volatiles. As early as 1729, the German-born August Sigmund Frobenius had described to England's Royal Society how to make aromatic essences with ether, the synthetic four-carbon chain that doesn't mix much with water and boils at body temperature (see page 436). Simply soak any aromatic in cold ether, then strain off the ether and let it evaporate. What's left is a nearly pure extract of the aromatic's carbon-chain molecules. Simple and ingenious again!

It took another century or so for French chemists to get to work on it, but this basic method—**solvent extraction**—gradually supplemented distillation in the growing fragrance industry, and largely replaced enfleurage. Hexane eventually became the standard solvent of choice; it's a gasoline-like six-carbon chain that chemists first found in petroleum and now manufacture from it, and it's cheaper, more stable, and less toxic than ether.

Solvent extraction of a flower or herb produces two useful fragrance materials. It removes fats and waxy materials along with the volatiles, so after the hexane is evaporated away, the initial extract may be viscous and even solid. It therefore came to be called the *concrete*. Mix the concrete with concentrated alcohol to wash out its small-chain volatiles, evaporate the alcohol from the wash, and you obtain the uncooked equivalent of an essential oil. This is called an **absolute** because it is absolved from the concrete, dissolved out of it. Like essential oils, concretes and absolutes represent tiny fractions of the original fresh flower: to extract a teaspoon of rose oil or absolute takes about forty-five pounds (twenty kilograms) of roses.

Absolutes, concretes, essential oils, tinctures: these were the basic materials of modern perfumery as it developed at the turn of the twentieth century. Today, the state of the art includes the use of novel solvents, so-called supercritical fluid extraction, with carbon dioxide gas pressurized into a semiliquid, and "fractional" and "molecular" distillations at very low pressures and temperatures, which allow individual volatile components of extracts to be separated and collected as **isolates**. All of these materials are often called "natural" because they originate with living plants and animals and reflect their qualities, in contrast to synthetic materials that are manufactured from plant wastes and cheap chemical feedstocks. The term can be misleading, however, because these "naturals" are also the product of sophisticated chemical science and technology. Often the more lifelike the perfumer wants a fragrance to be, the more high-tech the ingredients required!

Different extracts are like different artists' sketches of the same flower: each is a different likeness. To get an idea of how variable the volatile profiles of extracts can be, consider one aspect of the complex scent of a rose. Two of its more abundant volatiles are generically floral phenylethanol and sweet-rosy citronellol. Fragrance chemists found that the live flower on the plant emits about three times as much of the generic floral as the sweet-rosy—but that simply cutting the flower from the plant reverses that balance. Then of the several cut-flower extracts that they analyzed, the distilled essential oil reflected the cut-flower proportions, while the rose water hydrosol more resembled the blossom on the plant, as did the solvent-extracted absolute. The supercritical carbon dioxide extract balanced the two qualities more evenly. Extrapolate this variability to the other major rose volatiles and it's evident that different extracts of rose will all be recognizably rosy, but with their own nuances. The same is true of aromatic extracts in general. Perfumers can choose among many different versions for each ingredient in their compositions.

**ABUNDANCE OF TWO VOLATILES IN LIVING ROSES AND
ROSE EXTRACTS (PERCENTAGES OF TOTAL VOLATILES)**

Molecule	Rose, intact plant	Rose, cut	Rose water (hydrosol)	Rose essential oil	Rose absolute	Rose CO_2 extract
citronellol (rose, sweet)	18	60	15	30–40	7	30
phenylethanol (floral)	60	15	50	2	50	50

*Source: K. H. C. Başer, M. Kurkcuoglu, et al., Anadolu University, Turkey;
see Selected References, page 626.*

Enough about the ongoing quest to capture the world's pleasing smells. On now to the aromatics that have inspired it! The next half-dozen sections are a sampler of the materials commonly extracted for use in perfumery, a number of them making a return appearance from earlier chapters.

Fragrance materials from plants: flowers, fruits, leaves

Flowers are the inspiration and heart of perfumery, the archetype of transfixing olfactory beauty. Their fragility and transience make it challenging to capture their scents. Only a handful of true flower extracts are routinely available, and they're very expensive; synthetic approximations are much more common.

The three most prominent flowers in perfumery have been so from ancient times. The **rose** was prized throughout Europe and Asia and successfully distilled in early Islamic times. It's richly scented with a mix of floral and fruity monoterpenoids, fragmented tetraterpenoids, and the common floral phenolic phenylethanol (see page 238). Perfumers credit rose oils and absolutes as being harmonizing, unifying ingredients in their compositions, a quality that may come from their broad base of several hundred volatiles.

It was Persia's appreciation for scent and its trade with eastern Asia that brought jasmine vines and citrus trees to the Middle East. **Jasmine**, more effectively extracted with solvents or enfleurage than by distillation, shares almost nothing with rose: it's dominated by its own eponymous fatty-acid fragments, ketones and esters and lactones, with only a touch of the generic floral monoterpenoid linalool. It also stands out for what's sometimes called an animal or "dirty" note, from un-ethereal, down-to-earth indole and tarry-barnyardy cresol (see page 233). **Orange blossom**, or **neroli**, distillable like the rose, shares incongruous indole with jasmine but is otherwise fresh with terpenoids that it shares with lavender and conifer trees, and it has a fruity note from berry-like methyl anthranilate. Both rose and orange blossom waters found ongoing roles outside perfumery, as flavorings for baked goods and sweets in the Middle East and the onetime Moorish regions of southern Europe.

SOME FLOWERS USED IN FRAGRANCES

Flower	Component smells	Molecules
rose (*Rosa damascena* & *centifolia*)	floral, fruity, violet, sweet	damascenone, ionone, rose oxide, citronellol, nerol, geraniol, phenylethanol
jasmine (*Jasminum grandiflorum* & relatives)	floral, creamy, fruity, animal	jasmone, methyl jasmonates, jasmolactone, linalool, benzyl esters, indole, cresol
neroli, orange blossom (*Citrus aurantium*)	floral, fresh, lavender, warm, fruity, animal	linalool, linalyl acetate, myrcene, ocimenes, methyl anthranilate, indole
osmanthus (*Osmanthus fragrans*)	floral, violet, fruity, apricot, woody	linalool, ionones, g-decalactone, hexyl butyrate, citronellol
ylang-ylang (*Cananga odorata*; tropical Asia, Indian Ocean)	floral, balsamic, sweet, spicy, wintergreen, leather	linalool, geranyl acetate, germacrene, caryophyllene, methyl benzoate, benzyl esters, cresyl methyl ether
tuberose (*Polianthes tuberosa*)	floral, balsamic, sweet, fruity, wintergreen, animal	methyl & benzyl benzoates, methyl salicylate, methyl isoeugenol, methyl anthranilate, farnesene, indole

Later additions to the perfumer's roster of floral materials include Asian and American natives with distinctive versions of floweriness. **Osmanthus** flowers are rich in terpenoid fragments that it shares with stone fruits, especially apricots. **Ylang-ylang** (from the Tagalog for "wild") has balsamic and wintergreen qualities from a number of benzenoid-ring volatiles, and a leathery variation on cresol.

And **tuberose**, a small relative of agave long cultivated in Mexico, stands out for its collection of benzenoid rings, a touch of wintergreen, and a heavy, lingering sweetness.

The most important **fruits** in perfumery, dating back to the eighteenth-century Eau de Cologne, are species of citrus, which share a number of terpenoids with flowers and contribute fresh, bright qualities (see page 320). Their volatiles are concentrated in glands embedded in the fruit peels, from which they can be extracted by mechanical pressure as well as solvents. **Bergamot** is the fragrance fruit par excellence; its flesh is inedibly sparse and sour, but its peel is strongly scented with floral, lavender, and woody terpenoids. Bergamot is the defining aromatic in Earl Grey tea, a nineteenth-century English invention apparently modeled on expensive Chinese teas scented with jasmine. **Lemon** contributes its landmark bright aroma via the terpenoids neral and geranial, while **bitter orange or bigarade** carries resin and pine notes, and the Caribbean latecomer **grapefruit** has notes of wood and eucalyptus. Vanilla and black and pink peppercorns (see pages 292, 286) are among the dried-fruit spices sometimes extracted for use in fragrances.

SOME FRUITS USED IN FRAGRANCES

Fruit	Component smells	Molecules
bergamot (*Citrus bergamia*)	floral, lavender, lemon, pine	linalool, linalyl acetate, limonene oxide, neral, geranial, pinene
lemon (*Citrus limon*)	citrus, pine, turpentine, lemon, floral	limonene, pinene, terpinene, sabinene, neral, geranial, linalool
orange, sour (*Citrus aurantium*)	citrus, resinous, floral, pine	limonene, myrcene, linalool, pinene
grapefruit (*Citrus* x *paradise*)	fresh, green, floral, eucalyptus, woody, grapefruit	octanal, decenal, dodecanal, eucalyptol, nootkatone

Of the small plants valued as aromatics, only one provides perfumers with the typical smell of freshly crushed leaves: **violet leaf**, whose six-carbon green-leaf volatiles and similar nine-carbon fragments (see pages 154, 156) are preserved by solvent extraction. Other leafy materials include culinary herbs like the mints (see page 249), and a fragrance-specific handful that tend to echo flowers or citrus fruits through shared terpenoids. **Citronella**, familiar from insect repellents, gives a mix of lemon and rose, and the perfumer's **geranium** leaf, from a particular species of *Pelargonium*, is rosy with a fresh herbal-citrus background. The landmark **lavender** floral-herbal mix of linalool and its ester linalyl acetate is echoed by **clary sage** and by **petitgrain**, the leaves of the sour orange tree, but with added woody, earthy, herbal variations. (Fresh clary sage is notable for a sweaty thiol that doesn't survive in extracts.) **Patchouli**, the leaf of a Malayan member of the mint family, owes its earthy, woody qualities to **patchoulol** and other unique sesquiterpenoids, and a mustiness that may derive in part from the need to dry or slightly ferment it to extract it efficiently. Patchouli became prominent in nineteenth-century Europe when its moth-repellent leaves protected cashmere shawls in transit from India, and it had a second vogue in the back-to-the-earth counterculture of the 1960s. Perfumers also work with extracts of richly aromatic cured **tobacco** leaves (see page 448).

SOME LEAVES AND LICHENS USED IN FRAGRANCES

Material	Component smells	Molecules
violet leaf (*Viola odorata*)	green, leafy, floral, fatty, cucumber	C5–C9 aldehydes & alcohols, esp. nonadienol & nonadienal
citronella (*Cymbopogon* species; Asia)	lemon, rose	citronellal, citronellol, geraniol
geranium (*Pelargonium graveolens*; southern Africa)	rose, herbal, citrus	citronellol, geraniol, menthone, photocitrals
lavender, with flowering tops (*Lavandula angustifolia* & relatives)	floral, sweet, herbal, woody, green, fresh	linalool, linalyl acetate, terpinenol, caryophyllene, undecatrienes
clary sage, with flowering tops (*Salvia sclarea*; Europe)	floral, sweet, woody, earthy, ambergris	linalyl acetate, linalool, germacrene, caryophyllene, spathulenol, amber naphthofuran
petitgrain, sour orange leaf (*Citrus aurantium*)	floral, sweet, herbal, woody, green	linalool, linalyl acetate, ocimenes, myrcene, terpineol, isopropyl methoxypyrazine

Material	Component smells	Molecules
patchouli (*Pogostemon cablin*; S. Asia)	earthy, woody, camphor, floral	patchoulene, patchoulol, seychellene, bulnesene (sesquiterpenoids)
oakmoss, tree lichen (*Evernia prunastri, Pseudevernia furfuracea*)	green, seaweedy, earthy, woody, smoky, animalic	rare decorated phenol & benzoate rings (e.g., methoxymethylphenol, methyl dihydroxydimethyl benzoate)

By far the most unusual and fascinating plantlike material is **oakmoss**, actually not a moss but a *lichen*, one of the slow-growing survivalist creatures that form scaly patches on rocks and trees. Lichens are ancient and still mysterious associations between fungi and photosynthesizing cyanobacteria or algae. The fungal partners in oakmoss manufacture powerful chemical defenses, including a host of unique variations on benzoic and phenolic rings that provide smells of damp undergrowth and seaweed. During the Renaissance an oak lichen was known as an aromatic specific to the island of Cyprus (in French, *Chypre*); centuries later, along with similarly local labdanum resin, it helped define the perfume family now known as the *chypres*.

Fragrance materials from plants: resins, woods, roots

All the tree resins and woods of traditional incense (see pages 445–46) are also extracted and used in perfumery, where they lend their woody, sweet, balsamic notes to fragrance compositions. Two other resins are best known today in perfumery, though they too were burned on ancient altars in the Middle East. **Labdanum** is a sticky resin that accumulates on leaves and twigs of the Mediterranean shrub known as rock rose. It's unusually complex, with two dozen phenolics that provide woody, leathery, and animalic notes, and other volatiles reminiscent of hay, flowers, and fruits. **Galbanum** is the dried milky latex harvested from a Central Asian relative of garlicky asafetida, strangely dominated by intense green-vegetable qualities from volatiles it shares with seaweeds and green peas and capsicum peppers. It has sulfurous, woody, and animal aspects as well.

SOME RESINS USED IN FRAGRANCES

Resin	Component smells	Molecules
labdanum (*Cistus ladanifer* & *creticus*; Mediterranean)	woody, amber, leathery, animalic, fruity	pinene, amber naphthofuran, phenolics, sesquiterpenoids, lactones, raspberry ketone
galbanum (*Ferula galbaniflua*; W. Asia)	green, leafy, seaweedy, woody, musky	galbanolenes (undecatrienes), isobutyl methoxypyrazine, butyl & propenyl disulfides, C13–C16 rings

Of the traditional incense woods used in perfumery, **agarwood or oud** is especially prized for its richness, **sandalwood** for its exotic santalols and their animal side qualities, milky and even slightly urinous. Two other exotic woods are prominent fragrance materials: South American **rosewood** and Asian **ho wood**, a relative of cinnamon, both smell like flowers! They accumulate the flowery monoterpenoid linalool, and their extracts can substitute for more fragile floral materials. Then there are a couple of wood materials made by "destructive distillation": that is, by pyrolyzing wood, as if making tar, and collecting the volatiles. **Birch tar oil** and **cade oil** are rich in the phenolic volatiles that characterize woodsmoke, with persistent medicinal, tarry, and barnyard notes. Birch tar oil is also used to evoke leather; it was once a specialty of the Russian countryside prized for impregnating fine leather boots, and so came to be an expected aspect of their smell.

SOME SPECIAL WOOD MATERIALS USED IN FRAGRANCES

Wood material	Component smells	Molecules
rosewood (*Aniba rosaeodora*; S. America)	floral, rose, spicy	linalool, linalool oxides, geraniol, terpineol
ho wood, leaf (*Cinnamomum camphora* var. *linaloolifera*; SE Asia)	floral, rose	linalool, camphor, terpineol
birch tar (*Betula*; Eurasia)	smoky, tarry, medicinal	guaiacol, cresol, phenol
cade, juniper tar (*Juniperus oxycedrus*; Mediterranean)	smoky, tarry, woody	guaiacol, ethylguaiacol, cresol, cadinene

A handful of plant roots and rhizomes (underground stems) are dried for use in perfumery, and they carry woody smells from fifteen-carbon sesquiterpenoids, which are common tree-wood volatiles. But these underground organs are uniquely inventive with sesquiterpenoids, perhaps to defend themselves against the legions of microbes in the soils they inhabit. Shyobunones and acorenone and costols and zizaenones appear nowhere else in this book!

One of the most venerable underground aromatics is **nard or spikenard**, the rhizome of a plant native to the Himalayas and long prized in the distant Mediterranean. In an incident recorded in two of the New Testament gospels, a woman anointed Jesus with a small vessel of pure nard oil, and Judas remarked that it was worth a year's wages. "The house was filled with the fragrance of the perfume," reports John: woody, earthy, patchouli-like, somber rather than flowery. Similarly long-prized is **costus**, the root of a Central Asian thistle, an incense ingredient for thousands of years and used in Chinese joss sticks to this day. Among perfumers its essential oil is noted for an unusual animalic note sometimes likened to hair, a fur coat, or a wet dog. **Calamus or sweet flag** is the rhizome of a wetland grass with sweet, earthy, leathery side qualities that have been put to use in alcohols and tobaccos as well as perfumes, though care has to be taken to avoid its nonvolatile toxin asarone.

The most popular underground material today is **vetiver**, an intriguing Asian grass whose thin yard-long roots can have subtle floral and grapefruity sides to their woody earthiness. It's a component in many fragrances marketed to men. In India, wild *khus* roots are woven into shades and mats that are sprinkled with water to cool and scent the air. Vetiver oils are sometimes distilled at unusually high temperatures and pressures and have been found to contain more than three hundred mostly unusual terpenoids and their products, includ- ing **zizaenones**. This remarkable diversity appears to result in part from bacteria living in the vetiver roots, where they transform relatively few plant-produced molecules into a more diverse set, perhaps to fend off their own microbial competitors.

SOME ROOTS AND RHIZOMES USED IN FRAGRANCES

Material	Component smells	Molecules
nard (*Nardostachys jatamansi*; Himalayas)	heavy, woody, patchouli, spicy	patchoulol, valeranone, valeranal, aromadendrene
costus (*Saussurea costus* & *lappa*; C. & S. Asia)	woody, fatty, animal hair	dehydrocostus lactone, costunolide, costols, curcumene, selinene
calamus, sweet flag (*Acorus calamus*; Eurasia)	warm, woody, leather, spicy	shyobunones, acorenone, acoronone, camphene, pinene, camphor
vetiver (*Chrysopogon zizanoides*; India)	woody, earthy, floral, grapefruit, smoky	zizaenones, vetivol, vetivones, khusimol, kusimone, nootkatone, phenols & guaiacols
orris (*Iris pallida*, *I.* x *germanica*; Europe)	woody-floral, warm, violet, sweet	irones (C10 main chains, + rings)

The most conventionally pleasant underground aromatic is **orris root**, the rhizome of *Iris pallida* and a couple of close relatives, whose unusually thick—and prohibitively expensive—essential oil, orris butter, is produced by drying and aging several-year-old rhizomes, then grinding them, moistening with dilute acid, and finally distilling. The result is a floral-woody scent suggestive of violet flowers, apparently due to the slow breakdown of thirty-carbon triterpenoids into fragments called irones that are similar in structure and aroma to the distinctive violet ionones. As we'll see, triterpenoid breakdown also figures in the transformation of a smelly whale intestinal obstruction into exotic ambergris.

Fragrance materials from animals: creaturely and clinging

Today the most prized perfume ingredients are extracts of flowers and rare woods, but for many centuries and in many high cultures, they came from animals. In his ninth-century treatise on aromatics of the Persian and early Islamic empires, the Baghdad physician Yuhanna ibn Masawaih listed the five principal materials in order as musk, ambergris, agarwood, camphor, and saffron. Musk is a kind of deer secretion, and ambergris comes from whales. Arab perfumers also used civet, the paste with which distant relatives of the mongoose and hyena mark their territory; later on, Europeans adopted castoreum, the marking fluid sprayed by beavers.

Why would the wealthy choose to mark themselves, and even their food and drink, with the effluvia of other animals? As we've seen in chapter 5, the characteristic smells of animals' bodies come mainly from resident microbes breaking down animal protein machinery, rich in nitrogen and sulfur, into the exhaust fumes of metabolism: urinous amines and fecal skatole, stinky sulfides and thiols, horse-stable cresols. These simple waste products aren't especially pleasant. But they do get our attention, subliminally or overtly. A touch of musk or beaver is a way of asserting the wearer's animal presence by prestigious proxy rather than poor hygiene: a refined means of visceral communication.

And as one ingredient in a fragrance, mixed with a bunch of the plant kingdom's elaborately constructed volatiles, the degraded quality of animal volatiles might actually be an asset, precisely by adding an entirely different dimension to the overall smell. There's a natural precedent for this combination of the elaborated and the degraded: flowers like jasmine and gardenia that are called "exotic" or "dirty" for their atypical, nonflowery touches of indole, skatole, and cresol. In perfumes as well as flowers, these touches may accentuate the appeal of the floral volatiles by setting them off and boosting our attention to the mixture (see page 220).

But the particular animal materials long prized in perfumery bring something else to fragrances, something remarkable: namely several rare volatiles that aren't found in ordinary secretions and excreta and aren't the usual small waste-product molecules. They're as large as the plant kingdom's sesquiterpenoids, with between thirteen and seventeen carbons arranged in some kind of ring structure. The *macrocyclics* are big single rings, the carbon atoms forming a rough cross or stubby cigar shape. Some are *polycyclics*, with several smaller rings connected to one another. Like the woody sesquiterpenoids, these large molecules are heavy and not very volatile. Their long hydrocarbon stretches or multiple hydrocarbon rings encourage loose bonds with each other, with other volatile molecules, and with similar skin and hair oils. They're a hundred times less volatile than plain water, a million times less volatile than floral linalool—so their qualities tend to be the opposite of ethereal and heady. Instead of flying up at us, they draw us in, invite us to sniff harder to sense them more fully.

And their smells are nothing like the common protein breakdown products, the amines and sulfides and cresols. Instead they're what might be called *creaturely*: they suggest a living, warm-blooded presence itself, not its microbial posse. They can have a variety of secondary qualities, some subtly animalic but others sweet or floral or fruity. The fragrance chemist Philip Kraft describes the smell of the pure musk macrocyclic volatile, **muscone**, as "warm, sensual, sweet-powdery," with a "smooth, soft, and intimate *'skin-on-skin'* connotation" suggesting the skin of a baby—the smell of animal life as yet untouched by the exertion of living.

Why do some animal bodies go to the trouble of constructing these large molecules? The best evidence to date comes from the chin-gland marking secretions of European rabbits! Their polycyclics have the effect of slowing the release of more volatile components in the secretions: so those volatile signals persist longer on the animals and the objects they mark. In the terminology of the fragrance profession, the polycyclics are *tenacious* ingredients and act as *fixatives*, holding a perfume's volatiles on the skin and releasing them more gradually and evenly. Perfumers have also long observed that animal materials somehow enhance and harmonize the other ingredients in a composition. The self-described "skeptical chymist" Robert Boyle concurred more than three centuries ago, noting of the surprising power of faintly aromatic ambergris that "some things, that are not fragrant themselves, may yet much heighten the fragrancy of Odiferous bodies."

Prized as they've been for these and other reasons over more than a thousand years, animal materials are seldom used in commercial fragrances today. Nevertheless, it's still possible to track down small samples for sniffing. And while they themselves are rare, their disembodied creatureliness now permeates our everyday life, whether we wear perfume or not.

Animal fragrance: musk, ambergris, civet

The preeminent animal fragrance material is **musk**, the waxy abdominal secretion of rare Asian members of the deer family. *Moschus chrysogaster* and related species are small solitary animals with odd fanglike upper teeth, long hunted by the several civilizations that bordered *M. chrysogaster*'s Himalayan homeland. The word *musk* comes via Arabic from pre-Islamic Persia, where Chinese visitors reported men anointing their faces and beards with musk-scented oil, and where it was associated with sensuality—a sixth-century poem attributed to Imru' al-Qais describes stray grains of musk in his lover's rumpled bed. With the rise of Islam, musk's significance shifted from the sensual now to the spiritual hereafter. Muhammad prescribed that worshippers first purify their bodies by washing and perfuming themselves; he praised musk, and at his death his body was said to emit its scent. In the eternal garden paradise that Islamic tradition envisions for the righteous, soil and mountains and virginal maidens are said to be made of musk.

How did the smell of an animal's mating secretions become the smell of the earthly made holy? According to historian Anya King's fascinating *Scent from the Garden of Paradise*, the Arabs considered musk deer to be mysterious creatures living in distant high mountains and feeding on rare plants. They thought that, in this almost otherworldly existence, ordinary corruptible animal blood became refined into its incorruptible essence, into the solid musk grains that never spoil or stop emitting their special scent. Thus musk and its creaturely smell came to represent not the indulgence of our animal nature, but the aspiration to realize our nature's highest spiritualized possibility.

As a way of evoking the spiritual through smell, this is an intriguing alternative to the utter dematerialization represented by fuming incense. It found no purchase in Europe, however, where simple enjoyment of musk's smell and tenacity made it a prominent aromatic. It scented the gloves and perfumes and foods of the wealthy from the Renaissance through the eighteenth century, when it fell out of fashion, and then returned decades later as a supporting ingredient in composed perfumes.

Musk starts as a yellow, cloudy fluid that male deer secrete from a gland near the genitals some weeks before the mating period. It collects in a small pod, where it matures, probably influenced by microbes, into about an ounce of solid, dark-colored grains of waxes and steroids. The grains scent the genital area and attract the attention of females, and some of the volatiles enter the urine. Fragrance chemists have found that musk's special qualities are largely the product of two of its macrocyclic volatiles, muscone and muscopyridine, the latter a nitrogen molecule providing more of the animalic, urinous aspect.

Musk deer are in danger of being hunted to extinction, and international trade in Himalayan musk, the most prized, is now illegal. Small amounts are produced from captive animals.

Ambergris is the strangest of all fragrance materials, a startling demonstration of Hero Carbon's protean permutability. It begins as a stinking obstruction in the rectum of ocean-cruising sperm whales, and ends years later as sublime seashore jetsam, the finest emitting a smell like no other, with facets of the ocean and soil, exotic woods, incense, and tobacco. And it has a musklike ability to fix and enhance other fine odors.

The name *amber* comes via the Arabic *anbar* from a Somali word, so Arabs were likely introduced to it through traders along the Indian Ocean. By early Islamic times they were using it in incense, perfumes, and as a flavoring. China adopted it via trade, and Europe probably via the Crusades. The French suffix *-gris* distinguishes gray whale residues from a different jetsam amber, the yellowish fossilized tree resin. In Renaissance times ambergris was added along with musk to savory and sweet foods as well as to perfumes and royal anointing oils. Then with musk it fell out of favor, and has since been an expensive rarity.

The main ingredient in the making of ambergris is squid! Sperm whales consume them by the ton daily—and along with them countless hard, indigestible squid beaks and quills and eye lenses. The whales normally regurgitate these residues, but occasionally retain enough to form a mass that's too large to expel. The mass keeps accreting undigested food, as well as microbes, parasitic worms, and digestive fluids that include bile, the mixture of cholesterol-like lipids and colored remains of blood hemoglobin. It grows until the whale dies, sometimes from the obstruction itself, which can weigh a thousand pounds.

Freshly formed, ambergris is black with iron-rich bile, soft and stinking. But with months and years, both as older layers become buried in the growing mass and as the mass floats in the ocean after the whale's death, it becomes firmer and lighter in color and smell—a still mysterious improvement that continues in the beachcomber's closet and fragrance company's vault. Apparently microbes in the mass use the bile lipids to construct a waxy thirty-carbon triterpenoid called ambrein, some of which is then slowly broken down into a host of unusual volatile fragments—maybe by oxygen, or the bile's iron, or microbes, or some mix of these. The combination of **amber naphthofuran** and ambrinol, two polycyclic fragments, reproduces many of ambergris's facets, as well as its fixative powers.

Civet was not among ibn Masawaih's important aromatics, and seems to have been a relatively late and more frankly animalic addition to the fragrance repertoire. The civet, *Civettictis civetta*, is a small carnivore native to eastern Africa. Persians apparently first appreciated the secretions of its marking glands around the ninth century, began raising it, and Arab traders then disseminated it and the name (*zabad*) to China, India, and Europe. Renaissance exploration brought both the material and the animal, including related Asian species, directly to Europe, where it rose and fell in popularity along with musk and ambergris. Because both males and females constantly produce their waxy marking secretion and wipe it onto objects, they were kept captive and the secretion harvested by various means. A small amount of purportedly humane civet is still produced in Southeast Asia. It contains a number of animalic nitrogen-containing volatiles, some possibly microbial in origin, but its key component is **civetone**, a macrocyclic molecule similar to muscone.

SOME ANIMAL MATERIALS USED IN FRAGRANCES

Material and source	Component smells	Molecules
musk (Asian deer *Moschus moschiferus* & relatives)	musky, rich, animal, urinous	muscone (1-ring C16), muscopyridine (1-ring C16+N), androstenone (4-ring C19 steroid), cresol
ambergris (sperm & pygmy whales, *Physeter* & *Kogia* species)	exotic, woody, balsamic, earthy, mossy, ocean	amber naphthofuran (3-ring C16), ambrinol (2-ring C13), ionone variants
civet (African civet "cat" *Civettictis civetta* & Asian relatives)	musky, sweet, animal, fecal	civetone (1-ring C17), muscone, indole, skatole, ethyl & propyl amines
castoreum (N. Hemisphere beavers *Castor fiber* & *canadensis*)	warm, leathery, smoky, tobacco, amine-animalic	phenols, cresols, guaiacols, benzoic & cinnamic acids, benzaldehyde; N-containing castoramine & nupharidines & quinolines & pyrazines

continued

Material and source	Component smells	Molecules
hyraceum (African rock hyrax *Procavia capensis*)	animalic, aspects of musk, castoreum, tobacco, sandalwood	amines, benzamide & other N molecules, acids
beeswax (honeybee *Apis mellifera*)	sweet, fatty, honey, hay, floral	phenylethanol & phenylethyl acetate; benzoate esters, octanal, nonanal, decanal, linalool oxides
onycha, nakh, fire-dried (sea snail *Chicoreus ramosus* & others)	smoky, medicinal, seaweed, shellfish	phenol, cresol, chlorophenol, chlorocresol, dichlorophenol, pyridine

Animal fragrance: castoreum, hyraceum, beeswax, onycha

Among other and less prestigious animal materials that have been put to use in perfumery, **castoreum** derives from the marking secretion of the beaver, the large rodent that lives in northerly wetlands and chews down trees. It's the dried, waxy residue of a milky fluid that collects in two pouches toward the rear of both males and females and scents their urine. Castoreum has been used since ancient times as a medicine and was adopted for perfumery (and artificial vanilla flavors) in the nineteenth century. It lacks large fixative molecules but achieves similar effects with a unique blend of carbon rings, carbon-nitrogen rings, and nitrogen-sesquiterpenoid hybrids. Many of its barnyardy cresols, smoky guaiacols, and fruity-flowery benzenoids may derive from the beaver's lignin- and ring-rich diet of bark and twigs, while amines and leathery quinolines may be made by pouch microbes living on urine and sloughed skin. This mix explains why castoreum is suggestive of **leather**, which was traditionally made of animal hides stripped with urine and dung, toughened with phenolic-rich tree bark tannins and oak galls, and softened with animal oils.

Hyraceum is a new—and very old—fragrance material. It comes from a small rabbit-like mammal native to southern Africa, the hyrax, which inhabits caves and rocky overhangs. Hyraxes deposit their wastes in collective latrines; these dry out in the arid climate and have formed fossil beds several yards thick and tens of thousands of years old. Hyraceum is this fossilized waste, mainly residues of urine. Local peoples used it as medicine; around 1850, a German pharmacist noted its resemblance to castoreum, and it has since found occasional use in perfumery as a humane if nonrenewable substitute. Hyraceum seems to be rich in nitrogen-containing molecules, and one analysis tentatively identified a sixteen-carbon macrocyclic musk.

Beeswax is a different sort of animal aromatic: from the insect realm, lacking large-molecule fixatives, but enjoyable even before it's extracted! Bees build their honeycombs from a solid mass of large forty-carbon chains that they secrete from glands on their abdomens. Fittingly for the material of hexagonal honey cells filled with concentrated flower nectar, beeswax absolute contains many six-carbon benzenoid volatiles, along with a few floral and waxy medium-chain aldehydes familiar from citrus peels.

Onycha, from the Greek for "fingernail" (*nakh* from Sanskrit for "shell"), brings an unusual oceanic note to the perfumer's palette. It comes from large marine sea snails that block the opening in their shells with a hard thumbnail-like shield called the *operculum*. Information about the opercula themselves is scant, but when they're dried over a fire and their volatiles then dry-distilled, smoky phenols and cresols dominate, along with chlorinated versions from the abundant halogens in seawater that are reminiscent of the bromophenols in shellfish and whiskies (see pages 381, 402).

Plant musks and ambers: ambrette, angelica, star-struck pine

Fragrance lovers have long recognized that some plant materials share the warm and fixative qualities of musk and ambergris and civet, and can be substituted for them. As far back as early Islamic times, ibn Masawaih noted that labdanum resin from the shrubby rock rose resembles ambergris. The seventeenth-century chemist Robert Boyle found that seeds of a hibiscus relative called musk-seed indeed resembled musk. Modern fragrance chemists have verified those keen-nosed observations. They've cataloged both macrocyclic and polycyclic volatiles in these and a number of other plant materials—and some of the plant molecules are absolutely identical to the whale molecules.

These crossover molecules have led to some terminological confusion in the fragrance world. Professionals often refer to plant materials that carry macrocyclics and polycyclics, and the molecules themselves, as "musks" and "ambers," "musk" and "musky" now specifying a warm, tenacious quality, not the original complex animalic smell of musk-deer grains. And "amber" in perfumery has nothing to do with amber proper, the fossilized tree resin; it's short for "resembling an aspect of ambergris." (Fossil resin emits a faint mix of woody sesquiterpenoids and solventy benzene and naphthalene rings.) This fuzziness is nothing new: the *Oxford English Dictionary* lists nearly three dozen animals and plants with "musk" in their names, most of them simply because they have a pronounced or atypical smell (though muskrats do make a dozen musklike macrocyclics).

We've already sniffed labdanum and galbanum (see page 461), as well as cured tobacco (see page 448), which carries both several macrocyclic musks *and* the polycyclic amber naphthofuran of ambergris. **Ambrette** is now the usual name for what Boyle knew as musk-seed, which reflects the olfactory overlap between ambers and musks. It comes from an Asian relative of hibiscus and okra plants whose outer seed coat contains the sixteen-carbon **ambrettolide**, both musky and floral, along with fruity and unusual winey volatiles. And **angelica root**, from a Eurasian plant in the celery family, turns out to contain a half-dozen macrocyclics with between twelve and sixteen carbons, the most important being pentadecanolide, described as delicately animal, musky, and sweet.

**SOME PLANT MATERIALS NOTABLE PRIMARILY
FOR MUSKY OR AMBERY QUALITIES**

Material	Component smells	Molecules
ambrette seed (Asian *Abelmoschus moschatus*)	rich, sweet, floral, musky, fruity, winey	ambrettolide (1-ring C16), farnesyl acetate, pyrazines & pyridines
angelica root (Eurasian *Angelica archangelica*)	sweet, musky, woody	pentadecanolide (1-ring C15), other 1-ring C12–C16 macrocyclics, pinene, carene

There's also a musky material in the plant world that's too ephemeral for perfumery, but magical for the prepared nose. To become prepared, know that in order to understand why even the finest flower extracts don't smell exactly like the living flowers themselves, fragrance chemists in the 1970s took up the challenge of identifying the tiny quantities of volatiles released by living flowers into the nearby air, the "headspace." They eventually developed portable devices for capturing headspace volatiles and carried them all over the world to identify promising new aromatics. In the early 1990s, the pioneering fragrance hunter Roman Kaiser was on an expedition along the Mediterranean coast, and on a shrubby and wooded hill in Liguria had "one of the most impressive olfactory experiences" available in that aromatic region: "a very transparent, resinous, woody, musky scent" carried on the wind. Kaiser followed his nose to a mass of resin on a pine trunk exposed to the sun. He collected the resin's headspace volatiles for much of the afternoon, and sure enough was able to identify several different macrocyclic musks, including the pentadecanolide shared by angelica and tobacco. When he took a second sample around sunrise the next morning, the musks were absent.

Which suggests a field trip for the amateur smell explorer: an afternoon walk in the woods on a warm bright day. Look for conifer trunks or fallen limbs weeping resin, and get up close to scrape and sniff. Sample the fresh, the crusted, then hunt for patches warm from hours in the sun. And marvel at the convergence of Hero Carbon's companionable avatars in fanged Himalayan deer, squid-afflicted whales, the odd root and seed and shrub, and star-struck tree ooze.

New aromatics from chemists

For centuries deer musk was preeminent among perfume materials. Today it's vanishingly scarce. Yet we inhale its special quality every day as it emanates from soaps and shampoos and cosmetics and the laundry dryer. The story of how this came to pass—how singular, rare deer musk begat dozens of common chemical musks—distills the larger story of modern perfumery.

The chemist's nose has always been an important laboratory instrument, and the Musk Saga starts with accidental observations, mostly in Germany, that certain lab procedures involving nitrogen chemicals could create musky smells. As far back as the 1500s, the alchemist Paracelsus noted in his *Archidoxis* that the distilled mixture of an ammonia salt and sulfuric acid smelled like musk, a "noble odor." Two centuries later, Andreas Marggraf noticed "the smell of the strongest musk" when he distilled fossil amber and treated its essential oil with nitric acid. In the 1830s, several chemists repeated Marggraf's procedure with readily available coal tar and copal resins, and the resulting musky materials were put to commercial use in soaps.

Then in 1891, Albert Bauer synthesized the first pure "nitromusks," single benzene rings decorated with nitrous groups like **musk xylene**. Despite their small size, they had the creaturely and persistent qualities of the macrocyclics and polycyclics, and pleasingly sweet, floral, and "powdery" facets as well. Thirty years later, the pioneering perfumer Ernest Beaux included two of Bauer's musks in his iconic fragrance Chanel No. 5—along with tinctures of true deer musk and civet. The nitromusks weren't simply cheap substitutes; they were new materials that expanded the perfumer's range of olfactory effects.

These first artificial musks were the product of chemical reactions that happened to generate musky smells. A serious analysis of musk and civet themselves didn't come until early in the twentieth century, when German chemists were able to isolate their characteristic molecules. For the first time muscone and civetone could be smelled on their own. Then in 1926 a Croatian-Swiss chemist named Leopold Ružička, working for what would become the multinational fragrance company Firmenich, showed muscone and civetone to be macrocyclic rings of more than a dozen carbon atoms: molecular structures that at the time were thought to be impossible. Hence his 1939 Nobel Prize. Having determined their structures, Ružička showed how such rings could be constructed in the lab from simple carbon chains, and thereby how civet and musk volatiles could be manufactured without any animals.

Along the way, Ružička synthesized a structure not actually in deer musk but still musky-smelling, which he named *muscolide*. Initially this appeared to be another example of lab chemistry accidentally approximating the smell of musk, but it turned out to be lab chemistry *anticipating* the discovery of a natural musk. Nearly eighty years later, muscolide was identified in angelica root! Other fragrance chemists before and after Ružička have similarly anticipated the discovery of important natural volatiles, including the violet ionones and the amber naphthofuran of ambergris.

SOME SYNTHETIC VOLATILES
LATER FOUND IN NATURAL FRAGRANCE MATERIALS

Molecule synthesized	Smells	Found later in
anisaldehyde, 1845	acacia ("mimosa") flower, hay	acacia flower, 1903; vanilla, anise, fennel, dill, tarragon, basil, blackcurrant
piperonal (heliotropin), 1869	heliotrope flower, vanilla, cherry	vanilla, 1905; violet, blueberry, black pepper, tobacco
methyl anthranilate, 1887	fruity, floral	neroli, 1895; grape, strawberry
ionones, 1893	violet	violet flower, 1972
g-undecalactone, 1905	peach	gardenia, 1993
muscolide, 1928	musky	angelica, 2004
amber naphthofuran (Ambrox), 1950	ambergris	ambergris, 1990; labdanum, tobacco, sunstruck pine resin
methyl dihydrojasmonate (Hedione), c. 1960	fresh, floral, "transparent," "overripe lemons"	black tea, 1974; jasmine, osmanthus, lima sweet orange

Fast-forward to the present: chemists have since found dozens of macrocyclic and polycyclic molecules in plants and animals, and they've synthesized hundreds of new ones. Whereas the current international trade in natural musk grains and civet paste is measured in pounds per year, thousands of *tons* of manufactured musk molecules are poured yearly into personal and household products of all kinds. Most of them are used for their fixative, gradual-release properties and non-specific creaturely presence. Some of the synthetics, the "white musks," are used for a "hot-iron-on-fresh-laundry" quality—the opposite of the original musk's underside-of-wild-mountain-creature quality! The rare, strange, ambivalent animal material became a stepping-stone to the ubiquitous reassurance of domesticity and cleanliness.

The Musk Saga reflects the development of fragrance chemistry in general. In 1803, a German chemist named Kindt treated common turpentine with hydrochloric acid and obtained crystals of camphor, at the time an exotic import. In ensuing decades, other chemists synthesized cinnamon's cinnamaldehyde, the sweet-smelling coumarin of hay and tonka, vanilla's vanillin. Soon they were coaxing coal tar and other cheap raw materials into yielding many known flavor and fragrance molecules, but also novelty molecules that happened to simulate the smells of such things as heliotrope and mimosa flowers and leather. Since then, many avatars of Hero Carbon not known in nature have brought new possibilities to the perfumer's palette, among them the unextractable "watery" delicacy of lily of the valley, seashore freshness, and the particular sweetness of cotton candy. And fragrance biotechnologists are engineering microbe and plant cells to churn out valuable volatiles in vats. Fragrance chemistry is big business, with a dominant handful of multinational companies inventing, patenting, and trademarking new "captive" molecules every year.

SOME SYNTHETIC FRAGRANCE MOLECULES

Molecule	Smells
musks: nitro-, macrocyclic, polycyclic, linear, incl. **Galaxolide**, 1890s–	persistent, musky, freshly washed/pressed fabric
quinolines, c. 1880s–	leather
hydroxycitronellal, c. 1905	lily of the valley, sweet, waxy, green
Lilial, 1956 (butylphenyl methylpropional)	lily of the valley, watery, green
Calone, 1966 (methyl benzodioxepinone)	fresh, watermelon, marine, ozone
ethylmaltol, 1968	cotton candy, crème caramel
Vertofix, 1972 (methyl cedryl ketone)	woody, amber
Iso E Super, 1974 (octahydrotetramethyl acetonaphthone)	cedar, violet, amber

The success of synthetic fragrance chemistry brought both benefits and problems. Like natural molecules, non-naturals can be irritating, allergenic, or even toxic. Some are unusually stable, so when they're poured by the ton into care products, they end up accumulating in the environment and in animals, us included. At the same time, apart from their economic advantages, synthetic materials can evoke natural materials and ambient smells that actual extracts can't.

They can provide more reliable, humane, and sustainable approximations of rare and endangered sources. They're sometimes less irritating or allergenic than their more complex natural counterparts. And ironically, a judicious application of synthetics and highest-tech extracts can make it possible for a perfumer to be truer to the smells of a natural living material—a rose in the garden, a violet in the woods—than a time-honored essential oil can be.

Composing with volatiles: perfume notes

At last we arrive at perfumes themselves! Perfumers pluck aromatics from all over the planet, from suppliers low-tech and high-, and play freely with their disembodied extracts to make chimeras unknown in nature: of rose and jasmine, rose and vetiver root, rose and beaver, rose and a dozen other things—or several dozen. Most of these compositions evaporate from the world without leaving much of a trace, but connoisseurs recognize some as significant and influential achievements in the perfumer's art. We'll take a virtual sniff at just a handful; I've selected the details from the analyses of Philip Kraft, Robert Calkin and J. Stephan Jellinek, and the Fragrance Foundation.

First a brief description of perfume compositions in general. They're complex! A single plant extract may carry dozens to hundreds of volatile molecules, and perfumers may in turn blend dozens of these extracts together into a single fragrance. It's likely that no one, not even professionals, can identify each ingredient in a complex blend. Our perception of so many molecules is necessarily impressionistic. But it has some structure imposed on it by their different volatilities, by how quick they are to fly into the air and become depleted on the skin. In the perfumer's argot, "top" or "head" notes are immediately prominent but fade in less than an hour, while "middle" or "heart" notes persist for an hour or so, and "base" notes for hours. The initial proportions of these different notes, changes in their proportions as they evaporate at different rates, our exposure and adaptation to them, our attentiveness and active searching for components we expect: all of these factors can bring different qualities of a perfume into relief from moment to moment.

The volatility of perfume components depends on how large and heavy their molecules are and how strongly their chains and side groups are attracted to the skin's proteins and oils and to each other. Common head notes are the citrusy, pine, floral, fresh smells from ten-carbon monoterpenoids. Heart notes include woody smells from the larger fifteen-carbon sequiterpenoids, floral smells from terpenoid fragments, and fruity and balsamic smells from esters and benzenoid rings. Persistent and fixative base notes include woody and incense smells from sesquiterpenoids and olibanic acids, smoky and animalic smells from small but sticky phenolic rings, and the creaturely, enveloping smells of musks and ambers.

Some landmark perfumes

Modern perfume compositions have their roots in seventeenth-century France, when Louis XIV's court began to abandon the heavy and exotic musk-civet-ambergris oils of the day for perfumes scented with herbs and flowers and citrus fruits from southern Europe. One of them survives to this day and lent its name to an entire category of citrus-herbal compositions: **Eau de Cologne**, the *eau de vie*–based creation of an Italian distiller working in Germany around 1700. (Today, cologne can simply mean a dilute citrusy fragrance.) During the nineteenth century, the French essential-oil industry became established around Grasse in Provence, high-proof alcohol and other solvents came into use, and pharmacist-chemists throughout Europe manufactured volatile materials for goods of all kinds. The luxury perfume industry took shape, centered in Paris and on the creations of individual perfumers, composers in fragrance who from the beginning took advantage of the range of materials available to them, both extracted and manufactured.

The first notable perfume of the modern era is said to be Paul Parquet's **Fougère Royale**, "Royal Fern," of 1882. It was original and influential in two ways. Ferns were not an established fragrance material, so it was a "fantasy" or "abstract" composition. It juxtaposed cologne-like citrus and lavender with a vegetal earthiness that Parquet evoked with oakmoss extract and with synthetic coumarin: the first use of a synthetic volatile in a "fine" perfume. The composition was well received and gave its name to an entire genre of perfumes, the *fougères*. Parquet also anticipated what the gas chromatograph would reveal in 2010: coumarin is indeed an important volatile in many ferns!

Just a few years after Fougère Royale came Aimé Guerlain's **Jicky**, the oldest perfume in continuous production (with modifications) to this day, and one of the most influential for the range and number of its ingredients: three citrus extracts, four florals, six herbs, two each of woods and roots, plus civet, plus synthetic coumarin, vanillin, and nitromusks. The overall effect was a pyramid-like layering of fresh top notes, floral and woody heart notes, and animalic, sweet base notes, the structure by which many fine perfumes would be designed for the next century.

Then in 1921 arrived the renowned **Chanel No. 5**, composed by Ernest Beaux for the fashion designer Coco Chanel. Its essence is structured luxury: at the top, bergamot and rosewood; at the heart, sandalwood and a bouquet of rose, jasmine, ylang-ylang, and synthetic violet and lily of the valley; at the base, musk *and* civet *and* ambergris reinforced by synthetic coumarin *and* vanillin *and* nitromusks. Among perfumers it's also known for a large dose of synthetic medium-chain aldehydes, which on their own have waxy, citrus-peel qualities, but in the perfume are said to balance the rich florality: an effect that Beaux later wrote was inspired by his wartime experience of summer above the Arctic Circle, "when the lakes and rivers release a perfume of extreme freshness."

SOME LANDMARK PERFUMES

Perfume	Principal ingredients	Significance
Eau de Cologne, 1706	citrus & herb oils	defined citrus-herbal family
Fougère Royale, 1882	citrus, lavender, coumarin, oakmoss	defined a new "abstract" family
Jicky, 1889	bergamot, rosewood (linalool), lavender, sandalwood, heliotropin, coumarin, vanillin, civet	complex & structured composition
Chanel No. 5, 1921	bergamot, rosewood; rose, jasmine, methyl ionone, ylang-ylang, hydroxycitronellal, sandalwood; musk, nitromusks, ambergris, civet, vanillin, coumarin, C10–C12 aldehydes	complex, rich, luxurious; large dose of synthetic aldehydes
Eau Sauvage, 1966	bergamot, lemon, orange, Hedione, herb oils, citral, indole, oakmoss, vetiver, coumarin	introduces fresh "transparent" note from Hedione to eau de Cologne

continued

Perfume	Principal ingredients	Significance
White Linen, 1978	Hedione, hydroxycitronellal, nerol, Galaxolide, Vertofix, vetiver, aldehydes	mostly synthetics; dominant musk & aldehydes for fresh laundry effect
Trésor, 1990	Hedione, methyl ionone, heliotropin, Galaxolide, Iso E Super, vanillin	large & block proportions of synthetics; "hug me" combination
Kenzo pour Homme, 1991	bergamot, Calone, clary sage, juniper, cedar, vetiver, white musks	popularizes marine trend with synthetic Calone
Angel, 1992	bergamot, mandarin, patchouli, coumarin, vanillin, ethylmaltol	initiates gourmand trend with synthetic ethylmaltol

The next few decades brought a number of similarly complex compositions, including variations on citrus-labdanum-oakmoss "chypres" (from François Coty's **Chypre**, 1917) and incense-vanilla-based "orientals" (Jacques Guerlain's **Shalimar**, 1925), as well as the practice of using synthetic molecules to define a fragrance, and expensive natural extracts to enrich it with their complexity. With the 1960s and new waves—*nouvelles vagues*—rolling through the cinema and novels and French cuisine, perfumers also explored new materials and compositional structures, less layered and evolving less on the skin, more "blocky" and consistent. Edmond Roudnitska's **Eau Sauvage**, "Wild Water," added to an eau de Cologne citrus-herb mix the newly discovered Hedione and its fresh, bright, "transparent" volatility, which soon became a standard ingredient in modern perfumes. In the seventies, Sophia Grojsman's **White Linen** emphasized the pressed-laundry qualities of the synthetic musks, abetted by Hedione and aldehydes, while her 1990 **Trésor**, "Treasure," brought together large proportions of just a few synthetic molecules—musky, woody, floral, sweet—to create an influential combination whose paradoxically creaturely quality she described as "hug me"! Christian Mathieu's **Kenzo pour Homme**, "Kenzo for Men," helped popularize the new ambient-smell molecule Calone and the idea of beach and seashore fragrances. And Olivier Cresp's **Angel** helped establish the category of food-inspired "gourmand" perfumes with ethylmaltol, a stronger synthetic variant of natural maltol that's evocative of cotton candy and caramel.

These days the perfume world is faster-moving and more various than ever. It ranges from novelty scents and shopping-mall specials—simple, inexpensive, and formulated to satisfy consumer tastes—to the taste-making scent portfolios of established companies and perfumers, to artisanal or niche creations from independent perfumers following their own noses, scouring the globe for the best materials no matter what the cost, and producing limited editions for a select few. At the Osmothèque fragrance conservatory and museum in Versailles, visitors can smell originals or reconstructions of hundreds of perfumes, new and old. And of course there are non-European traditions and scents to get to know. Those of India, for example, where in the venerable fragrance center of Kannauj not far from the Taj Mahal, **mitti attar** is distilled from cakes of sun-baked clay directly into sandalwood-infused oil: a South Asian elevation of gaiaichor!

So although much of modern life has been deodorized or odorized with factitious freshness, there still are galaxies of glinting scent-filled flacons out there to sniff.

Listening to smells

The modern fragrance world is worth a visit even for smell explorers who aren't especially interested in fragrances. That's because it's home to a lively online subculture of enthusiasts with a collective interest in all things volatile, not just the flowery and musky. Along with reviews of new commercial scents, guides to making fragrances at home, suppliers for distillation rigs and solvents and other paraphernalia, there are also sources for many hundreds of aromatic materials, raw woods and resins and roots, distillates and isolates and synthetics—samples of nearly everything mentioned in this book, from rose oil to skatole. To smell and learn about dozens of them in a couple of hours, make a pilgrimage to the Aftel Archive of Curious Scents in Berkeley, California. I owe much of what I know about fragrance to many hours at perfumer Mandy Aftel's scent "organ," sniffing with her through the notes, trying out chords and discords, and imbibing her fascination with the natural materials they come from. Her Archive is a wonderful distillation of that experience.

The interest in aromatherapy, cannabis, and volatiles in general has also propelled the invention of devices that can be put to use as a sort of low-tech gas chromatograph. There are now small electrical vaporizers that control the temperature of a coin-size platform up to 400°F (200°C), just short of the point at which many organic materials begin to pyrolyze. With one of these advanced censers, you can heat an aromatic gently and progressively, just enough to shake its volatiles loose. Often it's possible to catch component smells as they come and go according to their different volatilities, and experience the composite nature of smells.

I've used a vaporizer to enjoy woods and resins, tobaccos and teas, soil and stones. It's not ideal for everything—frankincense has been found to emit its most distinctive qualities between 480°F and 590°F (250°C and 310°C)—but it can be revelatory. When Roman Kaiser had the opportunity to analyze a priceless sample of agarwood, he took a tiny fifty-milligram piece, heated it to 300°F, the equivalent of a medium-hot kitchen oven, and observed how its smell evolved over the course of twelve minutes. At first it was "sweet-balsamic, woody-floral," then more spicy, with "shades of vanilla and musk," then "deep noble woody," eventually with a "phenolic note of castoreum, sweet vanilla." Somehow the higher heat of a glowing coal orchestrates these notes into the chord that Kaiser describes as "the mother of all scents."

Temperature-controlled censing is the electronic version of a method for appreciating volatiles that's centuries old. Years ago I'd read in Liza Dalby's fine memoir of Japan, *East Wind Melts the Ice*, about a practice known as *monko*, whose name literally means "listening to incense." On my first visit to Japan, my friend Mio Kuriwaki arranged for a listening lesson at a Kyoto shop that has supplied the nearby imperial palace with aromatics for many generations. *Monko* has a long and somewhat obscure history that began in China. Today it's a fiddly but fascinating procedure centered on a teacup-size ceramic container filled with ash. You light a small piece of charcoal, bury the glowing coal shallowly in the ash, then set above it a flake of the mineral mica. You place a tiny chip of agarwood or another aromatic on the mica, cup your hands around the container near your nose, and sniff as volatiles are driven out by the indirect heat, without the distraction of generic smokiness. A *monko* session usually includes several materials, perhaps different grades or origins of agarwood, which you learn to distinguish and appreciate.

The name for this practice, incense *listening*, is an odd one, but in English it works as a usefully incongruous substitute for a word that we don't have. To *hear* a sound is simply to perceive it; to *listen* is to pay attention, to focus and try to comprehend. Listening to a smell means giving a moment and some energy to take it in more than once, to register its qualities and the feelings and thoughts that it stimulates. It can be exhilarating to listen to beautiful or strange aromatics. It's also exhaustingly inward to focus on the invisible and intangible, and rack the memory for precedents or comparables. But it's building a database, a nose and a sensory world, and building is work. It pays off as the world expands.

Chapter 18 · COOKED FOODS

....

> But on days when I was on my good behavior, they would bring out the waffle iron.
> Its rectangle would crush the fire of thorns burning red as the spikes of sword lilies.
> And soon the waffle would be in my smock, warmer to the fingers than to the lips.
> Yes, then indeed I was eating fire, eating its gold, its odor and even its crackling
> while the burning waffle crunched under my teeth.
>
> And it is always like that, through a kind of extra pleasure, like dessert, that fire
> shows its humaneness. It does not confine itself to cooking; it makes things crisp
> and crunchy. It puts the golden crust on the griddle cake; it gives a material form to
> festivity. As far back in time as we can go, gastronomic value has always taken
> precedence over nutritional value; and it is in joy and not in sorrow that humankind
> discovered its spirit. The conquest of the superfluous gives us a greater spiritual
> excitement than the conquest of the necessary. Humankind is a creation of desire,
> not a creation of need.
>
> · Gaston Bachelard, *The Psychoanalysis of Fire*, 1938

Our tour of the world's smells so far: Primeval molecules in interstellar space. The stink of life without oxygen. Chemical defenses and come-hithers of rooted plants. Exhaust fumes of roving animals. Kinky chains in cold-water rovers. Life's remnants in soil and smoke and solvents. Fragrances from mountain deer and chemistry labs. Now we come home to the most familiar and reliably appealing smells of all. Food and drink are the bits of the world that we savor most intimately. We fill the kitchen with their developing aromas, sniff over them at the table, bring them inside ourselves, and breathe their volatiles out for a last whiff even as they become part of us.

Animals eat to acquire the substance and energy of other living things. In several chapters above, we sample many of our favorite foodstuffs when they're alive or freshly harvested and smell of themselves: raw fruits and vegetables, herbs and spices, mushrooms, fish. Before eating them, though, we usually work them over into something else. We break them apart, combine them, heat them, encourage benign microbes to grow in them. These manipulations may have helped give birth to our species, and certainly fed its rise to dominance, because they make raw plant and animal tissues significantly more digestible and nourishing. The initially novel and superfluous practice of manipulating food has been so effective that it now appears to be a biological necessity for us.

Why would food work have caught on with our forebears in the first place? Its nutritional advantages probably wouldn't have been obvious. What *are* immediately obvious are easier chewing, and heightened flavors: stronger tastes and smells from the molecules that cooking both frees up and creates. Experiments have shown that chimpanzees and several other animals often choose unfamiliar cooked foods over familiar raw ones. Perhaps animal sensory systems interpret stronger flavors as signals of more accessible nourishment. Or perhaps strong flavors are more stimulating to the senses and more attractive—along with the crisp and crunchy and golden—simply for that reason. Or both! Nourishment and pleasure, need and desire: not that easy to distinguish, an entanglement both delightful and problematic.

If cooking helped make us human, it did so in part by helping us live more fully as animals. Cooked foods engage our senses and brain with the full range of stimuli that they evolved to detect and interpret, and that settled life has otherwise tended to narrow. The visual cues of colors, shapes, and consistencies; the sound effects of crisp, crunchy, slurpy; the tongue's exquisite sensitivity to gradations of solid and fluid, hot and cold; tastes sweet, sour, salty, savory, and bitter; smells that identify specific foods and their histories—all these perceptions can come into play in a single bite or sip. That moment is capable in turn of triggering associations, memories, thoughts, feelings—sometimes a sense of meaning. A rich encounter with the world and with ourselves, well worth listening to.

There are two main ways to transform raw foodstuffs. One is to ferment them with microbes, the planet's masters of *bio*-alchemy. We'll get to those sharp, heady, funky smells in the next chapter. Here we explore the very different smells of *pyro*-alchemy, cooking with heat. Fire became humane when humans learned to control its intensity, pause its headlong rush to smoke and ash, and explore the temperature range in which pyrolysis can make foodstuffs golden and aromatic.

Pyrolytic kitchen alchemy: odorless sugar, odorous caramel

Back at the beginning of chapter 2, I mentioned the magic of cooking caramel: begin with a single ingredient, white, sweet, odorless sugar crystals, then apply moderate heat, and you end up with a brown fluid that tastes sweet and sour and bitter all at once, in a kitchen filled with its landmark aroma. Having sniffed all around the planet since then, we can better appreciate how rich a representation of Hero Carbon's virtuosity that aroma is.

Table sugar is sucrose, a linked pair of decorated six-cornered carbon-oxygen rings. When heat energy breaks it apart slowly and gradually, not in a quick puff of smoke, it forms many different fragments. A few single rings are primarily responsible for the characteristic caramel smell. But they're accompanied by fragments that we've sniffed before in berries, peaches, coconuts, pineapple, citrus fruits, tomatoes, mushrooms, nuts, vanilla—and in hayfields and fallen leaves, in horse stables, in smoke. A lightly pyrolyzed golden caramel is fruity and buttery, a dark caramel more spicy and smoky.

SOME SMELLS OF CARAMEL

Component smells	Molecules	Also found in
caramel, cotton candy, fruity	maltol	raspberry, strawberry, katsura leaves (see page 346)
caramel, peanut	dihydromaltol	fermented milks, wine
caramel	dihydroxydimethyl furanone (diacetylformoin)	—
caramel, buttery	diacetyl	butter, cream
caramel, buttery	acetylpropionyl (pentanedione)	nuts, yogurt

Component smells	Molecules	Also found in
hay, woody, bready	furfural	sun-dried vegetation
sweet, almond	methyl furfural	berries, tomato
fatty, animal	hydroxymethyl furfural (HMF)	honey, pineapple, vanilla, sun-dried vegetation
sweet, hay, roasted	methyl furanone (angelica lactone)	bread, coffee, popcorn
coconut, creamy, fruity	g-nonalactone	coconut, peach
fruity, creamy, peach	g-decalactone	butter, peach, coconut
vanilla	vanillin	vanilla
smoky	guaiacol	smoke
barnyard, tar	cresol	tar, horse excreta
vinegar, sour	acetic acid	vinegar
cheesy, sweaty, sour	butyric, methylbutyric, octanoic & decanoic acids	fermented milks
floral, citrus peel	octanal	citrus peels
mushroomy	octenol	mushrooms
fruity	ethyl pentanoate (ester)	strawberry, apple, pineapple

All this—and plenty more—from essentially one molecule that isn't volatile itself! Caramel epitomizes the chemical creativity of cooking, the way it can transcend its ingredients and fill them to overflowing with volatile traces of the world at large.

The cook's plant and animal ingredients typically include thousands of different molecules, some with smells of their own, most capable of breaking into smellable fragments when they're hit with enough heat energy. Flavor chemists have documented dozens to thousands of volatiles in individual cooked foods. Fortunately, it seems that only a small fraction of them contribute significantly to the smells that we actually perceive.

Some of the most rigorous research on food aromas comes from a research group that's now part of the Leibniz Institute for Food Systems Biology near Munich. In 2014,

Peter Schieberle, Thomas Hofmann, and colleagues reviewed several decades of work and reported that only 230 molecules account for the majority of important volatiles in a wide range of foods, and that a given food aroma can be reasonably simulated with around a dozen of these key molecules in specific proportions. This "combinatorial code" might not include nuances that a connoisseur would notice, but it communicates a food's basic aromatic identity, just as a recognizable sketch does for a face. We'll encounter these key volatiles often as we sniff our way through the world of cooked and fermented foods.

Apart from the herbs and spices that we use specifically for their aromas, our raw ingredients consist mostly of nonvolatile, smell-less molecules. When these proteins, sugars, starches, fats, and membrane lipids are dismantled by the cook's judicious application of heat, all contribute to pleasing cooked aromas. They tend to generate common sets of volatiles, which I'll call "bouquets." These four bunches of related volatiles help define the generic "cooked" smell of most foods prepared with heat.

Cooked bouquets: fatty, sulfurous, sweet, nutty

Of the four main bouquets of cooked volatiles, the most common is also the least obvious. I call it the **fatty bouquet** because it arises from fats and oils and cell-membrane lipids. These long, nonvolatile carbon-hydrogen chains are the food molecules most vulnerable to alteration by oxygen and water and heat, even mild, nonpyrolytic cooking temperatures. Because all living things have membranes, all foods produce a fatty bouquet, even vegetables and fruits. It's a broad mix of starter-set molecules and smells that gives a kind of hazy volatile background, a generic breath of cooked-ness, sometimes with more specific facets. Chickens and other birds generate fatty bouquets dominated by the ten-carbon decadienal, whose smell is usually described as "deep-fried" because it's a key volatile produced by common frying oils manufactured from seeds. Certain oils whose carbon chains have more kinks generate fragments similar to key volatiles in sea creatures, so they smell fishy. The fats of animal milk and meats happen to generate the same lactones that give coconut and stone fruits their sweet-creamy qualities.

A FATTY BOUQUET OF COOKED SMELLS

Smells	Molecules	Favored by
cooked, fatty (a composite of green, mushroomy, cucumber-melon, fishy, metallic, fatty, waxy)	mix of C2–C10 alcohols, aldehydes, ketones (e.g., acetaldehyde, ethanol, hexanol, hexanal, octanal, octenol)	most foods
fried	decadienals	corn, cottonseed, sunflower, grapeseed oils; poultry
fishy	mix of C7, C8, C9 chains with double bonds (e.g., heptenal, heptadienal, octenone, octadienone, nonadienal)	seafoods; soy, canola, flaxseed (linseed) oils
coconut, creamy	lactones	animal fats

The other three cooked bouquets come from more rugged food molecules that are less vulnerable to attack by oxygen and water. They're mostly generated at boiling temperatures and above, when pyrolysis reactions among the chains themselves become significant. The reactions that generate the most volatiles involve sugars and the amino-acid building blocks of proteins. Their initial encounter with each other triggers a complex reaction cascade, the products of one stage successively reacting to form new sets of products and an increasingly rich aroma. At late stages of the cascade and temperatures well above the boil, they produce large aggregate molecules that give the typical brown color of toasted and roasted foods, so the entire cascade is often somewhat misleadingly referred to as the **browning reactions**. They're also known as the **Maillard reactions**, after the French chemist who first noted them a century ago, Louis-Camille Maillard.

One cooked bouquet produced in the early stages of Maillard pyrolysis is marked by the presence of sulfur, an element that all plants and animals carry in amino acids and various other molecules. The **sulfurous bouquet** contributes to the aroma of many boiled foods and is dominated by small volatiles familiar from hot springs and the ocean air, as well as less salubrious sources. Dimethyl sulfide, which is also emitted by saltwater creatures (see page 378) and so carries that humid, oceanic suggestion, is an important component of the smells of heated milk and many boiled vegetables (DMS production can be reduced by cooking vegetables in acidified water). Hydrogen sulfide and methanethiol can be unpleasantly sulfidic and putrid in other contexts, but the trace amounts found in cooked foods usually just add an appealing depth to the overall smell. Several disulfides and trisulfides are important notes in cooked onions and roasted meats, and the sulfur-containing aldehyde **methional** gives the interior of a freshly boiled potato its distinctive character.

A SULFUROUS BOUQUET OF COOKED SMELLS

Smells	Molecules	Also found in
boiled potato	methional (methyl thiopropanal)	persimmon, black truffles, rye
ocean air, boiled vegetables or milk	dimethyl sulfide, disulfide & trisulfide	ocean air, seaweeds, truffles, durian
boiled egg	hydrogen sulfide	hot springs, swamps, sewage
rotting vegetables	methanethiol	swamps, sewage

There's also a **sweet bouquet** of Maillard volatiles, which develops at temperatures high enough to turn foods light brown, usually during baking or frying. It consists mainly of carbon-oxygen rings altered from their carbohydrate originals, "sweet" in quality because some are defining components of browned-sugar caramel, and some, like **maltol**, are the same molecules made by plants to signal fruit ripeness. The furanones sotolon and furaneol, familiar from fenugreek and strawberries, are at the heart of the sweet bouquet. The carbon-oxygen chain diacetyl is a key volatile in butter, and phenolic-ring phenylacetaldehyde suggests honey.

A SWEET BOUQUET OF COOKED SMELLS

Smells	Molecules	Also found in raw
sweet, caramel	maltol	woods, berries, honey
sweet, woody, bready, brown	furfural, furan aldehyde	fruits, vanilla, tonka, hay
fenugreek, maple	sotolon (caramel furanone)	fenugreek, candy cap mushroom

Smells	Molecules	Also found in raw
fruity, caramel	furaneol (strawberry furanone)	strawberry, pineapple, mango, many fruits
buttery, sweet, caramel	diacetyl	strawberry, banana, butter
honey	phenylacetaldehyde	flowers, honey

The **nutty bouquet** of Maillard reaction products appears at temperatures high enough to turn foods dark brown, during toasting and roasting and grilling. These volatiles are mainly carbon rings that include nitrogen atoms, sulfur atoms, or both: drastic rearrangements of their original protein and carbohydrate sources. They come in many different configurations—pyrrolidines, thiophenes, thiazoles, thiolanes, others—but the pyrrolines and pyrazines are by far the most common. Toasty-roasty **alkyl pyrazines** trade the methoxy decoration of the vegetal pyrazines for simple methyl and ethyl groups.

A NUTTY BOUQUET OF COOKED SMELLS

Smells	Molecules	Also found in raw
nutty, toasty, popcorn, sweet	pyrrolines (C4, N1)	rice, pandan leaves
toasty, roasted, nutty, earthy	pyrazines (C4, N2)	potatoes
malty, cocoa, nutty, sweaty	methylbutanal	tomato, truffle

There's one especially important volatile in the nutty bouquet that figures in the smells of many more gently cooked foods: it's even formed during fermentations and in ripening tomatoes. The small branched-chain aldehyde **methylbutanal** includes neither sulfur nor nitrogen, and it breaks off from particular amino acids at much lower temperatures than the Maillard rings form. It's usually described as smelling like cocoa or barley malt (barley grain partly germinated, then toasted for making into a sweet syrup or beer), and sometimes it has a hint of the cheesy or sweaty quality of its cousin, methylbutyric acid.

Smells of cooking methods: boiled, toasted, grilled, smoked, fried

Cooking methods fall into two basic groups that apply very different temperatures to food and mark it with very different aromas. **Moist cooking methods** heat foods with liquid water or steam, which raise the food surface to the ordinary boiling point of water and no higher; pressure cooking gets a few tens of degrees above that. So the smells generated by **boiling, simmering, braising, steaming**, and **pressure cooking** are dominated by the food's intrinsic volatiles, and by the fatty and sulfurous bouquets generated by fragmentation of lipids and proteins. The smells of aromatic vegetables are intensified and rounded out; the smells of relatively non-aromatic raw meats become mainly fatty and sulfurous.

Dry cooking methods reach temperatures hundreds of degrees hotter than water or steam by means of heated oven air, infrared radiation from oven walls or glowing electrical elements or coals or flames, or hot oil or fat. They're able to evaporate much of the moisture at a food's surface and accelerate the pyrolysis cascade. Pyrolysis generates both small volatile molecules and large nonvolatile clusters of carbon rings that absorb light and color the food surface—light brown at first, darker as the reactions proceed, brown-black if the surface is charred. The surface color is a good indicator of flavor development and is even associated with the smells of some volatiles: the toasty, roasty furan, pyrroline, and pyrazine rings are often described as smelling "brown." Brown colors and flavors can also develop at lower temperatures in foods with substantial amounts of water—meat stocks and the grain mash for beer making are two examples—but only over long cooking times.

SOME CHARACTERISTIC SMELLS OF COOKING METHODS

Cooking methods	Component smells	Molecules
boiling, simmering, steaming	mild, cooked, sulfurous, sweet	aldehydes, sulfides
baking, toasting, roasting	sweet, toasted, roasted, buttery	furans, furanones, acetyl pyrroline, pyrazines, diacetyl

Cooking methods	Component smells	Molecules
frying	mild, fatty, cooked, sulfurous, sweet, toasted	aldehydes, sulfides, thiols, furans, acetyl pyrroline, pyrazines
grilling	sweet, roasted, charred	furans, furanones, pyrazines, guaiacol, cresols, skatole

The smells created by dry **baking, toasting, and roasting** are dominated by the sweet smells of carbohydrate breakdown and the rings of the nutty bouquet. Bread crusts and toast, toasted nuts, and the skin and crust of roasted chicken and beef are dominated by carbon-nitrogen rings. The very high temperatures of **grilling and broiling** can carry the pyrolysis of the food surface far enough to char it, dismantle the furans and pyrrolines and pyrazines, and produce the smoky and tarry breakdown products typical of smoldering wood (see page 411). Glowing **charcoal** emits its own petrochemical-like bouquet of further stripped-down BTEX carbon rings (see page 435) along with formaldehyde and other short-chain aldehydes, and sometimes ammonia. Food surfaces rich in nitrogen-containing proteins can develop tarry cresol and fecal skatole, which can contribute noticeably to the aroma of grilled meats and fish. **Smoking** applies pyrolysis volatiles to foods indirectly with the fumes of smoldering wood, so the foods themselves can be cooked gently. Some woods are famously better than others at producing aromatic rather than acrid fumes, yet surprisingly little is known about the chemical details; hickory wood apparently produces unusually abundant pyrazines. Chinese **tea-smoked** duck is made by scorching a mixture of tea leaves, sometimes twigs or sawdust but usually rice or flour, along with sugar and spices: the sugar contributes caramel-like furans and furanones; tea leaves, the phenol-related smoky and tarry rings; flour, sweet maltol and toasty pyrazines; and the whole mixture a couple of fruity branched chains. In Thailand, some desserts are flavored by being enclosed with a **scented candle,** *tian op,* whose wick smolders and fills the container with smoke. Here the key is the composition of the beeswax candle, which in addition to its long fragmentable chains is aromatized with incense materials like benzoin, frankincense or sandalwood, citrus peels, herbs, spices, essential oils, even musk!

SOME SMOKES APPLIED TO FOOD

Smoke source	Component smells	Molecules
woods	buttery, bready, caramel, smoky, clove, vanilla, tarry, roasted	diacetyl, furans, furanones, syringols, guaiacols, eugenol, vanillin, cresols, pyrazines
black tea + sugar + flour	sweet, bready, caramel, fruity, tarry, toasty	furans, furanones, maltol, methylbutyl acetate, heptanone, cresols, pyrazines
Thai scented candle, *tian op*	melon, citrus, orange, peel, waxy, smoky, vanilla, incense	heptanal, octanal, nonanal, decanal, octenal, guaiacol, vanillin, santalols

Uniquely among common cooking methods, **frying** heats foodstuffs with a foodstuff. We coat or immerse the food in a plant oil or animal fat, which unlike water can be heated well above the boiling point, thus boil moisture out of food surfaces, and encourage the reactions that turn the surfaces brown and flavorful. The long carbon chains of the oils and fats are also vulnerable to breakdown, and it's their fragments, mostly aldehydes, that give fried foods and kitchen air that characteristic smell. The decadienals seem to be key to the distinctive delicious-ness of fried foods, perhaps because they're rare enough in nature to stand out from the general fog of green-mushroomy-fruity-waxy aldehyde fragments. They're among the main fragments produced by linoleic acid, of which **safflower, sun-flower, grapeseed**, and **corn** oils have the highest proportions and so generate the strongest fatty-fried smell.

Appetizing though many of these aldehydes are, they're well known to be toxic to airway cells, and they contribute to respiratory diseases common among kitchen workers. This problem has been of special concern in China, where even at home foods may be stir-fried every day with temperatures and techniques that cause airborne oil droplets to combust spectacularly above the wok. The **"breath of the wok,"** *wok hei*, is often celebrated as the signature aroma of excellent stir-frying, but frequent exposure to the oil droplets and high levels of volatiles it generates—among them formaldehyde and other aldehydes, benzene and toluene—is thought to be responsible for the relatively high incidence of lung cancer among Chinese women.

SOME SMELLS OF HOT OILS AND FATS

Component smells	Molecules	Favored by
deep-fried	decadienals, heptenal, octenal, octadienal, nonadienals	safflower, sunflower, grapeseed, corn oils
plastic, fruity	nonenals, decenals	high-oleic soy, canola, olive oils
painty, fishy	heptenal, heptadienal, octenone, octadienone, nonadienal	soy, canola oils
acrid, choking	propenal (acrolein)	smoking-hot oils
scorched, acrid, choking, solvent	formaldehyde, acetaldehyde, benzene, toluene	oil droplet combustion

Aromatic plant oils and animal fats

Some cooking fats and oils are aromatic in their own right and flavor food surfaces as much as the generic fried aldehydes. Two are often manufactured from toasted dry seeds and carry signature toasted, roasted pyrazines. **Peanut oil**, from a South American member of the bean family, brings together pleasant fried decadienals with sweet, fruity, and vanilla notes. **Sesame oil**, from a seed native to India and possibly the first ever cultivated for its oil, is especially favored in Asia, the Middle East, and Africa. It's sold in both mild and more strongly roasted forms, the latter (like toasted sesame seeds) with an array of coffee-like, meaty sulfur volatiles and smoky guaiacol.

SOME AROMATIC COOKING OILS

Oil or fat	Component smells	Molecules
peanut (toasted)	fatty, fried, green, roasty, sweet, cooked apple, vanilla	nonenal, decadienals, pyrazines, ethyl methylbutyrate, damascenone, vanillin
sesame (toasted)	meaty, coffee, sulfurous, smoky, honey	furans, pyrazines, thiazoles, thiazolines, furfurylthiol, guaiacol, phenylacetaldehyde
olive, unrefined "extra virgin"	green apple, green leaves, grassy, fruity, tomato	hexenals, hexadienal, hexanal, hexenols, hexenyl acetate, pentenone

Olive oil comes not from dry seeds but from the fresh small fruits of a Mediterranean tree; it has been used from the times of ancient civilizations in lamps and skin care as well as cooking. Olives are richest in oil when they're ripe and purple-black and carry esters common to many ripe fruits. But at that stage they're soft, prone to damage and attack by microbes that can lend musty, winey-vinegary, and overripe or putrid smells to oils. Today, most high-quality olive oils are milled from green fruits that are just beginning to turn color—as Cato and Pliny observed in ancient times! Crushing causes the fruits' defensive enzymes to generate a dozen permutations of green-leaf aldehydes and alcohols, and some esters. That barrage of volatiles is trapped in the oils, which when fresh can provide an unparalleled study in green-leaf nuances: the notes range from cut grass to crushed leaves to tomato leaves, artichokes, green apples, green bananas, and immature almonds or walnuts. These notes gradually fade during storage, and they're quickly driven off during cooking, replaced by more generic fatty and fried bouquets.

Two other distinctive plant oils come from palm trees and evoke tropical cuisines. **Palm oil** is produced mainly from trees native to Africa and flavors dishes in regions of Brazil with historical ties to the slave trade. It's extracted from the palm's fruit pulp (less flavorful palm kernel oil comes from the seeds), and when unrefined has a deep red-orange color from carotenoid pigments. Its remarkable scent comes from fragments of those pigments that suggest wood and flowers and saffron and tobacco; in the fifteenth century, the Venetian traveler Alvise Cadamosto recorded its "pleasant flavor of violets."

SOME TROPICAL COOKING OILS

Oil or fat	Component smells	Molecules
palm, unrefined	woody, floral, sweet, resinous, saffron, tobacco	trimethyl cyclohexanone, trimethyl cyclohexenone, cyclocitral, ionol, linalool
coconut	coconut, creamy, fatty, sweet, nutty	octa-, deca-, dodeca-lactones, hexanal, nonanal, undecanone

Landmark **coconut oil** is extracted from the large fruit of an Asian palm and is prominent in South Asian cooking, in Western sweet dishes—and in sunscreens and tanning lotions. Its instantly recognizable smell is uniquely dominated by lactones (see page 280). They're present in the raw coconut flesh and its oil, and become increasingly abundant when the oil is heated. Most of them, and the smell of coconut itself, are described as having a sweet, fatty, creamy quality. In fact, the name *lactone* comes from the Latin word for "milk," whose fat-enriched cream also carries these molecules. A few other fruits, including pineapples and peaches, emit lactones along with the more typical fruit-ripening esters.

Sweet-creamy lactones are also a part of the appeal of various animal fats used in frying. We don't really know what lactones are doing in animal fats, but it seems to be a biochemical fluke that just happens to echo the plant ripeness signals. **Chicken fat** and **duck fat**, with abundant kinked carbon chains, both lead with pleasant fried-smelling decadienal, chicken fat supporting it with lactones, duck fat with mushroomy eight-carbon chains. **Lard**, rendered from the less kinked body fat of pigs, combines medium-length broken-chain aldehydes, waxy and citrus-peely, with mushroomy octenol and a coconutty lactone. **Beef tallow**, rendered from the even less kinked body fat of cows and steers, has the same waxy, citrus-peel notes as well as coconutty and peachy lactones, and long heating releases a signature beef-fat note, the branched thirteen-carbon aldehyde methyl tridecanal.

SOME ANIMAL FATS

Oil or fat	Component smells	Molecules
chicken fat	fried, sweet, fatty, fruity	decadienal, undecenal, dodecalactone, decalactone, nonenal, nonadienal
duck fat	fried, green, waxy, citrus, mushroomy, almond	decadienal, hexanal, nonanal, octenol & octenal, heptadienal, benzaldehyde
pork lard	waxy, citrus peel, green, mushroomy, coconut, creamy	decanal, heptenal, nonanal, hexanal, octenol, g-octalactone
beef tallow	waxy, citrus peel, floral, fried, green, sweet, creamy, coconut, peach, tallowy	decenal, undecenal, decadienal, nonanal, hexanal, nonenal, g-hexa-, octa-, nona-, undeca-lactones, methyl tridecanal
butter	butter, coconut, creamy, mushroomy, cowy, cheesy, barnyardy	diacetyl, d-decalactone, octenone, butyric acid, skatole
butter, heated	sweet, caramel, fried, mushroomy, coconut, creamy, fruity, barnyardy	furaneol, decadienal, octenone, d-octa-, deca-, dodeca-lactones, skatole
ghee	sweet, caramel, creamy, coconut, fruity	hydroxymethyl furfural, furanmethanol, maltol, d-hexa-, octa-, deca-, dodeca-lactones

In a category of its own is cow's-milk **butter**, which gets its aromas from a mild fermentation of the cream it's churned from (see page 564), along with the traces of proteins and sugars in residual droplets of milk. Fermentation microbes elevate the four-carbon diacetyl into its landmark buttery volatile, accompanied by multiple lactones, mushroomy octenone, and cheesy butyric acid. **Clarified butter**, heated to separate the fat molecules from the watery milk droplets, loses much of its diacetyl but gains sweet, fruity, coconutty rings. In the making of Indian **ghee**, the heating is prolonged to brown the milk solids, which generates a stronger caramel aroma.

Versatile reserves: milks, creams, eggs

We'll start our survey of cooked foods with some of the blandest and most versatile, including butter's parent. Milk and eggs are remarkable ingredients, first foods that nourish newborn mammals and unborn avians with concentrated protein and energy, and help the cook construct dishes that can be velvety or cakey or mostly air.

Milks are mixtures of myriad proteins, sugars, and several hundred different carbon chains, most of the latter bound up in fat molecules. But there are also volatile traces of the individual chains, and these can be amplified by the fat-breaking activity of enzymes in the milk and those carried by microbes. Fresh from the animal, milks have distinctive smells that vary according to the species and its feed. In common cow's milk, fat fragments and a couple of esters provide sweet, fruity, mushroomy notes, while a sulfide and nitrogen-containing indole come from proteins and fill in more animal qualities. **Pasteurized milk** has been heated at 140°F to 160°F (60°C to 70°C) to eliminate potentially harmful microbes, a mild cooking that brings a sulfurous oceanic note to the fore, along with coconutty decalactone. **Cream** is a fraction separated from milk with a fivefold higher concentration of fat droplets and the lactones they carry. When used to make rich whipped or ice creams, the aeration and agitation cause some fat breakage and a rise in the levels of cowy butyric and octanoic acids. **Skim milk**, with very little fat, has fewer volatiles and a mild aroma that may be quicker than full-fat milk to reveal the stale, cardboardy notes of oxidation (see page 536).

SOME MILKS AND CREAMS

Food	Component smells	Molecules
milk, cow, raw	fruity, sweet, sulfurous, waxy, mushroomy, musty, cowy	ethyl hexanoate & butyrate, dimethyl sulfone, nonanal, octenol, indole, butyric acid
milk, cow, pasteurized	sulfurous, grassy, waxy, mushroomy, musty, cowy, coconut	dimethyl sulfone, hexanal, nonanal, octenol, indole, butyric acid, d-decalactone

continued

Food	Component smells	Molecules
cream, cow	coconut, creamy, peach, cowy	deca-, dodeca-, dodeceno-lactones, skatole, butyric acid, dimethyl sulfide
milk & cream, cow, UHT	creamy, coconut, peach, sweet, grassy, toasty, vanilla	C8–C10 & C12 lactones, heptenal, acetyl pyrroline & thiazoline, vanillin
milk, sweetened condensed	coconut, creamy, peach, caramel, fruity, plastic	deca-, dodeca-, dodeceno-lactones, furanmethanol, heptanone, trimethylbenzene
milk, cow, nonfat dry	caramel, fenugreek, cowy, potato, fruity, coconut, vanilla, toasty	furaneol, sotolon, butyric acid, methional, aminoacetophenone, d-decalactone, vanillin, acetyl pyrroline & thiazoline
milk, goat, raw	goaty/stable, fecal, medicinal, fatty, floral, honey, vanilla	ethyloctanoic acid, skatole, propylphenol, decanoic acid, phenylacetic acid, vanillin
milk, goat, cooked	floral, honey, goaty/stable, fecal, flowery, coconut, creamy, peach, ocean	phenylacetic acid, ethyloctanoic acid, skatole, g-dodecenolactone, g-undecalactone, d-decalactone, DMS

UHT, or ultra-high-temperature, milks and creams have been heated drastically enough to last for a few months at room temperature. That extreme heat generates abundant lactones, but also toasty-roasty nitrogen and sulfur rings, and vanillin. **Sweetened condensed milk** relies for its stability in part on a large proportion of sugar and isn't as strongly heated or as toasty, though cooking it on its own quickly turns it into a delicious caramel sauce. **Nonfat dry milk**, deprived of its long carbon chains and both heated and dehydrated, develops a host of sweet, fruity, roasty, toasty rings, not much like the original animal fluid.

Goat and sheep milks stand apart thanks to their unusual branched-chain ethyloctanoic acid, which is largely unique to their fats and identified with them. Another important contributor to the goaty-sheepy note is fecal skatole, also a microbial product, and produced at higher levels on a diet of green forage that's relatively high in protein (see page 86). Cooking balances these qualities with floral phenylacetic acid and several nutty and fruity lactones.

Eggs contain the materials that an embryonic bird needs to develop into a chick capable of pecking its way through the shell. The tiny egg cell itself rides on the surface of the yellow yolk, a sphere rich in proteins and long-chain fat and membrane molecules. The surrounding sticky and colorless **albumen** is mainly water and proteins that are notably rich in sulfur-containing amino acids: the main source of the landmark volatile hydrogen sulfide, which provides the dominant smell of just-cooked whole eggs. *Not*, as is often said, microbially spoiled "rotten eggs," in which hydrogen sulfide takes a back seat to a putrid mix of short-chain acids and amines and thiols (intentionally fermented egg is a well-studied ingredient in deer repellents). Hydrogen sulfide is quite volatile and escapes cooked eggs quickly, so their aroma fades. With its diverse mix of raw materials, egg **yolk** develops a complex aroma that includes green, mushroom, and fatty, fried notes.

EGGS, COOKED AND "ROTTEN"

Food	Component smells	Molecules
egg, freshly cooked	sulfidic, rotten vegetable, fresh, vinegar	hydrogen sulfide, methanethiol, acetaldehyde, acetic acid
egg yolk, cooked	potato, green, toasted, mushroomy, honey, fried	methional, heptanal, acetyl pyrroline, octenone, phenylacetaldehyde, decadienal
egg, "rotten" (fermented)	cheesy, urinous, sulfurous, fruity	butyric & caproic acids, trimethylamine & other amines, hydroxyethanethiol, many esters
eggs, cooked & spoiled	rotten vegetable, sulfidic, cocoa, fresh	methanethiol, hydrogen sulfide, methylbutanal, acetaldehyde

Meats and meatiness

Meats, the edible muscles and organs of our fellow animals, fed the emergence of our species and remain among our most widely appealing foods. Even though many people avoid eating meat, today humankind slaughters tens of billions of food animals yearly. Meatiness is defined by a substantial texture, a savory taste, and characteristic aromas. When they analyzed the perfume emitted by beef as it roasts mouthwateringly in the oven, chemists working in Switzerland detected the presence of several *thousand* different carbon chains and rings, and a hundred sulfur molecules. The smell of cooked meat is a rich mix of all the basic cooked bouquets, augmented by a specifically meaty one that reflects the nature of animality itself.

Animals live by pursuing and consuming other forms of life, and the animal body is built largely of proteins, molecules that work actively to move and maintain it. Its tissues are therefore good sources of sulfur-containing amino acids and aromatic sulfurous fragments. Animals also depend on oxygen to generate the chemical energy that powers their muscles and organs, and they have specialized proteins for absorbing oxygen from the air and distributing it in their bodies: hemoglobin in blood and myoglobin in muscle and organs. Within these abundant proteins is a sub-molecule that does the actual oxygen-binding, and that molecule turns out to be largely responsible for the meatiness of cooked flesh.

The oxygen-binding component is *heme*, which gives animal blood and tissues their characteristic red color; the more active and energy-hungry the tissue, the darker it is with heme. At the center of each heme molecule is an atom of iron, an element whose ready sharing of electrons made it a valuable resource in the evolution of early life forms (see page 31) and made versions of heme ubiquitous among living things today. Animal tissues contain so much heme that when they're disrupted by cooking, the suddenly unmanaged reactivity of its iron causes chemical changes galore—including the generation of otherwise rare volatiles that help define meat flavor.

One meaty fragment develops before any cooking and contributes to the "bloody" smell of raw meat, an echo of our own heme-rich blood when we bite a lip or lick a cut. We usually describe the smell as "metallic" because it's similar to the smell left on our fingers when we handle coins, or in the air when we scrub a bare metal pan or sink. But metals themselves aren't volatile: we're actually smelling the fragments created when stray heme iron and other metals help oxygen attack the carbon chains in our cell membranes, skin oils, or dish soap. Polyunsaturated chains, kinked at several points by double bonds between adjacent carbon atoms, are the most susceptible, and we perceive the fragments that include a kink or two as metallic-smelling. In the case of blood and exposed surfaces of raw meat, the key fragment is a ten-carbon, one-kink aldehyde with an oxygen atom riding its middle: an **epoxy decenal**. Our hominid ancestors would have known that molecule and smell long before they paid much attention to rocks and ores, so for much of our prehistory, they may well have experienced metals as bloody-smelling.

SOME BLOODY, METALLIC, LIVERY, GAMY NOTES IN MEATS

Component smells	Molecules
bloody, metallic	epoxy decenal
metallic, mushroom	octenone, nonenone
geranium leaf, metallic	octadienone
metallic	hexenone, heptenone, nonenol

So heme-created epoxy decenal gives raw and rare meats their metallic note. More thorough cooking actually increases its levels, but also helps heme generate many other volatiles that muffle its presence. In addition to the generic cooked and cooking-method bouquets, a specifically **meaty bouquet** arises from unusual sulfurous products of amino acids and the vitamin thiamine. These furan sulfides and **furan thiols** variously evoke roasted meat or roasted onions (and also contribute these notes to coffee, unusual fruits, and cat urine!). For cooks who want to give a meaty flavor to meatless dishes, it's good to know that **yeast extract** made from microscopic fungi is even richer in these and related sulfur volatiles than is **beef extract**.

SOME SMELLS OF MEATS AND EXTRACTS, COOKING AND COOKED

Smell source or quality	Component smells	Molecules
beef while roasting	fatty, vegetal, sulfurous, potato, onion, meaty, roasted	20 aldehydes & ketones; abundant sulfur: methanethiol, methional, dimethyl trisulfide, methyl furanthiol, methyl thiophene, acetyl thiazoline, methyl mercaptopropanol
background cooked bouquets	fatty, vegetal, metallic; sulfurous; caramel, sweet, fruity; roasted, savory	broken-chain bouquet; hydrogen & methyl sulfides, methanethiol, methional; furans, furanones, lactones; pyrazines, thiazoles, thiolanes

continued

Smell source or quality	Component smells	Molecules
meaty	meaty, roasty, oniony	furanthiols, furan sulfides
beef extract	mushroomy, caramel/ fruity, roasty, sweaty	octenol, furaneol, trimethylpyrazine, methylbutyric acid
yeast extract	meaty, roasted, potato, mushroomy, sulfurous, toasted	methyl furanthiol & methyl dithiofuran, methional, octenone, dimethyl trisulfide, thiophenes, pyrazines

Meats from particular animals add their own signature volatiles to the basic meaty bouquet, which in stocks and sauces is generally richer when they're made at a long, slow simmer rather than a quick boil. For **beef** it's an unusual tallowy aldehyde derived from longer chains in its cell membranes; meats from older animals produce more of this methyl tridecanal, as does long stewing. **Pork** is unusually rich in sulfur-containing rings, perhaps because the tissue is rich in thiamine. Its "piggy," pigsty aspect comes from animalic volatiles made by microbes in the animals' guts and stored in their fat, while "boar taint" combines those qualities with the urinous smell of the male hormone androstenone (see page 101); male hogs are often castrated to minimize this. **Lamb and mutton** are marked by a sweaty-smelling branched acid generated by gut microbes and stored in the muscle fat. When animals are fed pasturage relatively rich in protein and lignin and other phenolic rings, the same process deposits barnyardy cresols and skatole, which combine with the methyloctanoic acid in an animalic "pastoral" flavor.

DISTINCTIVE SMELLS OF COOKED BEEF, PORK, LAMB, AND POULTRY

Meat	Distinctive smells	Molecules
beef	tallowy, beefy	methyl tridecanal, butyric acid
pork	sulfur, onion, roasted, porky; piggy	thianes, thiolanes, thiazines; skatole, androstenone, aminoacetophenone
lamb	sheepy, sweaty; animal	ethyloctanoic acid; skatole, cresol
chicken	fried, fatty, meaty, fruity, animal	decadienals, furanthiols, g-deca- & dodeca-lactones, cresol

Meat	Distinctive smells	Molecules
turkey	fried, mushroomy, sweet	decadienals, octenone, pentylthiophene
duck	fried, sweaty, mushroomy, coffee, sulfurous; roasted, caramel (liver)	decadienals, methylbutanal, octenol, furfurylthiol, dimethyl trisulfide; acetyl tetrahydropyridine

Chickens and other small birds work their muscles far less than heavy mammals, so they have a tenth or less the heme content and a mild cooked flavor. Their fatty skin boosts levels of deep-fried decadienal and sweet lactones. **Turkey** meat, little studied, seems to be distinguished in part by mushroomy and metallic octenone. **Ducks** are migratory birds capable of long-distance flying, so their flight-muscle breasts are richly endowed with heme iron and have a stronger, meatier flavor than chicken or turkey.

Of the inner animal organs that are commonly eaten, **hearts** and seed-grinding bird **gizzards** are hardworking muscles with plenty of heme and meaty flavor. Similarly, dark-red **livers** aren't muscles, but they are very active—they perform a host of essential biochemical services—and in fact contain plenty of thiamine, several times as much heme iron as muscle, and as much as ten times more of the highly kinked chains. So liver flavor tends to be strong, with metallic-smelling fragments especially prominent. By contrast, **foie gras**, the enlarged liver of ducks and geese that have been force-fed with grain, is nearly half fat, most of it not highly kinked. All the fat dilutes the heme and thiamine and kinked chains, a dilution reflected in the pale color and muted livery-ness.

Livery-ness is a particular flavor that some people who otherwise love meat can't abide. They find it in venison and other **game meats**—including grouse!—and in "grass-fed" beef. The difference in cooked smells between ordinary liver and foie gras parallels the difference between these meats and typical American **grain-fed beef**. Animals that live on open pastures take in a large proportion of multi-kinked carbon chains from whole plants, but are otherwise lean, so their metallic fragments are prominent. Feedlot cattle are given seeds and concentrates, relatively poor in multi-kinked chains, to speed their growth and fattening. That diet dilutes the multi-kinked chains in the animals' tissues, and in the cooked meat covers the metallic fragments with fatty, waxy, tallowy fragments and lactones. In Europe and South America, where pastured beef is the norm, the grain-fed flavor is considered abnormally fatty, rancid, and "off." In North America, where feedlot concentrates are the norm, the grass-fed flavor is considered livery and gamy, the grain-fed flavor nutty and beefy.

Why should livery, gamy, metallic flavors be more off-putting than any others? The answer may have to do with a vestigial alarm system inherited from our pre-culinary ancestors. The metallic-smelling molecule epoxy decenal turns out to be a landmark alerting volatile for many animals. Biologists have found it to be *the* key identifying volatile of blood, sufficient on its own to attract predator animals, repel prey— and trigger wariness in humans. No surprise, then, that it might sometimes detract from the pleasure of eating.

Force-feeding birds, castrating hogs, and grain-feeding cattle are all venerable versions of a kind of olfactory camouflage, a precooking on the hoof that emphasizes generically fatty smells over the specifically animal and bloody. They distance our eating from the smells of animal life and the violence of wounding and ending it; they help us forget that we are predatory animals taking pleasure in pieces of flesh that prefigure our own end.

Fish and fishiness

Meatiness in a food is a good thing. Fishiness is usually not, as we've seen (see page 384). Because most fish are cold-blooded and live in a cold, weight-supporting medium, they bear little resemblance to farm animals in their tissues and intrinsic volatiles. The same goes for their cooked aromas. Fish and shellfish muscles are relatively delicate, their texture best when gently cooked—and their flavor only slightly affected. As their usual pale color indicates, they contain little aroma-generating heme iron. Many fish have small patches of dark muscle near the fins and just under the skin, and these do have a stronger flavor than the white flesh just millimeters away.

The aromas of cooked fish come mainly from their cell-membrane and oil chains, which are highly kinked so that they stay fluid in cold waters (see page 383). The fishes' own enzymes break these chains into kinked fragments, mostly six, eight, and nine carbon atoms long, that have green, cucumbery, geranium-leaf, and melon qualities. Those are the volatiles of fresh fish, the **aquatic fatty bouquet**: not meaty, but not overtly fishy, instead vegetal and oceanic. Brief cooking amplifies these and adds sulfur-containing methional from amino-acid breakdown. According to one study, a recognizable sketch of cooked-fish smell can be created with a simple mixture of just two volatiles, geranium-leaf octadienone and cooked-potato methional.

When fish and shellfish are less fresh or cooked more thoroughly, the long exposure to oxygen or to heat energy breaks their kinked chains mainly into seven- and ten-carbon fragments—and this is the mix that smells distinctly fishy. Long cooking times and canning temperatures can develop some of the meaty furan sulfur volatiles; methyl furanthiol is known as the "fish thiol" for its importance in the smell of canned tuna.

SOME COMMON SMELLS IN COOKED FISH AND SHELLFISH

Smells	Molecules
aquatic fatty bouquet: green, fishy-fatty, mushroomy-metallic, geranium leaf, cucumber, melon, fried	hexenal, heptenal, heptadienal, octenone, octadienone, nonadienal, nonadienone, decadienals
cooked-fish bouquet: geranium leaf + cooked potato	methional + octadienone

Annoyingly, even fresh fish perfectly cooked, not at all fishy on the plate, can sometimes leave a fishy aftertaste in the mouth—if it's been preceded or followed by a sip of wine. Japanese chemists have identified the culprit as traces of ferrous iron commonly left in wines from metal winery equipment, which catalyze the breakage of kinked-chain residues as we eat. This emergent fishiness can be somewhat mitigated with an extra squeeze of lemon juice, whose citric acid binds iron.

Particular fish and shellfish have their own qualities. In freshwater **trout**, cooking develops potato and buttery qualities to fill out the vegetal broken-chain fragments. Cooking **ocean fish** like **cod** and **salmon** supplements their vegetal-fishy fragments with different proportions of potato, deep-fried, musty, and cocoa notes. Fatty Atlantic and king salmon can develop especially strong smells because they have twenty times the kinked-chain content of lean fish like cod, and so they generate more volatile fragments.

SOME COOKED FISH

Fish	Component smells	Molecules
trout	geranium leaf, cucumber, potato, buttery, earthy, musty	octadienone, nonadienal, methional, diacetyl, geosmin, methylisoborneol
salmon	geranium leaf, cucumber, cooked potato, fried, musty	octadienone, nonadienal, methional, decadienal, propional
cod	cooked potato, geranium leaf, cocoa, cucumber, fried	methional, octadienone, methylbutanal, nonadienal, decadienal
canned tuna	meaty	furanthiols, sulfides

Crustaceans—**shrimps**, **crabs**, **lobsters**, **crayfish** and their ilk—have a shared family note that's described as popcorn-like, even when they're boiled. The volatile responsible, acetyl pyrroline, develops because crustaceans balance the salinity of seawater by accumulating amino acids, among which proline readily generates pyrroline when heated. The digestive and reproductive organs of crustaceans—shrimp and crayfish heads, crab butter, lobster coral—are sometimes discarded but sometimes savored as the most flavorful parts, thanks to their relatively high levels of lipids and lipid-fragmenting iron.

SOME COOKED CRUSTACEANS

Crustacean	Component smells	Molecules
shrimp	popcorn, roasted nut, boiled potato, exposed seaweed	acetyl pyrroline, ethyl dimethylpyrazine, methional, bromophenols
crab	buttery, fishy, popcorn, boiled potato	diacetyl, pyrrolidine, acetyl pyrroline, methional
lobster	buttery, boiled potato, popcorn, sweaty, floral, animal	diacetyl, methional, acetyl pyrroline, methylbutyric & phenylacetic acids, skatole, aminoacetophenone

Of the members of the **mollusk** family, **octopus** and **squid** are most fishlike in being active swimmers and mostly muscle tissue. Because they're denser, cooks sometimes simmer them for hours, during which they can develop meaty thiols. Of the bivalves, **scallops** swim, so their meat is mainly muscle and resembles crustacean meats. More sedentary filter-feeding clams, mussels, and oysters collect and harbor plankton that balance ocean salinity with DMSP (see page 378), which cooking helps convert into ocean-air dimethyl sulfide: an especially appropriate transformation in clambakes at the beach, which often include similarly sulfidic cooked corn. **Clams** have a notably sweet smell, **mussels** potato and buttery notes, and cooked **oysters** retain a modified version of their original vegetal qualities (see page 393).

SOME COOKED MOLLUSKS

Mollusk	Component smells	Molecules
squid (long cooking)	cooked potato, cheesy, popcorn, fenugreek, roasted coffee, meaty	methional, butyric acid, acetyl pyrroline, sotolon, furfurylthiol, methyl furanthiol
scallops	boiled potato, toasty, ocean breeze, cooked cabbage, fishy	methional, pyrazines, dimethyl sulfide & disulfide
clams	sweet, toasted corn, nutty, ocean breeze	maltol, furaneol, acetyl thiazoline, acetyl thiazole, dimethyl sulfide
mussels	potato, butter, nutty, oniony, corn chip	methional, diacetyl, ethyl pyrazine, dimethyl trisulfide, acetyl thiazoline
oysters	whisky, green, waxy, mushroom, fatty, melon, ocean breeze	pentanol, pentenol, octanol, octenol, octadienol, benzaldehyde, lilac aldehyde, dimethyl sulfide

Companionable vegetables: tomatoes, carrots, celery, alliums

We shift now from animal flesh and its sometimes vegetal aroma notes, to vegetables and their sometimes meaty notes! To begin, several reliable companions in meat and fish dishes, and one in particular that provides a key aroma note to long-cooked stews.

The fruit of the **tomato** plant is the leading fresh vegetable in world production and among the most versatile in cooking, with its striking red color and rich, balanced flavor, the taste simultaneously sweet, tart, and savory, the smell a complex blend of planty qualities with an undertone of the sulfidic and cocoa-sweaty (see page 318). Heating eliminates the distinctive green tomato-leaf note while amplifying the sulfurous, and it encourages some dismantling of the fruit's copious terpenoid pigments, which brings cooked-apple damascenone to the fore and adds violet, lemon, and apricot notes. Prolonged or extreme heat in turn diminishes these in favor of simply vegetal dimethyl sulfide, the main volatile in cooked-down tomato paste.

COOKED TOMATOES, CARROTS, AND CELERY

Vegetable	Component smells	Molecules
tomato	cooked apple, floral, cooked potato, green, caramel, sweaty, mushroomy-metallic, sulfurous	damascenone, linalool, methional, hexenal, furaneol, methylbutanal, octenone, dimethyl trisulfide, dimethyl sulfide
carrot	floral, cooked apple, green, woody, pine, resinous	linden ether, ionone, damascenone, heptanal, nonenal, terpinolene, myrcene
celery, lovage	celery, herbal, sweet, fenugreek, cooked apple, floral, sweaty	phthalides, sotolon, damascenone, ionone, methylbutyric acid

Many main dishes in the European tradition begin with the preparation of a base of "aromatic" vegetables that brings together onions and/or garlic, carrots, and celery, all chopped and cooked in fat or oil. This mix is then combined with the main ingredient and a unifying liquid, a broth or wine or some crushed tomatoes. Like tomatoes, **carrots** are intensely colored with terpenoid-derived pigments, and when cooked emit some of the same fruity and floral fragments, with others that are woody and resinous. Its unusual and dominant linden ether is the main volatile in flowers of the linden tree, and in linden (*tilleul*) tea and honey. **Celery** and its close relatives, **celery root** and **lovage**, provide a complementary dimension with sweet-smelling furanone volatiles. Celery carries its distinctive phthalides, and all three produce fenugreek-like sotolon. In an amusing confusion of original and imitation, lovage (*Levisticum officinale*) is called the Maggi plant in northern

Europe because its cooked aroma resembles the long-standing Maggi brand of instant seasoning that simulates the aromatic-vegetable base.

An entirely different aromatic dimension comes from the sulfur virtuosos of the allium tribe, garlic and onions and their relatives. Though most vegetables emit ocean-air dimethyl sulfide when they're heated (plants are well stocked with its precursor derivative, the amino acid methionine), the allium defensive enzymes generate dozens of different sulfur molecules when the tissue is disrupted. Their copious sulfur volatiles can reinforce the sulfides and thiols of protein-rich meats, or substitute for them in meatless dishes.

Different cooking methods elicit very different versions of allium aroma. When garlic or onions are cooked intact or with an enzyme-inhibiting acid ingredient like tomatoes, the defensive enzymes are prevented from generating their full complement of volatiles, so the aroma is relatively subdued. When their tissues are disrupted by chopping or grating and then cooked slowly, the rich mix of defensive molecules forms and evolves, giving rise both to simpler methyl sulfides and more complex chains and rings.

COOKED ONIONS AND LEEKS

Vegetable	Component smells	Molecules
onion, sweated 212°F (100°C), 25 min.	oniony	propenyl & propyl sulfides, propenyl & propyl di- & tri-sulfides
onion, sautéed 300°F (155°C), 10 min.	oniony, toasty	propenyl & propyl sulfides, methylbutanal, pyrazines
onion, fried 260°F (130°C), 18 min.	oniony, fried, sweet	aldehydes, dimethyl sulfide, dimethyl di- & tri-sulfides, thiophenes, methional, diacetyl, furfural, acetyl furan
leek	green-fruity, leafy, oniony, sulfurous	methyl pentenal, hexanal, hexenal, propyl sulfides, propanethiol

According to a 2015 study from the University of Nantes, gently "**sweating**" chopped **onions** in a pan drives off the irritating volatiles into the kitchen air while giving the enzymes time to generate their oniony propenyl sulfides; quick **sautéing** reduces the sulfide output while adding malty and toasty notes from amino and sugar browning reactions; thorough **frying** amplifies the amounts of all but the propenyl sulfides and rounds out the aroma with abundant methyl sulfides, sweet furans, and buttery diacetyl. This savory-sweet effect is prominent in such things as French onion soup and the fried shallot garnishes popular in southern Asia.

Garlic behaves similarly with its complement of allyl sulfides. Cooking "tames" the strong, pungent quality of raw garlic by encouraging the conversion of stinging allicin (diallyl thiosulfinate) into the character-defining diallyl disulfide, and the generation of a mix of sulfides and rings. **Black garlic** is an unusual version, prepared by holding the intact bulbs at 140°F to 180°F (60°C to 80°C) for weeks, allowing both enzyme activity and thoroughgoing browning and other reactions, turning them black and enfolding the garlicky allyl sulfides in sweet-sour and roasty bouquets.

COOKED GARLIC

Preparation	Component smells	Molecules
sweated	garlicky, sulfurous	diallyl sulfide & di- & tri-sulfides, allyl methyl trisulfide, dithiane, butenal
fried	garlicky, sulfurous	diallyl disulfide, allyl methyl di- & tri-sulfides, vinyldithiins
black	garlicky, sweet, roasted, sweaty, vinegary	allyl methyl trisulfide, diallyl di- & tri-sulfides, furaneol, methylbutyric & acetic acids

How do the volatiles of aromatic vegetables and meats combine to give us the pleasures of something like a stew? In 2009 and 2011, German flavor chemists published analyses of the major volatiles in typical **meat stews** made by browning cubes of either beef or pork, then carrots, leeks, onions, and celery root, then adding water and simmering the combination for four hours. They found no evidence that the different materials reacted with each other to form novel volatiles. Instead, each ingredient made its own contribution to the stew liquid. Beef provided the characteristically tallowy methyl tridecanal and high levels of many volatiles, perhaps thanks to its abundant heme. The pork gave a more subdued aroma, with a distinct deep-fried note. The carrots provide terpenoids, notably floral ionone, and the celery root gives sweet-fenugreek sotolon.

Remarkably, the dominant note, described as "gravy-like," came not from the meats, but from the onions and leeks! The volatile responsible turned out to be a five-carbon, one-sulfur chain with a methyl decoration, a **mercaptomethyl pentanol**, **MMP** for short. It's formed by a sequence of reactions, the first caused by heat-sensitive onion enzymes, then ordinary chemical reactions that are accelerated by heat. So its production is encouraged by chopping or pureeing these alliums (but not garlic) well before cooking them to let the enzymes do their work, then cooking slowly for several hours.

BEEF AND PORK STEWS

Component smells	Contributed by	Molecules
tallow, beef fat	beef	methyl tridecanal
deep-fried	pork	decadienal
gravy-like, meaty, sweaty	leek, onion	mercaptomethyl pentanol
sulfurous	beef, pork, onion	methional, dimethyl trisulfide
fenugreek; caramel	celery root; all	sotolon, furaneol
floral, cooked fruit	carrot	ionone, damascenone
fatty, fried	beef, pork	aldehydes: nonanal, decanal, nonenal, decenal, undecenal . . .

Equally remarkably, modern humans have a variant of a single odorant receptor tuned and highly sensitive to MMP, which Neanderthals shared but other early hominids did not. Is it the case, as the scientists who discovered this speculate, that it reflects "an adaptation to our nutritional behavior toward onions across cultures"? A couple of intriguing clues: MMP smells meaty and brothy when very dilute, oniony and sweaty when more concentrated. The MMP receptor also responds to a related mercaptomethyl butanol and hexanol—which are volatile components of human sweat and perhaps the receptor's original targets. (*Mercapto*- and *sulfanyl* are near chemical synonyms; I repeat the volatile names used in the relevant studies.) In 2000, flavor chemists patented the use of these sweat volatiles to enhance meat

flavors in packaged foods (see page 121). Whatever evolutionary influences conspired to make modern humans so sensitive to sweat-adjacent MMP, it seems that
today when we smell good gravy, we're unconsciously registering a note of our aromatic selves.

Green vegetables

Many edible plant parts are green with photosynthetic chlorophyll and share a
"green" aspect to their aromas when cooked, but this isn't the simple cut-grass,
fresh smell of living plant tissues. Cooking inactivates the enzymes that produce
the green life volatiles, and it adds new fatty and sulfurous volatile bouquets and
particular notes. A 2000 study of **spinach** cooked for 30 and 120 minutes showed
how the aroma of cooked green vegetables can develop spicy, even meaty aspects.
Like its botanical relatives chard and beets, cooked spinach also emits earthy-
smelling geosmin (see page 348).

A number of green vegetables retain a distinctly vegetal greenness from their
methoxypyrazines (see page 181), which are stable to heat and can end up dominating a vegetable's cooked flavor. There are several common methoxypyrazines,
variously described as smelling earthy, musty, and vegetal; they're heavy and clinging compared to the GLVs. **Green beans and peas**, both seed pods of members of
the legume family, are dominated by them, with additional mushroomy notes and,
in the case of the beans, floral linalool. **Green capsicum peppers** combine a methoxypyrazine with a sulfurous thiazole, and as the dominant member of its "holy
trinity" of vegetable aromatics, lends the Louisiana Creole stew called gumbo a
special note. **Asparagus** is dominated by the sulfurous bouquet, ocean-air dimethyl
sulfide and potato methional; soil-blanched white asparagus is mild, while stalks
exposed to the sun have a strong vegetal-green note from a methoxypyrazine. (Eating asparagus also has sulfurous after-effects; see page 102.) **Celtuce**, also known as
stem or asparagus lettuce, is a variety of lettuce popular in China for its thick, long,
juicy-crunchy stem. It's oddly and wonderfully endowed with the same acetyl pyrroline that gives basmati rice and pandan leaves their distinctive aroma. Cooking
boosts that note above the vegetal methoxypyrazine background.

SOME MOSTLY GREEN VEGETABLES

Vegetable	Characteristic smells	Molecules
spinach, quick-cooked; stewed	green leaf, floral, sweet, earthy; fatty, smoky, clove, sulfurous, meaty	GLVs (hexenals & hexenols), cyclo-citral, ionone, safranal, geosmin; C12 aldehydes, vinylguaiacol, thiophenes, thiazoles
bean, green	vegetal green, geranium leaf, potato, mushroomy-metallic, nutty, floral	methoxypyrazines, octadienone, methional, octenone, acetyl pyrroline, linalool
pea, green	vegetal green, earthy, sulfurous, mushroomy-metallic, green fruit	methoxypyrazines, dipropyl disulfide, octenone, hexyl acetate
capsicum pepper, green	vegetal green, sulfurous	methoxypyrazine, butyl propylthiazole
asparagus	ocean air, vegetal green, potato, mushroomy	dimethyl sulfide, methoxypyrazine, methional, octanedione
celtuce, stem lettuce	popcorn, basmati rice, green, vegetal, earthy	acetyl pyrroline, methoxypyrazines, ethyl dimethylpyrazine
artichoke	mushroomy-metallic, green, waxy, honey, woody	octenone, hexenone, nonenal, decanal, phenylacetaldehyde, selinene, caryophyllene
okra	sweet, floral, cooked apple, woody, smoky, clove, fishy	acetyl furan, linalool, damascenone, caryophyllene, vinylguaiacol, eugenol, pyridine
cabbage family: bok choy, broccoli, brussels sprouts, cabbage, collards, kale, mustard greens	ocean, sulfurous, pungent	sulfides, isothiocyanates, nitriles, cyanides

Other green vegetables are less strongly green-flavored. **Artichokes**, the unopened flower bud of a large thistle, and **okra**, the immature seed pod of a hibiscus, stand apart with woody and floral terpenoids and smoky, spicy phenolics. A large group of vegetables shares a dominant sulfurousness with the alliums. The many members of the **cabbage** family, the **brassicas**, defend themselves with irri-

tating and aromatic isothiocyanates (see page 182). Heat encourages these molecules to react with others and form a variety of volatile products, including nitriles
and cyanides (with triple-bonded carbon-nitrogen pairs), as well as sulfides. Each
cabbage family member has its own shade of sulfurousness and pungency, **mustard
greens** being by far the most aggressive and **collards** and **kale** the least.

Vegetable miscellany

Here's a sampling of some other vegetables that aren't green or used as aromatics.
First, a few that stay close to or in the ground. A couple of underground tubers and
roots smell like the soil they're dug from, even when meticulously cleaned, because they synthesize volatiles similar to those emitted by food-hunting inhabitants of the soil, perhaps a way of disguising themselves. Some **potato** and **beet**
varieties make earthy and musty methoxypyrazines that persist when they're
cooked, and some beets also synthesize their own geosmin (see page 270).

When boiled and steamed, most vegetables generate mainly the fatty and sulfurous cooked bouquets, but **boiled potato** flesh also develops new nutty and earthy
pyrazines. **Potato skins** are rich in phenolic molecules that reinforce and toughen
their structure; when baked, they readily dry out and brown, become especially
nutty, and produce phenolic ring volatiles, smoky and spicy. **Fried potatoes** and **potato chips** are defined by the frying-oil volatile decadienal, potato-sulfurous methional, and nutty pyrazines. **Sweet potatoes** are unrelated to common potatoes and
aren't earthy. Varieties that are orange with carotenoid pigments generate floral, fruity
terpenoid fragments from them, along with sweet, caramel notes that are especially
evident after slow baking, a process that allows its enzymes time to convert starch
into reactive sugars.

Bamboo shoots, the moist growing tips of a tropical grass that eventually grows
to tree height, are unusual for their dominant phenolic relatives, hints of tree resin
and wintergreen. And **mushrooms**, the aboveground organs of underground fungi,
are rich in amino acids and become less obviously mushroomy when cooked, their
characteristic eight-carbon volatiles joined by a number of toasty, roasty, caramel,
and potato-sulfurous reaction products.

SOME GROUND-HUGGING VEGETABLES

Vegetable	Characteristic smells	Molecules
beet	earthy, metallic, clove-smoky, cooked apple, vegetal earthy	geosmin, methoxypyrazines, decenal, vinylguaiacol, damascenone
potato, baked	flesh: earthy, sulfurous, potato, sweaty, honey skin: nutty, roasted, smoky, clove	ethyl methylpyrazines, methional, methanethiol, dimethyl di- & tri-sulfides, methylbutanal, phenylacetaldehyde; c. 30 pyrazines, guaiacol & vinylguaiacol, eugenol
potato, sweet (*Ipomoea batatas*)	sweet, caramel, almond, floral, butter, honey	maltol, furfural, furan, pentylfuran, benzaldehyde, diacetyl, phenylacetaldehyde, geraniol, ionone
bamboo shoot	balsamic, sweet, wintergreen, woody, green, waxy	benzyl & methyl salicylates, cedrol, hexenal, heneicosane
mushroom, common, fried	sweaty, cooked potato, toasty, mushroomy-metallic, caramel, honey, roasted	methylbutanal, methional, acetyl pyrroline, octenone, furaneol, phenylacetaldehyde, pyrazines

Now a final half-dozen non-green, non-aromatic vegetables. **Cauliflower** is a pale oddity from the cabbage family, less endowed than its brassica sisters with sulfurous isothiocyanates but with buttery and mushroomy notes of its own. **Ripe capsicum peppers** lose much of the vegetal methoxypyrazine that defines their green stage and accumulate a range of unusual fruity sulfur volatiles, which cooking amplifies and supplements with floral terpenoids. **Eggplant** develops a distinctive mix of woody terpenoids, several sweet benzenoid rings, and a rare haylike sulfur molecule. Canned **winter squash**, including **pumpkin**, more severely heated than it would be in a kitchen, is dominated by sulfides and sweet, bready aldehydes. **Sweet corn** is so dominated by the ocean-air dimethyl sulfide that "cooked corn" is a term often used to describe the smell of that molecule. (A clam chowder made with corn and milk really brings the ocean to the table!) And **nori**, a red seaweed from the ocean, formed into thin sheets, dried, toasted, and used as a wrapper for Asian rice balls and sushi, retains its aquatic geranium-leaf and cucumber-oyster notes, but the drying and toasting supplement them with roasted, meaty, floral, sweet qualities.

SOME STALK, FRUIT, AND SEA VEGETABLES

Vegetable	Characteristic smells	Molecules
cauliflower	sulfurous, buttery, mushroomy-metallic	as for cabbage family, + diacetyl, octenone
capsicum pepper, red	sulfurous, fruity, metallic, vegetal green, floral, cooked apple	heptanethiol, heptanone & heptenone, isobutyl methoxypyrazine, linalool, damascenone, many C7 & C9 thiols
eggplant	pine, citrus, woody, almond, honey, sweet, dry hay	carene, bisabolene, benzaldehyde, phenylacetaldehyde, toluene, methyl thiopene carboxaldehyde
squash, winter, canned	sweaty, sulfurous, solvent, bready, sweet	methylbutanal, dimethyl sulfide & disulfide & trisulfide, methylpropanal, pentanal, furfural
corn, sweet	ocean, sulfurous, solvent, green	dimethyl sulfide, ethanethiol, acetone, ethanol, acetaldehyde
nori (seaweed), toasted	toasted, meaty, cheesy, geranium leaf, cucumber/oyster, floral, honey	pyrazines, thiazolyl ethanone, butyric & methylbutyric acids, octadienone, nonadienal, ionone, phenylacetaldehyde

Seeds: nuts, grains, beans

We've already sampled the innate raw qualities of nuts and grains in chapter 13; heating generally allows these to persist while adding generic cooked bouquets. **Nuts**, usually roasted or fried, smell like their raw selves with the addition of roasty pyrazines or deep-fried decadienal. **Pumpkin seeds** are unusual for their prominent sulfurous notes. Briefly boiled as is or re-formed to make noodles, **refined grains**, freed of their embryo and envelope, smell of themselves with the amplification of fatty cooked bouquets. **Whole-grain** foods, retaining the oil- and protein-rich embryo and protective seed coat, or bran, generate more volatiles and a stronger flavor than refined grains. From nonvolatile phenolic acids in the seed coat, boiling water generates volatile phenolics familiar from smoke and spices.

SOME COOKED SEEDS

Seeds	Component smells	Molecules
nuts, roasted	as on p. 280 + toasty, roasted, caramel, fatty, fried	as on p. 280 + pyrazines, furanthiols, furanones, decadienal
pumpkin seeds	malty, toasty, cooked potato & vegetable, spicy/smoky	methylbutanals, ethyl methylpyrazines, methional, dimethyl sulfide, vinylguaiacol
refined grains, boiled	as on p. 276 + cooked	as on p. 276 + fatty bouquet
whole grains, boiled	fatty, vanilla, smoky, clove	aldehydes, vanillin, guaiacol, vinylguaiacol
wild rice, boiled (*Zizania palustris*)	nutty, smoky, tealike, green	methylpyrazines, guaiacol, pyridine, hexanal, heptanal, nonanal

An unusually flavorful whole grain is **wild rice**, the seeds of a grass native to lakes and streams of the northern United States and southern Canada. It's traditionally harvested from the waters while still moist and maturing, before it "shatters" off the plant, then cured in heaps for some days before parching. Curing often leads to some bacterial fermentation, and parching is still sometimes done with wood fires. The resulting toasty, smoky, and tealike qualities are less prominent in modern mechanized production.

Three different forms of cooking give **corn or maize** very different qualities. Most include "popcorn-like" acetyl pyrroline; here the alternative "basmati rice–like" is more helpful. **Canned sweet corn** is dominated by seaside-vegetal dimethyl sulfide, acetyl pyrroline, and a corny thiazole. **Popcorn**, quickly cooked in a hot-air popper or microwave oven, emits hydrogen and dimethyl sulfides as the grains explode, and retains a mix of acetyl pyrroline and fatty, roast-coffee, spicy, and buttery notes (before any butter is added!). The long-simmered porridge **polenta** carries many of the same volatiles and a corny thiazole; when made from the whole grain, the spicy seed-coat note develops and then fades. The standouts among corn products are **hominy** and **nixtamal**, whole kernels cooked and steeped in an alkaline solution of mineral lime (calcium hydroxide). For hominy, the cooked kernels are dried, then cooked again whole or ground into grits or flour; nixtamal corn is ground wet into a dough to make **tortillas** and **tamales**. Alkaline cooking causes the amino acid tryptophan to break down and form the amino-acetophenone that, along with violet-like ionone, gives these foods their hauntingly floral-fruity note, reminiscent of Concord grapes and chestnut honey. **Corn chip** snacks add popcorn and generic fried volatiles.

SOME COOKED VERSIONS OF CORN

Corn preparation	Component smells	Molecules
canned sweet corn	ocean-air, corny, basmati, cocoa, clove	dimethyl sulfide, acetyl thiazole, acetyl pyrroline, ethyl dimethylpyrazine, vinylguaiacol
popcorn	basmati, fried, coffee, clove, vanilla, butter	acetyl pyrroline, decadienal, furfurylthiol, vinylguaiacol, vanillin, diacetyl
polenta	mushroom, waxy, basmati, ocean-air, sulfurous, honey, clove	octenol, nonanol, acetyl pyrroline, dimethyl sulfide & trisulfide, acetyl thiazole, phenylacetaldehyde, vinylguaiacol
corn tortillas (nixtamalized)	animal-floral, violet, cocoa, clove, mushroomy	aminoacetophenone, ionone, methylbutanal, vinylguaiacol, octenol
corn chips (nixtamalized)	fried, cocoa, coffee, animal-floral, basmati, clove	decadienals, methylbutanal, furfurylthiol, aminoacetophenone, acetyl pyrroline, vinylguaiacol, pyrazines

Wheat is the most important grain in the West, and much of it is consumed in the form of **bread**, a dough of ground grain shaped and cooked through, often after an aerating period of yeast fermentation. Our species has known the smells of bread for at least 14,000 years, from long before the beginnings of agriculture. Most bread doughs essentially toast on the outside and steam within, and in the process lose their yeast volatiles. The smell of common **white bread** from refined flour is a mix of fatty, sulfurous, and sweet bouquets inside the loaf, toasted and roasted in the browned crust. Phenolic bran molecules in stronger-flavored **whole-wheat bread** generate spicy volatiles throughout the loaf, but inhibit the formation of the crust's toasty-sweet pyrroline and furanone rings. **Sourdough breads** owe their sharp aroma and taste to vinegary acetic and other acids generated by bacteria that accompany the yeasts. Raw rye flour carries more aroma than wheat flour, including premade spicy, sweet, and buttery volatiles; traditional **whole-grain sourdough rye bread** has an especially rich flavor. **Pretzels** are one of a number of *Laugengebäcke,* or lye-baked goods, that apparently originated in Bavaria and get their dark brown color and distinctive flavor from dipping the shaped dough, before baking, in hot alkaline lye water. This simple treatment shifts the balance of browning reactions and their sweet and nutty bouquets (see pages 490–91), minimizing the formation of pyrazines and aldehydes and raising popcorn, caramel, buttery, and sweaty volatiles to the fore—with a minor floral echo of similarly alkaline nixtamal corn.

A last grain preparation: **roux** (French: "red") is a paste of flour and a cooking fat or oil, heated in a pan until it develops some color and aroma, then added to a liquid sauce, stew, or soup to thicken its consistency and contribute a background aroma. A lightly cooked blond roux carries mainly starter-set acids and ketones and a touch of almond-essence benzaldehyde; as it darkens with continued cooking, the roux develops sweet and nutty furans and pyrazines. The fat component is important: butter favors cheesy, floral, creamy-buttery volatiles, while polyunsaturated vegetable oil favors deep-fried and honey notes.

SOME BAKED BREADS AND ROUX

Bread versions	Component smells	Molecules
white bread crust	popcorn, cocoa, sweaty, potato, buttery, caramel, nutty	acetyl pyrroline, methylbutanals, methylbutyric acid, methional, diacetyl, furaneol, furfural, pyrazines
white bread interior	fatty, metallic, mushroomy, butter, cooked potato, honey, popcorn	decadienal, decenal, nonenal, octenone, diacetyl, methional, methylbutanol, phenylacetaldehyde, acetyl pyrroline
whole-wheat bread	caramel, woody, smoky, clove, solvent	vinylguaiacol, isomaltol, cyclopentanedione, octenol, esters
sourdough wheat bread	prominent vinegary, sweaty, solvent-fruity	acetic, methylpropionic, & methylbutyric acids, ethyl acetate

continued

Bread versions	Component smells	Molecules
sourdough rye bread	cheesy, spicy, vanilla, potato, fatty, fresh, vinegar, honey, floral	butyric & methylbutyric acids, sotolon, vanillin, methional, decadienal, hexanal, acetic & phenylacetic acids
pretzel	popcorn, caramel, honey, sweaty, buttery, floral	acetyl pyrroline, furaneol, phenylacetic & methylbutyric acids, diacetyl, aminoacetophenone
wheat flour roux	fatty, sweet, nutty	acids, aldehydes, ketones, furans, pyrazines;
with butter:	+ cheesy, floral, creamy;	nonanone, butyric acid, d-decalactone;
with vegetable oil:	+ deep-fried, honey	decadienal, phenylacetaldehyde

A last group of edible seeds, richer than the grains in protein and an important nutritional resource, comes from the legume family—dried beans, peas, lentils, and relatives. Their volatiles are relatively unexplored. Studies of **red kidney beans** have found their smell defined by a mix of sulfurous, earthy, mushroomy notes, along with a toasty pyrazine and a clove-like phenolic. **Soybeans** are unusually rich in oil and in highly kinked carbon chains, so when cooked they generate large numbers of fragments that define a fatty bouquet described as "beany"; there's also a distinct mushroomy note from stored octenol. Soybeaniness is not universally loved, and soybean breeding and manufacturing research have been aimed at reducing its levels in tofu, soy milk, and other products.

SOME BEANS

Seeds	Component smells	Molecules
red beans, boiled	cooked potato, sulfurous, earthy, toasty, mushroomy, clove	methional, ethyl & methyl sulfides, methylpyrazines, octenol, vinylguaiacol
soybeans, boiled & tofu	beany-fatty, mushroomy, almond, sulfurous	hexanal, hexenal, heptenal, octenal, octenol, nonanal, pentanal, decadienal, benzaldehyde, pentylfuran, dimethyl disulfide

Cooked fruits, syrups, bee-cooked honeys

As we've seen, our favorite fruits have already been concocted by their parent plants to be appealingly flavorful. We cook them mainly to prolong their otherwise brief edible lives, and in the process we change their original balance of tastes and aromas. Cooking usually intensifies sweetness and subdues distinctive esters and terpenoids; their cooked bouquets predictably bend toward the smells of cooked-sugar caramel. Sweet-hay furfural is an almost universal volatile in fruits cooked for any length of time.

SOME DRIED AND COOKED FRUITS

Dried fruit	Component smells	Molecules
prune (dried plum)	sweet, woody, floral, almond extract, fruity, honey	furfural, nonanal, benzaldehyde, ethyl cinnamate, phenylacetaldehyde, furaneol, methyl furfural
raisin (dried grape)	fatty, green, honey, cooked apple, fruity, roasted	nonenal, hexanal, phenylacetaldehyde, damascenone, ethyl hexanoate, pyrazines
cooked apple	fruity, apple, cooked apple, cocoa-sweaty, honey, almond essence, woody	acetate & butyrate & propionate esters, damascenone, methylbutanal, phenylacetaldehyde, benzaldehyde, furfural
strawberry preserves	strawberry, caramel, sweaty, cheesy, woody, cinnamon, peach, ocean air	furaneol, methylpropionic acid, butyric acid, methylbutyric acid, furfural, cinnamyl acetate, g-dodecalactone, dimethyl sulfide
sour orange marmalade	citrus, pine, floral, resinous, orange peel, fruity, woody	limonene, terpinenol, terpineol, linalool, myrcene, octanal, ethyl acetate, furfural

The simplest way to cook fruits is to let them dry out enough that microbes can no longer grow on them. **Prunes**, the dried version of plums, are the epitome of dried fruit, with sweet furans and furanones complementing the fruit's persistent almond, floral, and fruity notes. Grapes are converted into **raisins** by being slowly sun-dried, quickly air-dried at 160°F to 180°F (70°C to 80°C), or dried with the addition of sulfur dioxide, a common treatment that inhibits oxidation and browning reactions and thus preserves some of the fruit's natural color and aroma. When fruits are quickly simmered or boiled to make **fruit sauces** or **preserves**, more of the original ripening volatiles tend to survive, and the sweet brown bouquet is less prominent—though it becomes more so the longer preserves are kept in the pantry.

Fruit preserves are usually made with the pure crystalline sugar sucrose, which has been concentrated and refined from the internal fluids of sugarcane or sugar beet plants, and is mostly odorless. When the plant fluids are initially boiled down, nonsucrose sugars and amino acids react with each other, turn the fluids an ever deeper brown, and generate sweet and roasted bouquets. Refined white sugars are made by separating the sucrose crystals from the brown fluid. Because beets are dug from the ground and often piled for some time before processing, they sometimes carry unpleasant spoilage volatiles that can persist as traces in refined **beet sugar**. **Cane sugar** is made from aboveground stalks that are processed promptly, so it rarely suffers from off-aromas.

Unrefined or "raw" sugar crystals retain a film of the aromatic brown fluid, which is where the important furanone sotolon was identified in 1980. **Molasses** is that brown fluid repeatedly concentrated into a stronger-smelling black-brown syrup; it develops a prominent sulfurous bouquet, smoky and vanilla phenolic notes, and sharpness from volatile acids. **Sorghum syrup**, the cooked-down fluid of crushed sorghum stalks, has unusual fermented and solventy notes.

SOME SUGARS AND SYRUPS

Sweetener	Component smells	Molecules
refined beet sugar: trace off-odors	sour, cheesy, sweaty, earthy, burnt sugar	acetic, butyric, methylbutyric acids, geosmin, dimethylpyrazine
unrefined cane sugar	fenugreek, vanilla, sweet, coconut, smoky	sotolon, vanillin, maltol, g-nonalactone, vinylphenol & dimethoxyphenol
cane molasses	smoky, vanilla, sharp, potato, sweaty, caramel, buttery, ocean air, sulfurous	guaiacol, vanillin, acetic acid, methional, butyric & methylbutyric acids, furaneol, diacetyl, dimethyl sulfide & disulfide

Sweetener	Component smells	Molecules
sorghum syrup	green, whisky, sweaty, butter, toasted	acetaldehyde, methyl propanal & butanals, acetone, diacetyl, pyrazines
maple syrup	sweet, fruity, fenugreek, toasty, bready, vanilla, sharp	furans, furaneol, sotolon, pyrazines, maple lactone (cyclotene, methyl cyclopentenolone), vanillin, acetic acid
palm sugars	caramel, sweet, buttery, smoky, sharp	furans, furanones, furaneol, maltol, diacetyl, acetoin, furfural, acetic acid
malt syrup	cocoa, sweaty, mushroom, potato, vanilla, sweet, fruity, sulfurous	methylbutanal, octenone, methional, decadienal, vanillin, methylbutyric acid, furaneol, dimethyl sulfide

Sweet syrups and unrefined sugars are also made by boiling down the saps of maple, birch, and palm trees. North American **maple syrup** carries a rounded diversity of sweet and brown rings, including familiar sotolon and furaneol, and an unusual "maple lactone." **Malt syrup**, much used in Asia, is made by *malting* barley grain—allowing the seeds to germinate and activate the enzymes that convert their starch into sugar—then crushing the seeds in water, allowing the conversion to proceed, and boiling down the resulting sweet liquid. Thanks to barley's abundant protein, malt syrup has a prominent dose of methylbutanal from amino acid breakdown; that's why this molecule is often described as smelling malty, as well as cocoa-like and sweaty.

Many millions of years before our species first concentrated plant fluids into spoilage-resistant syrups, insects were doing just that with flower nectar. Bees slowly transform this fragile sweet reward into long-lived **honey**: drop by drop as individuals pump the gathered nectar in and out of their honey sacs, then by the honeycomb as workers fan the hive air. This process of controlled digestion and evaporation, more complex and nuanced than cooking's quick blasts of high heat, retains some volatiles from the original nectar and liberates others from nonvolatile storage forms. Honey's reputation as a sweet embalming fluid is apparently largely legendary, but it does at least preserve some scent from the brief life of flowers.

SOME HONEYS

Flower source	Component smells	Molecules
mixed	honey, floral, cinnamon aged: hay, tobacco, musty	phenylacetaldehyde, phenylethanol, phenylacetic acid, cinnamaldehyde; aged: hydroxymethyl furfural
orange	floral, herbal, cumin, grapey	linalool oxidation products, incl. lilac aldehydes, dill ether, menthenal; methyl anthranilate
chestnut	animal-floral, sweaty, bready	aminoacetophenone, benzoic acid & alcohol, methylbutyric acid, furans
linden	cooked apple, anise, floral, cooked potato	damascenone, anisaldehyde, linden ether, linalool, methional
buckwheat	malty, caramel, vanilla, floral, buttery, fruity	methylbutanal, butyrate esters, dimethyl sulfide, diacetyl, sotolon, damascenone
manuka	solventy, sweet, caramel, popcorn	dihydroxyacetone, acetyl formaldehyde (glyoxal), acetyl pyrroline, hydroxymethyl furfural
mixed forest honeydew; pine forest; oak forest	cooked apple, smoky, clove, almond, pine, coconut	damascenone, guaiacols, cresol, eugenol, benzaldehyde; pinene; oak lactone (methyl octalactone)

Most honeys contain sweet furan and floral phenolic rings. **Phenylacetaldehyde** is often described as honey-like on its own; it accumulates from the slow break-down of trace amino acids. Of common "monofloral" honeys from predominantly single flower types, **orange flower** resembles the perfumery ingredient neroli (see page 458), **chestnut** shares unusual animal-floral amino-acetophenone with Concord grapes and corn tortillas, strong **buckwheat** resembles malt syrup with its dominant methylbutanal and sulfides, and **manuka**, from a tree native to New Zealand and eastern Australia, has a unique solventy note. **Honeydew** is a honey-like concentrate that bees make by collecting the secretions of other insects that feed on plant fluids; it can have smoky, animal, and woody notes. As is true of dried fruits and preserves, honeys darken visibly with time as browning reactions proceed and obscure their more distinctive notes.

Pastries and cakes

At last we come to Gaston Bachelard's waffles and the pleasures of the superfluous. We cook most foods because we need their nourishment to live; we cook pastries and cakes because we want to indulge in sweet, rich tastes and textures and smells. Those pleasures are rooted in our biological need for energy, but they also motivate us for their own sake.

Sweet baked goods are mixtures of grain flours, sugars, and fats, shaped and cooked in an oven or pan hot enough to dry out and brown their surfaces. They develop a base of generic bouquets, fatty and sweet and toasted, that's pleasing on its own and a congenial background for the more specific aromas of other ingredients, fruits or nuts or spices or chocolate. I've been unable to find a published analysis of waffle volatiles—evidently not as beloved by flavor chemists as by Bachelard—but they're likely similar to the volatiles of basic pastries and cakes.

First, the simplest: a **puff pastry** baked from a dough containing only wheat flour, water, and a refined, odorless vegetable shortening. Once baked to dry brown flakiness, its main volatiles include fatty aldehydes with metallic, fried, and green notes, sweet-caramel furaneol, and popcorn-basmati acetyl pyrroline. When odorless vegetable fat is replaced with odorous butter, these are joined by the butter's coconutty decalactone, buttery diacetyl, and cheesy short-chain acids, all of them amplified by the cooking.

SOME PASTRIES AND CAKES

Baked good	Component smells	Molecules
puff pastry, margarine	fatty, metallic, fried, green, caramel, popcorn	epoxy decenal, decadienal, furaneol, nonenal, acetyl pyrroline
puff pastry, butter	coconut, caramel, fatty, cheesy, floral, potato, toasty, buttery	as above + d-decalactone, butyric & methybutyric acids, diacetyl

continued

Baked good	Component smells	Molecules
cookies, sponge cake crust (wheat flour, sugar, whole eggs, butter)	as for butter puff pastry + sweet, bready, cocoa, roasted	as for butter puff pastry + maltol, furfural, furans, methylbutanal, pyrazines
sponge cake interior	fatty, bready-whiskey, fruity, potato	lower furans, furanones & pyrazines than crust; fatty bouquet, pentanol, ethyl hexanoate, methional

Cookie comes from a Germanic word meaning "little cake," and **cakes and cookies** alike, including Bachelard's griddle cakes, generally add sugar and eggs to the flour and fat of simple pastries. Sugars contribute material for more extensive caramelization reactions and browning reactions with amino acids; eggs bring abundant proteins in the white, proteins and free amino acids in the yolk, which also adds long fragmentable fat chains and related molecules. Small thin cookies and cake surfaces dry out and develop the sweet, brown ring volatiles, maltol and the furans and furanones, along with cocoa-like methylbutanal and toasty pyrazines deriving from the amino acids. Some of these volatiles penetrate into the cake interior, but there the temperature never rises past the boiling point of water, and the aroma comes mainly from the fatty bouquet, potato-like methional, and a couple of fruity and alcohol-like notes.

Coffee and chocolate

To conclude this chapter on the smells of our cooked foods, two superfluities that often accompany pastries and cakes, two of our olfactory landmarks, and two exemplars of the dark arts of pyrolysis. Coffee and chocolate are both made from seeds of tree fruits, one native to Africa, the other to South America. Early human fans of the fruits discovered that cooking could transform their uninteresting kernels into something stimulating and delicious, and modern enthusiasts find endless nuances of flavor to enthuse over and market. Nuance mostly aside, here's what makes coffee coffee and chocolate chocolate.

There are two species of *Coffea* whose seeds are used to make **coffee**; they're now grown in the Americas and Asia as well as their African home. Both C. *arabica* and C. *robusta* trees attract animal dispersers with cherry-like fruits and protect the seeds inside with caffeine—a useful drug for us but toxic to many insect pests—and various phenolic-ring molecules. The green beanlike seeds carry embryo-nourishing protein, oils, starch, and sugars; they're roasted at temperatures ranging from 430°F to 500°F (220°C to 260°C), for two to ten or more minutes. The high heat turns the green beans brown to black, and fragments and reacts their resources into many hundreds of volatiles. Robusta beans tend to favor the production of smoky-spicy molecules; arabicas, the sweet and sulfurous.

Different "roasting profiles," or temperature-time paths, produce different flavor balances in the finished beans. Light roasts are sweet and bready from the predominance of furan rings, while the darkest roasts have the singed flavor of guaiacol and phenol from breakdown of defensive phenolics. Medium roasts emphasize the landmark aroma of coffee proper, which is largely created by sulfur-decorated furan rings and related thiols. **Furfurylthiol**, the **"coffee mercaptan,"** smells like roasted coffee and lends the same note to meats and toasted sesame seeds. Methyl furanthiol is a defining volatile in cooked meats, and some other coffee thiols also have an animal sulfurousness. Lesser contributors to medium-roast coffee aroma are nutty pyrazines, an earthy methoxypyrazine from the raw bean, phenolic-ring vanillin, and a couple of guaiacols with clove-like, spicy smells. Light- and medium-roast coffees can also carry flowery and fruity notes from the terpenoid linalool, and occasionally the decorated phenolic ring aptly known as the raspberry ketone. A side note for those who enjoy roasted **chicory root** as a caffeine-free coffee substitute: it shares many of coffee's prominent volatiles but adds its own unique note with the woody-peppery terpenoid rotundone (see page 170).

SOME SMELLS OF COFFEE

Component smells	Molecules
sweet, bready, brown, buttery	furans, furanones (furaneol, sotolon), maltol, diacetyl
coffee, roasted meat, sulfurous, catty, skunky	thiols: furfurylthiol, methanethiol, methyl furanthiol, mercaptomethylbutyl formate, methyl butenethiol . . .
sweaty, malty, cocoa	methylbutanals, methylpropanal

continued

Component smells	Molecules
nutty, roasty, earthy	pyrazines, methoxypyrazine
floral, fruity	acetaldehyde, propanal, linalool, damascenone, raspberry ketone (hydroxyphenyl butanone), octanal
clove, smoky, singed	ethyl & vinyl guaiacols, guaiacol, phenol, pyridine

The smells of coffee drinks depend on how they're prepared. The high pressure and concentration of espresso tend to emphasize the sulfurous and phenolic notes more than standard pour-over drip coffee does. And the smells are ever-changing. The character-defining thiols aren't very soluble in water, and they're unstable: so they're especially prominent when the roasted beans are freshly ground, less so in a finished beverage, and then they slowly disappear in the cup or carafe as they react with the brown pigment molecules. The overall aroma also changes as coffee cools down and its volatiles are released more slowly.

The smells of fluid coffee are ever evolving and dissipating, but the smells of solid **chocolate** are locked in, stable, slow to reveal themselves. The roasted cacao beans from which chocolate is made, superfinely ground with a substantial dose of sugar to balance their bitterness, are nearly half cocoa butter, a solid fat that melts in the mouth as we eat it, just below body temperature. The aromas it releases are very different from coffee's. Cacao beans oddly carry only traces of sulfur and sulfur-containing amino acids, so when roasted they develop just a hint of coffee's thiol volatiles. They're poorer in sugars than coffee beans and so generate fewer sweet furan rings. Moreover, they're far more gently roasted—at around 250°F to 285°F (120°C to 140°C) for around thirty minutes—so their abundant defensive phenolic molecules aren't fragmented into smoky and spicy volatile rings.

The distinctive smell of chocolate comes largely from methylbutanal, a branched amino-acid fragment that smells distinctly cocoa-like with a sweaty edge, and then from a bouquet of around twenty different pyrazines. (Cocoa powder is ground roasted cacao beans with much of the fat removed, a concentrate of chocolate's core flavor.) Some of the pyrazines smell like cocoa, some nutty, some musty and earthy, some burned. The overall aroma of a given chocolate depends on the relative proportions of these separate notes, and on a few other factors. A tart edge comes from abundant acetic and methylbutyric acids, while fruitiness comes from several esters. Some cacao varieties are prized for producing "fine" or "flavor" beans whose floral, herbal, fruity, or woody volatiles survive roasting. When beans are "alkalized" or "dutched" before roasting—treated with a carbonate solution that diminishes their acidity—they favor Maillard reactions that darken the color and shift the balance of pyrazines, and end up less sharp and fruity. In **milk chocolates**, made by mixing melted chocolate with dairy concentrates, the coconutty, buttery, and caramel dairy volatiles come to dominate, as they do in **white chocolate**, a blend of milk solids and cocoa butter with none of the browned cocoa-bean solids. And most versions of chocolate are flavored with vanilla, either an actual spice extract or a synthetic vanillin imitation.

SOME SMELLS OF DARK AND MILK CHOCOLATES

Component smells	Molecules
cocoa, malty, sweaty	methylbutanal, methylpropanal
cocoa, nutty, roasty, earthy	methyl- & methoxy-pyrazines, dimethyl dithiofuran, dimethyl trisulfide
sharp, vinegar, cheesy	acetic & methylbutyric acids
honey, sweet	phenylacetaldehyde, phenylacetic acid
caramel	sotolon, furaneol
floral	linalool, phenylethanol
fruity	methylbutyrate esters, methylbutyl acetate
herbal, woody, fruity	myrcene, ocimene; heptanol, heptanyl acetate, heptanone
fatty, creamy, coconut	g-octalactone, g- & d-decalactones
buttery	diacetyl
caramel	furaneol, sotolon

It takes patience to enjoy solid chocolate fully. Its volatiles are trapped throughout the cocoa fat and on the cocoa and sugar particles, and the volatiles escape into the air only as the fat melts. When the mouth's moisture wets the dry cocoa particles, new reactions take place and generate a fresh set of volatiles, notably additional methylbutanal. This is true of other dry foods as well, chips and crackers and bread crusts: the act of eating helps create their aromas. Flavor often blooms in the mouth.

How is it that cacao beans generate such a dark color and full aroma when cooked at temperatures that would barely color a coffee bean or peanut? By being primed to do so. I learned this firsthand around 1995, when I visited cacao and chocolate producers in Venezuela, smuggled a few cacao pods home, and made chocolate in my California kitchen. When I removed the seeds from the large woody pods, cleaned off the sweet pulp, and immediately roasted them, I ended up with hard, dry lumps that smelled like dried pinto beans, nothing like chocolate. But when I let the extracted seeds and pulp sit out for a few days as cacao producers do, until they smelled yeasty and vinegary, then cleaned and gently dried the seeds, they already had a faint chocolate smell. Brief "roasting" in the toaster oven amplified it.

The smells of the soured cacao pulp are sensible clues to the seeds' transformation. We now know that when the sugary pulp is exposed to the air—in rural Venezuela or suburban California—airborne microbes quickly colonize it. Yeasts consume its sugars and generate alcohol. Alcohol-consuming bacteria convert the alcohol into acetic acid, the essence of vinegar. The acetic acid penetrates the seeds, breaches compartments within their cells, and frees germination enzymes to dismantle the embryo's food reserves into amino acids and sugars. And these building-block molecules are prime ingredients for the pyrazine-making Maillard reactions, which are productive even at the relatively low roasting temperatures that allow some fruit and microbial volatiles to survive. In particular, the combination of enzymatic protein-breaking followed by moderate heat is what generates cocoa-defining methylbutanal.

So the key to successful cacao roasting is the invisible work of molecular machines: the microbial enzymes that turn sugars into alcohol and alcohol into acetic acid, plus the cacao enzymes that turn food reserves into aroma makers. We've encountered enzymes before as creators of green-leaf volatiles in plants and mushroomy eight-carbon chains in fungi, and we've met microbes as the source of most animal malodors. But enzymes and microbes have also given us wonderful alternatives to pyrolysis for making foods safe and nourishing and delicious. Curing and fermentation: the subjects of our final roundup of notable smells.

Chapter 19 · CURED AND FERMENTED FOODS

A poet once said, "The whole universe is in a glass of wine." . . . There are the things of physics: the twisting liquid which evaporates depending on the wind and weather, the reflections in the glass, and our imagination adds the atoms. The glass is a distillation of the earth's rocks, and in its composition we see the secrets of the universe's age, and the evolution of stars. What strange array of chemicals are in the wine? How did they come to be? There are the ferments, the enzymes, the substrates, and the products. There in wine is found the great generalization: all life is fermentation. . . . How vivid is the claret, pressing its existence into the consciousness that watches it!

· Richard Feynman, *The Feynman Lectures on Physics*, 1963

Surströmming: This celebrated (or should one say notorious?) kind of fermented herring is produced . . . in the north of Sweden. . . . Another fishery official recalled that as a young man he was in the harbour of the island of Ulvön on the August day when 200 barrels of surströmming were opened. As the smell billowed upwards, birds began to drop dead from the sky. Moreover, the wind carried the fumes over to a distant convoy of tugs hauling barges of limestone along the coast; whereupon every single tugmaster changed course for Ulvön.

· Alan Davidson, *North Atlantic Seafood*, 1979

The physicist Richard Feynman's poet of the wineglass is probably a convenient fiction. The closest candidate I've found is Robert Louis Stevenson, who did call fine wine "bottled poetry," and who did write verse—most memorably "Yo-ho-ho and a bottle of rum." But this saying sounds like a wine-loving professor's. I was a student at Caltech in the early 1970s and never missed a chance to imbibe Feynman's Queens-inflected delight in the wonders of the everyday. His exclamation above suggests the thrill of intense perception: the wine's vivid color *presses* its very existence through the eye and onto the mind drinking it in and imagining the agents that made it. Though Feynman oddly omits it, of course the next step is to sniff, then sip: aromas and flavors are vivid too! Smell and taste give us our most direct access to the work of Hero Carbon's agents of change, the ferments and enzymes.

If wine is bottled poetry, Swedish surströmming is barreled infernality. Alan Davidson read classics at Oxford and slipped a Lucretian tall tale into his fishy scholarship, attributing Lake Avernus deadliness to the fumes of herring ripened dockside (see page 23). I can testify that it is indeed advisable to open gas-swollen cans of putrid surströmming in the open air. When I've done so, no birds were harmed, but many flies and wasps were attracted—along with a few intrepid smell explorers. And that's the tale's truth: the smells of decomposition draw people as well as flies. Like wine aromas, they're the products of fermentation, and they've likely signified valuable nourishment for thousands of human generations.

Though the term now has a specific technical meaning in biochemistry, *fermentation* generally names a method of food preparation that relies not on heat but on microbes and their enzymes. The word comes from an ancient root that meant "to boil" or "to bubble." Heat a pot of crushed fresh grapes over a fire and it will bubble with water vapor. Let another pot of the same crushed grapes sit unheated for a few days and it will bubble too, this time with the metabolic exhaust of the microbes growing in it. Both bubblings are signs of the original food's physical and chemical transformation. In boiling, heat energy disorganizes plant and animal tissues and generates new volatile bouquets. In fermentation, microbes and their enzymes do the same.

It seems likely that spontaneous fermentation would have been an early and important supplement to the cooking with fire that nourished the early development of our species. Any meat not eaten or cooked within a day or two would gradually become easier to chew on its own as muscle enzymes broke it down from within, the enzymes of ubiquitous microbes from without. The starch and fiber of leftover pounded tubers would become more digestible. If fuel for fire was scarce, simply hiding surplus foodstuffs under rocks or soil or water would give enzymes time to make them more easily assimilated—and more flavorful.

This unmanaged version of fermentation is essentially partial *decomposition*, a relatively neutral synonym for *decay, spoilage, rot, putrefaction*. Today the words and smells suggest inedibility, the risk of illness, and disgust. But decomposing foods aren't always toxic, and in the early days of our species a tolerance for them— and experience of which were safe—would have been advantageous. Their strong smells would become associated with the satisfaction of hunger, and even sought out simply for their sensory stimulation. The sailor John Jewitt, held captive on Vancouver Island around 1800, reported that the peoples of Nootka Sound frequently ate salmon roe fresh but also let tubs of it ferment, and "they esteem it much more when it has acquired a strong taste . . . though scarcely any thing can be more repugnant to a European palate." To this day, "stink eggs" and "stink heads" are prized salmon preparations among the First Peoples of the Pacific Northwest, permafrost-ripened walrus and caribou are enjoyed across the Arctic— and surströmming has its tugmasters.

With the development of agriculture in more temperate regions than the far north, spontaneous versions of decomposition gave way to a variety of *managed* versions that could keep surplus foodstuffs reliably edible for weeks or months. *Cured* foods—from the Latin "to care for"—are protected from microbes by salting or drying, and are transformed mainly by their own enzymes. *Fermented* foods are made by inviting the assistance of particular benign microbes. They defend themselves and the food with weapons that are harmless and even attractive to us, mainly tart acids and relaxing alcohol.

In the processes of curing and fermentation, food and microbe enzymes generate volatile molecules and flavors unlike those of spontaneous decomposition. Today, when we're able to refrigerate and freeze and can foods to keep them from spoiling, fermented foods are prized less for their shelf life than for those superfluous flavors. The people who produce them go to extraordinary trouble to make them their own, so we have a gloriously unnecessary profusion of cheeses, sausages, teas, alcohols, and vinegars, not to mention condiments so intense that a spoonful flavors whole dishes. Fermented foods offer countless opportunities to marvel at human ingenuity and idiosyncrasy and desire. And at the range of microbial partners that we've recruited, from plant leaves and flowers and tree bark, from the soil and the oceans, from our animal companions, and from ourselves: a vast network whose connections we're just beginning to trace.

Unmanaged food chemistry:
staleness and rancidity

Before we get to the aromas of managed cures and fermentations, there are a few related landmarks worth sniffing at and understanding: namely, the all-too-familiar smells of *un*managed food storage, of staling and spoilage.

Foods are complex chemical mixes that naturally change with time. The smells that we call **stale** and **rancid** come from volatiles generated by the air's reactive oxygen, often with the collusion of a food's own enzymes. *Stale* comes from a root meaning "stand," and in medieval times was applied to beer or wine that had stood long enough to become clear and strong: stale ale was a good thing. *Rancid* comes from the Latin for "rotten" and came to be associated mainly with the smell of old fats and oils. These smells become progressively stronger, so they're a marker of the time that has passed since a food was harvested or prepared: valuable information for our scavenging ancestors, and for modern-day *Homo sapiens* as we scavenge in cabinets and refrigerators and check our finds to see if they smell "off."

Stale and rancid smells develop mainly when some combination of enzymes, oxygen, and light energy breaks the long carbon chains found in all foods into small volatile aldehydes and ketones. Many of the same molecules are among the fresh-smelling green-leaf or cucumber volatiles or the mushroomy fungal volatiles, or contribute to the pleasant background fatty bouquet of cooked foods. The wonderful smell of sun-dried laundry comes from a mix of straight-chain C5 to C10 aldehydes. However, *slow* oxidation generates more mixed bouquets, often including odd kinked chains, whose overall smells are very different. They're usually described as metallic, fishy, painty, and especially "cardboardy." The cardboard note was first described in dairy products in the 1940s and associated with the nine-carbon, one-kink nonenals, which were in fact later identified as the dominant volatiles in manufactured cardboard: cucumberiness out of place!

SOME STALE AND RANCID BOUQUETS

Composite smell	Component smells	Molecules
cardboardy	fatty, cucumber, green, fried, mushroom	nonenals, nonadienal, decadienal, octenol
metallic	mushroom, geranium leaf, metallic	octenone, octadienone
fishy	boiled fish, paint, mushroom, geranium leaf, cucumber	heptenal, heptadienal, octenone, octadienone, nonadienal
painty	fresh, solvent, green, mushroom, paint	acetaldehyde, propanal, pentenal, heptadienal

Rancidity in **cooking fats and oils** is especially obvious because these ingredients are full of vulnerable long carbon chains and often have little or no aroma of their own. Multi-kinked polyunsaturated oils are more susceptible to breakage than oils high in monounsaturated oleic acid, or in saturates, and they release different volatile fragments. Butter gets cheesy; olive oil and peanut oil get cardboardy and reminiscent of stale nuts; canola and soy oils get fishy; highly polyunsaturated flaxseed oil (aka linseed oil) gets painty. Manufacturers have developed more stable high-oleic soy and canola varieties for longer and more neutral shelf life. Antioxidant additives also slow oil rancidity, as does storage in cool and dark conditions.

SOME RANCID COOKING FATS AND OILS

Fat or oil	Component smells	Molecules
butter	cardboard, metallic	nonenals, hexanal, octenone, octadienone
olive oil	cardboard, fried, tallowy	nonenal, decadienal, octenal, decenal
standard canola, soy oils; high-oleic oils	painty, fishy; waxy, green, sweet, tallowy	propanal, hexenal, heptadienal, nonadienal, decatrienal; octanal, nonanal, decanal, decenal

Oil-rich **nuts** quickly go rancid when removed from the protection of their shells, their surfaces often damaged in the process. Walnuts and pine nuts are highest in polyunsaturates and most susceptible. **Grains** and the **flours, breads,** and

cereals made from them are much leaner than nuts, but develop similar off-smells. **Dried herbs** lose much of their aromatic identity and fade into hay. **Frozen vegetables** commonly have a stale flavor due to cell damage and uncontrolled chemical reactions as their tissue freezes and thaws. Exposure to bright light can stink up otherwise unspoiled, drinkable **milk** by helping one of its vitamins (B_2) attack a sulfur-containing amino acid and generate methanethiol. Light "skunks" **beer** by encouraging acids in hops to misbehave similarly. **Leftover meats** are compromised by the same iron-containing heme molecule that helped generate their deliciousness: it continues to fragment the meat lipids, especially when they're reheated, and generates the version of staleness known as WOF, **warmed-over flavor.** Rewarmed beef can smell livery because it may have ten times the freshly cooked levels of metallic epoxy decenal and green hexanal. Cured meats—bacon, hams, sausages—suffer less from WOF because the curing salt's nitrite inhibits oxidation.

SOME STALE AND RANCID FOODS

Food	Component smells	Molecules
nuts	green, waxy, citrus peel, cucumber	hexanal, octanal, nonanal, nonadienal
rice, grain flours, cereals, breads	green, fatty, waxy, citrus peel, beany	hexanal, heptanal, octanal, nonenal, pentylfuran
dried herbs	haylike, cocoa-sweaty	methyl nonadione; methyl propanal & butanal
frozen vegetables	green, solvent, fishy	hexenals, hexanal, pentanal, pentenal, heptenal
milk, lightstruck	rotting vegetable, cooked potato, sharp, acrid	methanethiol, methional, formic acid, acrolein
beer lightstruck	cardboard, green, sweet, cocoa-sweaty; sulfurous, skunky	nonenal, hexanal, acetaldehyde, furfurals, methylbutanals; methyl butenethiol
rewarmed meats & poultry	green, metallic, potato	hexanal, epoxy decenal, methional

The volatiles of rancidity and staleness are undesirable when they're in the foreground, but they're sometimes valued as background aromas or as tokens of desirable long aging. In southern France, oxidized cooking oils made from ripe olives are appreciated, and pieces of old ham fat are added to long-cooked stews. Many fans of Parmesan cheese save the hard and rancid rinds to simmer in broths and sauces.

Managed food chemistry: cured hams, anchovies, eggs

When well managed, the storage of raw animal tissues produces extraordinary flavors. It gives the tissue enzymes a chance to break down the abundant proteins and lipids into savory-tasting and aromatic fragments, and the fragments a chance to react and generate new tastes and smells. This intensification of flavor, along with the tenderizing effect, is the purpose behind *hanging* wild game animals for a few days, *aging* beef for a few weeks, and *curing* various meats for months.

The ingredient that makes extended curing possible is salt, an initial dose of which severely limits the growth of microbes on the food so that it can mature indefinitely without spoiling. The cured meat most commonly eaten as is, uncooked, is the large hind leg of the pig, the **ham**. Cured hams often do develop surface molds and yeasts, but their inner flavor comes from the chemistry of the meat itself, which is influenced by many factors: among them the pigs' breed, diet, and freedom to move, and age at slaughter; the amount of salt applied, with or without nitrates and nitrites; and the temperatures, humidities, and duration of curing. In general, the warmer or longer the cure, the more extensive the chemical changes and the more intense the final flavor.

Most hams share a core volatile bouquet that includes cooked and animalic notes—cocoa, potato, sweaty, rancid, barnyard—but also a prominent fruitiness from esters. Modern industrial **"city" hams** are generally injected with a salt-and-nitrite brine and "cured" in a matter of hours, then cooked at around 150°F (65°C); they have a mild aroma with a number of sulfur volatiles generated by the cooking. **Italian prosciutto** hams are typically cured for a year at cellar-like temperatures averaging around 60°F (15°C), and they develop a notably sweet, fruity quality that perhaps pushes cardboardy nonenal in the direction of its melon-cucumber quality in plants. Traditional **American country hams** from Virginia through the Carolinas, now largely eclipsed by industrial approximations, are cured for nine to twelve months at ambient temperatures that can reach 85°F (30°C) in the summer, and they have mushroomy, nutty, honeyed notes; even when unsmoked they often carry a whiff of smoky guaiacol. Chinese **Jinhua hams**, named for their home city near Shanghai, are cured for less than a year but go through periods of sun-drying and intentional mold growth and spend some weeks near 100°F (37°C); they have mushroomy and sulfurous notes and an intense meatiness.

The generally acknowledged gold standard for cured ham is *jamón ibérico de bellota*, acorn-fed Iberian hams: made in southwestern Spain from an indigenous breed of pig that lives actively in open oaklands eating vegetation and acorns, is slaughtered at a mature year or more of age, and cured at ambient temperatures for at least eighteen months, long enough to experience the heat of two Spanish summers. The very specific diet generates long-chain aldehydes with floral and waxy notes, and the curing conditions generate an especially intense aroma.

SOME CURED HAMS

Ham	Component smells	Molecules
most kinds	cocoa, sweaty, green-rancid, cooked potato, barnyard, berry & tropical fruits	methylbutanal, methylbutyric acid, hexanal, methional, cresol, ethyl methylbutyrate
industrial, cooked (nitrites, 1-day cure)	+ ocean/cooked corn, onion, meaty, buttery	+ dimethyl sulfide, ethanethiol, methyl thiopene, dimethyl dithiofuran, diacetyl
Italian prosciutto (Parma, San Daniele, 12 months, 60°F/15°C)	+ cucumber-melon, fenugreek, caramel, floral, honey, toasty	+ nonenal, sotolon, furaneol, phenylacetic acid, phenylacetaldehyde, acetyl pyrroline
American country (9–12 months, to 85°F/30°C)	+ mushroomy, toasty, honey, floral, peachy, caramel, smoky	+ octenone, acetyl pyrroline, phenylacetaldehyde, phenylethanol, nonalactone, furaneol, guaiacol
Chinese Jinhua (6–10 months; periods of sun-drying & intentional mold growth; to 100°F/37°C)	+ mushroomy, toasty, smoky, garlic, meaty, honey, peachy	+ octenone, acetyl pyrroline, guaiacol, dimethyl trisulfide, dimethyl dithiofuran, phenylacetaldehyde, nonalactone
Spanish jamón ibérico (24+ months, to 85°F/30°C)	+ meaty, mushroomy, cheesy, fruity, caramel, floral, waxy	+ methyl furanthiol, octenone, butyric acid, methylbutyrate, furaneol, octanal, nonanal

From massive hams to finger-size fish: cured **anchovies** are known for developing a savory flavor reminiscent of Iberian ham, much more than simple fishiness, and for that reason are a favorite ingredient in Mediterranean cooking (French anchoïade, Italian bagna cauda). These small fatty fish in the herring family are cured in 20–25 percent salt for some months, during which enzymes from their muscles, a portion of their gut, and the gut's resident microbes generate meaty flavor molecules and their precursors. More suggestive of oceanic origins are perishable **caviars**, fat-rich eggs of sturgeon and other fish that are only briefly and lightly salted and then refrigerated, giving a simpler and milder mix of aldehydes.

In a class by themselves are ancient-looking **century eggs**, whose original Chinese name is *pidan*, "leather egg." They're usually duck eggs, cured for months with a salty and strongly alkaline paste or brine made from ash, lye, or mineral lime. The extreme chemical conditions accomplish the fragmentation work that enzymes do in hams and anchovies. The result is startling: a solid, brown, transparent albumen, a jade-green semisolid yolk, and a smell that's powerfully ammoniacal and sulfidic when first opened, milder when cut, and fishier than anchovies thanks to the volatility of trimethylamine in alkaline conditions.

SOME CURED FISH, FISH EGGS, AND DUCK EGGS

Food	Component smells	Molecules
anchovies	cocoa, potato, fatty, geranium, melon, fried, beef fat	methylbutanal, methional, heptenal, octadienone, nonadienal, decadienal, epoxy decenal
caviar (sturgeon eggs, salted & matured)	fishy, cucumber, floral, waxy, deep-fried, cooked potato	heptenal, nonadienal, nonenal, nonanal, decadienals, methional
century duck egg (salt & alkali paste or solution, months)	ammonia & sulfidic; mushroom, fatty, oniony, fishy	ammonia & hydrogen sulfide; octenone, hexanal, nonenal, decadienal, diisopropyl & diethyl & dimethyl disulfides, trimethylamine

Managed food chemistry: tea leaves

Dried tea leaves don't usually find themselves cheek by jowl with cured pork and anchovies, but their aromas are also the product of managed, curated enzyme action, however brief. In fact, that brevity makes their transformation all the more magical. It's wonderful to experience firsthand—and easy enough to do so. *Camellia sinensis* is native to southern China, but plants are widely available and hardy enough that they're grown commercially in the northern United States. Get a plant, pick a couple of tender new leaves, and crush one gently. It'll smell like the green leaf it is, thanks to the green-leaf volatiles (see page 154). Then rub both between your palms a few times and let them sit for an hour or so. Smell again.

Perfume! Flowers and fruits conjured from green leaves. That's the enzyme alchemy that tea makers throughout East Asia have mastered and elaborated on to create hundreds of different variations. Teas are a case study in the human pursuit of nuances of sensation for their own sake. The following paragraphs can only hint at their diversity and complexities.

The preferred materials for tea making are the new leaves at a plant's growing tips. They're packed with enzymes and small molecules for building themselves, and because they're still tender and vulnerable to insects, they stockpile a wide range of chemical defenses: not only the usual green-leaf volatiles, but also terpenoids and benzenoids ready to be released by enzymes when the leaves sense stress or damage, including simple physical friction or flexing. Some aromas develop when the leaves are harvested and left to wither with water loss. Tossing or rubbing the leaves generates more. And applying heat to kill the enzymes and complete the drying drives off some volatiles, but also creates new ones.

To make **green tea,** the leaves are harvested, sometimes withered, but not handled before they're heated; they tend to have the least perfumed, sometimes strongly green smell. **Chinese green teas** are dried by pan-firing the leaves on a heated metal surface, which generates roasted and smoky notes. **Japanese green teas**, including the most common, **sencha**, are first steamed to inactivate enzymes, then rolled to break the leaf cells and encourage some chemical reactions, and finally heat-dried at a relatively low temperature, around 230°F (110°C), generating a green and meaty sulfur volatile: the cat ketone! **Matcha** and **gyokuro** green teas are made from leaves that have been shaded for several weeks before harvest; this causes the leaf cells to dismantle their abundant protein machinery for energy, in the process generating amino acids and eventually dimethyl sulfide, a contributor to their oceanic, vegetal, seaweedy quality.

For **oolong or wulong tea**, named from the Chinese for "black dragon," the leaves are withered, then annoyed several times over the course of hours by grabbing and rubbing and tossing them by hand, or bouncing them around in a rotating drum. This bruising sometimes supplements natural disturbance in the plantation: the special aromas of Taiwanese Oriental Beauty and Darjeeling second-crop teas are due in part to the nibbling of insects. The floral, jasmine-like smell of many oolongs comes in part from methyl jasmonate, a twelve-carbon ester and stress signal that damaged leaves emit to stimulate defenses in other leaves and plants (see page 162). Leaves of the Jin Xuan variety make a notable "milky"-flavored oolong, perhaps thanks to jasmolactone and g-decalactone.

For **black or red tea**, developed in nineteenth-century China and famously adopted in India (Assam, Darjeeling) and Sri Lanka (Ceylon), the leaves are withered, then crushed or cut to break their cells apart, mix up their contents, expose them to the air's oxygen, and encourage a riot of enzyme activity and chemical reactions. The overall effect is richly fruity, floral, and honeyed.

SOME VERSIONS OF CURED TEA LEAVES

Tea	Component smells	Molecules
green, Chinese: Longjing (unshaded leaves, withered, pan-fired dry)	floral, fruity, roasted, green capsicum, smoky, ocean air	geraniol, linalool, nerolidol, ionone, dihydroactinidiolide, methyl jasmonate, methyl & ethyl pyrazines, isopropyl methoxypyrazine, vinylphenol, dimethyl sulfide
green, Japanese: sencha (unshaded leaves, steamed, rolled, mild drying)	meaty, animal, floral, cooked apple, deep-fried, metallic, green	mercapto methylpentanone, methoxymethyl butanethiol, indole, methyl jasmonate, damascenone, decadienal, octadienone, nonadienal

continued

Tea	Component smells	Molecules
green, Japanese: matcha, gyokuro (shaded leaves, steamed, hot drying)	seaweed, sweaty, green, floral, mushroomy, metallic, cucumber, floral	dimethyl sulfide, methylbutanal, hexanal, heptenal, octenone, octadienone, nonadienal, ionone
oolong, wulong, Chinese (withered, tossed, heated, rolled/oxidized, heated dry)	floral, fruity, creamy, sweet, animal, violet, green, ocean air	phenylethanol, methyl jasmonate, jasmolactone, furaneol, indole, linalool, ionone, nerolidol, d-decalactone, hexanal, dimethyl sulfide
black: Chinese Keemun; Indian Assam, Darjeeling; Sri Lankan Ceylon (withered, crushed, oxidized, heated dry)	floral, citrus, rose, violet, honey, fruity, green, malty, nutty	linalool, geraniol, ionone, phenylacetaldehyde, damascenone, hexenol, methylbutanal, methylpyrazines
pu-erh (sun-withered, rolled, fermented)	musty, fallen leaves, floral, woody, sweet	di- & tri-methoxybenzenes, ionone, linalool oxides, terpineol, syringol

Finally a completely different preparation of camellia leaves. **Pu-erh teas** come from the southern Chinese province of Yunnan, where the leaves are partly sun-dried, lightly rolled and heated to retain some enzyme activity, then massed together into cakes and stored in warm humid conditions for years. They develop a brown color and smell reminiscent of fallen autumn leaves. The main volatiles are methoxybenzenes derived from the leaves' phenolic rings (gallic acid), the same molecules that give native Chinese rose species their faint smell (see page 239).

Though oolong and black teas have often been loosely described as fermented, they're better called oxidized. Pu-erh teas are truly fermented, transformed by a host of wild microbes, notably species of *Aspergillus* molds. They're a managed version of the decomposing autumn leaves they smell like. And for us they're a bridge to the realm of microbial food decompositions.

Unmanaged household microbes: soiled and spoiled

Back in chapter 1, I described mistaking the smell of a delicious cheese in my own kitchen for a stagnant sink drain. In Japanese restaurants, the smell of natto, fermented soybeans (see below, page 557) is used to describe a "dirty rag" odor that can develop in unused *oshibori* hand towels. Evidently intentional fermentations share volatiles with the malodors of the modern household! We and our homes are loaded with exploitable traces of organic matter, and the microbes living in, on, and alongside us have always scented human life, even in these plumbed and cleansered and refrigerated times. We've already encountered their smells on our own bodies' surfaces and cavities and residues (chapter 6). Wafting from things we surround ourselves with to keep the natural world's entropic forces at bay, they're the very breath of that entropy, and they remain unavoidable points of reference for the aromas of fermentation, even if only subliminally.

Whereas the stale smells of uncontrolled oxidation come from a few aldehydes, the stinkier smells of unmanaged microbes come from the more varied waste products of their metabolism as they plunder whatever organic matter they can find: *spoiled* originally meant "looted." The smells of the soiled and spoiled cluster into five unlovely volatile bouquets. Three are mostly from bacteria: **sour** and vomitous short-chain acids, **sweaty-animalic** branched-chain acids from protein breakdown, and stinky **sulfurous** molecules from the same. Molds produce their **mushroomy-musty** eight-carbon defenses and signals. And **solventy-fruity** alcohols, ketones, and short esters are made mostly by yeasts and molds.

SOME HOUSEHOLD MICROBIAL BOUQUETS

General smells	Household sources	Molecules
sour	kitchen sponges & cloths, dishwasher, vases & flowerpots	short-chain acids: acetic, propionic, butyric, hexanoic
sulfurous	kitchen drains & sponges, toilet	sulfur-containing: methanethiol, hydrogen sulfide, methyl sulfides

continued

General smells	Household sources	Molecules
musty-mushroomy	basement, bathroom, dank carpet	C8 chains, terpenoids: octenol, octenone, octadienone, methylisoborneol
sweaty-animalic	unwashed or damp laundry	short branched acids & aldehydes: methylbutyric & methylhexenoic acids, methylbutanals
solventy-fruity, lifting	spoiling foods	ketones, alcohols, esters: acetone, heptanone, ethanol, ethyl acetate

There are plenty of niches in the human-built environment where microbes thrive, on particles or films of food residues, on the remains of tiny insect cohabitants and shed flakes of our own skin, and even on the lipid-similar soaps we use to clean up. Bacteria dominate in moist spots and emit their typical small-molecule sour and sulfurous smelliness from the **kitchen sink and drain, dish cloths and sponges, refrigerators, shower stalls and curtains, damp carpets and basements.** (A quick fix for sourness: alkaline baking soda pulls off a hydrogen from the acid molecules, leaving them charged and nonvolatile.) Even the appliances that we use to clean up turn out to shelter extremophiles, domestic versions of the microbes that thrive in hot springs and mine wastes. **Dishwashers** harbor bacteria and yeasts on their gaskets; **washing machines** reduce microbial numbers on dirty clothes but spread plenty through the whole load, some of which survive dryers and laundry lines. When washed fabric is remoistened by humidity or perspiration, one of these, the bacterium *Moraxella osloensis*, generates the familiar sweaty smell of **dank towels and laundry** (from 4-methyl-3-hexenoic acid, cousin to our body odor's 3-methyl-2-hexenoic acid).

SOME SPOILED FOODS

Food	Component smells	Molecules
spoiled meats	sour, vomitous, sweaty, rancid, cabbagey, putrid, fruity, fatty, malty, cheesy;	butyric & methylbutyric acids, methylbutanal, hexanal, octenal, nonanal, dimethyl sulfide & disulfide & trisulfide, methyl thioacetate, ethyl acetate & butyrate;
spoiled fish	+ fishy, ammonia	+ dimethyl-, trimethyl- & methylbutyl-amines
spoiled milk, cream	sour, vomitous, sulfurous, solventy, fruity	acetic acid, ethanol, acetone, ethyl acetate & butyrate, butyric acid, dimethyl disulfide, methanethiol
moldy bread	mushroomy, musty, earthy, fruity	octenol, octanol, methylisoborneol, damascenone
moldy citrus fruit	solvent, furniture polish	ethanol, ethyl acetate, limonene
moldy tomato	sweaty, sulfurous, solventy	methylbutanal, dimethyl disulfide, toluene, xylene

When it comes to foods themselves, spoiling fruits and vegetables, breads, and meats all share a loss of their own intrinsic volatiles and the appearance of new, strong smells that are out of place at best. Fruitiness is delightful, but not coming from a piece of raw **meat**, which as it spoils starts out fruity-solventy but quickly progresses to putrid. Sometimes modern preservation methods actually make things worse. Raw milk fresh from the udder develops a sour, appley, buttery aroma, the wild beginnings of sour cream and yogurt and cheese; if it's pasteurized, the souring bacteria are killed off, and refrigeration gives an advantage to cold-tolerant bacteria (*Pseudomonas*) that eventually turn **pasteurized milk** foul with sulfurous, fruity, solventy products.

Mix all kinds of kitchen scraps together and tie them up in a plastic bag that traps moisture and heat: that's the recipe for a microbial paradise! Hence the smells familiar from garbage-piled city sidewalks on summer nights, which take me back to the battered metal cans it was my chore to tend decades ago. A breathtaking exuberance of sourness and fruitiness, sulfides and solvents.

Managed microbes: decomposition redeemed

Thousands of years ago, the pioneers of fermentation discovered ways to manage decomposition so that foodstuffs would remain edible longer than usual, and become stronger-flavored—but not with the random miasmas of staling and spoilage. They recruited a select corps of decomposers: microbes that limit their attack on food molecules, discourage their more destructive or toxic competitors from invading, and emphasize the more pleasant volatile bouquets, with the sulfurous, sweaty, and musty usually—but not always!—providing a relatively unobtrusive depth to the overall smell.

MICROBIAL BOUQUETS IN FERMENTED FOODS

Bouquets of smells & molecules	Prominent in
sour, sharp: short-chain acids	brined pickles, wines, beers, vinegars, sausages, soy sauce & miso, cheeses
solventy-fruity: alcohols, esters, ketones	
sulfurous: thiols, sulfides	cheeses, soy sauce & miso, fish sauces, shrimp pastes
sweaty-animalic, cocoa: branched acids & aldehydes	soy sauce, cheeses
musty-mushroomy: C8 chains, terpenoids	cheeses, sausages

The most important group of fermentation microbes is the **lactic acid bacteria**, or **LAB**, so named because their main defensive weapon is smell-less, pleasantly tart-tasting lactic acid. This group predates the first land plants and diversified with them; then some plant-dwellers specialized further to live on the animals that eat plants, from insects to dairy cows to ourselves. They're everywhere, and most of them are quick at consuming sugars and don't need oxygen, so LAB get a head start on vulnerable organic matter of all kinds, from trampled vegetation to spilled mother's milk. They're behind the spontaneous fermentation of our **vegetable pickles**, they're intentionally added to make **yogurt** and **cheeses** and raw sausages like **salamis**, and they play a role in **soy and fish sauces**. *Lactococcus lactis*, the main source of buttery diacetyl in dairy products, likely came from the green pasturage and udders of cattle, while *Lactobacillus helveticus*, which intensifies the flavor of many long-aged cheeses, apparently arrived in milk via the digestive tract of chickens. The original home of *Lactobacillus sanfranciscensis*, a bacterium that contributes to sourdough breads, is the gut and frass of *Tribolium* flour beetles!

SOME FERMENTATION BACTERIA

Bacteria	Foods	Main aroma contributions
lactic acid bacteria: *Lactobacillus, Lactococcus, Streptococcus, Leuconostoc, Pediococcus, Enterococcus, Tetragenococcus* . . .	yogurt, cheeses, sausages, soy & fish sauces, sourdough breads	solventy-fruity, sour, sulfidic, buttery (alcohols, esters, acetic & other acids, sulfides, diacetyl)
acetic acid bacteria: *Acetobacter, Gluconobacter, Gluconacetobacter, Komagataeibacter*	vinegars, kombucha, lambic beers	vinegar (acetic acid)
Propionibacterium freudenreichii	Swiss-style cheeses	sweaty (propionic & branched-chain acids)
Brevibacterium linens, aurantiacum	washed-rind cheeses	sulfidic, sweaty (sulfides, thiols, branched-chain acids)
Staphylococcus xylosus, carnosus, equorum	sausages	solventy-fruity, sulfidic, sweaty (alcohols, esters, sulfides, branched-chain aldehydes)
Bacillus subtilis	fermented beans & nuts, e.g., natto, dawadawa	sweaty, nutty (branched-chain acids, pyrazines)
Clostridium perfringens	salt-rising bread, Mediterranean flatbreads	sour, cheesy, vomit (butyric, propionic acids)
Halanaerobium praevalens	surströmming	sour, vomit, excremental (butyric & propionic acids, sulfides & thiols)

Among the other bacteria recruited for food fermentations, the **acetic acid bacteria**—so named because they defend themselves with vinegary acetic acid—ferment alcohols into **vinegars** and cooperate with yeasts to make **kombucha**; they're usually found growing on damaged grapes and other fruits. The *Propionibacterium* that gives **"Swiss" cheeses** like Emmental their distinctive sharp, sweaty aroma, is sister to other species that live on our skin; the *Brevibacterium* that contributes to the orange color and stinkiness of brine-washed **Époisses and Limburger cheeses** has close relatives on sweat-salty feet (see page 114) and in saltwater sediments and fish.

By far the most important fermenters from the fungus kingdom are the **yeasts**, which like bacteria live as single cells. First among them is *Saccharomyces cerevisiae*, which defends its turf with pleasingly psychoactive ethyl alcohol (see page 363). It also produces other alcohols, simple acids, and the acid-alcohol esters typical of ripe fruits, and it lends these aromas to **alcohols** and the **vinegars** and **distilled spirits** made from them. Related *Zygosaccharomyces* includes species that colonize partly dried or fermented fruits and tree sap; they contribute solvent, fruity, and caramel notes to condiments like **soy and fish sauces** and true **balsamic vinegars**. *Geotrichum*, an unusual yeast that forms moldlike filaments, spoils fruits but also gives some **moist cheeses** like Saint-Marcellin their "bloomy" white coat, and contributes sulfurous and medicinal volatiles.

SOME FERMENTATION YEASTS

Fungi	Foods	Main aroma contributions
Saccharomyces cerevisiae, exiguus, pastorianus, bayanus, uvarum . . .	breads, wines, beers, rice alcohols, vinegars	heady, alcoholic, fruity, floral (alcohols, esters)
Zygosaccharomyces rouxii, sapae, lentus	soy & fish sauces, balsamic vinegars, kombucha	heady, fruity, sweet caramel (alcohols, esters, furanones)
Brettanomyces/Dekkera bruxellensis, anomala . . .	beers, wines	spicy, smoky, barnyardy (phenols, acetic acid)
Debaromyces hansenii	sausages, moist-rind cheeses	fruity, sulfurous, ammonia (alcohols, esters, methional, thiols, ammonia)

Fungi	Foods	Main aroma contributions
Yarrowia lipolytica	cheeses	sulfidic, ammonia (sulfides, thiols, ammonia)
Candida milleri, humilis, versatilis, etchellsii . . .	sourdough breads, soy sauce	heady, alcoholic, fruity, floral (alcohols, esters)
Geotrichum candidum	bloomy-rind cheeses	fruity, floral, sulfurous, medicinal (esters, thiols & sulfides, thioesters, ketones, phenylethanol, phenol)

The multicellular, filament-forming **molds** that help ferment distinctive foods and drinks come from fungal clans that are widespread in nature, in soils and on vegetation, and include destructive and highly toxic members. The most familiar are the white penicillium species, one of which grows on the surface of moist **Camembert and Brie cheeses**, and another couple on the surface of **dry-cured sausages**. Blue-green penicillium species color the air channels poked into **blue cheeses** (molds require oxygen to grow). The white molds contribute mainly mushroomy-musty eight-carbon chains; the blues, fruity-solventy ketones.

Species of the molds *Aspergillus* and *Rhizopus* are essential ingredients in Asian multistage fermentations that produce intensely flavorful **soy pastes and sauces** and various **alcohols**. These molds secrete copious enzymes for digesting seed starch and proteins into sugars and amino acids, which bacteria and yeasts can then thrive on in a second fermentation. The initial mold stage contributes typical fungal eight-carbon mushroominess and fruity, solventy, and cocoa notes.

SOME FERMENTATION MOLDS

Fungi	Foods	Main aroma contributions
Penicillium camemberti, roqueforti, nalgiovense, salamii	bloomy-rind & blue-veined cheeses, sausages	mushroomy, fruity (octanol, octenol, alcohols, ketones)
Aspergillus oryzae, sojae, kawachii	soy sauce, miso, fermented black beans, rice alcohols	mushroomy, sweaty, fruity, alcoholic (octenol, methylbutanal, alcohols, acetone, diacetyl, ethyl acetate)
Rhizopus oligosporus, oryzae	tempeh (soybean cake)	solvents, sulfurous, rancid (ethanol, acetone, ethyl acetate, butanone, methylbutanol, dimethyl sulfide & trisulfide, methional, methylpropanal)

The traditional fermentation starter of mixed wild molds is known as **qu** (pronounced "choo") in China and **meju** in Korea; the later and more refined Japanese version, **koji**, is a pure culture of *Aspergillus*. This shift from spontaneous and complex fermentations to the more controlled and simplified has been a common one over the last century, especially in industrial-scale production. More recently, enthusiasts and manufacturers alike have been exploring new combinations of microbes and ingredients. In the tables that follow, wherever possible I've collected information for traditionally fermented foods, the flavor standard that industrial approximations aim for.

Fermented vegetables:
pickles, sauerkraut and kimchi, olives, capers

Pickled vegetables are among the most popular fermented—or, these days, pseudofermented—foods. For millennia they were an important means of preserving summer's perishable bounty for the winter; nowadays they're usually quick-punch, salty-sour condiments, often manufactured by simple acidification and preserved by pasteurization. True pickles have richer flavors. They're among the easiest fermented foods to make, and they offer a vivid demonstration of how simple microbial selection influences fermentation flavors. Smell explorers, do try this at home: chop up some vegetable trimmings, divide them between two cups or bowls, immerse one batch in plain water, the other in water plus a spoonful of salt. Cover them, leave in a warm place, and sniff them every day.

The unsalted bowl is an approximation of how a couple of Chinese fermented specialties are made: the **rotted amaranth** of Shaoxing, and the liquid used to marinate and flavor pieces of bean curd to make **stinky tofu**. With limited oxygen dissolved in the water, all kinds of anaerobic microbes extract energy from whatever materials they can, including nitrogen- and sulfur-containing amino acids. The result, as you'll smell, is a swampy set of sulfides and putrid cadaverine.

Most other pickles are made with salt precisely to avoid that swampiness. A number of lactic acid bacteria tolerate high salt levels better than most microbes, so the salt gives them an advantage over the amino-metabolizing anaerobes. The LAB get their energy from the plant carbohydrates and lipid chains, from which they generate preservative lactic acid and a mildly sharp mix of volatile acids and alcohols: the smells of your salted bowl.

SOME FERMENTED VEGETABLES AND PICKLES

Vegetable	Component smells	Molecules
rotted amaranth, mei xian cai geng (& stinky tofu: chou doufou)	ocean air, truffle, onion, gassy, garlic, animal, putrid	dimethyl, methylethyl & diethyl disulfides, indole, cadaverine
cucumber	sour, cheesy, musty, cucumbery, floral	hexenoic acids, heptanol, nonadienal, phenylethanol
cabbage: sauerkraut	sour, vinegary, solvents, sulfurous	acetic acid, ethanol, acetaldehyde, ethyl acetate, ethyl lactate, methyl butanol, dimethyl di- & tri-sulfides
cabbage + garlic + capsicum: kimchi	as above + garlicky	as above + diallyl disulfide, allyl methyl disulfide, methyl propyl disulfide

continued

Vegetable	Component smells	Molecules
olives, brined	solventy, vinegary, fruity	ethyl & other alcohols, acetic acid, ethyl acetate
olives, salt-cured	fruity, solventy, vinegary, sweaty	methyl methylbutyrate, hexenol, methylbutanol, ethyl propanoate, acetic & methylbutyric acids
capers	floral, piney, cabbagey, fruity	nerolidol, linalool, ionone, terpineol, methyl isothiocyanate, ethyl benzoate, methyl & ethyl octanoates

In **cucumber pickles**, typically fermented for a few weeks in a brine around 6 percent salt, the plant enzymes that produce the characteristic nine-carbon chains are suppressed, so cucumberiness becomes just one note in a mix of sour and heady fermentation volatiles. **Sauerkraut**, "sour cabbage," is made with half the salt and has a more diverse microflora and aroma, including sulfides evolved from the thiocyanate defenses that define the cabbage family (see page 182); Korean **kimchi**, the cabbage pickle that includes garlic and red pepper, greatly amplifies the sulfurous bouquet with garlicky allyl sulfides. **Brined olives** take months to ferment due to the waxy layer coating each fruit; yeasts end up dominating and giving them a pleasant mix of alcohols, acetic acid, and a solventy-fruity ester. The drier environment of **salt-cured olives** also favors yeasts and produces a range of branched-chain molecules and more diverse esters. (Canned olives are dominated by cooked-vegetable dimethyl sulfide.)

To me the most delightful fermented vegetable is the salted unopened flower bud of the **caper** bush, a distant relative of the cabbage family, which manages to be cabbagey and floral at the same time. The florality comes from two terpenoids, nerolidol and linalool, that scent the striking white and purple open flower, and another, violet-like ionone, somehow liberated by the fermentation. Several fruity esters provide backup.

Fermented seed foods:
breads, tempeh, stinky beans

Unlike perishable fresh vegetables, most seeds can be stored indefinitely as is. Grains and dried beans are generally fermented not to preserve them, but to make these dietary staples easier and more interesting to eat—or to produce alcohols, which we'll get to later.

The most familiar fermented grain foods are breads, usually aerated by the dominant growth of carbon dioxide–producing yeasts in doughs of moistened finely ground grain. Most bread flavor is generated during the cooking process (see page 520), which drives off many yeast volatiles. (**Packaged dry yeast** for bread-making has an intense smell generated from amino acids by the drying process, and include sulfides and nutty-popcorny pyrazines and pyrroles.) Doughs that include significant numbers of LAB and other bacteria develop volatiles that persist even after baking, notably acetic and other short-chain acids. Such **sourdough breads** come with many names and formulas, including Ethiopian **injera** made from tef, and south Indian **dosa** and **idli** made from mixed batters of rice and black gram beans.

Sourdough aromas are pleasantly balanced when the fermentation proceeds at a cool room temperature, but overheated starters and doughs often generate too much sweaty propionic acid and/or cheesy-vomitous butyric acid. Ironically, **supermarket bread aisles** often have a pervasive sour smell because manufacturers load their plastic-confined loaves with propionic acid salts to prevent mold growth.

Butyric cheesiness is actually prized in a few unusual breads. For their **salt-rising bread** and **eftazymo**, cooks in the eastern United States and the eastern Mediterranean have long selected bacteria as their chief dough raisers. They begin by pouring boiling water over ground grain or beans. This step instantly kills most microbes, yeasts included, but stimulates the tough dormant spores of survivalist bacteria to germinate. During the body-temperature fermentation, species of *Bacillus* secrete enzymes that break down the seeds' starch to sugars. Then harmless strains of *Clostridium perfringens*—other strains of which cause gas gangrene and food poisoning!—metabolize the sugars and amino acids, in the process generating leavening hydrogen gas as well as cheesy butyric acid and sulfurous methanethiol.

SOME FERMENTED SEED FOODS

Food	Component smells	Molecules
sourdough breads, injera	sour, vinegary, cheesy, sweaty	acetic, propionic, butyric, pentanoic acids
salt-rising bread, eftazymo (Greek self-leavened bread)	cheesy, vinegary, sulfurous	butyric & acetic acids, methanethiol
douzhi (Chinese mung bean broth)	sulfurous, cheesy, cucumber, clove, green, potato, barnyard, mushroom	dimethyl trisulfide, butyric acid, nonadienal, eugenol, hexenol, ethylphenol, methional, cresol, octenone
sufu (Chinese fermented tofu)	animal, sulfurous, disinfectant, cheesy, fruity; red sufu: above + yeasty	indole, dimethyl di-, tri-, & tetra-sulfides, phenol, butyric acid, methyl butyrate; + alcohols, esters
natto (Japanese sticky soybeans)	butter, toasted, coffee, mushroom, fruity, cheesy	diacetyl, di-, tri-, & tetra-methyl pyrazines, hydroxybutanone, octenol, methylpropionic & methylbutyric acids, ammonia
tempeh (Indonesian soybean cakes)	cooked potato, fresh, mushroom, nutty/popcorn, cabbage, honey	methional, methylpropanal, octenone, acetyl pyrroline, dimethyl trisulfide, phenylacetaldehyde

In Asia, protein-rich beans are an important staple food, and fermentations make them less monotonously beany. Some are not universally beloved. **Beijing douzhi or douzhir** is the protein-rich liquid left over after ground mung beans are treated to facilitate the extraction of their starch for noodle making, including a fermentation by wild lactic and acetic bacteria and species of *Clostridium*. A 2018 analysis noted that this official element of the city's "intangible cultural heritage," despite its "strong sulfurous, sour, cheese odor . . . is still enjoyed by a portion of the population of Beijing." I can testify: it's certainly tangible to the nose!

Sufu is a version of stinky tofu made by fermenting the soybean curd itself, not just soaking it in a prefermented brine. Its makers encourage several molds (*Actinomucor, Mucor, Rhizopus*) to grow on cubes of the curd, then let the mold enzymes do their breaking down during several months of brining. The tofu protein serves to generate prominent animalic volatiles along with a characteristic contribution from disinfectant-like, medicinal phenol. A dark red version of sufu owes its color and typical alcohol and ester notes to species of a yeast, *Monascus*, which is cultured separately on rice and then added to the aging brine.

Natto is the somewhat notorious Japanese version of a soybean food made throughout Asia; Africa produces similar fermentations from a range of other seeds. The cooked beans are wrapped loosely in microbe-laden straw or leaves, or inoculated with a culture of the *Bacillus subtilis* var. *natto* bacterium, and fermented at body temperature for a day or two. The microbes break down the abundant soy protein and leave the beans with a brown color, an intriguing sticky-slimy coating, and an aroma that combines cocoa and coffee qualities with sweaty and sulfidic notes and sometimes ammonia, along with toasty pyrazines that probably form spontaneously in the warmth as the proteins break down. Commercial bacterial strains that generate a relatively mild aroma include some collected from Chinese dust clouds that pass over Japan in the spring: they're said to produce "sky natto"!

Tempeh is an Indonesian fermented soybean cake, the least challenging of all the fermented soy foods. It's made much like natto, but the cooked beans are washed free of their seed coats and fermented with *Rhizopus* mold, whose mycelium filaments grow through the mass and bind it together into solid cakes. True to its fungal nature, the mold gives the beans a mushroomy-musty eight-carbon volatile along with vegetable, nutty, and honey notes.

Fermented seed condiments: soy sauce, miso, jiang

Fermented seed *foods* are flavorful preparations of nourishing but bland grains and beans. Fermented seed *condiments* are flavor concentrates for adding to bland foods of all kinds to satisfy our appetite for sensory stimulation. The peoples of Asia have led the way in developing and refining fermented seed products that "help the rice go down." One, the soy sauce made from soybeans, is now a landmark flavor over much of the planet. Its distinctive balance of volatiles and salty-savory taste make it and similar preparations unusually versatile. A condiment made and tasting very much like soy sauce, **murri**, was prepared from barley in the medieval Arab world. Today inventive chefs in many countries are applying Asian techniques to all kinds of raw materials, from local seeds to insects and dried yeast, while manufacturers rely increasingly on shortcuts to make inexpensive approximations.

There are many variations on fermented seed condiments, but most develop their flavors through the same general two-stage fermentation. They begin with the growth of molds on cooked seeds to make the starter qu or meju or koji, rich in enzymes and flavored with cooked-grain and fungal volatiles. Then the starter is combined with fresh batches of cooked soybeans and grains and a very salty brine. As the mixture ferments for days to months, salt-tolerant lactic acid bacteria generate preservative and volatile acids, and salt-tolerant yeasts generate alcohols, esters, and carbon-ring furanones and sulfur volatiles. An aging period of months to years gives time for the countless products and remains of seeds and microbes to react further with each other.

The finished pastes are usually sold mostly as is, while fluid sauces are usually pasteurized before bottling, at above-boiling temperatures for a few minutes, or at 140°F (60°C) for days; this cooking step deepens the color and flavor yet further. The results are strong-tasting concoctions, salty and savory and sour and sometimes sweet, with an aroma rich in microbial and cooked volatiles.

The core of **soy sauce** aroma comes from the reactions of the amino acids with each other and with traces of sugars and short-chain fragments, in part during fermentation and aging, in part from the final cooking. The simple **industrial approximation of soy sauce**, soybeans pressure-cooked with very strong hydrochloric acid, generates an aroma surprisingly similar to the fermented version, the same core blend of familiar amino-acid fragments—cocoa-like methylbutanal, potato-like methional—and some "brown"-smelling cooked molecules. However, only the involvement of yeasts generates the volatiles that give fermented soy sauces their greater breadth and depth—namely, heady alcohols and fruity esters, the tropical-fruit, roasted-coffee, and roasted-meat thiols, and the caramel-sweet sotolon and "shoyu furanone."

SOME SOY SAUCES AND MISO PASTES

Condiment	Component smells	Molecules
fermented soy sauce (steamed soy + roasted wheat, rice or soy koji, brined & fermented 6–8 months, pasteurized)	cocoa, potato, caramel, fenugreek, fruity, tropical fruity, lifting, smoky-spicy, roast meat, roast coffee	methylbutanal, methional, sotolon, shoyu (hydroxyethylmethyl) furanone, ethanol, ethyl methylpropionate, ethyl thiopropionate, other esters, ethylguaiacol, methyl furanthiol, furanmethanethiol
acid-hydrolyzed soy sauce	cocoa, potato, sour, rancid, roasted, smoky	methylbutanal, methional, phenylacetaldehyde, acetic, formic, & butyric acids, pyrazines, furans, sulfides, guaiacol
miso, white, yellow, red (steamed soybeans + rice koji, fermented + aged 6–12 months)	caramel, sweet, potato, floral, fruity, smoky, roast coffee	shoyu furanone, sotolon, methional, furfural, hexanol, phenylethanol, ethyl methylpropionate & butyrate, phenylethyl acetate, other esters, guaiacol, furanmethanethiol
miso, mame/hatcho (steamed soybeans, soy koji, fermented + aged 2–3 years)	cocoa, potato, caramel, honey, mushroom, smoky, sulfurous, roast coffee	methylbutanal, methional, shoyu furanone, phenylacetaldehyde, octenone, guaiacol, dimethyl trisulfide, furanmethanethiol

Japanese **miso**, the defining ingredient in its namesake soup, is the most familiar of the semisolid versions of fermented soybeans. Most misos are made with a starchy rice-based *Aspergillus* starter with no roasted grains or pasteurization, so they're lighter in color than soy sauce, usually lack the cocoa-like methylbutanal and instead are dominated by alcohols, esters, and furanones, fruity and caramel-smelling. The exception is **hatcho or mame miso**, made almost entirely of soybeans, with far less sugar and far more protein and amino acids, aged for two years or more, and very dark in its color and roasted savory flavor.

SOME OTHER FERMENTED SEED CONDIMENTS

Condiment	Component smells	Molecules
dou chi (whole black soybeans, steamed & fermented)	cocoa, fruity, nutty, mushroomy, floral, honey	methylbutanal, ethyl methylbutyrate, methylbutyl acetate, dimethylpyrazine, octenol, phenylethanol, phenylethyl acetate, phenylacetaldehyde
doubanjiang (steamed broad beans, crushed & fermented)	vinegar, almond extract, honey, toasty, caramel, floral, potato, cocoa	acetic acid, benzaldehyde, phenylacetaldehyde, dimethylpyrazine, shoyu furanone, phenylethyl acetate, methional, butanal
tianmianjiang (steamed wheat bread, molded & dried, brined & fermented)	malty, bready, toasted, mushroomy, sour, honey, smoky/clove	methylbutanal, furfural, acetyl furan, octenol, pentanoic acid, phenylacetaldehyde, ethylguaiacol
gochujang (chili powder, flour, salt, fermented with meju)	bell pepper, sulfurous, potato, apple, floral, buttery	himachalene, ionone, ethyl & vinyl guaiacols, isobutyl methoxypyrazine, methional, ethanol, methylpropanal, ethyl methylpropanoate, phenylethanol, ethyl lactate

Among other Chinese seed condiments, **dou chi**, or **fermented black beans**, are whole steamed soybeans salted and fermented with *Aspergillus* mold, with savory and mushroomy volatiles balancing a modest fruitiness; **doubanjiang**, **fermented broad bean paste**, made from the starchier, less protein-rich fava bean, doesn't develop as much of the methylbutanal cocoa note, and the Sichuan version that includes chili peppers carries their distinctive woody-floral mix of terpenoids. **Tianmianjiang**, or **fermented wheat paste**, is bean-free, made from steamed wheat-flour buns that are allowed to mold, then dried, brined, and fermented; it has bready and mushroomy qualities. Korean **gochujang**, **fermented chili paste**, has become more widely known and available over the last decade; it's made from a mixture of ground dried chilis and grain flour, rice or wheat, and smells both fruity and vegetal.

Fermented fish and meats: fish sauces, katsuobushi, surströmming, salami

Long before there were fermented seed sauces, there were fermented fish sauces. These were likely an early discovery in Southeast Asia that highly perishable small fish can be preserved and infused with flavor simply by salting or brining them and storing them for months. Though largely forgotten in the West until recently, similar condiments were much used among the ancient Greeks (*garon*) and Romans (*garum* and *liquamen*). Those made from mackerel innards became especially prized.

The Western **fish sauce** revival has come with the appreciation of Thai and Vietnamese cuisines. Their **nam pla** and **nuoc mam** are made by mixing anchovy-size fish with a third to half their weight in salt, fermenting in warm conditions for months, then drawing off the liquid and aging it further. The dominant microbes are lactic acid bacteria, with muscle and digestive enzymes from the fish helping to break down proteins and oils, and follow-on chemical reactions generating brown pigments and flavors. The result is a salty, tart, savory fluid with a rich meaty, cheesy, sulfurous aroma—only slightly fishy, because trimethylamine from saltwater fish (see page 386) isn't very volatile at the acidic pH of the sauce.

Pastes of fermented fish and shrimp are also common ingredients in Southeast Asian cooking. They're typically fermented more briefly, **fish paste** sometimes with grains to encourage yeast growth and alcohol-fruity aromas. **Shrimp pastes**, made from partly sun-dried shrimp, are distinctively rich in nutty, toasty pyrazines and a more prominent fishiness from trimethylamine, possibly due to a dominance of protein-digesting microbes over lactic acid bacteria and the resulting alkaline pH.

The most elaborately produced fermented fish is without doubt the Japanese **katsuobushi**, which lends its savory taste and complex aroma to **dashi**, the basic Japanese broth and cooking liquid. Skipjack tuna fillets are simmered until firm, then briefly and repeatedly smoked over several days, then inoculated with *Aspergillus*, *Penicillium*, and *Eurotium* molds and fermented in a wooden box for some weeks, then sun-dried, then scraped clean—and this fermentation-drying cycle is repeated several times! The finished fillets are so hard and dense that they make a metallic ringing sound when struck; they have a distinctive mix of smoky, roasted, fishy, and sulfurous smells, with mushroomy and cheesy background notes from the molds.

There are relatively few examples of fermented animal flesh eaten as is, not as a condiment, maybe because it's not easy to eat a large portion of such strong-flavored stuff! Three exceptions that prove the rule are ammoniacal Icelandic shark and Korean skate, **hákarl** and **hongeo-hoe** (see page 387), and **surströmming**, the herring whose sulfurous, fishy, excremental fumes are allegedly deadly to birds. Surströmming is fermented in both barrel and increasingly swollen can by bacteria from saltwater sediments, and then eaten in chunks—but with bland potatoes and strong drink to wash it down.

The standout example of a beloved fermented meat is the dry-cured sausage typified by **salami**: chopped meat, usually pork, salted and stuffed raw into protein casings made from animal intestines or approximations thereof, fermented for weeks or months—from within by lactic acid bacteria, and from without by a coating of *Penicillium* molds and *Debaromyces* (and other) yeasts. Along with the fungal coating, nitrate and nitrite salts limit fat oxidation and rancidity and allow the meatier amino-acid and sulfur derivatives more prominence. The bacteria provide sour and cheesy acids, plus cocoa and potato notes; the fungi, their trademark musty mushroominess; the bacteria and yeasts, alcohols and esters and their fruitiness.

SOME FERMENTED MEATS AND FISH

Condiment or food	Component smells	Molecules
fish sauces (Thai nam pla, Vietnamese nuoc mam)	sulfurous, cocoa-sweaty, cheesy, fishy	methanethiol, methylpropanal & butanal, dimethyl trisulfide, methional, butyric & methylbutyric acids, trimethylamine
fish paste (Thai patis, Filipino bagoong)	cocoa-sweaty, almond extract, geranium leaf, metallic, mushroomy, sulfurous, fruity	methylbutanals, benzaldehyde, butanal, pentenal, octadienone, pentenol, dimethyl di- & tri-sulfides, ethyl acetate & butyrate, butyl butyrate

Condiment or food	Component smells	Molecules
shrimp paste (Thai kapi, Filipino belachan)	nutty, roasted, fishy, sulfidic, cocoa, sweaty, cheesy, buttery	dimethyl, dimethylethyl & trimethylethyl pyrazines, trimethylamine, dimethyl di- & tri-sulfides, methylbutanal & butyric acid, diacetyl
preserved skipjack tuna fillets (Japanese katsuobushi)	smoky, roasted, sulfidic, meaty, fishy, cucumber, metallic, mushroomy, cheesy, sweaty	guaiacols, phenols, cresols, pyrazines, furans, dimethyl sulfide & disulfide, thiofurans, trimethylamine, heptenal, nonenal, nonadienal, octenol, octadienone, butyric & methylbutyric acids
surströmming	fishy, rot, fecal, cheesy, vomit, disinfectant, fruity	trimethylamine, di- & tri-sulfides, butyric acid, phenol, butyrate esters
salami	sharp, cheesy, toasty, cooked potato, mushroomy, fruity	acetic, butyric, methylbutyric acids, methional, acetyl pyrroline, octenone, octenol, ethyl butyrate, ethyl methylpropionate

Fermented milks: yogurt, sour cream, butter, cheeses

Milk, the fluid first food of newborn mammals, is a versatile blank slate for fermentation. When transformed into the solid concentrate called **cheese**, it can support diverse communities of bacteria and yeasts and molds, and develop in flavor for months or years. Dairying peoples have coaxed it into hundreds of variations: rewarding territory for the smell explorer.

Milk fermentations share a few defining volatiles, and one of them highlights the central role of fermentation within the animal body. Four decades ago, food scientists reported sketching a recognizable caricature of cheddar cheese with just three molecules. Two were mildly sulfurous methional and buttery diacetyl. The third was butyric acid, which I've often described as "cheesy"—not an enlightening term when sampling cheeses!

Butyric acid is an important molecule in animal life. Bacteria produce this four-carbon chain from residual plant materials in the animal digestive system, where it both suppresses harmful microbes and feeds the cells of the intestinal lining (see page 97). Mammals supply this beneficial acid to their newborns by bundling it in their milk's fat molecules, to be unbundled by the infant's stomach enzymes. Microbial enzymes can do the same unbundling. Hence butyric acid's shifting qualities: it can suggest cheese, but also infant spit-up, or spoiled milk, or diapers, or manure. Here, best simply to let it stand for itself: *butyric* names the sour, sharp facet that many cheeses share with these other less pleasant animal by-products.

First a few noncheese dairy fermentations. **Yogurt** is soured and thickened by culturing scalded milk with lactic acid bacteria. Its aroma is that of cooked milk with a sharp and fresh edge, the freshness from acetaldehyde, whose green-apple quality is prominent in the watery whey when whole yogurt is strained. Preindustrial versions of **crème fraîche** and **sour cream** were the flavorful result when freshly collected raw milk was left overnight for its fat globules to rise and form the cream layer, and wild lactic bacteria grew spontaneously and lightly acidified it. **Butter** was churned from that acidified cream and carried the same volatiles, **diacetyl** most prominently, so diacetyl became the landmark milk-fat volatile. Today sour creams are made by culturing pasteurized cream with lactic acid bacteria; mass-produced butter is usually flavored with an extract of cultured milk or an aroma concentrate. Less familiar **smen** is a North African butter that's salted and sealed in jars to ripen for months or years, in the process losing diacetyl and gaining a cheese-like collection of acids and aldehydes.

SOME FERMENTED MILKS AND CREAMS

Product	Component smells	Molecules
yogurt	fresh, green, buttery, sour, ocean air	acetaldehyde, diacetyl, acetic & butyric acids, dimethyl sulfide
sour cream, crème fraîche	buttery, coconut, potato, meaty, ocean air, green apple, vinegar	diacetyl, acetoin, d-decalactone, methyl furanthiol, methional, dimethyl sulfide, acetaldehyde, acetic acid

Product	Component smells	Molecules
butter, cultured	butter, coconut, cheesy	diacetyl, d-decalactone, butyric acid
"buttermilk" (cultured skim milk)	coconut, sweet, buttery, vanilla, metallic	g- & d-decalactones & octalactones, diacetyl, vanillin, epoxy decenal, acetic acid, acetaldehyde, dimethyl sulfide
smen (ripened butter)	butyric, sweaty, fruity, green	butyric, pentanoic, hexanoic acids, ethyl methylpropionate, heptenal

Now a set of representative **cheeses**. Their flavors come from some combination of the following factors: the milk's own volatiles and enzymes, plus its own wild microbes if it's used raw, not pasteurized; an added enzyme preparation, rennet, that coagulates the proteins and forms solid curds; added "starter" lactic acid bacteria that acidify the milk promptly; added "adjunct cultures" of flavor-enhancing microbes; and after the curds are formed into a cheese, other microbes that are encouraged to grow on its surfaces during a period of maturing, or ripening.

I've sorted a dozen cheese styles into several groups. The first is flavored primarily by the milk, rennet, and starter bacteria. Fresh **goat cheese** has an overt goaty element from the animal's distinctive short-chain acids (see page 86), along with coconutty and grapey volatiles. **Mozzarella** may be made from cow's milk or richer water buffalo milk, the standard in its Italian homeland and productive of a stronger and more animalic aroma. Genuine Greek **feta**, brined or heavily salted, highlights sheep milk's short-chain acids; it's sharp, sour, butyric, sweaty, with mollifying fruity esters.

SOME FRESH AND BRINED CHEESES

Cheese	Distinctive smells	Molecules
goat, fresh	goaty, waxy, sharp, butyric, sweaty, buttery, mushroomy, potato, fecal, coconut, grape	hexanoic, decanoic, ethyl- & methyl-octanoic acids, diacetyl, skatole, g-octalactone, d-dodecalactone, aminoacetophenone
mozzarella: cow	fruity, winey, floral	ethyl & methyl butyrates, methyl butanol, phenylacetaldehyde, nonanal, octenol
water buffalo	mushroomy, floral, mothball/animal, fresh	octenol, nonanal, indole, hydroxybutanone, methyl butenol, octanone, heptanal
feta, sheep's milk	sweaty, butyric, fruity	hexanoic, decanoic, butyric acids & ethyl esters, nonanal, nonanone

A second group is importantly flavored by microbes that flourish only after the cheeses have been formed. **Camembert** and **Brie** have "bloomy" rinds, their surfaces white with a matte coat of *Penicillium* mold and *Geotrichum* yeasts that contribute mushroomy chains, fruity esters, stinky sulfur volatiles, and sometimes ammonia from extreme protein breakdown, evident when a cheese is first unwrapped or if it's overripe. **Blue cheeses** have colored veins where air channels have been poked into the curd so a different *Penicillium* mold can grow; they're made from a variety of milks and mold strains, but all owe their shared fruity-solvency smell to ketones, especially heptanone. The notoriously stinky sticky-rind **Époisses**, **Munster**, **Limburger**, and their ilk are regularly smeared—"washed"—with brine to encourage the growth of salt-tolerant yeasts and bacteria. The bacteria break down surface proteins into a mix of sweaty branched acids, sulfides and thiols, and medicinal phenol. The smell can quickly fill a room, but the inner cheese, and the experience of eating, is milder and more balanced.

SOME SURFACE- AND VEIN-RIPENED CHEESES

Cheese	Distinctive smells	Molecules
bloomy rind: Camembert, Brie	mushroomy, sweaty, buttery, sulfurous	octenol, octenone, undecanone, diacetyl, phenylethyl acetate, d-decalactone, methanethiol, dimethyl sulfide, butyric & methylbutyric acids
blue-veined: Roquefort, Gorgonzola, Stilton	buttery, sweaty, fruity-spicy, solvent, green, mushroomy, fruity	diacetyl, methylbutyric acids, heptanone & nonanone, octenol, ethyl hexanoate
washed rind: Époisses, Munster, Limburger	sweaty, sulfurous, floral, chemical, fruity	hexanoic & methylbutyric acids, phenylethanol, dimethyl disulfide, phenol, indole, acetophenone, ethyl acetate, methanethiol, methyl thioacetate & propionate & butyrate

A third group includes larger cheeses (to 130 pounds, 60 kilograms, or more) whose curds are often "cooked" to around 120°F (50°C) to remove more moisture, so they end up drier, firmer, and longer-lived than bloomy and sticky-rind cheeses. They're flavored mainly by wild, starter, and sometimes adjunct bacteria, along with surviving rennet and milk enzymes, as well as by the cooking step and by slow chemical reactions that take time to generate another layer of volatiles, including caramel-like sotolon and furaneol. Dutch **Gouda** is notable for its sweet, nutty, meaty mix of esters, sotolon, and a key beef-fat aldehyde (see page 504). **Emmental**, the familiar hole-filled Swiss cheese, gets its sweaty-sweet smell from two unusual bacteria: *Propionibacterium freudenreichii*, which produces both propionic acid and the carbon dioxide gas that forms the holes, and *Lactobacillus helveticus*, whose high enzyme activity leads to the formation of abundant fruity esters and furaneol. The rinds of **Comté** and **Gruyère** cheeses from the Jura Mountains are washed with brine during the initial period of maturing, and they develop a rich aroma with meaty, oniony, roasted, earthy, and sweet aspects. Swiss **Appenzeller** is notably strong in sweaty-foot branched acids.

Cheddar is the world's most imitated cheese. Made in the traditional way from raw milk on farms in its home region in the west of England, it's intensely flavored, earthy and vegetal and animalic. Approximated in factories worldwide for mass consumption, it's mainly buttery, gently sulfurous, mushroomy, and caramel-sweet, nowadays with the help of adjunct *Lactobacillus helveticus*. "Sharp" cheddars are aged long enough for butyric and other short-chain acids to become evident.

SOME SEMIHARD CHEESES RIPENED FOR MONTHS

Cheese	Distinctive smells	Molecules
Gouda	fruity, sweet, sweaty/cheesy, beefy, caramel	ethyl butyrate & hexanoate, hexanoic acids, methyl tridecanal, g-dodecalactone, sotolon
Emmental	sharp, butyric, sweaty, fruity, potato, caramel	acetic, propionic, & methylbutyric acids, methylbutanal, ethyl propanoate, furaneol
Comté, Gruyère, Appenzeller	savory, beefy, roasted onion, sweaty, fruity, honey, floral, earthy	methanethiol, dimethyl trisulfide, methylbutanal, nonenal, phenylacetaldehyde, methylbutyric & methylpentanoic acids, ethyldimethyl & diethylmethyl pyrazines
cheddar: farmhouse	soil, green capsicum, cooked potato, barnyard, coconut, cheesy, sweaty, rose, fruity, fenugreek	isopropyl & isobutyl methoxypyrazines, methional, cresol, d-dodecalactone, butyric & methylbutyric & acetic acids, phenylethanol, ethyl octanoate, damascenone, sotolon;
mild	buttery, fatty, cooked potato, popcorn, sulfidic, mushroomy, caramel	diacetyl, heptenal, methional, acetyl pyrroline, dimethyl trisulfide, octenone, octadienone, furaneol

The fourth group is a set of hard, dry cheeses that can develop in flavor for a year or two. **Pecorino romano** is a sheep's milk cheese coagulated with the protein- and fat-digesting enzymes in an extract of lamb stomach, so it's rich in butyric and sweaty fatty acids. **Parmesan**, also Italian in origin and the saltiest and driest of the group, at its best as genuine Parmiggiano Reggiano and aged for two years, is simultaneously sharply butyric and honeyed and fruity, often reminiscent of pineapple.

SOME HARD CHEESES RIPENED MONTHS TO YEARS

Cheese	Distinctive smells	Molecules
pecorino romano (sheep's milk, lamb rennet)	sour, butyric, sweaty, fruity	butyric, hexanoic, octanoic, acetic, propionic acids, ethyl butyrate, hexyl acetate, butanone
Parmesan	sharp, butyric, fruity, honey, sulfurous, nutty	acetic, butyric acids, ethyl butyrate & hexanoate, phenylacetaldehyde, diacetyl, dimethyl trisulfide, dimethylpyrazines
rinds of mimolette, tomme de Savoie cironée (*ciron* = mite)	lemony, honey, gasoline, fatty-nutty	neral, dehydrocitral, phenylacetaldehyde, tridecane, acaridial

To conclude, a couple of cheeses whose intriguing spoor I learned to track in French countryside farmers' markets. The dry rinds of aged **mimolette** are allowed to develop colonies of barely visible mites, which leave behind a telltale powder and a characteristic strong smell from several signaling molecules. These include lemony and gasoline-like chains and unusual **acaridial**, whose smell its discoverer Walter Leal described to me as "surprisingly good" for the pheromone of a spider relative! Cheese mites are generally considered to be a pest, but I and others appreciate their contribution to the experience of **tomme de Savoie** and mimolette, and the soft German **Milbenkäse** is named for and features them.

Fruit alcohols: wines

In the realm of food and drink, **alcohols** are the intoxicating and flavorful result when yeasts ferment the sugar in sweet fluids into ethyl alcohol. Humans have been making alcoholic beverages for something like seven thousand years from such various materials as ripe fruits, plant and tree saps, cooked grains or vegetables, and sugar-refining residues. In the following sections we'll sample some familiar modern alcohols and related products.

Wines are yeast-fermented plant juices, grape juice above all. They're named for the grapevine, and grape wines are by far the most sophisticated and widely produced. They apparently originated in the prehistoric Caucasus, then passed into the ancient Middle East and Mediterranean, and from there to western and northern Europe, the Americas, Australia, and nearly every temperate climate friendly to *Vitis vinifera* and its sister species. The myriad molecules of fermented grape juice generate combined sensations of smell, taste, and touch that can be intense, subtle, changing, fleeting, lingering: deliciously absorbing. They invite repeated sniffing and sipping, contemplation, sharing and comparing perceptions with others. For connoisseurs who taste systematically and develop an expert palate, wine flavors can be revelatory of the grapes and microbes and processes that made them.

These wonderfully superfluous pleasures, and the demand for wines that provide them, have inspired an immense body of knowledge and lore about wine flavors. The next few pages are meant to give the smell explorer a few guideposts to sniff for in wines themselves, and in the ongoing conversation about them.

Wine aromas are endlessly diverse, but they share a common vinous bouquet generated by *Saccharomyces* yeasts. It includes two of the five common microbial bouquets, the solventy-fruity and the sour. The major volatile is ethyl alcohol, ethanol, around 10 percent of the liquid, which is accompanied by traces of other alcohols, several short-chain acids, and ethanol-acid esters. The combination of acetic acid and ethyl acetate is known as "volatile acidity," an essential aspect of vinosity but a sign of deterioration when it becomes too prominent.

SOME COMPONENTS OF GENERAL VINOUS BOUQUET

Smell qualities	Molecules
alcohol, solvent	ethanol
solvent, cocoa, whiskey, floral	other alcohols: methylbutanol, phenylethanol
fruity, solvent	esters: ethyl & phenylethyl acetates, ethyl butyrate, ethyl methylbutyrate . . .
sharp, vinegar, sweaty	acids: acetic, butyric, hexanoic, octanoic
fresh, green apple, pungent	aldehydes: acetaldehyde
cooked apple, violet	terpenoid fragments: damascenone, ionone

Grapes themselves have a typically fruity endowment of esters and terpenoid fragments. Some grape varieties carry other molecules that help make their wines distinctive. Oddly, those defining volatiles and smells are often barely detectable in the raw grape juice! I experienced this phenomenon firsthand during the 2007 harvest at Fox Run Vineyards in upstate New York, when winemaker Peter Bell invited me to taste the juice of freshly crushed Riesling grapes. It was generically fruity, little more. Then he gave me some actively fermenting juice: strongly citrusy and floral. As the pioneering oenologist Émile Peynaud put it, "Wine smells more of fruit than the grape itself."

It's likely that all grape varieties store some defensive chemicals bound to non-volatile molecules, sugars or amino acids, to be released by the fruit's enzymes when it's damaged. Yeasts have similar enzymes—perhaps to be able to exploit the non-volatiles, perhaps to attract insect transportation by amplifying fruit signals (see page 364). So fermentation greatly enhances the native aromas of many varieties. White **Gewürztraminer**, **Muscat**, and **Riesling** grapes release citrus, floral, and specifically roselike monoterpenoids, Riesling also a petroleum-like naphthalene structure **TDN** (see page 420); the red **Syrah** grape, a black-peppery sesquiterpene. Many white grape varieties, but **Sauvignon Blanc** most prominently, release a range of sulfur-containing thiols with tropical fruit and blackcurrant or cat-urine smells. Sauvignon Blanc can also have a vegetal facet, thanks to the same methoxypyrazines found in fresh peas and beans; the red **Bordeaux** varieties sometimes share this as well.

SOME DISTINCTIVE WINE AROMAS DERIVED FROM GRAPES

Smell qualities	Main grape sources	Molecules
floral, citrus, rose, lychee	Muscat, Gewürztraminer, Riesling	linalool, geraniol, rose oxide
kerosene	Riesling	TDN (trimethyl dihydronaphthalene)
coconut, sweet	Gewürztraminer	wine lactone
black pepper	Syrah	rotundone
passion fruit, grapefruit	Sauvignon Blanc & others	mercaptohexyl acetate, mercaptohexanol
boxwood, blackcurrant, cat urine	Sauvignon Blanc & others	mercaptomethyl pentanone

continued

Smell qualities	Main grape sources	Molecules
vegetal, green	Cabernet Sauvignon & Franc, Merlot, Sauvignon Blanc	methoxypyrazines
"foxy": sweet, strawberry	Concord, muscadine, scuppernong	aminoacetophenone

The primary contributors to wine aroma are the main yeast fermentation and the grapes themselves. Several other secondary contributors may come into play, depending on how the grapes and wine are handled. **Lactic acid bacteria** sometimes contribute buttery diacetyl. ***Brettanomyces* yeasts** metabolize protective phenolic rings in the grape skins and generate debatably desirable leather, "sweaty saddle," barnyard, clove, and smoke notes. Traces of **sulfur** and **copper** from traditional vineyard sprays, and **sulfur dioxide** used to control the fermentation, can react to form volatiles with sulfurous, smoky struck flint, and mineral qualities (see page 369).

An important factor in the flavor of many wines is temporary storage in contact with **wood barrels, staves, or chips** whose surfaces are often "toasted" with a gas flame to pyrolyze the wood components and add sweet, spicy, nutty aromas (see page 411). Barrel storage goes back many centuries, but the connoisseurship of oak species and toasting degrees only a few decades. Furans from the wood and sulfur compounds from the yeasts can sometimes react to form thiols that smell of roast coffee, roasted meat, and smoky struck flint.

SOME WINE AROMAS DERIVED FROM BARRELS, MICROBES, AND REACTIONS

Smell qualities	Main sources	Molecules
buttery	lactic acid bacteria	diacetyl
barnyard, "sweaty saddle," clove, smoky	*Brettanomyces* yeasts	ethylphenol, guaiacol, catechol; vinylphenol, vinylguaiacol
sulfurous, struck flint, mineral	yeast sulfur molecules, added sulfur dioxide, dissolved copper & iron	hydrogen sulfide, methanethiol & ethanethiol, sulfanes

CURED AND FERMENTED FOODS 573

Smell qualities	Main sources	Molecules
woody: smoky, clove, vanilla, coconut, stable	toasted oak: barrels, barrel staves, chips	guaiacol, eugenol, vanillin, oak (whiskey) lactone, ethyl & vinyl phenols & guaiacols
roasted coffee or meat; struck flint, smoke	barrels + chemical reactions	furfurylthiol, methyl furanthiol; benzenemethanethiol
aged: green apple, honey, potato	slow oxidation & other chemical reactions	acetaldehyde, phenyl-acetaldehyde, methional
fenugreek, caramel		sotolon
sulfurous, vegetal		DMS, methanethiol
coconut		lactones
dried fruit		methyl nonadione

Oxygen and the passage of time are responsible for chemical reactions among the many grape, microbe, and barrel molecules, whose overall effect is called **aging**. Some wines are thought to be at their finest after years in the bottle. Initially harsh branched alcohols are converted to milder aldehydes; eventually fruity and floral aromas fade as esters split back into acids and alcohols, and alcohols and thiols (sulfur alcohols) are oxidized to aldehydes. Aged aromas resemble honey, dried fruits, and the sulfurousness of cooked and canned vegetables; nutty lactones often increase as well, and eventually the caramel note of the furanone sotolon.

Sweet-smelling sotolon is a key volatile in several unusual kinds of wine. **Sherries** from southern Spain are white wines aged in partly empty barrels; oxidation of alcohols to aldehydes, and the chemical reaction of acetaldehyde to produce sotolon, generate much of the distinctive nutty-sweet aroma. **Madeira**, from the Portuguese island, is held for months at 115°F (45°C) to approximate the extreme conditions of shipping wine to the Americas and Asia centuries ago. The result is facets of honey, cooked apple, and bread.

SOME WINE AROMAS DERIVED FROM SPECIAL HANDLING: SHERRY, MADEIRA, NOBLE ROT

Wine & smell qualities	Main sources	Molecules
sherry: fresh green, cocoa/ sweaty, buttery, fenugreek/caramel	flor yeast, oxidation	diethoxyethane, methylbutanals, acetaldehyde, methional, diacetyl, sotolon
Madeira: honey, cooked apple, bready, toasty, fenugreek/caramel	*estufagem* (115°F/45°C for 3 months), oxidation	phenylacetaldehyde, damascenone, furfurals, guaiacol, sotolon
Sauternes*, Tokaji Aszú, Trockenbeerenauslese: honey, peach & apricot, caramel, fenugreek * + grapefruity, citrus, onion	*Botrytis* mold, grape defenses, chemical reactions	phenylacetaldehyde, g-nonalactone & g-decalactone, furaneol, shoyu furanone, sotolon * + sulfanyl hexanol & pentanol & heptanol, methyl sulfanyl butanal

The most remarkable of the unusual furanone-touched wines begin with what's been called the fungal equivalent of Dr. Jekyll and Mr. Hyde, the destructive gray-rot mold *Botrytis cinerea*. When autumn vineyard conditions are just right, it can become the more restrained "**noble rot**," which perforates the grape skins, stimulates their chemical defenses, leaves its own chemical marks, and encourages unusual yeasts to join in. In short, it ferments the grapes alive, transforming them on the vine. These botrytized grapes are then fermented with yeasts to make Hungarian **Tokaji Aszú**, French **Sauternes**, and German/Austrian **Trockenbeerenauslese**, all sweet with sugars and complementary honey and fruit aromas; the Sauvignon Blanc grapes in Sauternes lend sulfurous grapefruit, blackcurrant, and even bacon-like notes.

SOME AROMA DEFECTS IN WINES

Defect and smells	Main sources	Molecules
cork taint: musty	cork disinfectant + fungus	trichloroanisole; geosmin, methylisoborneol, methoxydimethyl pyrazine
smoke taint: ashtray, tarry	fires near vineyard	guaiacol, methyl & vinyl guaiacols, cresols
ladybug taint: vegetal	vineyard infestation	methoxypyrazines
fruit-fly taint: waxy, floral	female fruit-fly in wineglass	undecenal

Wines can end up with undesirable smells, which include an excess of volatile acidity in the vinous bouquet or of the sweaty-saddle volatiles produced by *Brettanomyces*. Musty, dank **cork taint** is caused by fungal growth on chlorine-disinfected cork stoppers. In **smoke taint**, grapes exposed to smoky air detoxify the phenols and guaiacols by bonding them to nonvolatile sugars; later, grape and yeast enzymes release them during fermentation, as do mouth microbe enzymes during drinking! **Ladybug taint** is a green-pepper, asparagus smell that results when too many of those helpful insects end up crushed along with the grapes. Then there's **fruit-fly taint**: if you see a fly overcome by the bouquet of an open glass or carafe, fish it out quickly! Female drosophila carry a waxy-floral aldehyde as a pheromone to attract males, and can leave a detectable off-odor in the glass within minutes.

Grain alcohols: beers

Like wines, beers are yeast-fermented plant matter, beloved since the Stone Age for the mind-altering alcohol supplied by *Saccharomyces cerevisiae*. In fact, the name of this workhorse yeast means "sugar fungus of beer," christened as it was in the nineteenth century by scientists in beer-making northern Europe, not the grape-friendly south. Unlike sugary wine grapes, the **barley** and other dry grains that support beer fermentations lock their sugars up in the tightly packed long chains that we call starch. So how do alcohol-loving humans get a sugar-loving yeast to ferment grains? By cooking the grains to unpack their starch chains, then enlisting enzymes to break them into their free sugars—and in a variety of ingenious ways. These include the application of human saliva, the preparation of moldy grains in the Chinese starter qu, and most direct of all, simply moistening a portion of raw grains, or *malting* them, so that they begin to sprout and generate enzymes for turning their *own* starch into metabolizable sugars.

Freshly malted, grassy-smelling grains are "kilned," or heated to dry them out for later use and to generate flavor. Low kilning temperatures preserve enzyme activity and produce a mild flavor; high temperatures sacrifice enzymes for a darker color and stronger roasted flavor. **Barley malts and malt extracts** are also used as ingredients in candies and sweet drinks; they have a distinctive smell due largely to the methylbutanals, which for this reason are often described as "malty."

To brew a batch of beer, some combination of ground-up malt and raw grain is mixed with water at around 150°F (65°C), hot enough to unpack the grain starch and speed the enzymes' work. After some time, usually an hour or less, the resulting sweet grain juice, or *wort*, is drawn off and boiled for an additional few hours with another ingredient: **hops**, the not very flowery-looking female flowers and seed clusters of a vine closely related to cannabis. This step drives off many of the malt and cooked-wort volatiles with the water vapor, but extracts new dimensions of flavor from the hops.

Hops were apparently introduced to brewing in medieval Bavaria to slow spoilage of the finished beer. They gradually caught on everywhere, and today they often dominate beer aroma. Like cannabis buds, they're richly endowed with terpenoids, and in addition carry defensive phenolic ring molecules, called the alpha and beta acids, that suppress spoilage bacteria and produce pleasantly bitter molecules during the wort boiling. Today brewers can choose among many different hop varieties and use them fresh or dried or aged; they often add some late in brewing to retain volatiles that wort boiling drives off. Hop volatiles in beer range from floral and resinous terpenoids to fruity esters, citrusy and blackcurrant-catty thiols, and leafy, sweaty aldehydes and acids.

SOME BEER AROMAS DERIVED FROM HOPS

Smells given to finished beer by hops	Molecules
cooked apple, rose	damascenone
floral, rose, violet	linalool, geraniol, ionone
citrus fruit, apple, pineapple	ethyl methylbutyrates & methylpropanoate & pentanoate

Smells given to finished beer by hops	Molecules
green, leafy	hexanal, hexenal, hexenol, nonadienal
sulfidic	dimethyl trisulfide
fruity, blackcurrant, catty, grapefruit	mercaptomethyl pentanone, mercaptopentanol, mercaptohexanol
sweaty	methylbutyric & hexenoic acids
resinous, spicy	myrcene, humulene

After the wort and hops are boiled together, the liquid is cooled and strained and fermented for some weeks. Yeasts give beers a basic cerevisial bouquet quite similar to the vinous bouquet of wines, but with different proportions of alcohols and esters, usually little volatile acidity, often a touch of clove-smoky vinylguaiacol from phenolic rings that grains use to reinforce their cell walls and seed coats, and one or two sweet, caramel furanones. In **dark beers**, brewed with strongly kilned malts, caramel furanones and maltol and the phenolic rings, some smoky and tarry from the pyrolyzed grains, are especially prominent. The bouquet of **stale beer,** familiar from bars and long-forgotten bottles, is a mix of cardboardy and solventy aldehydes that are present in the fresh beer bound to other molecules and slowly released with time. They're sometimes joined by a skunky-smelling sulfur molecule that exposure to light generates from hop thiols.

There are three traditional methods of fermenting grain juice into beer, and they produce three very different kinds of flavors. The mildest is the most recent and the most popular worldwide. **Lager** beers were developed several hundred years ago in Germany, where brewers intent on fermenting and storing beers at cool cellar temperatures (*lagern* means "to store") singled out an unusual yeast that worked well in those conditions and settled conveniently to the bottom of the fermentation tank. That yeast is now recognized as a domesticated hybrid of the standard beer yeast with its own name, *Saccharomyces pastorianus*, and it's found nowhere in the natural world except as a brewery escapee. Bottom-fermented lager beers have a subdued blend of fruity, flowery, solventy, and sulfurous fermentation volatiles.

Before the development of lagering, beer wort would be fermented at warm room temperatures by the addition of *Saccharomyces cerevisiae*, which would collect at the tank surface. The combination of warmth and ready access to oxygen results in more active microbial metabolism and a greater quantity and diversity of volatiles in these top-fermented beers, often called **ales**, which are characterized by a strongly fruity aroma and little or no sulfurousness. A distinctive version is Bavarian **Weissbier** or **Hefeweizen**, which is made with a large portion of **wheat** and particular yeasts, *Saccharomyces delbrueckii* and others, that are copious producers of a banana-like ester. They also metabolize the grain seed-coat phenolic rings into clove/medicinal/barnyard rings—which are unwelcome notes in many other beers, as the nearly identical brett flavor is in many wines. Most domesticated beer yeasts have nonfunctional genes for generating the rings, but some brewers now seek out strains still designated as POF+, or phenolic off-flavor positive, precisely to obtain those not-necessarily-off flavors.

SOME MALTS AND BEERS

Malt or beer	Distinctive smells	Molecules
malt	cocoa, sweaty, cooked potato, fried	methylbutanals, methional, decadienal
bottom-fermented		
lager, light	fruity, malty, floral, ocean air, smoky-clove, caramel	damascenone, ethyl butyrate & hexanoate, methylbutanol, phenylethanol, dimethyl sulfide, vinylguaiacol, furaneol
lager, dark	caramel, sweet, fruity, sweaty	as for light lager, except: lower esters, much higher furaneol, phenylethanol, methylbutyric acid, maltol
top-fermented		
pale ale	fruity, malty, floral, caramel	as for light lager, except: 2 to 20 times higher esters; little dimethyl sulfide
stout (dark malt)	sweet, caramel, clove, tarry	maltol, ethylguaiacol, cresols

Malt or beer	Distinctive smells	Molecules
wheat	clove, barnyardy, medicinal, sweaty, banana	vinylguaiacol & phenol, methylbutyric acid, methylbutyl (isoamyl) acetate
wild-fermented lambic	spicy, clove, medicinal, fruity, sour, vinegary, ocean air, sweaty	ethyl & vinyl guaiacols, ethylphenol, methyl cinnamate, acetic acid, dimethyl sulfide, hexanoic & methylbutyric acids
stale lightstruck, skunky	cardboard, sweet, bready, solvent; sulfurous	nonenal, hydroxymethyl furfural, ethoxymethyl furan; methyl butenethiol

The third traditional beer fermentation method is spontaneous, or "wild," letting the microbes enter the wort from the environment. The **lambics** of Belgium are made with a mix of malted barley and wheat. The boiled wort is cooled overnight in large open pans, then slowly fermented in microbe-laden wooden vats and then barrels for a year or more, during which more than two thousand different fungi and bacteria come and go. Species of *Saccharomyces* are gradually replaced by the same *Brettanomyces* yeasts that give barnyard and other phenolic notes to wines. Lactic and acetic acid bacteria also contribute volatile acidity and a winelike aspect that would be a mark of spoilage in other styles of beer. Some lambics are made frankly fruity by the addition of fruits or fruit syrups, cherry and raspberry among the most common. Today lambic-like "sour" beers are made with a variety of methods.

Grain alcohols: Chinese rice wines, Japanese sake

Even with the availability of everything-included beer brewing kits, it still takes several weeks for the smell explorer to experience the everyday magic of bland cereality transformed into fragrant alcohol. Not so with Asian alcohols! The simplest require just a few days and a manufactured domestic edition of the Chinese qu starter culture. Local versions of these ingenious starters are found throughout Asia and house a tremendous diversity of microbes. In the West, the easiest to find are dry, marble-size Shanghai or wine "yeast balls" (*jiuqu* or *jiuyao*), which in addition to yeasts carry molds to digest the grain starch into sugars, and lactic bacteria that discourage spoilage microbes and supply tartness. To make fresh rice wine, you cook and cool a few cups of glutinous rice, mix in a ground-up yeast ball, and leave it in an airtight container in a warm place. After a few days, the rice exudes a clear liquid: **fresh rice wine**, known as **laozao** or **jiuniang** in China, **khao mahk** in Thailand, **makgeolli** in Korea. . . . It hits most of the mouth's pleasure buttons, simply and delightfully: sweet, tart, savory, fruity, and lightly boozy with a typical yeast bouquet of heady alcohols and esters.

A much stronger rice wine in both flavor and alcohol (8 to 18 percent) is the Chinese **yellow wine**, **huangjiu**, made in essentially the same way but over a much longer time. The qu starter for rice wines is usually made from wheat fermented at warmer-than-body temperatures, 120°F (50°C) and above, and is usually dominated by *Aspergillus* and *Rhizopus* molds, with several yeasts and a dozen or more bacteria accompanying. A large proportion of qu is added to the steamed rice and provides a distinctive qu quality, described as yeasty and bready, to the finished wine. This second fermentation traditionally lasts several months around 75°F (25°C) in open earthenware vessels, followed by several years of aging and then the mild cooking of pasteurization. The result balances fruity esters with clove, potato, mushroom, and floral volatiles (regulations specify minimum levels of rosy phenylethanol), while aging brings sweet vanillin and the soy-sauce-like mix of methylbutanal and sotolon. Modern streamlined production methods tend not to generate the strong sauce and vanilla notes.

SOME ASIAN RICE WINES

Alcohol	Component smells	Molecules
laozao: Chinese fresh rice wine	solvent, floral, fruity, vinegar, butyric	methylbutanol, phenylethanol, phenylethyl acetate, ethyl acetate & butyrate, acetic & propionic & butyric acids
huangjiu: Chinese yellow rice wine	clove, cooked potato, mushroom, sulfidic, fruity, floral, vanilla, cocoa, caramel, sweaty, almond extract	vinylguaiacol, methional, octenone, dimethyl trisulfide, ethyl butyrate & hexanoate, phenylethanol, vanillin, methylbutanal, sotolon, butyric & methylbutyric acids, benzaldehyde

Alcohol	Component smells	Molecules
sake: Japanese rice wine	alcohol, fruity, pineapple, apple, banana, grassy, sweaty	methylbutanol, ethyl hexanoate & octanoate, methylbutyl acetate, hexanoic & octanoic acids;
aged	fenugreek, cocoa	sotolon, methylbutanal

The most delicate of all grain wines is **nihonshu** or **sake**, a Japanese offshoot of Chinese rice wine that became especially refined in temples and shrines beginning in the twelfth century, then both streamlined and further refined in the twentieth. Instead of a complex wild starter, it's made with a pure and select culture of the mold *Aspergillus oryzae*, the koji, and specific strains of yeast that have been selected for their brewing qualities. The main fermentation takes place on rice grains that have been "polished" to remove not only the seed coat and germ, but the outer oil- and mineral-rich layers as well—at least a third and sometimes more than half the mass of the original grain. Sake rice is therefore mostly starch, largely a blank volatile canvas against which the yeasts' creativity can be appreciated.

Since the nineteenth century, sake fermentation temperatures have been kept low, between 45°F and 65°F (8°C and 18°C), originally by brewing only in the winter: conditions that favor the production of fruity esters. There are many possible variations on production conditions, degrees of rice polishing, acidification with spontaneous or pure lactic acid bacterial cultures or the simple addition of lactic acid, yeast strains, and additions of concentrated alcohol, sugars, and other ingredients. Of the many types of sake, those called *junmai* are made from koji, rice, and water only, while *honjozo* sakes can include some added alcohol. The "special" and "very special" categories, *ginjo* and *daiginjo*, are defined by the use of rice grains milled to less than half their original weight, fermentation at the lowest temperatures for four or five weeks, and a distinctively fruity-floral aroma, *ginjo-ka*. Ordinary *junmai* and *honjozo* sakes are fermented at warmer temperatures for three weeks, and are more likely to have cereal, cocoa, and sometimes mushroomy or sweaty notes. Most sake is aged for just six months and consumed over the next year or so. Relatively rare **aged sake**, matured for at least three years, develops *jukusei-ka*, "ripened smell," by accumulating sotolon and methylbutanal along with sulfur volatiles, which give it sherry, Madeira, and more savory soy-sauce notes—aspects that it shares with Chinese yellow wines.

Vinegars and kombucha

If you leave an unfinished bottle of wine or beer or sake out for a few days, it turns sour. Manage the same conditions for a few weeks or more, and you transform a spoiled drink into the intensely flavored, versatile liquid called *vinegar*. The name comes via French from the Latin *vinum*, "wine," and *acer*, "sharp" or "sour." Both of our words *acid* and *acetic* come from that second root, and acetic acid is the molecule that defines vinegar, making up around 5 percent of the liquid. There are many different acids in our food and drink, all of them sour to the taste, but not all are volatile. Acetic acid is, and it smells sour and pungent as well as specifically vinegary. Vinegars have many uses in the kitchen, especially in sauces and condiments, and are also enjoyed in drinks simply diluted with water or as a component of fruit-syrup mixtures called shrubs.

Vinegars are the product of fermentation by acetic acid bacteria, which add oxygen to yeast's defensive two-carbon ethyl alcohol, H_3CCH_2OH, turning it into their own two-carbon weapon, acetic acid, H_3CCOOH. They can be made from a wide range of starting materials, from fruits and grains to the sweet saps of palm trees and sugarcane, young coconut juice, even food wastes. Access to the air's oxygen is essential to the fermentation, which takes months when the alcohol is simply left in a partly filled barrel, just a day or two if it's constantly aerated. Slow fermentations tend to generate more alcohol-acid esters that help fill out the acetic aroma.

Distilled or **spirit vinegar** is the simplest and cheapest because it's made on an industrial scale with nearly pure distilled alcohol, diluted and supplemented with minimal nutrients for the acetic acid bacteria. Ordinary **wine vinegar** can have floral, buttery, even cheesy facets, while true **sherry vinegar** from Jerez, matured and sometimes also fermented in the solera system of partly filled wood barrels, carries a wider range of fruity esters, with a fenugreek-caramel note from sotolon that develops in months to years (for *riserva* vinegars) of aging. The apple origins of **cider vinegar** come through in a couple of distinctive esters, while **malt vinegar**, a favorite in beer-loving England, distinguishes itself with branched-chain volatiles, cocoa-ish and sweaty and fruity.

Balsamic vinegars are sweet and milder than ordinary vinegars, made in northern Italy for many centuries, and known since the eighteenth by the name that indicates a soothing balm, not a sharp accent. Preindustrial balsamic vinegars were made by first simmering fresh grape juice for several days to reduce it to a brown syrup, sweet and tart with concentrated sugars and acids, then letting spontaneous alcoholic and acetic fermentations struggle simultaneously in wooden barrels over the course of *years*. The result is a thick, intensely flavored but only mildly acetic syrup, nearly black from extensive sugar caramelization and reactions between sugars and amino acids (see page 489), with caramel and woody aromas.

When this local specialty of Modena and Reggio Emilia became popular in the twentieth century, the method traditional in these cities was streamlined with cultured yeast and acetic bacteria. But *aceto balsamico di Modena tradizionale* must still be fermented and aged for at least twelve years in a sequence of barrels made of different woods. It's an expensive condiment, used by the drop. Several officially approved Modenese *balsamico* approximations are mixtures of cooked grape syrup with ordinary wine vinegar and a caramel coloring, aged for a few months in wooden barrels. They have a roughly similar set of volatiles but a stronger acetic smell, and they're used more freely in dressings and sauces.

SOME VINEGARS AND KOMBUCHA

Vinegar	Component smells: acetic +	Molecules: acetic acid +
"spirit," "distilled"	solvent, fruity, fresh, cocoa-sweaty	ethyl acetate, ethanol, acetaldehyde, acetone, methyl propanal & butanal
red wine vinegar	vinegar, cheesy, floral, fruity, buttery, caramel	ethyl acetate, ethanol, methylbutyric acid, phenylethanol, diacetyl, butyric acid
sherry	fruity, sweaty, clove, fenugreek, solvent	methylbutyl acetate, ethyl methylbutyrate & methylpropionate, methylbutyric acid, ethylguaiacol, sotolon, ethyl acetate
cider vinegar	cooked apple, solvent, floral, fruity	diethyl succinate, ethyl acetate, phenylethanol, butanol, butyl acetate, ethyl propionate

continued

Vinegar	Component smells: acetic +	Molecules: acetic acid +
malt vinegar	cheesy, buttery, solvent, cocoa, fruity, banana	methylpropionic acid, acetoin, acetaldehyde, methylbutanol, methylbutyl acetate
balsamic vinegar: *tradizionale di Modena*, 12 years, c. 2% acetic acid	buttery, sweet, caramel, cheesy, floral, honey	diacetyl, furfural, acetyl furan, maltol, cyclotene, vanillin, butyric & methylbutyric acids, phenylethanol & phenylethyl acetate
common, c. 6% acetic acid	more acetic & sour, less sweet & caramel	higher acids & alcohols, lower furans
Chinese vinegars:		
Zhenjiang (rice)	solvent, bready, floral, cheesy, cocoa-sweaty, almond extract, nutty	ethanol, ethyl acetate, furfural, phenylethanol, phenylacetaldehyde, methylbutyric acid, methylbutanal, benzaldehyde, tetramethyl pyrazine
Shanxi, aged (sorghum, or millet, buckwheat, wheat)	boiled potato, vanilla, buttery, nutty, cheesy, coconut, smoky, sulfurous, honey	methional, vanillin, diacetyl, pyrazines, methylbutyric acid, g-nonalactone, guaiacol, dimethyl trisulfide, phenylacetaldehyde
kombucha	solvent, fresh, green apple	ethanol, acetaldehyde

Chinese vinegars round out acetic acid with distinctive sets of volatile companions. Production commences with the usual qu-started alcoholic fermentation of cooked grains, usually sorghum or rice. To encourage the growth of oxygen-requiring acetic acid bacteria, the fermentation containers are left open to the air, or the mash is regularly mixed with grain hulls that trap bubbles on their surfaces. Salt is sometimes added to stop the fermentation, and a final heating step pasteurizes the liquid and encourages further flavor-deepening chemical reactions.

Two contrasting examples of Chinese vinegars: **Zhenjiang** or **Chinkiang vinegar** is from the south and made with glutinous rice, like yellow wine, and it has a winey aroma rich in esters and alcohols, together with toasty furfural and pyrazines. **Shanxi vinegar**, from the north, is based on sorghum. After more than a month of fermentation, the liquid is heated at 185°F (85°C) for several *days*, then left to age and concentrate outdoors for a year in large open ceramic jars, evaporating more water than acetic acid in the summer, and freezing water for easy removal in the cold winter. The result is a viscous, nearly black vinegar, nutty and woody and sulfurous.

Also apparently originating in northern China is the vinegary beverage known as **kombucha**. The basic version is black or green tea sweetened with 5 to 10 percent table sugar and fermented with a mixed microbial starter that includes a variety of yeasts and lactic and acetic bacteria. This semisolid floating "SCOBY," or symbiotic culture of bacteria and yeasts, generates a mix of acetic and some nonvolatile acids, along with alcohol, acetaldehyde, and carbon dioxide gas that gives the liquid a fresh, tingly quality.

Distilled spirits: brandies, whiskeys, baijius, tequilas, rums

To conclude this chapter, the sublime collaboration between Earth's masters of bio- and pyro-alchemy: yeast-fermented liquids that have been concentrated into their highly intoxicating and flavorful essences by distillation, the controlled evaporation and condensation of alcohol and its companion volatiles (see page 453). There's evidence that alcohol was first distilled from fermented grapes and sugarcane in ancient India, many centuries before *spiritus vini*, the active breathlike "spirit" of wine, in the medieval West. (Central Asian peoples may have been the first to concentrate alcohols, by simply leaving them outdoors to freeze some of the water and draining off the alcohol-enriched remainder.) Cultures across the planet have since gone to remarkable efforts not only to make alcoholic spirits from *any* fermentable material, but also to wrap alcohol's strong, solventy pungency in other volatiles that make it more pleasant to drink.

So there are many varieties of spiritous experience for the smell explorer to seek out and savor. To appreciate their aromas—a practice sometimes called *nosing*—connoisseurs will often dilute spirits with an equal volume of water. Because they contain so many two-carbon alcohol molecules, carbon-chain volatiles tend to nestle in among them and stay tucked in the liquid. Adding the no-carbon, volatile-unfriendly water molecules disrupts the nestling and forces more carbon chains into the air where we can smell them.

Depending on which of the many kinds of still designs distillers use, they have varying degrees of control over how hot the fermented material gets and which of the volatiles that accompany alcohol in the vapors get captured and included in the distillate. Some volatiles come from the original plant material, most from the fermentation yeasts, and some from the "cooking" of the distillation process. Various alcohols and esters form the basic spiritous bouquet, along with sweet and smoky aromatics if the spirits are stored in barrels. With an alcohol content of 20 to 40 percent or more, most distillates are immune to microbial spoilage, so they're easily stored for decades, during which their strange arrays of chemicals can react with each other and their flavors evolve.

SOME COMMON COMPONENTS OF SPIRITOUS BOUQUET

Smell qualities	Molecules
alcohol, solvent	ethanol
solvent, cocoa, floral	other alcohols: methylbutanols, phenylethanol
fruity, solvent	esters: ethyl acetate, ethyl butyrate, ethyl methylbutyrate . . .
floral, cooked apple	terpenoid fragment: damascenone
wood, toasted/charred: vanilla, coconut, smoky, spicy, clove	vanillin, whiskey (oak) lactone, guaiacol & ethylguaiacol, eugenol

A couple of spirits begin as anonymous distillates, more than 90 percent alcohol and intentionally neutral in flavor. **Vodka** (Russian: "little water") is made by diluting neutral spirits to a solution that's around 40 percent alcohol and 60 percent water, with just traces of other volatiles unless it's flavored. **Gin** (from the Dutch for "juniper") is neutral spirits redistilled or otherwise aromatized with juniper berries and some combination of coriander seed, citrus peels, and other aromatic seeds and roots.

SOME RELATIVELY SIMPLE DISTILLED SPIRITS

Spirit	Distinctive smells	Molecules
vodka (unflavored)	alcohol, solvent	methylbutanols, ethyl decanoate, methylbutyl octanoate, limonene, geranyl acetone, styrene, toluene
gin (neutral spirits, juniper & other aromatics)	pine, resinous, citrus, woody	pinene, myrcene, limonene, terpinene, terpineol, geranyl acetate, linalool, cadinene, caryophyllene, humulene

Brandy (from German: "burnt wine") names spirits distilled from fermented fruits, most commonly grapes. The best-known brandies are **Cognac** and **Armagnac** from southwest France, made from nondescript grapes and aged in barrels, richly endowed with fruity, floral yeast volatiles and woody sweetness. Cognacs aged for decades mellow in taste and texture and develop a prized aromatic quality known unpromisingly as **rancio**, with esters split and the acids oxidized to floral-cheesy, mushroomy, fatty ketones. Some distinctive kinds of Peruvian **pisco** are made using floral grape varieties whose terpenoids persist in the distilled brandy. Similarly, **apple and pear brandies** retain esters that characterize the original fruits and their ciders and perries, with the added support of cooked-apple damascenone formed during the distillation.

SOME SPIRITS DISTILLED FROM FRUITS

Spirit	Distinctive smells	Molecules
brandy, grape: Cognac	strongly fruity, floral, cocoa, woody	damascenone, methylpropanal, many ethyl esters, diacetyl, vanillin, vinyl & ethyl guaiacol;
aged rancio character	mushroomy, blue cheese, creamy, waxy, coconut	heptanone, nonanone, undecanone, propyl octanoate
brandy, grape: pisco aromatico (Italia, Moscatel, Torontel . . .)	floral, perfumed, fresh, honey, caramel	linalool, hexanol, phenylacetaldehyde, furaneol

continued

Spirit	Distinctive smells	Molecules
brandy, apple	cooked apple, sweaty, woody	damascenone, many butyrate & butyl & methylbutyl esters, methylbutyric acid, ethyl phenol & guaiacol
brandy, pear	pear, sweaty, fatty	ethyl decadienoates, methylbutyric acid, nonenal

Grain-based beers and rice wines aren't as estery as fruit wines, and the same is true of their distillates. **Whisky**, from the Gaelic words for *aqua vitae*, "water of life," is the Scots spelling for the distillate of fermented malt and barley, which concentrates malt's cocoa-sweaty branched-chain alcohols and aldehydes. When the malted barley grains have been dried with the heat from peat-fired kilns, the distillate retains traces of the smoky, sometimes seaweedy pyrolysis volatiles (see pages 381, 402). Whereas most Scotch whiskies are aged in used barrels, **American whiskeys**, made with corn or rye or a mixture of the two, are generally aged in newly charred oak barrels that contribute much of their sweet, vanilla, and coconut aromas.

SOME SPIRITS DISTILLED FROM GRAINS

Spirit	Distinctive smells	Molecules
whisky, malt (malted barley, other grains)	cocoa, sweaty, fruity, oily	methylbutanals, methylbutanols, hexanol, butanols, ethyl laurate
peat-kilned malt	+ smoky, clove, leathery, tarry, horse stable, seacoast-medicinal	+ guaiacol, methyl & ethyl guaiacols, ethylphenol, cresols, bromophenols
whiskey, corn or rye	cooked apple, vanilla, coconut, peach, clove, smoky	damascenone, vanillin, g-nonalactone, oak lactone, g-decalactone, eugenol, guaiacol, ethylguaiacol
awamori (rice)	mushroomy, sulfurous	octenol, methyl thioacetate, dimethyl trisulfide
shochu, buckwheat	sweet, resinous, waxy, onion, sweaty	methyl cinnamate, ethyl octanoate, methionol, methylbutyric acid

Undeniably different from most other grain spirits are a couple of versions of Japanese **shochu**, which is enjoyed without long aging. The predecessor of shochu was apparently **awamori**, made on the island of Okinawa using long-grain rice, pot stills that came via trade from Thailand, and an unusual black strain of koji *Aspergillus* that contributes mushroomy and sulfurous touches. Rice-based shochus carry the usual spiritous bouquet, but **buckwheat shochu** stands out with a resinous-fruity benzenoid ester.

Mainland Asian grain spirits span a much greater range of flavors than whiskeys. Chinese *baijiu*—"clear" or "white" alcohol, usually made from sorghum— involves complex, sometimes recursive production methods. These include "solid-state" processing, in which the cooked grains are mixed with the powdered qu without submerging and diluting them in water, then covered and fermented for weeks in a moist heap, and finally shoveled into a basket and suspended over steaming liquid, whose vapor percolates through and pulls the trapped alcohol and volatiles out with it.

Of the dozen or so baijiu styles, here's a sampler of three. The **sauce aroma** type (e.g., Moutai or Maotai, the most widely available baijiu brand) involves a hot-cultured qu (150°F/65°C) and a thrice-repeated cycle of hot fermentation (120°F/50°C) and steam distillation. The end result: volatiles typical of toasting and roasting, with an overall aroma resembling soy sauce. The **strong aroma** type is defined by high levels of the fruity ester ethyl hexanoate. Its main fermentation takes place in cellar chambers lined with "pit mud," a mix of wet soil and fermentation residues teeming with bacteria, which are the source of copious hexanoic and other starter-set acids that give rise to fruity esters, but also persist as sweaty notes. The rice-based *chi xiang* ("chi aroma") or *zhi xiang* ("fat aroma") baijiu has a fatty quality from aldehydes that develop when the spirit is aged for three weeks in contact with a large piece of cooked pork!

Baijius are traditionally aged in clay jars, not wood, sometimes only briefly, and they're fiery. When I nearly choked on my first sip, the Chengdu chef Yu Bo explained that what many Westerners call harshness the Chinese prize as "exciting." I've had many sips since and find the ingenuity of the process and diversity of aromas pretty exciting too.

**SOME CHINESE SPIRITS DISTILLED FROM
SORGHUM AND OTHER GRAINS**

Baijiu type	Distinctive smells	Molecules
sauce aroma	roasted, cocoa, meaty, sulfurous	pyrazines, methylbutanal, furfurylthiol, methanethiol, dimethyl trisulfide
strong aroma	fruity, sweaty, barnyardy	ethyl hexanoate & copious esters, hexanoic acid, cresol, skatole
fatty aroma (*chi xiang* aka *zhi xiang*)	fruity, fatty, floral, coconut, sweaty	ethyl esters, octenal, nonenal, phenylethanol, g-nonalactone, butyric & hexanoic acids

A last batch of spirits: those distilled from materials other than fruits and grains. Mexican **tequila** and **mezcal** start with massive fructose-rich stems of agave plants, steamed many hours for tequila, roasted in wood-fired ovens for mezcal. They share resinous terpineol, and mezcal adds toasted and smoky notes from the oven.

SOME SPIRITS DISTILLED FROM PLANT ROOTS AND STEMS

Spirit	Distinctive smells	Molecules
tequila (steamed *Agave tequilana*); mezcal (roasted *A. angustifolia* & others)	floral, resinous, waxy, buttery, sulfurous + bready, caramel, burnt, medicinal, smoky, tarry	damascenone, linalool, terpineol, ethyl decanoate, diacetyl, dimethyl disulfide + methyl furfural, furanmethanol, phenol, guaiacol, cresols
rhum agricole, cachaça (raw cane juice, *Saccharum officinarum*)	fruity, cooked apple, floral, fatty, ocean air, vegetal	ethyl methylbutyrate, damascenone, phenylethanol, nonadienal, dimethyl sulfide
rum (molasses, cooked cane juice)	vanilla, sweet, fruity, cocoa, buttery, coconut, clove, cinnamon	vanillin, ethyl butyrate & methylbutyrate, damascenone, methylbutanal, diacetyl, oak lactone, allylguaiacol

The sucrose-rich sap of sugarcane, anciently fermented and distilled in India, today gives us the main commercial spirits of Central and South America. **Agricole rums** from the French Caribbean, and Brazilian **cachaças**, are made by fermenting raw cane juice, and they share the original juice's fatty and vegetal/ocean aspects. The more common **rums** developed on early British plantations are made from by-products of sugar refining, mainly molasses (see page 524), which lends some of its own intense aroma to the distillate. Seventeenth-century rums were notoriously harsh with "hogo," from the French *haut goût*, "strong flavor"; recently the term has been redefined as a strong fruity aroma, achieved in part by encouraging putrid-smelling bacterial fermentations in refining wastes. When distilled along with fermented molasses, their unpleasant short-chain acids join with the alcohol to form pleasant esters: a modern Caribbean version of the Chinese pit-mud alchemy that flavors strong-aroma baijiu!

A heady footnote to fermentations, flowers, and fruits

Many fermented foods, from wine to cheese to soy sauce and hogo rums, owe part of their delightfulness to esters, the volatiles that largely define the general quality of fruitiness. Traces of the acid and alcohol ingredients that make esters are produced by most living things in the course of their basic metabolism, but prolific alcohol and ester production is the specialty of yeasts. While alcohols are an effective weapon against microbial competition, the biological function of esters remains enigmatic. They may be a by-product of metabolism, or a way of making both acids and alcohols less toxic to the cells that produce them, or programmed signals for attracting mobile insects—any or all of the above.

In chapter 14, I mention the demonstrated role of yeast alcohols and esters in attracting fruit flies to ripe fruits, with its implication that insects are partly responsible for the pleasurably intensified fruity-floral smells of wine and beer (see page 364). There's now evidence that yeasts and their insect carriers may have helped set a precedent for what flowers and fruits themselves have evolved to smell like.

In a 2018 study, the Swedish biologist Paul Becher and his colleagues noted that the single-cell yeast clan separated from the filament-making remainder of Kingdom Fungi around 300 million years ago. This was after insects had appeared on Earth, but well before the arrival of the flowering plants. Becher's group studied the volatile emissions of nine widely different yeast species, some of them the cause of human yeast infections, some of them with ancestry that has been traced back 150 million years. All of them produced two alcohols, floral phenylethanol and cocoa-like methylbutanol, and most of them also produced ethyl alcohol and the solventy-fruity ester ethyl acetate. Furthermore, in laboratory tests, all species, even the human disease yeasts, attracted fruit flies. The fact that these traits for volatile production are so widely shared suggests that they're ancient, going back to common ancestors near the beginnings of the yeast clan—and long predating the evolution of flowers and fruits. So, conclude Becher et al., "yeast-insect communication may have contributed to the evolution of insect-mediated pollination in plants." That is, flowers may emit insect-attracting alcohols and esters in part because yeasts had done so first.

Maybe so: but plants had almost surely been aiming volatiles at insects long before the yeasts did, mostly to repel and confuse them, sometimes to attract them for spore dispersal. In chapter 7, I noted Florian Schiestl's finding that plants have likely influenced the behavior of insects by co-opting their chemical language, which is diverse enough to have produced lemony ants and floral butterflies and tropical-fruity toe-biting bugs. Benzenoid rings are also important in flower scents, and insects apparently made them before flowering plants did. Perhaps the yeasts' insect appeal did prompt plants to add alcohols and esters to their floral vocabulary, or to give them more prominence. Perhaps this estery nudge somehow carried through to the fruits that develop from flower tissues. And perhaps other microbes participated in these developments. The lactic acid bacteria scent our vegetable and milk fermentations with alcohols and esters, and they're an even more ancient lineage than Kingdom Fungi, some of them also vegetation denizens like yeasts and dependent on animals for dispersal (see page 548).

There are still plenty of fascinating questions to ponder in the evolution of yeast and flower and fruit scents. What seems certain today is that some of the osmocosm's greatest delights, Earth's nectars and ambrosia, are the joint concoction of microbes, insects, plants, and our fellow mammals. Fermented foods and drinks are humankind's contribution to this ancient and intricate web—an intangible, inaudible, largely invisible set of relationships that we're able to perceive and savor through our sense for the essential.

Conclusion · MY SECOND GROUSE

In 2006, a year after the gobsmacking lunch that I describe in the prologue, I returned to Spain and England to check up on the gastronomic avant-garde. Once again I found myself in London during grouse season, so of course I went back to St. John for another smack! This time I dined with a friend, a Japanese neuroscientist. Yuki doesn't study perception, but she knows and loves food. We ordered and ate with gusto, beginning with marrow bones and lamb tongue. The wine was an earthy Cahors, the preliminaries delicious, and the grouse . . . powerful. Once again the flavor was intensely meaty, almost too much. But this time— happily!—I wasn't struck dumb. Our conversation was all over the place and nonstop.

A chef friend once pointed out to me, apropos of avant-garde novelty, that a new dish is new only once. The same goes for a classic dish newly tasted.

Surprises get our attention. Neurobiologists have found that the brain does much of its work without our being consciously aware of it, drawing on past experience to anticipate what sensations and perceptions the next moment will bring. When some aspect of the actual moment matches prediction, the brain is effectively on top of what's going on and doesn't bother to pay much attention to it. But if reality suddenly fails to match prediction—when there's something new or unexpected—the brain focuses conscious attention on the discrepancy, to evaluate it and act accordingly.

I figure that, when I had my first bite of grouse, my brain predicted the familiar flavor of domestic duck or squab. But the wild bird was so different and so strong that the surprise momentarily hijacked my wits away from the conversation and everything else: *Stop! Don't swallow! You sure this isn't bad?* Then a year later, the first grouse was in my experiential database, and my brain anticipated intense funk. Expectation satisfied, wits intact, conversation uninterrupted.

It was the wallop of my first grouse that propelled me to visit its homeland, get schooled in grouse life and after-life, and eventually undertake this survey of the greater osmocosm. Nothing along the way has struck me speechless again, but years of deliberate attention to smells have brought smaller surprises galore, and with them, almost daily, some mix of fresh experience and understanding and wonderment. The experiences: the many smells themselves, their nuances and intriguing echoes. The understanding: the backstories, the secret lives that Proust and Sartre intuited in smells, the otherwise unapparent workings and relationships of things. The wonderment: at the dense web of interdependencies woven by protean Hero Carbon among its needy yet resourceful avatars, moment by moment, eon on eon—and at the fact that we can sense its traces intimately, in a moment's sniff! Wonderment as well at the achievement of the human mind in coming to know all this and constantly discovering more.

I hope that this collection of descriptions and stories will nudge you to sniff out actual flying molecules in your own life, and venture beyond familiar pleasures to smells that usually make us hold our breath. Many of these turn out to be signs of living things working hard to stay alive, and they can be more than simply repellent. When the food writer and China expert Fuchsia Dunlop tasted the swampy fermented specialties of Shaoxing, southwest of Shanghai, she found stalks of rotted amaranth "putrescent and wildly exciting at the same time." Years ago I witnessed the slaughter of a young Easter goat, and I still vividly remember its inner organs, filled with its last grassy meal and hot and stinking with life, turned out steaming into the winter air. Stupefaction from grouse funk, wild excitement from putrescence, essential animality unforgettably exposed: amid all the managed smells of the modern world, sometimes what's most affecting is a whiff of the realities of life and death.

What meaning can there be in those managed and manufactured smells? It's true that technology and commerce have deodorized much of modern life and reodorized it with synthetic odor clichés—citrus for "freshness," pine for "cleanness." True also that supermarket foods are often stripped-down industrial simulations of originals once made from a host of other living things. Rather than tasting of the world that made and sustains us, many manufactured foods taste only of what their makers think they can sell us.

All true: but there's another facet for the smell explorer to find in the commodified aisles of the osmocosm. It took eons of dealmaking with animals and microbes for plants to come up with our favorite volatile molecules. Synthetic flavors and fragrances arrived when organic chemists discovered that they could conjure the same molecules from pine bark and petroleum in a few hours. That's remarkable in its own way! So yes, let's savor natural aromatics for the qualities that inspired imitation in the first place, and for their roles in the lives of plants. But also appreciate synthetics for making scarce and pricy aromas more broadly affordable—and as an impressive joint achievement of Hero Carbon and human chemists, partners in the probing of matter's possibilities.

To conclude, a few last words about smelling through the world with and without this book.

The tables in the chapters above list component smells to "listen" for when smelling things actively and attentively and searchingly. Unlike tasting notes for wine and coffee and the like, which offer the subjective impressions of individual people doing the tasting, these component smells are the sensations triggered by molecules that are objectively present in a given thing, according to at least one published chemical analysis and usually several.

Even so, the tables only indicate notes that you *may* perceive in the overall smells of things. They're not prescriptions for what you *should* perceive. Smell perception simply isn't prescribable, for reasons that would fill another book. Here are three in brief: we all have different noses and brains and experiential databases; it's challenging even for trained professionals to identify volatile notes in mixtures; and sometimes volatile mixtures generate a new quality that isn't just a simple sum of their components. To experience these complications firsthand, sit down with friends, share a bottle of cola, and see who can distinguish its half-dozen defining spice extracts. Not easy! Curry spice blends are even more challenging.

SOME INGREDIENTS IN COLA AND CURRY FLAVORS

Flavor	Component spices
cola	vanilla, citrus peel, cinnamon, coriander, nutmeg, orange flower
curry	cumin, coriander, fenugreek, cinnamon, clove, fennel, cardamom, ginger, black pepper, turmeric

So sometimes we get a recognizable whiff of the volatiles that are objectively there in things, and sometimes we don't. But the real reward of active smelling isn't the momentary satisfaction of identifying notes. Above all there's the smelling itself! Our senses exist to be stimulated, our perceptual powers to be exercised, and it's exciting, deeply fulfilling even, to engage and experience them consciously. When we nose an intriguing flower or finger a leaf or sip a cola, and take the time to sniff repeatedly and searchingly for component smells, we experience their qualities more fully than when we smell with brain on autopilot. Even if we don't find the subsmells, we notice what we actually *do* smell. We can notice how a smell changes from sniff to sniff as our predictive brain responds to its persistence. We can ask ourselves what the smell reminds us of, why it intrigues us, whether it evokes a memory or emotion. And long after our smell receptors have released the volatiles that were momentarily a physical part of us, the tangle of perception and thought and feeling remains, part of the wondrous inner web that will help shape our future experiences.

Perfumers have a fine term—French, of course!—for the scent of a perfume as it lingers in the air after its wearer passes by: *sillage*, the trail left by a moving body. It's used as well to mean animal tracks, the wake of a boat, a jet contrail. Movement is relative, so imagine the volatile wakes swirling around us as our bodies pass through a world where nearly everything is releasing molecules of itself. When we get in the habit of listening to smells, we get better at noticing these invisible ephemeral trails and what they might lead back to, the many volatile givens and gifts that fill our lives.

Acknowledgments

....

This book has been more than a decade in the making, and it's a pleasure to think back and name the many people who've helped me out along the way.

To begin, my thanks to Fergus Henderson and Trevor Gulliver for that life-changing lunch in 2005, and to Ben Weatherall and his family for a revelatory couple of weekends in grouse country, off-duty London chefs and costume dinners and backyard fireworks included! I'm also grateful to Heston Blumenthal for organizing the tour that took me to St. John in the first place, and for our many other scouting trips and chats about the inner experience of eating.

There's no better way to explore the osmocosm than doing some actual smelling with other knowledgeable smellers, but that's not always easy. For the chance to know firsthand both the tedium and the panic of gas chromatography-olfactometry, I'm indebted to Arielle Johnson, and also to Pat Brown and Celeste Holz-Schietinger. For the experience of near-fresh giant water bugs I thank Pim Techamuanvivit; of scented Thai candles, Leela Punyaratabandhu. For the guided appreciation of rare aromatics and classic perfumes, my gratitude to Mandy Aftel and Victoria Frolova; of olive oils, to Darrell Corti, Alexandra Devarenne, Paul Vossen, and Pablo Voitzuk; of grapes and wines, to Peter Bell and Amy Albert; of spirits, to Audrey Saunders and Tony Conigliaro.

For the invitation to dig ripe truffles in Oregon, I thank Charles Lefevre and Leslie Scott; to delve into foods and incense in Japan, Kumiko Ninomiya and Mio Kuriwaki; to savor "stinky" vegetable fermentations in China, Fuchsia Dunlop; to swoon over hams in Spain, Jorge Ruiz and Carlos Tristancho; to trace cheese flavors in England, Randolph Hodgson and Jamie Montgomery and Joe Schneider and Bronwen Percival, and in Comté, Jean-Louis Carbonnier; to follow baijiu production from mud pit to cup in China, Derek Sandhaus, Bill Isler, and Don Lee.

I'm also grateful to a number of experts in matters olfactory for sharing their knowledge and insights, from the overarching to the very particular. It's been a privilege to join conversations about food smells and human evolution with Richard Wrangham and Rachel Carmody, and about flavor chemistry and perception with Terry Acree, Gary Beauchamp, Nadia Berenstein, Paul Breslin, Jeannine Delwiche, Susan Ebeler, Stuart Firestein, Jean-Xavier Guinard, Arielle Johnson, Marcia Pelchat, Gordon Shepherd, Dana Small, and Andrew Taylor. For answering my questions about beaver secretions, my thanks to Dietland Müller-Schwarze; about cilantro, to Keith Cadwallader; about Oregon truffles and Chinese alcohols, to Michael Qian; about Chinese stinky vegetables, to Luping Liu; about cheese mites, to Michael Heethoff and Walter Leal; about wines, to Jancis Robinson and Andrew Waterhouse and Hildegarde Heymann; about sake, to Izumi Motai and Chris Pearce; and about ancient authors and etymologies, to Richard Thomas.

I can't overstate the importance to this project of my affiliation with the University of California, Davis, and the access it has given me to scholarly resources that are otherwise unaffordable for most people, including taxpayers who help subsidize them. I owe this boon to successive department chairs of Food Science and Technology, James Seiber, Michael McCarthy, and Linda Harris. Lee Meddin helped me navigate the Davis system, and Clare Hasler-Lewis made me a campus regular early on by involving me with the Robert Mondavi Institute. My thanks to all.

Another person at Davis to whom I'm especially grateful: Alice Phung, a chemist and this book's molecule maker, quick on the draw and imperturbable no matter how many tweaks and variations I asked for.

I'm much obliged to a number of people who read and commented on draft versions of this book. In the initial stages, Rebecca Saletan helped me get my bearings in outer space. Victoria Frolova made important suggestions for the fragrance chapter. Gary Beauchamp, Sharon Long, and Lubert Stryer read several chapters, and John McGee marked up many. Arielle Johnson cast her sharp flavor-chemist's eye over nearly the entire manuscript; she and Tom Pold brainstormed titles with me, and Tom schooled me in publishing logistics. Mandy Aftel and Foster Curry read every last word of every chapter as I finished it, and spent hours with me going over Foster's meticulous annotations.

Mandy: my thanks to you for our years of friendship and writerly mutual aid, for sharing with me your astounding collection of aromatics and books and artifacts, and for our many hours together sniffing and talking, sniffing and talking.

At Penguin Press, I'm deeply grateful to Ann Godoff for her patience with a manuscript both long overdue and a swerve from what I'd originally proposed, and for mapping out in great detail how to make it much better. My thanks as well to Amanda Dewey for her artful integration of text and drawings, and to Casey Denis, Gary Stimeling, and their colleagues for meticulous copyediting and production work in difficult circumstances.

At The Wylie Agency, I thank Andrew Wylie for his heartening support and invaluable counsel as the years rolled by, and Tracy Bohan for finding good publishing homes overseas.

For their interest and encouragement and various shared olfactory adventures, thanks to my sister Joan and brother Michael, and to friends Dave Arnold, Nastassia Lopez, Devaki Bhaya and Arthur Grossman, Bronwen and Francis Percival, Shirley Corriher, Yuki Goda, Daniel Patterson, Mark Alfenito, and Sarah Wally. In addition to their putting up with much smell talk on visits home, I'm grateful to my son John for helping me get a whiff of several lab reagents and to my daughter Florence for bringing black walnuts and truly horsey "horse chestnuts" from her weekend stints at the ranch.

Finally, for arranging for me to make tea in Uji and learn about sake yeasts and Wagyu grades in San Francisco, for her fellow writer's understanding and tolerance of my getting stuck at the desk, for her buoyant spirit, and for her love, I thank my partner Elli Sekine. And for Elli and family and friends I thank my lucky stars.

Selected References

....

I got hooked on research and writing as a student in the 1970s, when I was trying to figure out how John Keats managed to become a great poet before he'd reached my age. He died of tuberculosis at twenty-five, in 1821. I spent months in the stacks of my university's main library getting to know the now obscure books that Keats read, taking notes with pen and paper. The yellowed pages and fragile bindings scented my way back to eighteenth-century ways of seeing the world. Their smells sometimes reminded me of my childhood visits to the creaky mansion that housed our city library, the rooms and corridors crammed with books, and with unsettling intimations of people and places and times way beyond my grasp. Old libraries and books still take me back, but nowadays they also draw me in, to remember how papermakers, bookmakers, air, and time cook up that sharp, smoky sweetness from the remains of plants and trees.

SOME SMELLS OF OLD BOOKS

Component smells	Sources	Molecules
vinegar, sharp	cellulose (paper fiber)	acetic, propionic, butyric acids
bready, sweet	cellulose (paper fiber)	furfural
green, fatty, rancid, waxy, citrus peel	lipids (paper fiber, bindings, finger oils)	hexanal, heptanal, octanal, nonanal, decanal
almond essence	lignin (wood pulp)	benzaldehyde
vanilla	lignin (wood pulp)	vanillin
smoky	lignin (wood pulp)	guaiacol

I don't spend much time in libraries anymore. I've done most of the work for this book at home, reading and writing virtual pages on monitors and keyboards and only occasionally noticing their volatile presence, the faint smells of warm circuit-board resins. Such an ordinary accompaniment to the extraordinary access that those circuits have opened to the sweep of human life and knowledge! During ten years of osmocosmic immersion they've brought me several thousand scientific reports and dozens of books, the work of generations of researchers in many different fields. The following pages credit only a small fraction of these, weighted toward recent publications that provide links back to their predecessors. I hope you'll at least scan through them, simply to get a sense of the range and depth of knowledge that our clan has accumulated about life in a world of flying molecules. Natural histories of the ancients, modern cosmochemistry, geochemistry, atmospheric chemistry, marine chemistry, body chemistry, food and fragrance chemistry, industrial and occupational chemistry—on and on: astounding!

For my quotations from non-English texts, I've listed accessible English editions; the translations from French and Latin are my own. To cite as many sources as possible, I've abbreviated them freely. Full journal titles shouldn't be difficult to decipher: for example, "*J Agric Food Chem*" is shorthand for the *Journal of Agricultural and Food Chemistry*. If you're interested in a closer look at a particular study, all you have to do is enter a couple of author names and title words into a search engine like Google Scholar, and the correct link should pop up on the first page of results. Full texts may be behind subscription paywalls, but summaries are almost always accessible, and these usually include the key results. Don't let unfamiliar jargon get in the way of your curiosity: just skim until you find what you're looking for. Follow your nose.

SOME GENERAL BOOKS ABOUT SMELLS

Classen, C., D. Howes, et al. (1994). *Aroma: The Cultural History of Smell*. Routledge.

Drobnick, J., ed. (2006). *The Smell Culture Reader*. Berg.

Gilbert, A. N. (2008). *What the Nose Knows: The Science of Scent in Everyday Life*. Crown.

Kaiser, R. (2006). *Meaningful Scents around the World: Olfactory, Chemical, Biological, and Cultural Considerations*. Wiley.

Büttner, A., ed. (2017). *Springer Handbook of Odor*. Springer.

PREFACE: MY FIRST GROUSE

Proust, M. (1921–22). *Sodome et Gomorrhe*. In *À la recherche du temps perdu* (1954), vol. 2, 738. Gallimard. Trans. C. K. S. Moncrieff and T. Kilmartin (1981), *Remembrance of Things Past*, vol. 2, *Cities of the Plain*, 764. Random House.

Dobson, A., and P. Hudson (1995). The interaction between the parasites and predators of red grouse *Lagopus lagopus scoticus*. *Ibis* 137:S87–S96.

Högstedt, G. (2014). Prolonged aerial chase of willow grouse *Lagopus lagopus* by common raven *Corvus corax*. *Ornis Nor* 37:15.

Martínez-Padilla, J., S. M. Redpath, et al. (2014). Insights into population ecology from long-term studies of red grouse. . . . *J Anim Ecol* 83:85–98.

Hudson, P. J., A. P. Dobson, et al. (2002). Parasitic worms and population cycles of red grouse. In *Population Cycles: The Case for Trophic Interactions*, ed. A. Berryman, 109–30. Oxford Univ. Press.

MacDiarmid, H. (1948). "My heart always goes back to the North." *Poetry* 72(4):175–79.

INTRODUCTION: A SENSE
FOR THE ESSENTIAL

Sartre, J.-P. (1947). *Baudelaire*. Trans. M. Turnell (1950), 174. New Directions.

Atala, A. (2013). *D.O.M.: Rediscovering Brazilian Ingredients*. Phaidon.

Shepherd, G. M. (2004). The human sense of smell: Are we better than we think? *PLoS Biol* 2(5):e146.

Shepherd, G. M. (2013). *Neurogastronomy*. Columbia Univ. Press.

Gibson, J. J., and L. Carmichael (1966). *The Senses Considered as Perceptual Systems*. Houghton Mifflin.

Latour, B. (2004). How to talk about the body? The normative dimension of science studies. *Body & Soc* 10(2–3):205–29.

Delon-Martin, C., J. Plailly, et al. (2013). Perfumers' expertise induces structural reorganization in olfactory brain regions. *NeuroImage* 68:55–62.

Gilbert, A. N. (2008). *What the Nose Knows: The Science of Scent in Everyday Life*. Crown.

CHAPTER 1. AMONG THE STARS

Serres, M. (1985). *Les Cinq Sens*. Trans. M. Sankey and P. Cowley (2008), *The Five Senses*, 163. Continuum.

Reeves, H. (1981). *Patience dans l'Azur*. Trans. R. A. Lewis and J. S. Lewis (1984), *Atoms of Silence*. MIT Press.

McGuire, B. A. (2018). 2018 Census of interstellar, circumstellar, extragalactic, protoplanetary disk, and exoplanetary molecules. *Astrophys J Suppl Ser* 239:17.

Ehrenfreund, P., and J. Cami (2010). Cosmic carbon chemistry: From the interstellar medium to the early earth. *Cold Spring Harbor Perspect Biol* 2: a002097.

Aponte, J. C., J. P. Dworkin, et al. (2014). Assessing the origins of aliphatic amines in the Murchison meteorite. . . . *Geochim Cosmochim Acta* 141:331–45.

Jenniskens, P., M. D. Fries, Q.-Z. Yin, and Sutter's Mill Meteorite Consortium (2012). Radar-enabled recovery of the Sutter's mill meteorite. . . . *Science* 338:1583–87.

Sandford, S. A., J. Aléon, et al. (2006). Organics captured from comet 81P/Wild 2 by the Stardust spacecraft. *Science* 314(5806):1720–24.

Pizzarello, S., and E. Shock (2010). The organic composition of carbonaceous meteorites. . . . *Cold Spring Harbor Perspect Biol* 2:a002105.

CHAPTER 2. PLANET EARTH, EARLY LIFE,
STINKING SULFUR

Lucretius (~50 BCE). *De Rerum Natura*. Trans. A. Esolen (1995), *On the Nature of Things*, book 6, lines 739–49, 807–12, 817–18. Johns Hopkins Univ. Press.

Serres, M. (1985). *Les Cinq Sens*. Trans. M. Sankey and P. Cowley (2008), *The Five Senses*, 164. Continuum.

Arndt, N. T., and E. G. Nisbet (2012). Processes on the young earth and the habitats of early life. *Ann Rev Earth Planet Sci* 40:521–49.

Chirouze, F., G. Dupont-Nivet, et al. (2012). Magnetostratigraphy of the neogene Siwalik group in the far eastern Himalaya. . . . *J Asian Earth Sci* 44: 117–35 [black salt].

Dodd, M. S., D. Papineau, et al. (2017). Evidence for early life in Earth's oldest hydrothermal vent precipitates. *Nature* 543:60–64.

Jo, S.-H., K.-H. Kim, et al. (2013). Study of odor from boiled eggs over time using gas chromatography. *Microchem J* 110:517–29.

Sleep, N. H. (2010). The Hadean-Archaean environment. *Cold Spring Harbor Perspect Biol* 2:a002527.

Sleep, N. H., and D. K. Bird (2008). Evolutionary ecology during the rise of dioxygen in the Earth's atmosphere. *Philos Trans Roy Soc B Biol Sci* 363:2651–64.

Rothschild, L. J., and R. L. Mancinelli (2004). Life in extreme environments. *Nature* 409:1092–1101.

Canfield, D. E., M. T. Rosing, et al. (2006). Early anaerobic metabolisms. *Philos Trans Roy Soc B Biol Sci* 361:1819–36.

Ballard, R. D. (1977). Notes on a major oceanographic find. *Oceanus* 20:35–44.

Tobler, M., C. N. Passow, et al. (2016). The evolutionary ecology of animals inhabiting hydrogen sulfide-rich environments. *Ann Rev Ecol Evol Syst* 47: 239–62.

Roldan, A., N. Hollingsworth, et al. (2015). Bio-inspired CO_2 conversion by iron sulfide catalysts. . . . *Chem Commun* 51:7501–4.

Kump, L. R., A. Pavlov, and M. A. Arthur (2005). Massive release of hydrogen sulfide to the surface ocean and atmosphere during intervals of oceanic anoxia. *Geology* 33:397.

CHAPTER 3. LIFE'S STARTER SET

Reeves, H. (1981). *Patience dans l'Azur*. Trans. R. A. Lewis and J. S. Lewis (1984), *Atoms of Silence*, 64. MIT Press.

Atkins, P. (2003). *Atkins' Molecules*. Cambridge Univ. Press.

CHAPTER 4. ANIMAL BODIES

Problemata (~200 BCE). Trans. E. S. Forster (1984), *Problems*, 13.4, in *The Complete Works of Aristotle*, ed. J. Barnes, 1410. Princeton Univ. Press.

Serres, M. (1985). *Les Cinq Sens*. Trans. M. Sankey and P. Cowley (2008), *The Five Senses*, 164. Continuum.

Animal life

Gould, S. J., ed. (1993). *The Book of Life*. Norton.

Hillis, D. M., H. C. Heller, et al. (2020). *Life: The Science of Biology*. Macmillan.

Brieger, L. (1885). *Weitere Untersuchungen über Ptomaine*. Berlin.

Amoore, J. E., L. J. Forrester, et al. (1975). Specific anosmia to 1-pyrroline: The spermous primary odor. *J Chem Ecology* 1(3):299–310.

Animal death and decomposition

Rosier, E., S. Loix, et al. (2015). The search for a volatile human specific marker in the decomposition process. *PLoS One* 10:e0137341.

Hussain, A., L. R. Saraiva, et al. (2013). High-affinity olfactory receptor for the death-associated odor cadaverine. *Proc Natl Acad Sci USA* 110:19579–84.

Liberles, S. D. (2015). Trace amine-associated receptors: Ligands, neural circuits, and behaviors. *Curr Opinion Neurobiology* 34:1–7.

Kalinová, B., H. Podskalská, et al. (2009). Irresistible bouquet of death—how are burying beetles . . . attracted by carcasses. *Naturwissenschaften* 96: 889–99.

Animal excrements

Fossey, D. (1983). *Gorillas in the Mist*. Houghton Mifflin, 46.

Lowe, J., S. Kershaw, et al. (1997). The effect of *Yucca schidigera* extract on canine and feline faecal volatiles. . . . *Res in Vet Sci* 63:67–71.

Mansourian, S., J. Corcoran, et al. (2016). Fecal-derived phenol induces egg-laying aversion in drosophila. *Curr Biol* 26:2762–69.

Albuquerque, T. A. de (2012). Diversity and effect of the microbial community of aging horse manure on stable fly. . . . PhD dissertation, Kansas State Univ.

Blanes-Vidal, V., M. N. Hansen, et al. (2009). Characterization of odor released during handling of swine slurry. *Atmos Enrivon* 43:2997–3005.

Yasuhara, A. (1987). Identification of volatile compounds in poultry manure. . . . *J Chromatogr A* 387:371–78.

Shahack-Gross, R. (2011). Herbivorous livestock dung . . . archaeological significance. *J Archaeological Sci* 38:205–18.

Bogaard, A., R. Fraser, et al. (2013). Crop manuring and intensive land management by Europe's first farmers. *Proc Natl Acad Sci USA* 110:12589–94.

Wright, D. W., D. K. Eaton, et al. (2005). Multidimensional gas chromatography-olfactometry for . . . malodors from confined animal feeding operations. *J Ag Food Chem* 53:8663–72.

Dalton, P., E. A. Caraway, et al. (2011). A multi-year field olfactometry study near a concentrated animal feeding operation. *J Air & Waste Management Assoc* 61:1398–1408.

Cushman, G. T. (2013). *Guano and the Opening of the Pacific World: A Global Ecological History*. Cambridge Univ. Press.

Wet dog, paws

Doty, R. L., and I. Dunbar (1974). Attraction of beagles to conspecific urine, vaginal and anal sac secretion odors. *Physiol Behav* 12:825–33.

Hardham, J. M., K. W. King, et al. (2008). Transfer . . . description of *Odoribacter denticanis* sp. nov., isolated from the crevicular spaces of canine periodontitis patients. *Int J Systematic and Evol Microbiol* 58:103–9.

Allaker, R. P. (2010). Investigations into the microecology of oral malodour in man and companion animals. *J Breath Res* 4:017103.

Carrier, C. A., J. L. Seeman, et al. (2011). Hyperhidrosis in naïve purpose-bred beagle dogs. . . . *J Am Assoc for Lab Animal Sci* 50(3):396–400.

Brouwer, E., and H. J. Nijkamp (1953). Occurrence of two valeric acids . . . in the hair grease of the dog. *Biochemical J* 55(3):444–47.

Young, L., P. Pollien, et al. (2002). Compounds responsible for the odor of dog hair coat. *World Small Animal Vet Assoc Congress Proceedings*.

Immortal in mortality

Calasso, R. (2010). *L'Ardore*. Trans. R. Dixon (2014), *Ardor*. Farrar, Straus and Giroux. 41–42 [*Satapatha Brahmana*].

CHAPTER 5. ANIMAL SIGNALS

Müller, F. (1878). Blumen der Luft. *Kosmos* 3, 187.

Müller-Schwarze, D. (2006). *Chemical Ecology of Vertebrates*. Cambridge Univ. Press.

Ferrero, D. M., J. K. Lemon, et al. (2011). Detection and avoidance of a carnivore odor by prey. *Proc Natl Acad Sci USA* 108:11235–40.

Wernecke, K. E. A., D. Vincenz, et al. (2015). Fox urine exposure induces avoidance behavior in rats and activates the amygdalar olfactory cortex. *Behavioural Brain Res* 279:76–81.

Cats, dogs, skunks, beavers

Starkenmann, C., Y. Niclass, et al. (2015). Odorant volatile sulfur compounds in cat urine. . . . *Flavour Fragr J* 30:91–100.

Miyazaki, M., T. Miyazaki, et al. (2018). The chemical basis of species, sex, and individual recognition using feces in the domestic cat. *J Chem Ecology* 44:364–73.

Doty, R. L., and I. Dunbar (1974). Attraction of beagles to conspecific urine, vaginal and anal sac secretion odors. *Physiol Behav* 12:825–33.

Preti, G., E. L. Muetterties, et al. (1976). Volatile constituents of dog and coyote anal sacs. *J Chem Ecology* 2:177–86.

Wood, W. F., B. G. Sollers, et al. (2002). Volatile components in defensive spray of the hooded skunk. . . . *J Chem Ecology* 28(9):1865–70.

Müller-Schwarze, D. (2011). *The Beaver: Its Life and Impact.* Comstock.

Goat cheese, lamb meat, wool
Homily of Pope Francis for Chrism Mass, March 28, 2013. http://www.vatican.va/content/francesco/en/homilies/2013/documents/papa-francesco_20130328_messa-crismale.html.
Birch, E. J., T. W. Knight, et al. (1989). Separation of male goat pheromones responsible for stimulating ovulatory activity in ewes. *New Zealand J Agric Res* 32:337–41.
Murata, K., S. Tamogami, et al. (2014). Identification of an olfactory signal molecule that activates the central regulator of reproduction in goats. *Curr Biol* 24:681–86.
Walkden-Brown, S. W., B. J. Restall, et al. (1994). Effect of nutrition on seasonal patterns . . . on sebaceous gland volume and odour in Australian cashmere goats. *Reproduction* 102:351–60.
Iwata, E., T. Kikusui, et al. (2003). Substances derived from 4-ethyl octanoic acid account for primer pheromone activity for the "male effect" in goats. *J Vet Medical Sci* 65:1019–21.
Kaffarnik, S., Y. Kayademir, et al. (2014). Concentrations of volatile 4-alkyl-branched fatty acids in sheep and goat milk and dairy products. . . . *J Food Sci* 79:C2209–14.
Siefarth, C., and A. Büttner (2014). The aroma of goat milk. . . . *J Agric Food Chem* 62:11805–17.
Carunchiawhetstine, M. E., Y. Karagul-Yuceer, et al. (2003). Identification and quantification of character aroma components in fresh chevre-style goat cheese. *J Food Sci* 68:2441–47.
Ha, J. K., and R. C. Lindsay (1991). Volatile alkylphenols and thiophenol in species-related characterizing flavors of red meats. *J Food Sci* 56:1197–1202.
Madruga, M., I. Dantas, et al. (2013). Volatiles and water- and fat-soluble precursors of Saanen goat and cross Suffolk lamb flavour. *Molecules* 18:2150–65.
Jover, E., M. Ábalos, et al. (2003). Volatile fatty acids as malodorous compounds in wool scouring water and lanolin. *Environ Technol* 24:1465–70.
Lisovac, A., and D. Shooter (2003). Volatiles from sheep wool and the modification of wool odour. *Small Ruminant Res* 49:115–24.

Insects
Attygalle, A. B., and E. D. Morgan (1984). Chemicals from the glands of ants. *Chem Soc Reviews* 13: 245–78.
d'Ettorre, P. (2016). Genomic and brain expansion provide ants with refined sense of smell. *Proc Natl Acad Sci USA* 113:13947–49.
Millar, J. G. (2004). Pheromones of true bugs. In *The Chemistry of Pheromones and Other Semiochemicals II*, ed. S. Schulz, 37–84. Springer.
Eisner, T., M. Eisner, et al. (2005). *Secret Weapons: Defenses of Insects, Spiders, Scorpions, and Other Many-Legged Creatures.* Harvard Univ. Press.
Kiatbenjakul, P., K.-O. Intarapichet, et al. (2015). Characterization of potent odorants in male giant water bug. . . . *Food Chem* 168:639–47.

CHAPTER 6. THE HUMAN ANIMAL
Problemata (~200 BCE). Trans. E. S. Forster (1984), *Problems*, 4.24, 13.8, 13.7, in *The Complete Works of Aristotle*, ed. J. Barnes, 1356, 1411–12. Princeton Univ. Press.
Stoddart, D. M. (1990). *The Scented Ape: The Biology and Culture of Human Odour.* Cambridge Univ. Press.

Human volatiles and microbiomes
Pathak, A. K., P. K. Sinha, et al. (2013). Diabetes—a historical review. *J Drug Delivery Ther* 3:83–84.
Crofford, O. B., R. E. Mallard, et al. (1977). Acetone in breath and blood. *Trans Am Clinical and Climatological Assoc* 88:128.
Costello, B. de Lacy, A. Amann, et al. (2014) A review of the volatiles from the healthy human body. *J Breath Res* 8:014001.
Amann, A., B. de Lacy Costello, et al. (2014). The human volatilome: Volatile organic compounds (VOCs) in exhaled breath, skin emanations, urine, feces and saliva. *J Breath Res* 8:034001.
Ding, T., and P. D. Schloss (2014). Dynamics and associations of microbial community types across the human body. *Nature* 509:357–60.
Gill, S. R., M. Pop, et al. (2006). Metagenomic analysis of the human distal gut microbiome. *Science* 312:1355–59.

Human excrement
Jiang, T., F. L. Suarez, et al. (2001). Gas production by feces of infants. *J Pediatric Gastroenterol Nutr* 32:534–41.
Parrett, A. M., and C. A. Edwards (1997). In vitro fermentation of carbohydrate by breast fed and formula fed infants. *Archives of Disease in Childhood* 76:249–53.
Fukuda, S., H. Toh, et al. (2011). Bifidobacteria can protect from enteropathogenic infection through production of acetate. *Nature* 469:543–47.
Brouns, F., B. Kettlitz, et al. (2002). Resistant starch and "the butyrate revolution." *Trends Food Sci Technol* 13:251–61.
Marquet, P., S. H. Duncan, et al. (2009). Lactate has the potential to promote hydrogen sulphide formation in the human colon. *FEMS Microbiol Letters* 299:128–34.
Moore, J. G., L. D. Jessop, et al. (1987). Gas-chromatographic and mass-spectrometric analysis of the odor of human feces. *Gastroenterology* 93:1321–29.
Sato, H., H. Morimatsu, et al. (2002). Analysis of mal-

odorous substances of human feces. *J Health Science* 48:179–85.

Russell, W. R., L. Hoyles, et al. (2013). Colonic bacterial metabolites and human health. *Curr Opinion Microbiol* 16:246–54.

Garner, C. E., S. Smith, et al. (2007). Volatile organic compounds from feces and their potential for diagnosis of gastrointestinal disease. *FASEB J* 21:1675–88.

Gas and urine

Franklin, B. (~1780). *To the Royal Academy. The Bagatelles from Passy*. Text and facsimile. Eakins Press.

Glowka, W. (1985). Franklin's perfumed proposer. *Studies in American Humor* 4(4):229–41.

Magendie, F. (1816). Note sur les gaz intestinaux de l'homme sain. *Ann Chimie Physique* 2:292–96.

Suarez, F. L., J. Springfield, and M. D. Levitt (1998). Identification of gases responsible for the odour of human flatus and evaluation of a device [charcoal cushion] purported to reduce this odour. *Gut* 43:100–104.

Tangerman, A. (2009). Measurement and biological significance of the volatile sulfur compounds hydrogen sulfide, methanethiol and dimethyl sulfide. . . . *J Chromatog B* 877:3366–77.

Wahl, H. G., A. Hoffmann, et al. (1999). Analysis of volatile organic compounds in human urine. . . . *J Chromatog A* 847:117–25.

Breath and mouth

Mochalski, P., J. King, et al. (2013). Blood and breath levels of selected volatile organic compounds. . . . *Analyst* 138:2134.

Corrao, S. (2011). Halitosis: New insight into a millennial old problem. *Internal and Emergency Medicine* 6:291–92.

van den Velde, S., M. Quirynen, et al. (2007). Halitosis associated volatiles in breath of healthy subjects. *J Chromatog B* 853:54–61.

Porter, S. R., and C. Scully (2006). Oral malodour (halitosis). *Brit Med J* 333:632–35.

Ross, B. M., S. Babay, et al. (2009). The use of selected ion flow tube mass spectrometry to detect and quantify polyamines in headspace gas and oral air. *Rapid Comms Mass Spectrom* 23:3973–82.

Peynaud, E. (1987). *The Taste of Wine: The Art and Science of Wine Appreciation*, trans. M. Schuster, 56–57. Macdonald Orbis.

Starkenmann, C., B. Le Calvé, et al. (2008). Olfactory perception of cysteine-S-conjugates from fruits and vegetables. *J Agric Food Chem* 56:9575–80.

Cerny, C., and R. Guntz-Dubini (2013). Formation of cysteine-S-conjugates in the Maillard reaction of cysteine and xylose. *Food Chem* 141:1078–86.

Skin

Wood, A. P., and D. P. Kelly (2010). Skin microbiology, body odor, and methylotrophic bacteria. In *Handbook of Hydrocarbon and Lipid Microbiol*, ed. K. N. Timmis, 3203–13. Springer.

Mochalski, P., K. Unterkofler, et al. (2014). Monitoring of selected skin-borne volatile markers of entrapped humans. . . . *Analytical Chemistry* 86:3915–23.

Dormont, L., J.-M. Bessière, et al. (2013). Human skin volatiles: A review. *J Chem Ecology* 39:569–78.

Steeghs, M. M. L., B. W. M. Moeskops, et al. (2006). On-line monitoring of UV-induced lipid peroxidation products from human skin. . . . *Int J Mass Spectrometry* 253:58–64.

Haze, S., Y. Gozu, et al. (2001). 2-Nonenal newly found in human body odor tends to increase with aging. *J Investigative Dermatology* 116:520–24.

Gallagher, M., C. J. Wysocki, et al. (2008). Analyses of volatile organic compounds from human skin. *British J Dermatology* 159:780–91.

Hara, T., A. Kyuka, et al. (2015). Butane-2,3-dione: The key contributor to axillary and foot odor. . . . *Chem & Biodiversity* 12:248–58.

Kimura, K., Y. Sekine, et al. (2016). Measurement of 2-nonenal and diacetyl emanating from human skin surface. . . . *J Chromatogr B* 1028:181–85.

Mitro, S., A. R. Gordon, et al. (2012). The smell of age: Perception and discrimination of body odors of different ages. *PLoS One* 7:e38110.

Head and feet

Labows, J. N., K. J. McGinley, et al. (1979). Characteristic gamma-lactone odor production of the genus *Pityrosporum* [*Malessezia*]. *Appl and Environ Microbiol* 38:412–15.

Goetz, N., and D. Good (1988). Detection and identification of volatile compounds evolved from human hair and scalp. . . . *J Soc Cosmet Chem* 39(1):1–13.

Brouwer, E., and H. J. Nijkamp (1952). Volatile acids in the secretion products (hair grease) of the skin. *Biochemical J* 52:54–58.

Benn, C. D. (2002). *Daily Life in Traditional China: The Tang Dynasty*. Greenwood Press.

James, A. G., D. Cox, et al. (2013). Microbiological and biochemical origins of human foot malodour. *Flavour Fragr J* 28:231–37.

Knols, B. G. J., and R. De Jong (1996). Limburger cheese as an attractant for the malaria mosquito *Anopheles gambiae*. *Parasitol Today* 12(4):159–61.

Vagina and semen

Huggins, G. R., and G. Preti (1976). Volatile constituents of human vaginal secretions. *Am J Obstet Gynecol* 126:129–36.

Stumpf, R. M., B. A. Wilson, et al. (2013). The primate vaginal microbiome: Comparative context and implications for human health and disease. *Am J Phys Anthropol* 152:119–34.

Butler, S. (2009). The scent of a woman. *Arethusa* 43:87–112 [Catullus].

Amoore, J. E., L. J. Forrester, et al. (1975). Specific

anosmia to 1-pyrroline: The spermous primary odor. *J Chem Ecology* 1(3):299–310.

Ross, B. M., S. Babay, et al. (2009). The use of selected ion flow tube mass spectrometry to detect and quantify polyamines in headspace gas and oral air [and semen]. *Rapid Comms Mass Spectrom* 23:3973–82.

Longo, V., A. Forleo, et al. (2018). HS-SPME-GC-MS metabolomics approach for sperm quality evaluation. . . . *Biomed Physics & Eng Express* 5(1):015006.

Armpits

Huysmans, J.-K. (1880). *Croquis Parisiens*. Vaton.

James, A. G., C. J. Austin, et al. (2013). Microbiological and biochemical origins of human axillary odour. *FEMS Microbiol Ecology* 83:527–40.

Troccaz, M., N. Gaïa, et al. (2015). Mapping axillary microbiota responsible for body odours. . . . *Microbiome* 3:3.

Minhas, G. S., D. Bawdon, et al. (2018). Structural basis of malodour precursor transport in the human axilla. *eLife* 7.

Natsch, A., J. Schmid, et al. (2004). Identification of odoriferous sulfanylalkanols in human axilla secretions. . . . *Chem & Biodiversity* 1:1058–72.

Acknowledging our emanations

McBurney, D. H., S. Streeter, and H. A. Euler (2012). Olfactory comfort in close relationships: You aren't the only one who does it. In *Olfactory Cognition: From Perception and Memory to Environmental Odours and Neuroscience*, ed. G. M. Zucco, R. S. Herz, B. Schaal, 59–72. John Benjamins.

Hasegawa, Y., M. Yabuki, et al. (2004). Identification of new odoriferous compounds in human axillary sweat. *Chem & Biodiversity* 1:2042–50.

Lundström, J. N. (2009). Neural processing of body odor. *Perfumer Flavorist* 34(4):49–51.

Dunlop, F. (2011). Global menu: Kicking up a stink. *Financial Times*, May 20.

CHAPTER 7. SWEET SMELLS OF SUCCESS

Milton, J. (1674). *Paradise Lost*, Book 5, lines 291–300.

Greening and scenting of Earth

Zahnle, K., N. Arndt, et al. (2007). Emergence of a habitable planet. *Space Sci Revs* 129:35–78.

Nies, D. H. (2014). Systematics of life, its early evolution, and ecological diversity. In *Ecological Biochemistry*, ed. G.-J. Krauss and D. H. Nies, 49–75. Wiley.

Harholt, J., Ø. Moestrup, et al. (2016). Why plants were terrestrial from the beginning. *Trends Plant Sci* 21:96–101.

Kenrick, P., and P. R. Crane (1997). The origin and early evolution of plants on land. *Nature* 389:33–39.

Morris, J. L., M. N. Puttick, et al. (2018). The time-scale of early land plant evolution. *Proc Natl Acad Sci USA* 115:E2274–83.

Wickett, N. J., S. Mirarab, et al. (2014). Phylotranscrip-tomic analysis of the origin and early diversification of land plants. *Proc Natl Acad Sci USA* 111:E4859–68.

Weng, J.-K., R. N. Philippe, et al. (2012). The rise of chemodiversity in plants. *Science* 336:1667–70.

Plant inventions for support and self-defense

Novaes, E., M. Kirst, et al. (2010). Lignin and biomass: A negative correlation for wood formation and lignin content. . . . *Plant Physiol* 154:555–61.

Renault, H., D. Werck-Reichhart, et al. (2019). Harnessing lignin evolution for biotechnological applications. *Curr Opinion Biotechnol* 56:105–11.

Labandeira, C. C., and E. D. Currano (2013). The fossil record of plant-insect dynamics. *Ann Rev Earth and Planetary Sci* 41:287–311.

Labandeira, C. C. (2006). The four phases of plant-arthropod associations in deep time. *Geologica Acta* 4(4):409–38.

Bar-On, Y. M., R. Phillips, et al. (2018). The biomass distribution on Earth. *Proc Natl Acad Sci USA* 115:6506–11.

Plant inventions for sex and dispersal

Pellmyr, O., and L. B. Thien (1986). Insect reproduction and floral fragrances: Keys to the evolution of the angiosperms? *Taxon* 35:76–85.

Dudareva, N., and E. Pichersky, eds. (2006). *Biology of Floral Scent*. CRC Press.

Lovisetto, A., F. Guzzo, et al. (2012). Molecular analyses of mads-box genes trace back to gymnosperms the invention of fleshy fruits. *Mol Biol Evol* 29:409–19.

Rodríguez, A., B. Alquézar, et al. (2013). Fruit aromas in mature fleshy fruits as signals of readiness for predation and seed dispersal. *New Phytologist* 197:36–48.

Guimarães, P. R., M. Galetti, et al. (2008). Seed dispersal anachronisms: Rethinking the fruits extinct megafauna ate. *PLoS One* 3:e1745.

Nevo, O., D. Razafimandimby, et al. (2018). Fruit scent as an evolved signal to primate seed dispersal. *Sci Adv* 4:eaat4871.

Cronberg, N. (2012). Animal-mediated fertilization in bryophytes—parallel or precursor to insect pollination in angiosperms? *Lindbergia* 35:76–85.

Schiestl, F. P., and S. Dötterl (2012). Evolution of floral scent and olfactory preferences in pollinators: Coevolution or pre-existing bias? *Evolution* 66:2042–55.

"Chemical" and medicinal plant molecules

Sarret, M., P. Adam, et al. (2017). Organic substances from Egyptian jars of the Early Dynastic period (3100–2700 BCE): Mode of preparation, alteration processes and botanical (re)assessment of "cedrium." *J Archaeol Sci Reports* 14:420–31.

Chen, W., I. Vermaak, et al. (2013). Camphor—a fumigant during the Black Death and a coveted fragrant wood in ancient Egypt and Babylon—a review. *Molecules* 18:5434–54.

Earley, L. S. (2004). *Looking for Longleaf: The Fall and Rise of an American Forest.* Univ. North Carolina Press.

Vernon, L. F. (2018). From surgical suite to fresh breath: The history of Listerine®. *Int J Dentistry and Oral Health* 4(3):1–7.

Irritating, soothing, pleasing plant molecules

Cometto-Muñiz, J. E., and C. Simons (2015). Trigeminal chemesthesis. In *Handbook of Olfaction and Gustation,* ed. R. L. Doty, 1089–1112. Wiley.

Startek, J. B., T. Voetsm, et al. (2019). To flourish or perish: Evolutionary TRiPs into the sensory biology of plant-herbivore interactions. *Pflügers Archiv—Eur J Physiol* 471:213–36.

Kerman, F., A. Chakirian, et al. (2011). Molecular complexity determines the number of olfactory notes and the pleasantness of smells. *Sci Rep* 1:206.

Khan, R. M., C.-H. Luk, et al. (2007). Predicting odor pleasantness from odorant structure: Pleasantness as a reflection of the physical world. *J Neuroscience* 27:10015–23.

Keller, A., R. C. Gerkin, et al. (2017). Predicting human olfactory perception from chemical features of odor molecules. *Science* 355:820–26.

CHAPTER 8. PLANT VOLATILE FAMILIES: GREEN, FRUITY, FLOWERY, SPICY

Thoreau, H. D. (1862). Wild apples. *Atlantic Monthly* 10(61):513–26.

General plant volatile production

Weng, J.-K., R. N. Philippe, et al. (2012). The rise of chemodiversity in plants. *Science* 336:1667–70.

Gang, D. R. (2005). Evolution of flavors and scents. *Ann Rev Plant Biol* 56:301–25.

Pott, D. M., S. Osorio, et al. (2019). From central to specialized metabolism: An overview. . . . *Front Plant Sci* 10:835.

Schwab, W., R. Davidovich-Rikanati, et al. (2008). Biosynthesis of plant-derived flavor compounds. *Plant J* 54:712–32.

Dudareva, N., F. Negre, et al. (2006). Plant volatiles: Recent advances and future perspectives. *Crit Revs Plant Sci* 25:417–40.

Kreis, W., and J. Munkert (2019). Exploiting enzyme promiscuity to shape plant specialized metabolism. *J Exp Bot* 70:1435–45.

Green-leaf volatiles

Engelberth, J., H. T. Alborn, et al. (2004). Airborne signals prime plants against insect herbivore attack. *Proc Natl Acad Sci USA* 101:1781–85.

Hatanaka, A. (1993). The biogeneration of green odour by green leaves. *Phytochemistry* 34:1201–18.

Matsui, K. (2006). Green leaf volatiles: Hydroperoxide lyase pathway. . . . *Curr Opinion Plant Biol* 9:274–80.

Esters, lactones, jasmonoids

Duchesne, A. N. (1766). *Histoire Naturelle des Fraisiers.* Paris.

Li, G., H. Jia, et al. (2014). Emission of volatile esters . . . during ripening of "Pingxiangli" pear fruit. . . . *Scientia Horticulturae* 170:17–23.

Beekwilder, J., M. Alvarez-Huerta, et al. (2004). Functional characterization of enzymes forming volatile esters from strawberry and banana. *Plant Physiol* 135:1865–78.

Maga, J. A., and I. Katz (1976). Lactones in foods. CRC *Crit Revs Food Sci Nutr* 8:1–56.

Heil, M., and J. Ton (2008). Long-distance signalling in plant defence. *Trends Plant Sci* 13:264–72.

Wang, J., D. Wu, et al. (2019). Jasmonate action in plant defense against insects. *J Exp Bot* 70:3391–3400.

Zhu, Z., and R. Napier (2017). Jasmonate—a blooming decade. *J Exp Bot* 68:1299–1302.

Terpenoids

Lange, B. M. (2015). The evolution of plant secretory structures and emergence of terpenoid chemical diversity. *Ann Rev Plant Biol* 66:139–59.

Chen, F., D. Tholl, et al. (2011). The family of terpene synthases in plants. . . . *Plant J* 66:212–29.

Pichersky, E., and R. A. Raguso (2018). Why do plants produce so many terpenoid compounds? *New Phytologist* 220:692–702.

Furanones, phenolics, benzenoids

Slaughter, J. C. (2007). The naturally occurring furanones: Formation and function from pheromone to food. *Biol Revs* 74:259–76.

Tohge, T., M. Watanabe, et al. (2013). The evolution of phenylpropanoid metabolism in the green lineage. *Crit Revs Biochem Mol Biol* 48:123–52.

Widhalm, J. R., and N. Dudareva (2015). A familiar ring to it: Biosynthesis of plant benzoic acids. *Molecular Plant* 8:83–97.

Nitrogen and sulfur volatiles

Routray, W., and K. Rayaguru (2018). 2-Acetyl-1-pyrroline: A key aroma component of aromatic rice and other food products. *Food Reviews Int* 34:539–65.

Murray, K. E., and F. B. Whitfield (1975). The occurrence of 3-alkyl-2-methoxypyrazines in raw vegetables. *J Sci Food Agric* 26:973–86.

Iranshahi, M. (2012). A review of volatile sulfur-containing compounds from terrestrial plants. . . . *J Ess Oil Res* 24:393–434.

Block, E. (2010). *Garlic and Other Alliums: The Lore and the Science.* Royal Society of Chemistry.

CHAPTER 9. MOSSES, TREES, GRASSES, WEEDS

Agatharchides (~150 BCE). *On the Erythraean Sea,* trans. S. M. Burstein. In A. Dalby (2000), *Dangerous Tastes: The Story of Spices,* 33. Univ. California Press.

Keller, H. (1908). *The World I Live In*, 65. Century.

Liverworts, mosses, horsetails, ferns

Cronberg, N. (2012). Animal-mediated fertilization in bryophytes—parallel or precursor to insect pollination in angiosperms? *Lindbergia* 35:76–85.

Asakawa, Y. (2004). Chemosystematics of the Hepaticae. *Phytochemistry* 65:623–69.

Saritas, Y., M. M. Sonwa, et al. (2001). Volatile constituents in mosses (*Musci*). *Phytochemistry* 57: 443–57.

Fons, F., D. Froissard, et al. (2013). Volatile composition of six horsetails. *Nat Product Comms* 8:509–12.

Fons, F., D. Froissard, et al. (2010). Biodiversity of volatile organic compounds from five French ferns. *Nat Product Comms* 5:1655–58.

Froissard, D., F. Fons, et al. (2011). Volatiles of French ferns and "fougère" scent in perfumery. *Nat Product Comms* 6:1723–26.

Kessler, M., E. Connor, and M. Lehnert (2015). Volatile organic compounds in the strongly fragrant fern genus *Melpomene*. . . . *Plant Biol* 17:430–36.

Jiao, J., Q.-Y. Gai, et al. (2013). Ionic-liquid-assisted microwave distillation coupled with headspace single-drop microextraction followed by GC-MS for the rapid analysis of essential oil in *Dryopteris fragrans*. *J Separation Sci* 36:3799–3806.

Cedar, pine, spruce, redwood

Epic of Gilgamesh (~1200 BCE). F. N. Al-Rawi and A. R. George (2014). Back to the cedar forest: The beginning and end of Tablet V of the standard Babylonian Epic of Gilgameš. *J Cuneiform Studies* 66(1):69–90.

Fleisher, A., and Z. Fleisher (2000). The volatiles of the leaves and wood of Lebanon cedar. . . . *J Ess Oil Res* 12:763–65.

Paoli, M., A.-M. Nam, et al. (2011). Chemical variability of the wood essential oil of *Cedrus atlantica manetti* from Corsica. *Chem & Biodiversity* 8:344–51.

Adams, R. P. (1987). Investigation of *Juniperus* species of the U.S. for new sources of cedarwood oil. *Ec Bot* 41(1):48–54.

Krauze-Baranowska, M., M. Mardarowicz, et al. (2002). Antifungal activity of the essential oils from some species of the genus *Pinus*. *Zeitschrift für Naturforschung* C 57:478–82.

Kurose, K., D. Okamura, and M. Yatagai (2007). Composition of the essential oils from the leaves of nine *Pinus* species and the cones of three of *Pinus* species. *Flavour Fragr J* 22:10–20.

Yu, E. J., T. H. Kim, et al. (2004). Aroma-active compounds of *Pinus densiflora* (red pine) needles. *Flavour Fragr J* 19:532–37.

Andersen, N. H., and D. D. Syrdal (1970). Terpenes and sesquiterpenes of *Chamaecyparis nootkatensis* leaf oil. *Phytochemistry* 9:1325–40.

von Rudloff, E. (1981). The leaf oil terpene composition of incense cedar and coast redwood. *Canadian J Chemistry* 59:285–87.

von Rudloff, E., M. S. Lapp, et al. (1988). Chemosystematic study of *Thuja plicata*. . . . *Biochem Systematics Ecol* 16:119–25.

Frankincense, myrrh, balsams

Langenheim, J. H. (2003). *Plant Resins: Chemistry, Evolution, Ecology, and Ethnobotany*. Timber Press.

Ben-Yehoshua, S., C. Borowitz, et al. (2012). Frankincense, myrrh, and balm of Gilead. . . . *Hort Reviews* 39:1–76.

Cerutti-Delasalle, C., M. Mehiri, et al. (2016). The (+)-cis- and (+)-trans-olibanic acids: Key odorants of frankincense. *Angewandte Chemie* 128:13923–27.

Niebler, J., and A. Büttner (2016). Frankincense revisited, Part I: Comparative analysis of volatiles in commercially relevant *Boswellia* species. *Chem & Biodiversity* 13:613–29.

Niebler, J., and A. Büttner (2015). Identification of odorants in frankincense. . . . *Phytochemistry* 109:66–75.

Niebler, J., K. Zhuravlova, et al. (2016). Fragrant sesquiterpene ketones as trace constituents in frankincense volatile oil. . . . *J Natural Products* 79:1160–64.

Hanuš, L. O., T. Rezanka, et al. (2005). Myrrh—*Commiphora* chemistry. *Biomed Papers* 149:3–28.

Burger, P., A. Casale, et al. (2016). New insights in the chemical composition of benzoin balsams. *Food Chem* 210:613–22.

Custódio, D. L., and V. F. Veiga-Junior (2012). True and common balsams. *Rev Brasil Farmacogn* 22: 1372–83.

Copals, amber

Case, R. J., A. O. Tucker, et al. (2003). Chemistry and ethnobotany of commercial incense copals . . . of North America. *Economic Botany* 57:189–202.

Gramosa, N. V., and E. R. Silveira (2005). Volatile constituents of *Copaifera langsdorffii* from the Brazilian northeast. *J Ess Oil Res* 17:130–32.

Feist, M., I. Lamprecht, et al. (2007). Thermal investigations of amber and copal. *Thermochimica Acta* 458:162–70.

Santiago-Blay, J. A, and J. B. Lambert (2017). Plant exudates and amber: Their origin and uses. *Arnoldia* 75(1):2–13.

Cinnamon, eucalypts, bay laurels, avocado

Ravindran, P. N., K. Nirmal Babu, et al., eds. (2004). *Cinnamon and Cassis: The Genus Cinnamomum*. CRC Press.

Woehrlin, F., H. Fry, et al. (2010). Quantification of flavoring constituents in cinnamon. . . . *J Agric Food Chem* 58:10568–75.

Kim, L., I. E. Galbally, et al. (2011). BVOC emissions from mechanical wounding of leaves and branches of *Eucalyptus sideroxylon* (red ironbark). *J Atmospheric Chemistry* 68:265–79.

Tabanca, N., C. Avonto, et al. (2013). Comparative investigation of *Umbellularia californica* and *Laurus nobilis* leaf essential oils. . . . *J Agric Food Chem* 61:12283–91.

Niogret, J., N. D. Epsky, et al. (2013). Analysis of sesquiterpene distributions in the leaves, branches, and trunks of avocado. . . . *Am J Plant Sci* 4:922–31.

Sandalwood, agarwood,
palo santo, oak, cherry

Hasegawa, T., H. Izumi, et al. (2012). Structure-odor relationships of α-santalol derivatives with modified side chains. *Molecules* 17:2259–70.

Chen, H., Y. Yang, et al. (2011). Comparison of compositions and antimicrobial activities of essential oils from chemically stimulated agarwood, wild agarwood and healthy *Aquilaria sinensis* trees. *Molecules* 16:4884–96.

Ishihara, M., T. Tsuneya, et al. (1993). Components of the volatile concentrate of agarwood. *J Ess Oil Res* 5:283–89.

Naef, R. (2011). The volatile and semi-volatile constituents of agarwood . . . *Flavour Fragr J* 26:73–87.

Yukawa, C., Y. Imayoshi, et al. (2006). Chemical composition of three extracts of *Bursera graveolens*. *Flavour Fragr J* 21:234–38.

Fernández de Simón, B., E. Esteruelas, et al. (2009). Volatile compounds in acacia, chestnut, cherry, ash, and oak woods. . . . *J Agric Food Chem* 57:3217–27.

Martínez, J., E. Cadahía, et al. (2008). Effect of the seasoning method on the chemical composition of oak heartwood to cooperage. *J Agric Food Chem* 56:3089–96.

Mugwort, sage, yarrow, stinking goosefoot,
creosote bush

Barney, J. N., A. G. Hay, et al. (2005). Isolation and characterization of allelopathic volatiles from mugwort. . . . *J Chem Ecology* 31:247–65.

Borek, T. T., J. M. Hochrien, et al. (2006). Composition of the essential oil of white sage, *Salvia apiana*. *Flavour Fragr J* 21:571–72.

Amoore, J. E., and L. J. Forrester (1976). Specific anosmia to trimethylamine: The fishy primary odor. *J Chem Ecology* 2:49–56.

Jardine, K., L. Abrell, et al. (2010). Volatile organic compound emissions from *Larrea tridentata* (creosote bush). *Atmos Chem Phys* 10:12191–206.

Rhew, R. C., B. R. Miller, et al. (2008). Chloroform, carbon tetrachloride and methyl chloroform fluxes in southern California ecosystems. *Atmos Environ* 42:7135–40.

Strobel, G., S. K. Singh, et al. (2011). An endophytic/pathogenic *Phoma* sp. from creosote bush . . . and its VOCs. *FEMS Microbiol Letters* 320:87–94.

Grasses, clovers, sweetgrass

Brilli, F., L. Hörtnagl, et al. (2012). Qualitative and quantitative characterization of volatile organic compound emissions from cut grass. *Environ Sci Technol* 46:3859–65.

Mayland, H. F., R. A. Flath, et al. (1997). Volatiles from fresh and air-dried vegetative tissues of tall fescue. . . . *J Agric Food Chem* 45:2204–10.

Ruuskanen, T. M., M. Müller, et al. (2011). Eddy covariance VOC emission and deposition fluxes above grassland using PTR-TOF. *Atmos Chem Phys* 11:611–25.

Quijano-Celis, C. E., J. A. Pino, et al. (2010). Chemical composition of the leaves essential oil of *Melilotus officinalis* (L.) Pallas from Colombia. *J Ess Oil Bearing Plants* 13:313–15.

Tava, A. (2001). Coumarin-containing grass: Volatiles from sweet vernalgrass. . . . *J Ess Oil Res* 13(5): 367–70.

Isoprene, forest emissions

Pollastri, S., T. Tsonev, et al. (2014). Isoprene improves photochemical efficiency and enhances heat dissipation in plants. . . . *J Exp Bot* 65:1565–70.

Gershenzon, J. (2008). Insects turn up their noses at sweating plants. *Proc Natl Acad Sci USA* 105:17211–12.

Guenther, A., T. Karl, et al. (2006). Estimates of global terrestrial isoprene emissions. . . . *Atmos Chem Phys* 6(11):3181–3210.

Guenther, A., P. Zimmerman, et al. (1994). Natural volatile organic compound emission rate estimates for U.S. woodland landscapes. *Atmos Environ* 28:1197–1210.

Kidd, C., V. Perraud, et al. (2014). Integrating phase and composition of secondary organic aerosol from the ozonolysis of α-pinene. *Proc Natl Acad Sci USA* 111:7552–57.

Jain, S., J. Zahardis, et al. (2014). Soft ionization chemical analysis of secondary organic aerosol from green leaf volatiles emitted by turf grass. *Environ Sci Technol* 48:4835–43.

Kirstine, W., I. Galbally, et al. (2002). Air pollution and the smell of cut grass. In *Conference Proceedings: 16th International Clean Air Conference*, Christchurch, New Zealand.

CHAPTER 10. FLOWERS

Aristotle (~325 BCE). *Sense and Sensibilia*, 5. Trans. J. I. Beare in *The Complete Works of Aristotle* (1984), ed. J. Barnes, 704–5. Princeton Univ. Press.

Pliny (~75 CE). *Natural History*, book 21, ch. 1. Trans. J. Bostock and H. T. Riley in *Pliny the Elder: The Natural History* (1855), https://www.perseus.tufts.edu/hopper/text?doc=Perseus:text:1999.02.0137.

Genders, R. (1994). *Scented Flora of the World*. Hale.

Goody, J. (1993). *The Culture of Flowers*. Cambridge Univ. Press.

Flower evolution, not-so-flowery throwbacks
Raguso, R. A. (2004). Flowers as sensory billboards. . . . *Curr Opinion Plant Biol* 7:434–40.
Raguso, R. A. (2008). Wake up and smell the roses: The ecology and evolution of floral scent. *Ann Rev Ecol Evol Systematics* 39:549–69.
Jürgens, A., S. Dötterl, et al. (2006). The chemical nature of fetid floral odours. . . . *New Phytologist* 172:452–68.
Zhang, X., K. Chingin, et al. (2018). Deciphering the chemical origin of the semen-like floral scents in three angiosperm plants. *Phytochemistry* 145:137–45.

Mixed floral volatiles, syndromes, signals
Dudareva, N. A., and E. Pichersky. (2006). *Biology of Floral Scent*. CRC Press.
Muhlemann, J. K., A. Klempien, et al. (2014). Floral volatiles: From biosynthesis to function. *Plant Cell Environ* 37:1936–49.
Kessler, D., C. Diezel, et al. (2013). *Petunia* flowers solve the defence/apparency dilemma of pollinator attraction by deploying complex floral blends. *Ecol Lett* 16:299–306.
Raguso, R. A. (2016). More lessons from linalool: Insights gained from a ubiquitous floral volatile. *Curr Opinion Plant Biol* 32:31–36.
Knudsen, J. T., R. Eriksson, et al. (2006). Diversity and distribution of floral scent. *Botan Rev* 72:1–120.
Grabenhorst, F., E. T. Rolls, et al. (2007). How pleasant and unpleasant stimuli combine in different brain regions: Odor mixtures. *J Neurosci* 27:13532–40.
Grabenhorst, F., E. T. Rolls, et al. (2011). A hedonically complex odor mixture produces an attentional capture effect in the brain. *NeuroImage* 55:832–43.

Flowers of the ancients
Kaiser, R. (2006). *Meaningful Scents around the World: Olfactory, Chemical, Biological, and Cultural Considerations*, 114–21 [lilies, lotus]. Wiley.
Moraga, A. R., J. L. Rambla, et al. (2009). Metabolite and target transcript analyses during *Crocus sativus* stigma development. *Phytochemistry* 70:1009–16.
Li, H. L. (1951). *The Garden Flowers of China*. Ronald Press.
Kaiser, R. (1993). *The Scent of Orchids: Olfactory and Chemical Investigations*. Elsevier.
Li, S., L. Chen, et al. (2012). Identification of floral fragrances in tree peony. . . . *Scientia Horticulturae* 142:158–65.
Kaneko, S., J. Chen, et al. (2017). Potent odorants of characteristic floral/sweet odor in Chinese chrysanthemum flower tea infusion. *J Agric Food Chem* 65:10058–63.
Zheng, C. H., T. H. Kim, et al. (2004). Characterization of potent aroma compounds in *Chrysanthemum coronarium* L. (Garland). . . . *Flavour Fragr J* 19:401–5.

European wild and garden flowers
Hammami, I., N. Kamoun, et al. (2011). Biocontrol of *Botrytis cinerea* with essential oil and methanol extract of *Viola odorata* L. flowers. *Arch Appl Sci Res* 5:44–51.
Surburg, H., M. Guentert, et al. (1993). Volatile compounds from flowers: Analytical and olfactory aspects. In *Bioactive Volatile Compounds from Plants*, ed. R. Teranishi, R. G. Buttery, et al., 168–86 [lily of the valley]. American Chemical Society.
Rohrig, E., J. Sivinski, et al. (2008). A floral-derived compound attractive to the tephritid fruit fly parasitoid *Diachasmimorpha longicaudata*. *J Chem Ecology* 34:549–57 [alyssum].
Dobson, H. E. M., J. Arroyo, et al. (1997). Interspecific variation in floral fragrances within the genus *Narcissus* (Amaryllidaceae). *Biochem Systematics Ecol* 25:685–706.
Mookherjee, B. D., R. W. Trenkle, et al. (1989). Live vs. dead, Part II: A comparative analysis of the headspace volatiles. . . . *J Ess Oil Res* 1:85–90 [narcissus].
Chen, H.-C., H.-S. Chi, et al. (2013). Headspace solid-phase microextraction analysis of volatile components in *Narcissus tazetta* var. *chinensis* Roem. *Molecules* 18:13723–34.
Brunke, E.-J., F.-J. Hammerschmidt, et al. (1993). Flower scent of some traditional medicinal plants. In *Bioactive Volatile Compounds from Plants*, ed. R. Teranishi, R. G. Buttery, et al., 282–96 [chamomile]. American Chemical Society.
Omidbaigi, R., F. Sefidkon, et al. (2003). Roman chamomile oil: Comparison between hydro-distillation and supercritical fluid extraction. *J Ess Oil Bearing Plants* 6:191–94.
An, M., T. Haig, et al. (2001). On-site field sampling and analysis of fragrance from living lavender (*Lavandula angustifolia* L.) flowers. . . . *J Chromatogr A* 917:245–50.
Lis-Balchin, M. T. (2012). Lavender. In *Handbook of Herbs and Spices*, ed. K. V. Peter, 329–47. Elsevier.
Clery, R. A., N. E. Owen, et al. (1999). An investigation into the scent of carnations. *J Ess Oil Res* 11:355–59.
Nimitkeatkai, H., Y. Ueda, et al. (2005). Emission of methylbutyric acid from *Gypsophila paniculata* L. . . . *Scientia Horticulturae* 106:370–80 [baby's breath].

European vine, shrub, and tree flowers
Porter, A. (1999). Floral volatiles of the sweet pea *Lathyrus odoratus*. *Phytochemistry* 51:211–14.
Schlotzhauer, W. S., S. D. Pair, et al. (1996). Volatile constituents from the flowers of Japanese honeysuckle (*Lonicera japonica*). *J Agric Food Chem* 44:206–9 [includes European *Lonicera caprifolium*].

Adumitresei, L., I. Gostin, et al. (2009). Chemical compounds identified in the leaf glands of *Rosa agrestis savi* and *Rosa rubiginosa* L. *Analele Ştiinţifice ale Universităţii "Al. I. Cuza" din Iaşi: Matematică* (Romania), 55(1):39–45.

Antonelli, A., C. Fabbri, et al. (1997). Characterization of 24 old garden roses from their volatile compositions. *J Agric Food Chem* 45:4435–39.

Kreck, M., S. Püschel, et al. (2003). Biogenetic studies in *Syringa vulgaris* L. . . . lilac aldehydes and lilac alcohols. *J Agric Food Chem* 51:463–69.

Li, Z.-G., M.-R. Lee, et al. (2006). Analysis of volatile compounds emitted from fresh *Syringa oblata* flowers. . . . *Analytica Chimica Acta* 576:43–49.

Naef, R., A. Jaquier, et al. (2004). From the linden flower to linden honey—volatile constituents of linden nectar, the extract of bee-stomach and ripe honey. *Chem & Biodiversity* 1(12):1870–79.

Asian flowers

Johnson, T. S., M. L. Schwieterman, et al. (2016). Lilium floral fragrance. . . . *Phytochemistry* 122:103–12.

Kong, Y., M. Sun, et al. (2012). Composition and emission rhythm of floral scent volatiles from eight lily cut flowers. *J Am Soc for Hort Sci* 137:376–82.

McCulloch, M. (2015). Fragrance removal in *Lilium* L. subdivision *Orientalis* (Oriental lily). University of Minnesota Department of Horticultural Science.

Bera, P., J. N. R. Kotamreddy, et al. (2015). Interspecific variation in headspace scent volatiles composition of four commercially cultivated jasmine flowers. *Natural Product Res* 29:1328–35.

Christensen, L. P., H. B. Jakobsen, et al. (1997). Volatiles emitted from flowers of γ-radiated and nonradiated *Jasminum polyanthum* Franch. in situ. *J Agric Food Chem* 45:2199–2203.

Rout, P. K., S. N. Naik, et al. (2010). Composition of absolutes of *Jasminum sambac* L. flowers. . . . *J Ess Oil Res* 22:398–406.

Ishikawa, M., T. Honda, et al. (2004). "Aqua-space®," a new headspace method for isolation of natural floral aromas. . . . *Biosci Biotech Biochem* 68:454–57 [gardenia].

Mookherjee, B. D., R. W. Trenkle, et al. (1989). Live vs. dead, Part II: A comparative analysis. . . . *J Ess Oil Res* 1:85–90 [osmanthus].

Perriot, R., K. Breme, et al. (2010). Chemical composition of French mimosa absolute oil. *J Agric Food Chem* 58:1844–49.

Zhang, H., and W. N. Setzer (2013). The floral essential oil composition of *Albizia julibrissin* growing in northern Alabama. *Am. J Ess Oils Nat Prod* 1(2):41–42.

African and American flowers

Azuma, H., L. B. Thien, et al. (1999). Floral scents, leaf volatiles and thermogenic flowers in Magnoliaceae. *Plant Species Biol* 14:121–27.

Barreto, A. S., G. D. Feliciano, et al. (2014). Volatile composition of three floral varieties of *Plumeria rubra*. *Int J Curr Microbiol Appl Sci* 3(8):598–607.

Ao, M., B. Liu, et al. (2013). Volatile compound in cut and un-cut flowers of tetraploid *Freesia hybrida*. *Natural Product Res* 27:37–40.

Fu, Y., X. Gao, et al. (2007). Volatile compounds in the flowers of freesia parental species and hybrids. *J Integrative Plant Biol* 49:1714–18.

Ogunwande, I. A., and N. O. Olawore (2006). The essential oil from the leaves and flowers of "African marigold," *Tagetes erecta* L. *J Ess Oil Res* 18:366–68.

Tankeu, S. Y., I. Vermaak, et al. (2013). Essential oil variation of *Tagetes minuta*. . . . *Biochem Systematics Ecol* 51:320–27.

Reverchon, E., and G. Della Porta (1997). Tuberose concrete fractionation by supercritical carbon dioxide. *J Agric Food Chem* 45:1356–60.

Rodyoung, A., K. Sa-nuanpuag, et al. (2015). Volatile releasing patterns of tuberose flowers. . . . *Acta Horticulturae* 307–11.

Kays, S. J., J. Hatch, et al. (2005). Volatile floral chemistry of *Heliotropium arborescens* L. "Marine." *HortScience* 40:1237–38.

Raguso, R. A., C. Henzel, et al. (2003). Trumpet flowers of the Sonoran desert: Floral biology of *Peniocereus* cacti and sacred *Datura*. *Int J Plant Sci* 164:877–92.

Suzuki, S. (1995). Development of new fragrance products by headspace method. *J Japan Oil Chemists' Soc* 44(4):274–82 [cereus, freesia].

Roses through history

Fougère-Danezan, M., S. Joly, et al. (2015). Phylogeny and biogeography of wild roses. . . . *Annals of Botany* 115:275–91.

Adumitresei, L., I. Gostin, et al. (2009). Chemical compounds identified in the leaf glands of *Rosa agrestis savi* and *Rosa rubiginosa* L. *Analele Ştiinţifice ale Universităţii "Al. I. Cuza" din Iaşi: Matematică* (Romania) 55(1):39–45.

Picone, J. M., A. Clery Robin, et al. (2004). Rhythmic emission of floral volatiles from *Rosa damascena semperflorens* cv. Quatre Saisons. *Planta* 219:468–78.

Rusanov, K., N. Kovacheva, et al. (2011). Traditional *Rosa damascena* flower harvesting practices evaluated through GC/MS. . . . *Food Chem* 129:1851–59.

Scalliet, G., F. Piola, et al. (2008). Scent evolution in Chinese roses. *Proc Natl Acad Sci USA* 105:5927–32.

Mookherjee, B. D., R. W. Trenkle, et al. (1989). Live vs. dead, Part II: A comparative analysis. . . . *J Ess Oil Res* 1:85–90 [hybrid tea rose].

Cherri-Martin, M., F. Jullien, et al. (2007). Fragrance heritability in hybrid tea roses. *Scientia Horticulturae* 113:177–81.

Caissard, J.-C., V. Bergougnoux, et al. (2006). Chemical and histochemical analysis of "Quatre Saisons

Blanc Mousseux," a moss rose of the *Rosa* × *damascena* group. *Ann Bot* 97:231–38.

CHAPTER 11. EDIBLE GREENS AND HERBS

Evelyn, J. (1699). *Acetaria: A Discourse of Sallets*. Facsimile (1982). Prospect Books.

Hui, Y. H., ed. (2010). *Handbook of Fruit and Vegetable Flavors*. Wiley.

Peter, K. V., ed. (2012). *Handbook of Herbs and Spices*. Elsevier.

Kitchen greens

Charron, C. S., D. J. Cantliffe, et al. (1996). Photosynthetic photon flux, photoperiod, and temperature effects on emissions of (Z)-3-hexenal, (Z)-3-hexenol, and (Z)-3-hexenyl acetate from lettuce. *J Am Soc for Hort Sci* 121:488–94.

Goetz-Schmidt, E. M., and P. Schreier (1986). Neutral volatiles from blended endive (*Cichorium endivia*, L.). *J Agric Food Chem* 34:212–15.

Flamini, G., P. L. Cioni, et al. (2003). Differences in the fragrances of pollen, leaves, and floral parts of garland (*Chrysanthemum coronarium*). . . . *J Agric Food Chem* 51:2267–71.

Umano, K., Y. Hagi, et al. (2000). Volatile chemicals identified in extracts from leaves of Japanese mugwort. . . . *J Agric Food Chem* 48:3463–69.

Chin, H.-W., Q. Zeng, et al. (1996). Occurrence and flavor properties of sinigrin hydrolysis products in fresh cabbage. *J Food Sci* 61:101–4.

Breme, K., P. Tournayre, et al. (2010). Characterization of volatile compounds of Indian cress absolute. . . . *J Agric Food Chem* 58:473–80.

Masanetz, C., H. Guth, et al. (1998). Fishy and hay-like off-flavours of dry spinach. *Zeitschrift für Lebensmitteluntersuchung und -Forschung A* 206:108–13.

Shim, J.-E., and H. Hee Baek (2012). Determination of trimethylamine in spinach, cabbage, and lettuce. . . . *J Food Sci* 77:C1071–76.

Dregus, M., and K.-H. Engel (2003). Volatile constituents of uncooked rhubarb (*Rheum rhabarbarum* L.) stalks. *J Agric Food Chem* 51:6530–36.

Dregus, M., and H.-G. Schmarr, et al. (2003). Enantioselective analysis of methyl-branched alcohols and acids in rhubarb stalks. *J Agric Food Chem* 51:7086–91.

Götz-Schmidt, E. M., and P. Schreier (1988). Volatile constituents of *Valerianella locusta*. *Phytochemistry* 27(3):845–48.

Delort, E., A. Jaquier, et al. (2012). Volatile composition of oyster leaf. . . . *J Agric Food Chem* 60:11681–90.

Mhamdi, B., W. A. Wannes, et al. (2009). Volatiles from leaves and flowers of borage. . . . *J Ess Oil Res* 21:504–6.

Herbs in the mint family

Karousou, R., M. Balta, et al. (2007). "Mints," smells and traditional uses in Thessaloniki (Greece) and other Mediterranean countries. *J Ethnopharmacol* 109:248–57.

Kelley, L. E., and K. R. Cadwallader (2018). Identification and quantitation of potent odorants in spearmint oils. *J Agric Food Chem* 66:2414–21.

Kokkini, S., and D. Vokou (1989). *Mentha spicata* (Lamiaceae) chemotypes growing wild in Greece. *Economic Botany* 43:192–202.

Teixeira, M. A., and A. E. Rodrigues (2014). Coupled extraction and dynamic headspace techniques for the characterization of essential oil and aroma fingerprint of *Thymus* species. *Ind & Eng Chem Res* 53:9875–82.

Figuérédo, G., P. Cabassu, et al. (2006). Studies of Mediterranean oregano populations. . . . *Flavour Fragr J* 21:134–39.

Baher, Z. F., M. Mirza, et al. (2002). The influence of water stress on plant height, herbal and essential oil yield and composition in *Satureja hortensis* L. *Flavour Fragr J* 17:275–77.

Karousou, R., C. Efstathiou, et al. (2012). Chemical diversity of wild growing *Origanum majorana* in Cyprus. *Chem & Biodiversity* 9:2210–17.

Flamini, G., P. L. Cioni, et al. (2002). Main agronomic-productive characteristics of two ecotypes of *Rosmarinus officinalis* L. and chemical composition of their essential oils. *J Agric Food Chem* 50:3512–17.

Stešević, D., M. Ristić, et al. (2014). Chemotype diversity of indigenous Dalmatian sage (*Salvia officinalis* L.) populations in Montenegro. *Chem & Biodiversity* 11:101–14.

Nitta, M., H. Kobayashi, et al. (2006). Essential oil variation of cultivated and wild perilla analyzed by GC/MS. *Biochem Systematics Ecol* 34:25–37.

Peer, W. A., and J. H. Langenheim (1998). Influence of phytochrome on leaf monoterpene variation in *Satureja douglasii*. *Biochem Systematics Ecol* 26:25–34.

Basil and "Pesto alla Genovese"

Vieira, R. F., and J. E. Simon (2006). Chemical characterization of basil (*Ocimum spp.*) based on volatile oils. *Flavour Fragr J* 21:214–21.

Bernhardt, B., L. Sipos, et al. (2015). Comparison of different *Ocimum basilicum* L. gene bank accessions. . . . *Ind Crops Products* 67:498–508.

Elementi, S., R. Neri, et al. (2006). Biodiversity and selection of "European" basil (*Ocimum basilicum* L.) types. *Acta Horticulturae* 99–104.

Amadei, G., and B. M. Ross (2012). Quantification of character-impacting compounds in *Ocimum basilicum* and "Pesto alla Genovese." . . . *Rapid Commun in Mass Spectrom* 26:219–25.

Miele, M., R. Dondero, et al. (2001). Methyleugenol in *Ocimum basilicum* L. cv. Genovese Gigante. *J Agric Food Chem* 49:517–21.

Fischer, R., N. Nitzan, et al. (2011). Variation in essential oil composition within individual leaves of sweet basil (*Ocimum basilicum* L.) is more affected by leaf position than by leaf age. *J Agric Food Chem* 59:4913–22.

Herbs in the celery family

Kurobayashi, Y., E. Kouno, et al. (2006). Potent odorants characterize the aroma quality of leaves and stalks in raw and boiled celery. *Biosci Biotech Biochem* 70:958–65.

Bylaitė, E., J. P. Roozen, et al. (2000). Dynamic deadspace-gas chromatography-olfactometry analysis of different anatomical parts of lovage (*Levisticum officinale* Koch.) at eight growing stages. *J Agric Food Chem* 48:6183–90.

Lisiewska, Z., W. Kmiecik, et al. (2007). Content of basic components and volatile oils in green dill. . . . *J Food Quality* 30:281–99.

Stefanini, M. B., L. C. Ming, et al. (2006). Essential oil constituents of different organs of fennel. . . . *Rev Bras Pl Med* 8:193–98.

Başer, K. H. C., N. Ermin, et al. (1998). The essential oil of *Anthriscus cerefolium* (L.) Hoffm. (chervil) growing wild in Turkey. *J Ess Oil Res* 10:463–64.

Cadwallader, K. R., D. Benitez, et al. (2005). Characteristic aroma components of the cilantro mimics. In *Natural Flavors and Fragrances*, ed. C. Frey and R. Rouseff, 117–28. American Chemical Society.

Masanetz, C., and W. Grosch (1998). Key odorants of parsley leaves. . . . *Flavour Fragr J* 13(2):115–24.

Ulrich, D., T. Bruchmüller, et al. (2011). Sensory characteristics and volatile profiles of parsley. . . . *J Agric Food Chem* 59:10651–56.

Herbs in the daisy family

Obolskiy, D., I. Pischel, et al. (2011). *Artemisia dracunculus* L. (Tarragon): A critical review. . . . *J Agric Food Chem* 59:11367–84.

Khalilov, L. M., E. A. Paramonov, et al. (2001). . . . Composition of vapor isolated from certain species of *Artemisia* plants. *Chemistry of Natural Compounds* 37(4):339–42 [wormwood].

Mastelić, J., O. Politeo, et al. (2008). Contribution to the analysis of the essential oil of *Helichrysum italicum* . . . *Molecules* 13:795–803 [curry plant].

Meshkatalsadat, M. H., J. Safaei-Ghomi, et al. (2010). Chemical characterization of volatile components of *Tagetes minuta* L. . . . *Digest J Nanomaterials and Biostructures* 5(1):101–6 [huacatay].

Other herbs from the Americas

Calvo-Irabién, L. M., V. Parra-Tabla, et al. (2014). Phytochemical diversity of the essential oils of Mexican oregano. . . . *Chem & Biodiversity* 11:1010–21.

Shahhoseini, R., A. Estaji, et al. (2013). The effect of different drying methods on the content and chemical composition of essential oil of lemon verbena. . . . *J Ess Oil Bearing Plants* 16(4):474–81.

Blanckaert, I., M. Paredes-Flores, et al. (2012). Ethnobotanical, morphological, phytochemical and molecular

evidence for the incipient domestication of epazote . . . in a semi-arid region of Mexico. *Genetic Resources and Crop Evolution* 59:557–73.

McBurnett, B. G., A. A. Chavira, et al. (2006). Analysis of *Piper auritum*: A traditional Hispanic herb. In *Hispanic Foods*, ed. M. H. Tunick and E. González de Mejía, 67–76 [hoja santa]. American Chemical Society.

Buttery, R. G., L. C. Ling, et al. (1987). Tomato leaf volatile aroma components. *J Agric Food Chem* 35:1039–42.

Herbs from East Asia

Avoseh, O., O. Oyedeji, et al. (2015). *Cymbopogon* species. . . . *Molecules* 20:7438–53 [lemongrass].

Agouillal, F., Z. M. Taher, et al. (2017). A review of . . . *Citrus hystrix* DC. *Biosciences, Biotechnol Res Asia* 14:285–305 [makrut lime].

Jiang, L., and K. Kubota (2001). Formation by mechanical stimulus of the flavor compounds in young leaves of Japanese pepper (*Xanthoxylum piperitum* DC). *J Agric Food Chem* 49:1353–57.

Starkenmann, C., L. Luca, et al. (2006). Comparison of volatile constituents of *Persicaria odorata* (Lour.). . . . *J Agric Food Chem* 54:3067–71 [rau ram].

Steinhaus, M. (2015). Characterization of the major odor-active compounds in the leaves of the curry tree *Bergera koenigii* L. . . . *J Agric Food Chem* 63:4060–67.

Wakte, K. V., R. J. Thengane, et al. (2010). Optimization of HS-SPME conditions for quantification of 2-acetyl-1-pyrroline and study of other volatiles in *Pandanus amaryllifolius* Roxb. *Food Chem* 121: 595–600.

Cosmopolitan herbs

Mnayer, D., A.-S. Fabiano-Tixier, et al. (2014). Chemical composition, antibacterial and antioxidant activities of six essentials oils from the Alliaceae family. *Molecules* 19:20034–53.

Nikolić, M., T. Marković, et al. (2013). Chemical composition and biological activity of *Gaultheria procumbens* L. essential oil. *Ind Crops Prods* 49:561–67 [wintergreen].

Brendel, S., T. Hofmann, and M. Granvogl (2019). Characterization of key aroma compounds in pellets of different hop varieties. . . . *J Agric Food Chem* 67:12044–53.

Andre, C. M., J.-F. Hausman, et al. (2016). *Cannabis sativa*: The plant of the thousand and one molecules. *Frontiers in Plant Sci* 7:19.

CHAPTER 12. EDIBLE ROOTS AND SEEDS: STAPLES AND SPICES

Pliny (~75 CE). *Natural History*, book 12, ch. 14. Trans. J. Bostock and H. T. Riley in *Pliny the Elder: The Natural History* (1855), https://www.perseus.tufts.edu/hopper/text?doc=Perseus:text:1999.02.0137.

Saul, H., M. Madella, et al. (2013). Phytoliths in pottery reveal the use of spice in European prehistoric cuisine. *PLoS One* 8:e70583.

Kraft, K. H., C. H. Brown, et al. (2014). Multiple lines of evidence for the origin of domesticated chili pepper, *Capsicum annuum*, in Mexico. *Proc Natl Acad Sci USA* 111:6165–70.

Hui, Y. H., ed. (2010). *Handbook of Fruit and Vegetable Flavors*. Wiley.

Peter, K. V., ed. (2012). *Handbook of Herbs and Spices*. Elsevier.

Parthasarathy, V. A., ed. (2008). *Chemistry of Spices*. Centre for Agriculture and Bioscience International.

Underground vegetables

McKenzie, M., and V. Corrigan (2016). Potato flavor. In *Advances in Potato Chemistry and Technology*, ed. L. Kaur and J. Singh, 339–68. Elsevier.

Wang, Y., and S. J. Kays (2001). Effect of cooking method on the aroma constituents of sweet potatoes (*Ipomoea batatas* [L.] Lam.). *J Food Quality* 24:67–78.

Bach, V., U. Kidmose, et al. (2012). Effects of harvest time and variety on sensory quality and chemical composition of Jerusalem artichoke (*Helianthus tuberosus*) tubers. *Food Chem* 133:82–89.

Ulrich, D., T. Nothnagel, et al. (2015). Influence of cultivar and harvest year on the volatile profiles of leaves and roots of carrots. . . . *J Agric Food Chem* 63:3348–56.

Neffati, M., and B. Marzouk (2009). Roots volatiles and fatty acids of coriander. . . . *Acta Physiologiae Plantarum* 31:455–61.

Sharopov, F. (2015). Phytochemistry and bioactivities of selected plant species with volatile secondary metabolites. PhD dissertation, Ruperto-Carola Univ., Heidelberg, 45–47 [parsnip].

Maher, L., and I. L. Goldman (2018). Endogenous production of geosmin in table beet. *HortScience* 53:67–72.

Pungent and aromatic roots and bulbs

Masuda, H., Y. Harada, et al. (1996). Characteristic odorants of wasabi (*Wasabia japonica matum*), Japanese horseradish, in comparison with those of horseradish (*Armoracia rusticana*). In *Biotechnology for Improved Foods and Flavors*, ed. G. R. Takeoka, R. Teranishi, et al., 67–78. American Chemical Society.

Block, E. (2010). *Garlic and Other Alliums*. Royal Society of Chemistry.

Pang, X., J. Cao, et al. (2017). Identification of ginger (*Zingiber officinale* Roscoe) volatiles and localization of aroma-active constituents by GC–Olfactometry. *J Agric Food Chem* 65:4140–45.

Singh, G., I. P. S. Kapoor, et al. (2010). Comparative study of chemical composition and antioxidant activity of fresh and dry rhizomes of turmeric. . . . *Food Chem Toxicol* 48:1026–31.

Jirovetz, L., G. Buchbauer, et al. (2003). Analysis of the essential oils of the leaves, stems, rhizomes and roots of the medicinal plant *Alpinia galanga* from southern India. *Acta Pharm Zagreb* 53(2):73–82.

Wagner, J., M. Granvogl, et al. (2016). Characterization of the key aroma compounds in raw licorice. . . . *J Agric Food Chem* 64:8388–96.

Safaralie, A., S. Fatemi, et al. (2008). Essential oil composition of *Valeriana officinalis* L. roots cultivated in Iran. *J Chromatogr A* 1180:159–64.

Grains, beans, peanut

Maeda, T., J. H. Kim, et al. (2008). Analysis of volatile compounds in polished-graded wheat flours. . . . *Eur Food Res Technol* 227:1233–41.

Nordlund, E., R.-L. Heiniö, et al. (2013). Flavour and stability of rye grain fractions in relation to their chemical composition. *Food Res Int* 54:48–56, https://doi.org/10.1016/j.foodres.2013.05.034.

Janeš, D., D. Kantar, et al. (2009). Identification of buckwheat (*Fagopyrum esculentum* Moench) aroma compounds with GC–MS. *Food Chem* 112:120–24.

Janeš, D., H. Prosen, et al. (2012). Identification and quantification of volatile aroma compounds of tartary buckwheat (*Fagopyrum tataricum* Gaertn.) and some of its milling fractions. *J Food Sci* 77:C746–51.

McGorrin, R. J. (2019). Key aroma compounds in oats and oat cereals. *J Agric Food Chem* 67:13778–89.

Cramer, A.-C. J., D. S. Mattinson, et al. (2005). Analysis of volatile compounds from various types of barley cultivars. *J Agric Food Chem* 53:7526–31.

Bradbury, L. M. T., R. J. Henry, et al. (2016). Flavor development in rice. In *Biotechnology in Flavor Production*, ed. D. Havkin-Frenkel and N. Dudai, 221–42. Wiley.

Yang, D. S., R. L. Shewfelt, et al. (2008). Comparison of odor-active compounds from six distinctly different rice flavor types. *J Agric Food Chem* 56:2780–87.

Asikin, Y., Kusumiyati, et al. (2018). Alterations in the morphological, sugar composition, and volatile flavor properties of petai (*Parkia speciosa* Hassk.) seed during ripening. *Food Res Int* 106:647–53.

Roland, W. S. U., L. Pouvreau, et al. (2017). Flavor aspects of pulse ingredients. *Cereal Chemistry J* 94:58–65.

Oomah, B. D., L. S. Y. Liang, et al. (2007). Volatile compounds of dry beans. . . . *Plant Foods Hum Nutr* 62:177–83.

Chetschik, I., M. Granvogl, et al. (2008). Comparison of the key aroma compounds in organically grown, raw West African peanuts (*Arachis hypogaea*) and in ground, pan-roasted meal produced thereof. *J Agric Food Chem* 56:10237–43.

Nuts

Cadwallader, K., and S. Puangpraphant (2008). Flavor and volatile compounds in tree nuts. In *Tree Nuts*, ed. F. Shahidi and C. Alasalvar, 109–25. CRC Press.

Kim, Y. K., K. N. Chung, et al. (1986). Volatile components of pinenut. *Korean J Food Sci Technol* 18(2):105–9.

Sonmezdag, A. S., H. Kelebek, et al. (2018). Pistachio oil (*Pistacia vera* L. cv. Uzun): Characterization of key odorants. . . . *Food Chem* 240:24–31.

Franklin, L. M., and A. E. Mitchell (2019). Review of the sensory and chemical characteristics of almond (*Prunus dulcis*) flavor. *J Agric Food Chem* 67:2743–53.

Burdack-Freitag, A., and P. Schieberle (2012). Characterization of the key odorants in raw Italian hazelnuts (*Corylus avellana* L. var. Tonda Romana) and roasted hazelnut paste. . . . *J Agric Food Chem* 60:5057–64.

Lee, J., L. Vázquez-Araújo, et al. (2011). Volatile compounds in light, medium, and dark black walnut and their influence on the sensory aromatic profile. *J Food Sci* 76:C199–204.

Gong, Y., A. L. Kerrihard, et al. (2018). Characterization of the volatile compounds in raw and roasted Georgia pecans by HS-SPME-GC-MS. *J Food Sci* 83:2753–60.

Wang, W., H. Chen, et al. (2019). Effect of sterilization and storage on volatile compounds, sensory properties and physicochemical properties of coconut milk. *Microchemical J* 153:104532.

Celery family spices

Sowbhagya, H. B., P. Srinivas, et al. (2010). Effect of enzymes on extraction of volatiles from celery seeds. *Food Chem* 120:230–34.

Bailer, J., T. Aichinger, et al. (2001). Essential oil content and composition in commercially available dill cultivars in comparison to caraway. *Ind Crops Prods* 14:229–39.

Besharati-Seidani, A., A. Jabbari, et al. (2005). Headspace solvent microextraction: A very rapid method for identification . . . of Iranian *Pimpinella anisum* seed. *Analytica Chimica Acta* 530:155–61.

Hammouda, F., M. Saleh, et al. (2014). Evaluation of the essential oil of *Foeniculum vulgare* Mill (fennel). . . . *African J Tradit Compl Altern Meds* 11:277.

Sriti, J., T. Talou, et al. (2009). Essential oil, fatty acid and sterol composition of Tunisian coriander fruit different parts. *J Sci Food Agric* 89:1659–64.

Hashemi, P., M. Shamizadeh, et al. (2009). Study of the essential oil composition of cumin seeds. . . . *Chromatographia* 70:1147–51.

Azizi, M., G. Davareenejad, et al. (2009). Essential oil content and constituents of black zira (*Bunium persicum* [Boiss.] B. Fedtsch.) from Iran during field cultivation (domestication). *J Ess Oil Res* 21:78–82.

Singh, G., P. Marimuthu, et al. (2004). Chemical, antioxidant and antifungal activities of volatile oil of black pepper [ajwain] and its acetone extract. *J Sci Food Agric* 84:1878–84.

Degenhardt, A., M. Liebig, et al. (2012). Novel insights into flavor chemistry of asafetida. In *Recent Advances in the Analysis of Food and Flavors*, ed. S. Toth and C. Mussinan, 167–75. American Chemical Society.

Other Eurasian spices

Mebazaa, R., A. Mahmoudi, et al. (2009). Characterisation of volatile compounds in Tunisian fenugreek seeds. *Food Chem* 115:1326–36.

Zawirska-Wojtasiak, R., S. Mildner-Szkudlarz, et al. (2010). Gas chromatography, sensory analysis and electronic nose in the evaluation of black cumin (*Nigella sativa* L.) aroma quality. *Gas Chromatography* 56:11

Ozturk, I., S. Karaman, et al. (2014). Aroma, sugar and anthocyanin profile of fruit and seed of mahlab (*Prunus mahaleb* L.) *Food Analytical Methods* 7: 761–73.

Macchia, M., L. Ceccarini, et al. (2013). Studies on saffron (*Crocus sativus* L.) from Tuscan Maremma (Italy). *Int J Food Sci Technol* 48(11):2370–75.

Pungent spices: mustards and peppers

Dai, R., and L.-T. Lim (2014). Release of allyl isothiocyanate from mustard seed meal powder. . . . *J Food Sci* 79:E47–53.

Liu, L., G. Song, et al. (2007). GC–MS Analysis of the essential oils of *Piper nigrum* L. and *Piper longum* L. *Chromatographia* 66:785–90.

Jagella, T., and W. Grosch (1999). Flavour and off-flavour compounds of black and white pepper (*Piper nigrum* L.), I–III, *Eur Food Res Technol* 209(1):16–31.

Ma, Y., Y. Wang, et al. (2019). Sensory characteristics and antioxidant activity of *Zanthoxylum bungeanum* pericarps. *Chem & Biodiversity* 16:e1800238 [Sichuan pepper].

Jiang, L., and K. Kubota (2004). Differences in the volatile components and their odor characteristics of green and ripe fruits and dried pericarp of Japanese pepper. . . . *J Agric Food Chem* 52:4197–4203.

Martín, A., A. Hernández, et al. (2017). Impact of volatile composition on the sensorial attributes of dried paprikas. *Food Res Int* 100:691–97.

Lawrence, B. M. (2016). Pink pepper fruit and leaf oils. *Perfumer & Flavorist* 41:5, 56–60.

Asian aromatics: cinnamon, clove, cardamom

Ravindran, P. N., K. Nirmal Babu, et al., eds. (2004). *Cinnamon and Cassia: The Genus* Cinnamomum. CRC Press.

Chaieb, K., H. Hajlaoui, et al. (2007). The chemical composition and biological activity of clove essential oil. . . . *Phytotherapy Res* 21:501–6.

Howes, M.-J. R., G. C. Kite, et al. (2009). Distinguishing Chinese star anise from Japanese star anise. . . . *J Agric Food Chem* 57:5783–89.

Gochev, V., T. Girova, et al. (2012). Low temperature extraction . . . seeds from cardamom (*Elettaria cardamomum* [L.] Maton). *J BioScience & Biotechnol* 1(2):135–39.

Aromatics from the Americas: allspice, annatto, tonka, vanilla

García-Fajardo, J., M. Martínez-Sosa, et al. (1997).

Comparative study of the oil and supercritical CO_2 extract of Mexican pimento. . . . *J Ess Oil Res* 9: 181–85.

Galindo-Cuspinera, V., M. B. Lubran, et al. (2002). Comparison of volatile compounds in water- and oil-soluble annatto (*Bixa orellana* L.) extracts. *J Agric Food Chem* 50:2010–15.

Bajer, T., S. Surmová, et al. (2018). Use of simultaneous distillation-extraction . . . for characterisation of the volatile profile of *Dipteryx odorata* (Aubl.) Willd. *Industrial Crops and Products* 119:313–21.

Brunschwig, C., S. Rochard, et al. (2016). Volatile composition and sensory properties of *Vanilla × tahitensis* bring new insights for vanilla quality control. *J Sci Food Agric* 96:848–58.

Belanger, F. C., and D. Havkin-Frenkel, eds. (2018). *Handbook of Vanilla Science and Technology*. Wiley.

CHAPTER 13. FRUITS

Bartram, W. (1791). *Travels through North and South Carolina.* . . .

Wallace, A. R. (1869). *The Malay Archipelago: The Land of the Orang-utan and the Bird of Paradise: A Narrative of Travel, with Studies of Man and Nature*, vol. 1, 74–75. Macmillan.

Bunyard, E. A. (1929). *The Anatomy of Dessert*. Reprint (2006), 4. Modern Library.

Corner, E. J. H. (1964). *The Life of Plants*, 218. World Publishing.

Hui, Y. H., ed. (2010). *Handbook of Fruit and Vegetable Flavors*. Wiley.

Fruit volatiles

Nevo, O., and Ayasse, M. (2019). Fruit scent: Biochemistry, ecological function, and evolution. In *Co-Evolution of Secondary Metabolites*, ed. J.-M. Merillon and K. G. Ramawat, 1–23. Springer.

Guth, H. (1996). Determination of the configuration of wine lactone. *Helv Chim Acta* 79:1559–71.

McGorrin, R. J. (2011). The significance of volatile sulfur compounds in food flavors: An overview. In *Volatile Sulfur Compounds in Food*, ed. M. C. Qian, X. Fan, et al., 3–31. American Chemical Society.

Tian Shan and pome fruits

Browning, F. (1998). *Apples*. North Point Press.

Dzhangaliev, A. D. (2003). The wild apple tree of Kazakhstan. *Hort Revs* 29:63–303.

Dzhangaliev, A. D., T. N. Salova, et al. (2003). The wild fruit and nut plants of Kazakhstan. *Hort Revs* 29:305–71.

Mabberley, D. J., and B. E. Juniper (2009). *The Story of the Apple*. Timber Press.

Spengler, R. N. (2019). Origins of the apple: The role of megafaunal mutualism in the domestication of *Malus* and rosaceous trees. *Frontiers Plant Sci* 10:617.

Aprea, E., M. L. Corollaro, et al. (2012). Sensory and instrumental profiling of 18 apple cultivars. . . . *Food Res Int* 49:677–86.

Brown, S. K. (2016). Breeding and biotechnology for flavor development in apple. . . . In *Biotechnology in Flavor Production*, ed. D. Havkin-Frenkel and N. Dudai, 264–80. Wiley.

Sugimoto, N., P. Forsline, et al. (2015). Volatile profiles of members of the USDA Geneva Malus core collection. . . . *J Agric Food Chem* 63:2106–16.

Donno, D., G. L. Beccaro, et al. (2012). Application of sensory, nutraceutical and genetic techniques to create a quality profile of ancient apple cultivars. *J Food Quality* 35:169–81.

Suwanagul, A., and D. G. Richardson (1997). Identification of headspace volatile compounds from different pear (*Pyrus communis* L.) varieties. *Acta Horticulturae* 475:605–24.

Li, G., H. Jia, et al. (2012). Characterization of aromatic volatile constituents in 11 Asian pear cultivars. . . . *African J Agric Res* 7(34):4761–70.

Tateo, F., and M. Bononi (2010). Headspace-SPME analysis of volatiles from quince whole fruits. *J Ess Oil Res* 22:416–18.

Stone fruits, figs, persimmons

Petersen, M. B., and L. Poll (1999). The influence of storage on aroma, soluble solids, acid and colour of sour cherries. . . . *Eur Food Res Technol* 209:251–56.

Hayaloglu, A. A., and N. Demir (2016). Phenolic compounds, volatiles, and sensory characteristics of twelve sweet cherry (*Prunus avium* L.) cultivars grown in Turkey. *J Food Sci* 81:C7–18.

Lozano, M., M. C. Vidal-Aragón, et al. (2009). Physicochemical and nutritional properties and volatile constituents of six Japanese plum (*Prunus salicina* Lindl.) cultivars. *Eur Food Res Technol* 228:403–10.

Sabarez, H. T., W. E. Price, et al. (2000). Volatile changes during dehydration of d'Agen prunes. *J Agric Food Chem* 48:1838–42.

Wang, Y., F. Chen, et al. (2010). Effects of germplasm origin and fruit character on volatile composition of peaches and nectarines. In *Flavor and Health Benefits of Small Fruits*, ed. M. C. Qian and A. M. Rimando, 95–117. American Chemical Society.

Greger, V., and P. Schieberle (2007). Characterization of the key aroma compounds in apricots. . . . *J Agric Food Chem* 55:5221–28.

King, E. S., H. Hopfer, et al. (2012). Describing the appearance and flavor profiles of fresh fig (*Ficus carica* L.) cultivars. *J Food Sci* 77:S419–29.

Mujić, I., M. Bavcon Kralj, et al. (2012). Changes in aromatic profile of fresh and dried fig. . . . *Int J Food Sci Technol* 47:2282–88.

Beaulieu, J. C., and R. E. Stein-Chisholm (2016). HS-GC–MS volatile compounds recovered in freshly pressed "Wonderful" cultivar and commercial pomegranate juices. *Food Chem* 190:643–56.

Mayuoni-Kirshinbaum, L., and R. Porat (2014). The

flavor of pomegranate fruit: A review. *J Sci Food Agric* 94:21–27.

Wang, Y., D. Hossain, et al. (2012). Characterization of volatile and aroma-impact compounds in persimmon. . . . *Flavour Fragr J* 27:141–48.

Berries, kiwifruit, grapes

Hummer, K. E., N. Bassil, et al. (2011). *Fragaria*. In *Wild Crop Relatives: Genomic and Breeding Resources*, ed. C. Kole, 17–44. Springer.

Schwieterman, M. L., T. A. Colquhoun, et al. (2014). Strawberry flavor: Diverse chemical compositions, a seasonal influence, and effects on sensory perception. *PLoS One* 9:e88446.

Ulrich, D., D. Komes, et al. (2007). Diversity of aroma patterns in wild and cultivated *Fragaria* accessions. *Genetic Resources and Crop Evolution* 54:1185–96.

Ulrich, D., and K. Olbricht (2014). Diversity of metabolite patterns and sensory characters in wild and cultivated strawberries. *J Berry Res* 4:11–17.

Du, X., and M. Qian (2010). Flavor chemistry of small fruits: Blackberry, raspberry, and blueberry. In *Flavor and Health Benefits of Small Fruits*, ed. M. C. Qian and A. M. Rimando, 27–43. American Chemical Society.

Zhu, J., F. Chen, et al. (2016). Characterization of the key aroma volatile compounds in cranberry. . . . *J Agric Food Chem* 64:4990–99.

Jung, K., O. Fastowski, et al. (2017). Analysis and sensory evaluation of volatile constituents of fresh blackcurrant (*Ribes nigrum* L.) fruits. *J Agric Food Chem* 65:9475–87.

Mouhib, H., and W. Stahl (2014). From cats and blackcurrants: Structure and dynamics of the sulfur-containing cassis odorant cat ketone. *Chem & Biodiversity* 11:1554–66.

Hempfling, K., O. Fastowski, et al. (2013). Analysis and sensory evaluation of gooseberry (*Ribes uva crispa* L.) volatiles. *J Agric Food Chem* 61:6240–49.

Garcia, C. V., S.-Y. Quek, et al. (2012). Kiwifruit flavour: A review. *Trends Food Sci Technol* 24:82–91.

Nieuwenhuizen, N. J., A. C. Allan, et al. (2016). The genetics of kiwifruit flavor and fragrance. In *The Kiwifruit Genome*, ed. R. Testolin, H.-W. Huang, et al., 135–47. Springer.

Sun, Q., M. J. Gates, et al. (2011). Comparison of odor-active compounds in grapes and wines from *Vitis vinifera* and non-foxy American grape species. *J Agric Food Chem* 59:10657–64.

Yang, C., Y. Wang, et al. (2011). Volatile compounds evolution of three table grapes. . . . *Food Chem* 128:823–30.

Cucumbers and melons

Schieberle, P., S. Ofner, et al. (1990). Evaluation of potent odorants in cucumbers (*Cucumis sativus*) and muskmelons (*Cucumis melo*). . . . *J Food Sci* 55:193–95.

Gonda, I., Y. Burger, et al. (2016). Biosynthesis and perception of melon aroma. In *Biotechnology in Flavor Production*, ed. D. Havkin-Frenkel and N. Dudai, 281–305. Wiley.

Aubert, C., and M. Pitrat (2006). Volatile compounds in the skin and pulp of Queen Anne's pocket melon. *J Agric Food Chem* 54:8177–82.

Liu, Y., C. He, et al. (2018). Comparison of fresh watermelon juice aroma characteristics of five varieties. . . . *Food Res Int* 107:119–29.

Vergauwen, D., and I. De Smet (2019). Watermelons versus melons: A matter of taste. *Trends Plant Sci* 24:973–76.

Tomatoes, chilis, and relatives

Davidovich-Rikanati, R., Y. Sitrit, et al. (2016). Tomato aroma: Biochemistry and biotechnology. In *Biotechnology in Flavor Production*, ed. D. Havkin-Frenkel and N. Dudai, 243–63. Wiley.

Mayer, F., G. Takeoka, et al. (2002). Aroma of fresh field tomatoes. In *Freshness and Shelf Life of Foods*, ed. K. R. Cadwallader and H. Weenen, 144–61. American Chemical Society.

Kreissl, J., and P. Schieberle (2017). Characterization of aroma-active compounds in Italian tomatoes. . . . *J Agric Food Chem* 65:5198–5208.

Xu, Y., and S. Barringer (2010). Comparison of tomatillo and tomato volatile compounds. . . . *J Food Sci* 75(3):C268–73.

Berger, R. G., F. Drawert, et al. (1989). The flavour of cape gooseberry. . . . *Zeitschrift für Lebensmitteluntersuchung und -Forschung* 188:122–26.

Yilmaztekin, M. (2014). Characterization of potent aroma compounds of cape gooseberry (*Physalis peruviana* L.) fruits. . . . *Int J Food Properties* 17:469–80.

Forero, M. D., C. E. Quijano, et al. (2009). Volatile compounds of chile pepper (*Capsicum annuum* L. var. *glabriusculum*) at two ripening stages. *Flavour Fragr J* 24:25–30.

Luning, P. A., T. de Rijk, et al. (1994). Gas chromatography, mass spectrometry, and sniffing port analyses of volatile compounds of fresh bell peppers. . . . *J Agric Food Chem* 42:977–83.

Naef, R., A. Velluz, et al. (2008). New volatile sulfur-containing constituents in . . . red bell peppers. . . . *J Agric Food Chem* 56:517–27.

Starkenmann, C., and Y. Niclass (2011). New cysteine-*S*-conjugate precursors of volatile sulfur compounds in bell peppers. . . . *J Agric Food Chem* 59:3358–65.

Pino, J., E. Sauri-Duch, et al. (2006). Changes in volatile compounds of habanero chile pepper (*Capsicum chinense* Jack. cv. Habanero) at two ripening stages. *Food Chem* 94:394–98.

Citrus fruits

Mabberley, D. J. (2004). Citrus (Rutaceae): A review of recent advances in etymology, systematics and medical applications. *Blumea* 49:481–98.

Wu, G. A., J. Terol, et al. (2018). Genomics of the origin and evolution of Citrus. *Nature* 554:311–16.

Dugo, G. (2010). *Citrus Oils: Composition, Advanced Analytical Techniques, Contaminants, and Biological Activity.* CRC Press.

Porat, R., S. Deterre, et al. (2016). The flavor of citrus fruit. In *Biotechnology in Flavor Production*, ed. D. Havkin-Frenkel and N. Dudai, 1–31. Wiley.

Lota, M.-L., D. de Rocca Serra, et al. (2002). Volatile components of peel and leaf oils of lemon and lime species. *J Agric Food Chem* 50:796–805 [terpenoid percentages in peel oils].

Verzera, A., A. Trozzi, et al. (2005). Essential oil composition of *Citrus meyerii* Y. Tan. and *Citrus medica* L. cv. Diamante and their lemon hybrids. *J Agric Food Chem* 53:4890–94 [terpenoid percentages in peel oils].

Arena, E., N. Guarrera, et al. (2006). Comparison of odour active compounds . . . between hand-squeezed juices from different orange varieties. *Food Chem* 98:59–63.

Selli, S., and H. Kelebek (2011). Aromatic profile and odour-activity value of blood orange juices obtained from Moro and Sanguinello. . . . *Industrial Crops and Products* 33:727–33.

Tomiyama, K., H. Aoki, et al. (2012). Characteristic volatile components of Japanese sour citrus fruits: yuzu, sudachi and kabosu. *Flavour Fragr J* 27:341–55.

Agouillal, F., Z. Taher, et al. (2017). A review of . . . *Citrus hystrix* DC. *Biosci, Biotechnol Res Asia* 14: 285–305 [makrut lime].

Dugo, G., and I. Bonaccorsi, eds. (2013). *Citrus Bergamia: Bergamot and Its Derivatives.* CRC Press.

Subtropical and tropical fruits

Amira, E. A., F. Guido, et al. (2011). Chemical and aroma volatile compositions of date palm (*Phoenix dactylifera* L.) fruits at three maturation stages. *Food Chem* 127:1744–54.

El Arem, A., E. B. Saafi, et al. (2012). Volatile and nonvolatile chemical composition of some date fruits (*Phoenix dactylifera* L.) harvested at different stages of maturity. *Int J Food Sci Technol* 47:549–55.

Aurore, G., C. Ginies, et al. (2011). Comparative study of free and glycoconjugated volatile compounds of three banana cultivars from French West Indies: Cavendish, Frayssinette and Plantain. *Food Chem* 129:28–34.

Bugaud, C., E. Deverge, et al. (2011). Sensory characterisation enabled the first classification of dessert bananas. *J Sci Food Agric* 91:992–1000.

Munafo, J. P., J. Didzbalis, et al. (2014). Characterization of the major aroma-active compounds in mango (*Mangifera indica* L.) cultivars. . . . *J Agric Food Chem* 62:4544–4551.

Sung, J., J. H. Suh, et al. (2019). Relationship between sensory attributes and chemical composition of different mango cultivars. *J Agric Food Chem* 67:5177–88.

Mahattanatawee, K., P. R. Pérez-Cacho, et al. (2007). Comparison of three lychee cultivar odor profiles. . . . *J Agric Food Chem* 55:1939–44.

Obenland, D., S. Collin, et al. (2012). Influence of maturity and ripening on aroma volatiles and flavor in "Hass" avocado. *Postharvest Biol Technol* 71:41–50.

Fuggate, P., C. Wongs-Aree, et al. (2010). Quality and volatile attributes of attached and detached "Pluk Mai Lie" papaya during fruit ripening. *Scientia Horticulturae* 126:120–29.

Ulrich, D., and Wijaya, C. H. (2010). Volatile patterns of different papaya (*Carica papaya* L.) varieties. *J Appl Bot Food Qual* 83:128–32.

Porto-Figueira, P., A. Freitas, et al. (2015). Profiling of passion fruit volatiles. . . . *Food Res Int* 77:408–18.

Shaw, G. J., P. J. Ellingham, et al. (1983). Volatile constituents of feijoa. . . . *J Sci Food Agric* 34:743–47.

Steinhaus, M., D. Sinuco, et al. (2009). Characterization of the key aroma compounds in pink guava. . . . *J Agric Food Chem* 57:2882–88.

Wijayaa, C., I. Silamba, et al. (2014). Correlation between flavor profile and sensory acceptance of two pineapple cultivars and their new genotype. In *Flavour Science*, ed. V. Ferreira and R. Lopez, 325–29. Elsevier.

Tokitomo, Y., M. Steinhaus, et al. (2005). Odor-active constituents in fresh pineapple. . . . *Biosci Biotechnol Biochem* 69(7):1323–30.

Ginkgo, vanilla, durian

Del Tredici, P. (2007). The phenology of sexual reproduction in *Ginkgo biloba*: Ecological and evolutionary implications. *Botanical Review* 73:267–78.

Parliment, T. H. (1995). Characterization of the putrid aroma compounds of *Ginkgo biloba* fruits. In *Fruit Flavors*, ed. R. L. Rouseff and M. M. Leahy, 276–79. American Chemical Society.

Householder, E., J. Janovec, et al. (2010). Diversity, natural history, and conservation of vanilla (Orchidaceae) in Amazonian wetlands of Madre de Dios, Peru. *J Botanical Res Institute of Texas* 4:227–43.

Lubinsky, P., M. Van Dam, et al. (2006). Pollination of vanilla and evolution in Orchidaceae. *Lindleyana* 75(12):926–29.

Gigant, R., S. Bory, et al. (2016). Biodiversity and evolution in the *Vanilla* genus. In *The Dynamical Processes of Biodiversity: Case Studies of Evolution and Spatial Distribution*, ed. O. Grillo and G. Venora, 1–26. InTech.

Li, J.-X., P. Schieberle, et al. (2017). Insights into the key compounds of durian (*Durio zibethinus* L. "Monthong") pulp odor. . . . *J Agric Food Chem* 65:639–47.

Nakashima, Y., P. Lagan, et al. (2008). A study of fruit-frugivore interactions in two species of durian (*Du-*

rio, Bombacaceae) in Sabah, Malaysia: Short Communications. *Biotropica* 40:255–58.

CHAPTER 14. THE LAND: SOIL, FUNGI, STONE

Pliny (~75 CE). *Natural History*, book 17, ch. 3. Trans. J. Bostock and H. T. Riley in *Pliny the Elder: The Natural History* (1855), https://www.perseus.tufts.edu/hopper/text?doc=Perseus:text:1999.02.0137.

Bear, I. J., and R. G. Thomas (1964). Nature of argillaceous odour. *Nature* 201(4923):993–95.

Bear, I. J., and R. G. Thomas (1966). Genesis of petrichor. *Geochimica et Cosmochimica Acta* 30(9): 869–79.

Fungi and streptomycetes

Naranjo-Ortiz, M. A., and T. Gabaldón (2019). Fungal evolution: Major ecological adaptations and evolutionary transitions. *Biological Reviews* 94(4): 1443–76.

Brundrett, M. C. (2002). Coevolution of roots and mycorrhizas of land plants. *New Phytologist* 154: 275–304.

Seipke, R. F., M. Kaltenpoth, et al. (2012). Streptomyces as symbionts: An emerging and widespread theme? *FEMS Microbiol Reviews* 36:862–76.

Leaf drop, litter, compost

Purahong, W., T. Wubet, et al. (2016). Life in leaf litter: Novel insights into community dynamics of bacteria and fungi during litter decomposition. *Mol Ecol* 25:4059–74.

Ramirez, K. S., C. L. Lauber, et al. (2010). Microbial consumption and production of volatile organic compounds at the soil-litter interface. *Biogeochem* 99:97–107.

Isidorov, V., Z. Tyszkiewicz, et al. (2016). Fungal succession in relation to volatile organic compounds emissions from Scots pine and Norway spruce leaf litter-decomposing fungi. *Atmos Environ* 131:301–6.

Leff, J. W., and N. Fierer (2008). Volatile organic compound (VOC) emissions from soil and litter samples. *Soil Biol Biochem* 40:1629–36.

Franich, R. A. (1992). Macrocyclic lactones in radiata pine forest floor litter. *Phytochemistry* 31:2532–33.

Tiefel, P., and R. G. Berger (1993). Seasonal variation of the concentrations of maltol and maltol glucoside in leaves of *Cercidiphyllum japonicum*. *J Sci Food Agric* 63:59–61.

Kutzner, H. J. (2000). Microbiology of composting. In *Biotechnology*, ed. H. J. Rhem and G. Reed, 2nd ed., vol. 11c, 35–100. Wiley.

Goldstein, N. (2008). Getting to know the odor compounds [in composting]. *BioCycle* 43(7):51–53.

Zhang, H., G. Li, et al. (2016). Influence of aeration on volatile sulfur compounds (VSCs) and NH_3 emissions during aerobic composting of kitchen waste. *Waste Management* 58:369–75.

Soil and geosmin

Crowther, T. W., J. van den Hoogen, et al. (2019). The global soil community and its influence on biogeochemistry. *Science* 365:eaav0550.

Paul, E. A. (2016). The nature and dynamics of soil organic matter. . . . *Soil Biol Biochem* 98:109–26.

Bennett, J. W., R. Hung, et al. (2012). Fungal and bacterial volatile organic compounds: An overview and their role as ecological signaling agents. In *Fungal Associations*, ed. B. Hock, 373–93. Springer.

Peñuelas, J., D. Asensio, et al. (2014). Biogenic volatile emissions from the soil. *Plant Cell Environ* 37: 1866–91.

Thaysen, A. C. (1936). The origin of an earthy or muddy taint in fish. *Annals Appl Biol* 23:99–104.

Gerber, N., and H. A. Lechevalier (1965). Geosmin, an earthy-smelling substance isolated from actinomycetes. *Appl Environ Microbiol* 13(6):935–38.

Gerber, N. N. (1977). Three highly odorous metabolites from an actinomycete: 2-isopropyl-3-methoxypyrazine, methylisoborneol, and geosmin. *J Chem Ecology* 3:475–82.

Buttery, R. G., and J. A. Garibaldi (1976). Geosmin and methylisoborneol in garden soil. *J Agric Food Chem* 24:1246–47.

Watson, S. B. (2003). Cyanobacterial and eukaryotic algal odour compounds: Signals or by-products? *Phycologia* 42:332–50.

Mold and mushroom volatiles

Dickschat, J. S. (2017). Fungal volatiles—a survey from edible mushrooms to moulds. *Nat Prod Rep* 34:310–28.

Tribe, H. T., E. Thines, et al. (2006). Moulds that should be better known: The wine cellar mould, *Racodium cellare* Persoon. *Mycologist* 20:171–75.

Fraatz, M. A., and H. Zorn (2011). Fungal flavours. In *The Mycota*, ed. M. Hofrichter, vol. 10, 249–68. Springer.

Pudil, F., R. Uvira, et al. (2014). Volatile compounds in stinkhorn. . . . *Eur Scientific J* 10(9):163–71.

Jung, M. Y., D. E. Lee, et al. (2019). Characterization of volatile profiles of six popular edible mushrooms. . . . *J Food Sci* 84:421–29.

Misharina, T. A., S. M. Muhutdinova, et al. (2010). Formation of flavor of dry champignons. . . . *Appl Biochemistry and Microbiol* 46:108–13.

Chen, C.-C., and S.-E. Liu, et al. (1986). Enzymic formation of volatile compounds in shiitake mushroom (*Lentinus edodes* Sing.). In *Biogeneration of Aromas*, ed. T. H. Parliment and R. Croteau, 176–83. American Chemical Society.

Wu, C.-M., and Z. Wang (2000). Volatile compounds in fresh and processed shiitake mushrooms. . . . *Food Sci Technol Res* 6:166–70.

de Pinho, P. G., B. Ribeiro, et al. (2008). Correlation between the pattern volatiles and the overall aroma of wild edible mushrooms. *J Agric Food Chem* 56: 1704–12.

Fons, F., S. Rapior, et al. (2003). Volatile compounds in the *Cantharellus, Craterellus* and *Hydnum* genera. *Cryptogamie Mycologie* 24(4):367–76 [chanterelle, black trumpet].

Cho, I. H., S. M. Lee, et al. (2007). Differentiation of aroma characteristics of pine-mushrooms (*Tricholoma matsutake* Sing.) of different grades. . . . *J Agric Food Chem* 55:2323–28.

Taşkin, H. (2013). Detection of volatile aroma compounds of *Morchella*. . . . *Notulae Botanicae Horti Agrobotanici Cluj-Napoca* 41:122.

Tietel, Z., and S. Masaphy (2018). Aroma-volatile profile of black morel (*Morchella importuna*) grown in Israel. *J Sci Food Agric* 98:346–53.

Truffles

Zambonelli, A., ed. (2016). *True Truffle (Tuber spp.) in the World: Soil Ecology, Systematics and Biochemistry.* Springer.

Buzzini, P., C. Gasparetti, et al. (2005). Production of volatile organic compounds (VOCs) by yeasts isolated from the ascocarps of black (*Tuber melanosporum* Vitt.) and white (*Tuber magnatum* Pico) truffles. *Archives of Microbiol* 184:187–93.

Splivallo, R., and S. E. Ebeler (2015). Sulfur volatiles of microbial origin are key contributors to human-sensed truffle aroma. *Appl Microbiol and Biotechnol* 99:2583–92.

Vahdatzadeh, M., A. Deveau, et al. (2015). The role of the microbiome of truffles in aroma formation. . . . *Appl and Environ Microbiol* 81:6946–52.

Talou, T., M. Doumenc-Faure, et al. (2001). Flavor profiling of 12 edible European truffles. In *Food Flavors and Chemistry*, ed. A. M. Spanier, F. Shahidi, et al., 274–80. Royal Society of Chemistry.

Yeasts

Péter, G., M. Takashima, et al. (2017). Yeast habitats: Different but global. In *Yeasts in Natural Ecosystems: Ecology*, ed. P. Buzzini, M.-A. Lachance, et al., 39–71. Springer.

Becher, P. G., G. Flick, et al. (2012). Yeast, not fruit volatiles, mediate *Drosophila melanogaster* attraction, oviposition and development. *Functional Ecology* 26:822–28.

Saerens, S. M. G., F. R. Delvaux, et al. (2010). Production and biological function of volatile esters in *Saccharomyces cerevisiae*. *Microbial Biotechnol* 3:165–77.

Christiaens, J. F., L. M. Franco, et al. (2014). The fungal aroma gene ATF1 promotes dispersal of yeast cells through insect vectors. *Cell Reports* 9:425–32.

Dweck, H. K. M., S. A. M. Ebrahim, et al. (2015). Olfactory proxy detection of dietary antioxidants in *Drosophila*. *Curr Biol* 25:455–66.

Smith, B. D., and B. Divol (2016). *Brettanomyces bruxellensis*, a survivalist prepared for the wine apocalypse and other beverages. *Food Microbiol* 59:161–75.

Steensels, J., L. Daenen, et al. (2015). *Brettanomyces* yeasts—from spoilage organisms to valuable contributors to industrial fermentations. *Int J Food Microbiol* 206:24–38.

Wet soil and stone, gunflint, swamps

Placella, S. A., E. L. Brodie, et al. (2012). Rainfall-induced carbon dioxide pulses result from sequential resuscitation of phylogenetically clustered microbial groups. *Proc Natl Acad Sci USA* 109:10931–36.

Starkenmann, C., C. J.-F. Chappuis, et al. (2016). Identification of hydrogen disulfanes and hydrogen trisulfanes in H_2S bottle, in flint, and in dry mineral white wine. *J Agric Food Chem* 64:9033–40.

Tominaga, T., G. Guimbertau, et al. (2003). Contribution of benzenemethanethiol to smoky aroma of certain *Vitis vinifera* L. wines. *J Agric Food Chem* 51:1373–76.

Kreitman, G. Y., J. C. Danilewicz, et al. (2017). Copper(II)-mediated hydrogen sulfide and thiol oxidation to disulfides and organic polysulfanes and their reductive cleavage in wine. . . . *J Agric Food Chem* 65:2564–71.

Brown, K. A. (1985). Sulphur distribution and metabolism in waterlogged peat. *Soil Biol Biochem* 17:39–45.

Wajon, J. E., B. V. Kavanagh, et al. (1988). Controlling swampy odors in drinking water. *J Am Water Works Assoc* 80:77–83.

Westermann, P. (1994). The effect of incubation temperature on steady-state concentrations of hydrogen and volatile fatty acids during anaerobic degradation in slurries from wetland sediments. *FEMS Microbiol Ecol* 13:295–302.

CHAPTER 15. THE WATERS: PLANKTON, SEAWEEDS, SHELLFISH, FISH

Steinbeck, J. (1962). *Travels with Charley: In Search of America.* Viking Press.

Clark, E. (1964). *The Oysters of Locmariaquer.* Pantheon.

Life in the waters

Swenson, H. (1983). *Why Is the Ocean Salty?* U.S. Geological Survey.

Brodie, J., C. X. Chan, et al. (2017). The algal revolution. *Trends Plant Sci* 22:726–38.

Open-ocean sulfurousness

Yancey, P. H. (2005). Organic osmolytes as compatible, metabolic and counteracting cytoprotectants in high osmolarity and other stresses. *J Exp Biol* 208:2819–30.

Giordano, M., and Prioretti, L. (2016). Sulphur and algae: Metabolism, ecology and evolution. In *The Physiology of Microalgae*, ed. M. A. Borowitzka, J. Beardall, et al., 185–209. Springer.

Haas, P. (1935). The liberation of methyl sulphide by seaweed. *Biochemical J* 29(6):1297–99.

Johnston, A. W., R. T. Green, et al. (2016). Enzymatic breakage of dimethylsulfoniopropionate—a signature molecule for life at sea. *Curr Opinion Chemical Biol* 31:58–65.

Seymour, J. R., R. Simo, et al. (2010). Chemoattraction to dimethylsulfoniopropionate throughout the marine microbial food web. *Science* 329:342–45.

Savoca, M. S., M. E. Wohlfeil, et al. (2016). Marine plastic debris emits a keystone infochemical for olfactory foraging seabirds. *Sci Advances* 2:e1600395.

Seashore halogens: chlorine, bromine, iodine volatiles

Paul, C., and G. Pohnert (2011). Production and role of volatile halogenated compounds from marine algae. *Nat Prod Rep* 28:186–95.

Küpper, F. C., L. J. Carpenter, et al. (2013). In vivo speciation studies and antioxidant properties of bromine in *Laminaria digitata* reinforce the significance of iodine accumulation for kelps. *J Exp Bot* 64:2653–64.

Chung, H. Y., W. C. Joyce Ma, et al. (2003). Seasonal distribution of bromophenols in selected Hong Kong seafood. *J Agric Food Chem* 51:6752–60.

Fuller, S. C., D. C. Frank, et al. (2008). Improved approach for analyzing bromophenols in seafood. . . . *J Agric Food Chem* 56:8248–54.

Liu, M., P. E. Hansen, et al. (2011). Bromophenols in marine algae and their bioactivities. *Marine Drugs* 9:1273–92.

Seafood aromas, fresh and stale

Shahidi, F., and K. R. Cadwallader, eds. (1997). *Flavor and Lipid Chemistry of Seafoods.* American Chemical Society.

Nollet, L. M. L., and F. Toldrá, eds. (2009). *Handbook of Seafood and Seafood Products Analysis.* CRC Press.

Haard, N. F., and B. K. Simpson, eds. (2000). *Seafood Enzymes: Utilization and Influence on Postharvest Seafood Quality.* CRC Press.

Ma, J., I. M. Pazos, et al. (2014). Microscopic insights into the protein-stabilizing effect of trimethylamine N-oxide (TMAO). *Proc Natl Acad Sci USA* 111:8476–81.

Summers, G., R. D. Wibisono, et al. (2017). Trimethylamine oxide content and spoilage potential of New Zealand commercial fish species. *New Zealand J Marine and Freshwater Res* 51:393–405.

Fish

Kawai, T., and M. Sakaguchi (1996). Fish flavor. *Crit Revs Food Sci and Nutrition* 36:257–98.

Ólafsdóttir, G., and R. Jónsdóttir (2009). Volatile aroma compounds in fish. In *Handbook of Seafood and Seafood Products Analysis*, ed. L. M. L. Nollet and F. Toldrá, 97–117. CRC Press.

Benanou, D., F. Acobas, et al. (2003). Analysis of off-flavors in the aquatic environment. . . . *Analyt Bioanalyt Chem* 376:69–77.

Howgate, P. (2004). Tainting of farmed fish by geosmin and 2-methyl-iso-borneol. . . . *Aquaculture* 234:155–81.

Liu, S., T. Liao, et al. (2017). Exploration of volatile compounds causing off-flavor in farm-raised channel catfish (*Ictalurus punctatus*) fillet. *Aquaculture Int* 25:413–22.

Shellfish

Pennarun, A.-L., C. Prost, et al. (2002). Identification and origin of the character-impact compounds of raw oyster *Crassostrea gigas. J Sci Food Agric* 82:1652–60.

Pennarun, A.-L., C. Prost, et al. (2003). Comparison of two microalgal diets. 2. Influence on odorant composition and organoleptic qualities of raw oysters. . . . *J Agric Food Chem* 51:2011–18.

van Houcke, J., I. Medina, et al. (2016). Biochemical and volatile organic compound profile of European flat oyster (*Ostrea edulis*) and Pacific cupped oyster (*Crassostrea gigas*). . . . *Food Control* 68:200–207.

Kube, S., A. Gerber, et al. (2006). Patterns of organic osmolytes in two marine bivalves, *Macoma balthica* and *Mytilus* spp. . . . *Marine Biol* 149:1387–96 [mussels].

Baek, H. H., and K. R. Cadwallader (1997). Character-impact aroma compounds of crustaceans. In *Flavor and Lipid Chemistry of Seafoods*, ed. F. Shahidi and K. R. Cadwallader, 85–94. American Chemical Society.

Whitfield, F. B., F. Helidoniotis, et al. (1995). *Effect of Diet and Environment on the Volatile Flavour Components of Crustaceans.* CSIRO and Fisheries Res and Development Corporation.

Rodríguez-Bernaldo De Quirós, A., J. López-Hernández, et al. (2001). Comparison of volatile components in fresh and canned sea urchin (*Paracentrotus lividus*, Lamarck) gonads. . . . *Eur Food Res Technol* 212:643–47.

Seaweeds and sea salt

Laudan, R. (1996). *Food of Paradise: Exploring Hawaii's Culinary Heritage.* Univ. Hawai'i Press.

Mouritsen, O. G., and J. D. Mouritsen (2013). *Seaweeds: Edible, Available, and Sustainable.* Univ. Chicago Press.

Güven, K. C., E. Sezik et al. (2013). Volatile oils from marine macroalgae. In *Natural Products*, ed. K. G. Ramawat and J.-M. Mérillon, 2883–2912. Springer.

Whitfield, F. B., F. Helidoniotis, et al. (1999). Distribution of bromophenols in species of marine algae from eastern Australia. *J Agric Food Chem* 47:2367–73.

Balbas, J., N. Hamid, et al. (2015). Comparison of . . . volatile composition between commercial and New Zealand made wakame from *Undaria pinnatifida. Food Chem* 186:168–75.

Blouin, N. A., J. A. Brodie, et al. (2011). Porphyra: A marine crop shaped by stress. *Trends Plant Sci* 16:29–37.

Shu, N., and H. Shen (2012). Identification of odour-active compounds in dried and roasted nori (*Porphyra yezoensis*). . . . *Flavour Fragr J* 27:157–64.

Miyasaki, T., H. Ozawa, et al. (2014). Discrimination of excellent-grade "nori," the dried laver *Porphyra* spp. . . . *Fisheries Sci* 80:827–38.

Burreson, B. J., R. E. Moore, et al. (1976). Volatile halogen compounds in the alga *Asparagopsis taxiformis* (Rhodophyta). *J Agric Food Chem* 24:856–61.

Le Pape, M.-A., J. Grua-Priol, et al. (2004). Optimization of dynamic headspace extraction of the edible red algae *Palmaria palmata* and identification of the volatile components. *J Agric Food Chem* 52:550–56.

Sánchez-García, F., A. Mirzayeva, et al. (2019). Evolution of volatile compounds and sensory characteristics of edible green seaweed (*Ulva rigida*). . . . *J Sci Food Agric* 99:5475–82.

Silva, I., M. A. Coimbra, et al. (2015). Can volatile organic compounds be markers of sea salt? *Food Chem* 169:102–13.

Influences of ocean volatiles

Le Chêne, M. (2012). Algues vertes, terrain glissant. *Ethnologie française* 42:657 [toxic sea lettuce].

Biester, H., D. Selimović, et al. (2006). Halogens in pore water of peat bogs. . . . *Biogeosci* 3:53–64.

Bendig, P., K. Lehnert, et al. (2014). Quantification of bromophenols in Islay whiskies. *J Agric Food Chem* 62:2767–71.

CHAPTER 16. AFTER-LIFE: SMOKE, ASPHALT, INDUSTRY

Faraday, M. (1861). *The Chemical History of a Candle.* Reprint (1960), 13, 27. Viking.

Robinson, V. (1937). Coal-tar contemplations. *Scientific Monthly* 45:354–56.

Fire and life

Bowman, D. M. J. S., J. K. Balch, et al. (2009). Fire in the earth system. *Science* 324:481–84.

Pausas, J. G., and J. E. Keeley (2009). A burning story: The role of fire in the history of life. *BioScience* 59:593–601.

Lenton, T. M., T. W. Dahl, et al. (2016). Earliest land plants created modern levels of atmospheric oxygen. *Proc Natl Acad Sci USA* 113:9704–9.

Glasspool, I. J., D. Edwards, et al. (2004). Charcoal in the Silurian as evidence for the earliest wildfire. *Geology* 32:381.

Bond, W. J., and A. C. Scott (2010). Fire and the spread of flowering plants in the Cretaceous. *New Phytologist* 188:1137–50.

He, T., J. G. Pausas, et al. (2012). Fire-adapted traits of *Pinus* arose in the fiery Cretaceous. *New Phytologist* 194:751–59.

Nelson, D. C., G. R. Flematti, et al. (2012). Regulation of seed germination and seedling growth by chemical signals from burning vegetation. *Annu Rev Plant Biol* 63:107–30.

Bowman, D. M. J. S., J. Balch, et al. (2011). The human dimension of fire regimes on Earth. *J Biogeography* 38:2223–36.

MacDonald, K. (2017). The use of fire and human distribution. *Temperature* 4:153–65.

Wrangham, R. (2009). *Catching Fire: How Cooking Made Us Human.* Basic Books.

Pyrolysis

Ciccioli, P., M. Centritto, et al. (2014). Biogenic volatile organic compound emissions from vegetation fires. *Plant Cell Environ* 37:1810–25.

Alén, R., E. Kuoppala, et al. (1996). Formation of the main degradation compound groups from wood and its components during pyrolysis. *J Analyt Appl Pyrolysis* 36:137–48.

Yokelson, R. J., R. Susott, et al. (1997). Emissions from smoldering combustion of biomass. . . . *J Geophys Res* 102:18865–77.

Branca, C., P. Giudicianni, et al. (2003). GC/MS characterization of liquids generated from low-temperature pyrolysis of wood. *Ind Eng Chem Res* 42:3190–3202.

Czerny, M., R. Brueckner, et al. (2011). . . . odor qualities and odor detection thresholds of volatile alkylated phenols. *Chemical Senses* 36:539–53.

Tar, charcoal, pitch, turpentine

Mazza, P. P. A., F. Martini, et al. (2006). A new Palaeolithic discovery: Tar-hafted stone tools in a European Mid-Pleistocene bone-bearing bed. *J Archaeol Sci* 33:1310–18.

Schenck, T., and P. Groom (2018). The aceramic production of *Betula pubescens* (downy birch) bark tar using simple raised structures: A viable Neanderthal technique? *Archaeol Anthropol Sci* 10:19–29.

Zilhão, J. (2019). Tar adhesives, Neandertals, and the tyranny of the discontinuous mind. *Proc Natl Acad Sci USA* 116:21966–68.

Antal, M. J., and M. Grønli (2003). The art, science, and technology of charcoal production. *Ind Eng Chem Res* 42:1619–40.

Pliny (~75 CE). *Natural History,* book 15, ch. 7. Trans. J. Bostock and H. T. Riley in *Pliny the Elder: The Natural History* (1855), https://www.perseus.tufts.edu/hopper/text?doc=Perseus:text:1999.02.0137.

Koller, J., U. Baumer, et al. (2005). Herodotus' and Pliny's embalming materials identified on ancient Egyptian mummies. *Archaeometry* 47:609–28.

Earley, L. S. (2004). *Looking for Longleaf: The Fall and Rise of an American Forest.* Univ. North Carolina Press.

Hennius, A. (2018). Viking Age tar production and outland exploitation. *Antiquity* 92:1349–61.

Bitumen, petroleum, coal

Connan, J. (1999). Use and trade of bitumen in antiquity and prehistory: Molecular archaeology reveals secrets of past civilizations. *Phil Trans Roy Soc Lond B* 354:33–50.

Carter, R. (2006). Boat remains and maritime trade in the Persian Gulf during the sixth and fifth millennia BC. *Antiquity* 80:52–63.

Nissenbaum, A., and S. Buckley (2013). Dead Sea asphalt in ancient Egyptian mummies—why? *Archaeometry* 55:563–68.

Murali Krishnan, J., and K. Rajagopal (2003). Review of the uses and modeling of bitumen from ancient to modern times. *Appl Mechanics Reviews* 56:149–214.

Hatcher, P. G., and D. J. Clifford (1997). The organic geochemistry of coal: From plant materials to coal. *Org Geochem* 27:251–74.

Rasmussen, B. (2005). Evidence for pervasive petroleum generation and migration in 3.2 and 2.63 Ga shales. *Geology* 33:497.

Vu, T. T. A., B. Horsfield, et al. (2013). The structural evolution of organic matter during maturation of coals and its impact on petroleum potential and feedstock for the deep biosphere. *Organic Geochem* 62:17–27.

Illuminating gas, kerosene, coal tar, single volatiles

Pagel, W. (2002). *Joan Baptista Van Helmont: Reformer of Science and Medicine*. Cambridge Univ. Press.

Matthews, W. (1827). *An Historical Sketch of the Origin, Progress, and Present State of Gas-lighting*. London.

Smalley, E. V. (1883). Striking oil. *Century Magazine* 26(3):323–39.

Faraday, M. (1815). On new compounds of carbon and hydrogen, and on certain other products obtained during the decomposition of oil by heat. *Proc Roy Soc London* 2:248–49.

Kaiser, R. (1968). "Bicarburet of hydrogen": Reappraisal of the discovery of benzene in 1825 with the analytical methods of 1968. *Angewandte Chem Int Ed*, Engl 7:345–50.

Sneader, W. (2005). *Drug Discovery: A History*. Wiley.

Reichenbach, K. (1835). *Das Kreosot in Chemischer, Physischer und Medicinischer Beziehung*, 14.

Runge, F. (1834). *Über einige Produkte der Steinkohlendestillation. Annalen der Physik und Chemie*, 31:65–78.

von Hofmann, A. W. (1849). Introduction, xlv–lxiii. *Reports of the Royal College of Chemistry and Research Conducted in the Laboratories in the Years 1845-6-7*.

Plastics and solvents

Powers, V. (1993). *The Bakelizer: A National Historic Chemical Landmark*. National Museum of American History, Smithsonian Institution, American Chemical Society.

Bruno, P., M. Caselli, et al. (2008). Monitoring of volatile organic compounds in non-residential indoor environments. *Indoor Air* 18:250–56.

Burdack-Freitag, A., A. Heinlein, et al. (2017). Material odor emissions and indoor air quality. In *Springer Handbook of Odor*, ed. A. Büttner, 563–84. Springer.

Kataoka, H., Y. Ohashi, et al. (2012). Indoor air monitoring of volatile organic compounds . . . In *Advanced Gas Chromatography*, ed. M. A. Mohd, 161–84. InTech.

Chang, J. C. S., R. Fortmann, et al. (2002). Air toxics emissions from a vinyl shower curtain. In *Proceedings: Indoor Air 2002*, 542–47.

Curran, K., and M. Strlič (2015). Polymers and volatiles: Using VOC analysis for the conservation of plastic and rubber objects. *Studies in Conservation* 60:1–14.

Coles, R., ed. (2003). *Food Packaging Technology*. Blackwell.

Bravo, A., J. H. Hotchkiss, et al. (1992). Identification of odor-active compounds resulting from thermal oxidation of polyethylene. *J Agric Food Chem* 40: 1881–85.

Sanders, R. A., D. V. Zyzak, et al. (2005). Identification of 8-nonenal as an important contributor to "plastic" off-odor in polyethylene packaging. *J Agric Food Chem* 53:1713–16.

Delaunay-Bertoncini, N., F. W. M. van der Wielen, et al. (2004). Analysis of low-molar-mass materials in commercial rubber samples. . . . *J Pharmaceut Biomed Anal* 35:1059–73.

Faber, J., and K. Brodzik (2017). Air quality inside passenger cars. *AIMS Environ Sci* 4:112–33.

Van Lente, R., and S. J. Herman (2001). The smell of success—exploiting the leather aroma. Soc. Automotive Eng. Technical Paper 2001-01-0047.

McDonald, B. C., J. A. de Gouw, et al. (2018). Volatile chemical products emerging as largest petrochemical source of urban organic emissions. *Science* 359: 760–64.

Tar-barrel toxins, ethereality

Lowe, D. (2007). Wake up and smell the solvents, https://blogs.sciencemag.org/pipeline /archives/2007/02/18/wake_up_and_smell_the _solvents.

Pott, P. (1775). *Chirurgical Observations Relative to the Cataract, the Polyplus of the Nose, the Cancer of the Scrotum, the Different Kinds of Ruptures, and the Mortification of the Toes and Feet*. London.

Cook. J. W., C. L. Hewett, et al. (1933). The isolation of a cancer-producing hydrocarbon from coal tar. *J Chem Soc* 1933:395–405.

Lim, S. K., H. S. Shin, et al. (2014). Risk assessment of volatile organic compounds benzene, toluene, ethylbenzene, and xylene (BTEX) in consumer products. *J Toxicology and Environ Health*, Part A, 77(22–24):1502–21.

Hubbard, T. D., I. A. Murray, et al. (2016). Divergent Ah receptor ligand selectivity during hominin evolution. *Mol Biol Evol* 33:2648–58.

Johnston, F. H., S. Melody, et al. (2016). The pyrohealth transition: How combustion emissions have shaped health through human history. *Phil Trans Roy Soc B* 371:20150173.

Frobenius, A. S. (1729). An account of a *spiritus vini aethereus*, together with several experiments tried therewith. *Phil Trans Roy Soc* 36:283–89.

Priesner, C. (1986). *Spiritus aethereus*—Formation of ether and theories on etherification from Valerius Cordus to Alexander Williamson. *Ambix* 33(2): 129–52.

CHAPTER 17. FRAGRANCES

Needham, J., and Lu Gwei-djen (1974). *Science and Civilisation in China*, vol. 5, *Chemistry and Chemical Technology*, part 2, *Spagyrical Discovery and Invention*. Cambridge Univ. Press, 154.

Dickinson, E. (1863). Essential Oils—are wrung. Emily Dickinson Archive, https://www.edickinson.org/editions/1/image_sets/236034.

Pickenhagen, W. (2017). The history of odor and odorants. In *Springer Handbook of Odor*, ed. A. Büttner, 1–12. Springer.

Trujillo, R. (2018). Tide: The history of the smell of clean in North America. *Perfumer Flavorist* 43(11): 28–32.

Incense and smudging

Schafer, E. H. (1985). *The Golden Peaches of Samarkand: A Study of Tang Exotics*. Univ. California Press.

Baum, J. M. (2013). From incense to idolatry: The reformation of olfaction in late medieval German ritual. *The Sixteenth Century J* 44(2):323–44.

Kanafani-Zahar, A. (1983). *Aesthetics and Ritual in the United Arab Emirates: The Anthropology of Food and Personal Adornment among Arabian Women*. American Univ. Beirut.

Niebler, J., and A. Büttner (2015). . . . A new approach to identify thermally generated odorants in frankincense. *J Analyt Appl Pyrolysis* 113:690–700.

Naef, R. (2011). The volatile and semi-volatile constituents of agarwood. . . . *Flavour Fragr J* 26:73–87.

Ishihara, M., T. Tsuneya, et al. (1993). Components of the agarwood smoke on heating. *J Ess Oil Res* 5:419–23.

Takeoka, G. R., C. Hobbs, et al. (2010). Volatile constituents of the aerial parts of *Salvia apiana* Jepson. *J Ess Oil Res* 22:241–44.

Borek, T. T., J. M. Hochrien, et al. (2006). Composition of the essential oil of white sage, *Salvia apiana*. *Flavour Fragr J* 21:571–72.

Cohen, R., K. G. Sexton, et al. (2013). Hazard assessment of United Arab Emirates (UAE) incense smoke. *Sci of the Total Environment* 458–60:176–86.

Lin, T.-C., G. Krishnaswamy, et al. (2008). Incense smoke: Clinical, structural and molecular effects on airway disease. *Clin Mol Allergy* 6:3.

Tobacco

Proctor, R. N. (2012). The history of the discovery of the cigarette–lung cancer link: evidentiary traditions, corporate denial, global toll. *Tobacco Control* 21:87–91.

Hahn, B. (2011). *Making Tobacco Bright: Creating an American Commodity, 1617–1937*. Johns Hopkins Univ. Press.

Baker, R. R., and L. J. Bishop (2004). The pyrolysis of tobacco ingredients. *J Analyt Appl Pyrolysis* 71: 223–311.

Perfetti, T., and A. Rodgman (2011). The complexity of tobacco and tobacco smoke. *Beiträge zur Tabakforschung Int / Contributions to Tobacco Res* 24:215–32.

Xiang, Z., K. Cai, et al. (2014). Analysis of volatile flavour components in flue-cured tobacco. . . . *Anal Methods* 6:3300.

Leffingwell, J. (1999). Basic chemical constituents of tobacco leaf and differences among tobacco types. In *Tobacco: Production, Chemistry, and Technology*, ed. D. L. Davis and M. T. Nielson. Blackwell.

Leffingwell, J. C., E. D. Alford, et al. (2013). Identification of the volatile constituents of Cyprian Latakia tobacco. . . . *Leffingwell Rep* 5(2):1–29.

Leffingwell, J., and E. Alford (2005). Volatile constituents of Perique tobacco. *J Environ Agric Food Chem* 4:1–6.

Rustemeier, K., R. Stabbert, et al. (2002). Evaluation of the potential effects of ingredients added to cigarettes. Part 2: Chemical composition of mainstream smoke. *Food Chem Toxicol* 40:93–104.

Taylor, H., D. Winter, et al. (1999). . . . odorous volatile organic compounds (VOCs) from concentrated aged sidestream smoke (SSS). *Beiträge zur Tabakforschung Int / Contributions to Tobacco Res* 18:175–87 [residues on cloth].

Bazemore, R., C. Harrison, et al. (2006). Identification of components responsible for the odor of cigar smoker's breath. *J Agric Food Chem* 54:497–501.

Frauendorfer, F., M. Christlbauer, et al. (2014). Elucidation of ashtray odor. In *Flavour Science*, ed. V. Ferreira and R. Lopez, 47–51. Elsevier.

Cannabis, moxa

ElSohly, M. A., ed. (2007). *Marijuana and the Cannabinoids*. Springer.

Moir, D., W. S. Rickert, et al. (2008). A comparison of mainstream and sidestream marijuana and tobacco cigarette smoke produced under two machine smoking conditions. *Chem Res Toxicol* 21:494–502.

Deng, H., and X. Shen (2013). The mechanism of moxibustion: Ancient theory and modern research. *Evidence-Based Compl Alt Med* 2013:1–7.

Wheeler, J., B. Coppock, et al. (2009). Does the burning of moxa (*Artemisia vulgaris*) in traditional Chinese medicine constitute a health hazard? *Acupunct Med* 27:16–20.

Capturing volatiles: distillation, alcohol, solvents

Genders, R. (1972). *Perfume through the Ages*. Putnam.

Morris, E. T. (1984). *Fragrance: The Story of Perfume from Cleopatra to Chanel*. Scribner.

McHugh, J. (2012). *Sandalwood and Carrion: Smell in Indian Religion and Culture*. Oxford Univ. Press.

Zohar, A., and E. Lev (2013). Trends in the use of perfumes and incense in the Near East after the Muslim conquests. *J Royal Asiatic Soc* 23:11–30.

Kanafani-Zahar, A. (1983). *Aesthetics and Ritual in the United Arab Emirates: The Anthropology of Food and Personal Adornment among Arabian Women*. American Univ. Beirut.

Forbes, R. J. (1956). *Studies in Ancient Technology*, vol. 3. Brill Archive.

Levey, M. (1959). *Chemistry and Chemical Technology in Ancient Mesopotamia*. Elsevier.

Belgiorno, M. R. (2016). *The Perfume of Cyprus: From Pyrgos to François Coty*. Ermes.

Castel, C., X. Fernandez, et al. (2009). Perfumes in Mediterranean antiquity. *Flavour Fragr J* 24:326–34.

Naves, Y. R., G. Mazuyer, et al. (1947). *Natural Perfume Materials: A Study of Concretes, Resinoids, Floral Oils and Pomades*. Reinhold.

Başer, K. H. C., and G. Buchbauer, eds. (2015). *Handbook of Essential Oils: Science, Technology, and Applications*. CRC Press.

Başer, K. H. C., M. Kurkcuoglu, et al. (2003). Turkish rose oil research: Recent results. *Perfumer Flavorist* 28(2):34–43.

Kurkcuoglu, M., and K. H. C. Başer (2003). Studies on Turkish rose concrete, absolute, and hydrosol. *Chem Nat Compounds* 39(5):457–64.

Plant fragrance materials
See Naves et al. and Başer and Buchbauer under "Capturing volatiles" (above); also see references for chapters 9–13.

Arctander, S. (1960). *Perfume and Flavor Materials of Natural Origin*. Self-published.

Swamy, M., and U. Sinniah (2015). A comprehensive review . . . of *Pogostemon cablin* Benth. *Molecules* 20:8521–47 [patchouli].

Joulain, D., and R. Tabacchi (2009). Lichen extracts as raw materials in perfumery: Oakmoss. *Flavour Fragr J* 24:49–61.

Froissard, D., F. Fons, et al. (2011). Volatiles of French ferns and "fougère" scent in perfumery. *Nat Product Comms* 6(11):1723–26.

Weyerstahl, P., H. Marschall, et al. (1998). Constituents of commercial labdanum oil. *Flavour Fragr J* 13(5):295–318.

Miyazawa, N., A. Nakanishi, et al. (2009). Novel key aroma components of galbanum oil. *J Agric Food Chem* 57:1433–39.

Chauhan, R. S., M. C. Nautiyal et al. (2017). Effect of post-harvest drying methods on the essential oil composition of *Nardostachys jatamansi* DC. *J Ess Oil Bearing Plants* 20:1090–96.

Liu, Z. L., Q. He, et al. (2012). Essential oil composition and larvicidal activity of *Saussurea lappa* roots. . . . *Parasitol Res* 110:2125–30 [costus].

Marongiu, B., A. Piras, et al. (2005). Chemical composition of the essential oil and supercritical CO_2 extract of *Commiphora myrrha* (Nees) Engl. and of *Acorus calamus* L. *J Agric Food Chem* 53:7939–43.

Belhassen, E., J.-J. Filippi, et al. (2015). Volatile constituents of vetiver: A review. *Flavour Fragr J* 30:26–82.

Del Giudice, L., D. R. Massardo, et al. (2008). The microbial community of vetiver root and its involvement into essential oil biogenesis. *Environ Microbiol* 10:2824–41.

Brenna, E., C. Fuganti, et al. (2003). . . . the ten isomers of irone. *Comptes Rendus Chimie* 6:529–46 [orris].

Mookheriee, B. D., and R. A. Wilson (1990). Tobacco constituents—their importance in flavor and fragrance chemistry. *Perfumer & Flavorist* 15(1):27–49.

Animal fragrance materials, plant musks
Levey, M. (1961). Ibn Māsawaih and his treatise on simple aromatic substances: Studies in the history of Arabic pharmacology. *J Hist Med Allied Sci* 16:394–410.

King, A. H. (2017). *Scent from the Garden of Paradise. Musk and the Medieval Islamic World*. Brill.

Kraft, P. (2004). Aroma chemicals, IV: Musks. In *Chemistry and Technology of Flavors and Fragrances*, ed. D. J. Rowe, 143–68. Blackwell.

Hayes, R. A., B. J. Richardson, et al. (2003). To fix or not to fix: The role of 2-phenoxyethanol in rabbit, *Oryctolagus cuniculus*, chin gland secretion. *J Chem Ecology* 29(5):1051–64.

Boyle, R. (1676). Experiments and observations about the mechanical production of odours. In *Works of the Honourable Robert Boyle* (London, 1772), vol. 4, 272.

Clarke, R. (2006). The origin of ambergris. *Lat Am J Aquat Mamm* 5(1):7–21.

Rowland, S. J., and P. A. Sutton (2017). Chromatographic and spectral studies of jetsam and archived ambergris. *Natural Product Res* 31:1752–57.

Panten, J., H. Surburg, et al. (2014). Recent results in the search for new molecules with ambergris odor. *Chem & Biodiversity* 11:1639–50.

Dannenfeldt, K. H. (1985). Europe discovers civet cats and civet. *J History Biology* 18:403–31.

Tang, R., F. X. Webster, et al. (1993). Phenolic compounds from male castoreum of the North American beaver, *Castor canadensis*. *J Chem Ecology* 19:1491–1500.

Prinsloo, L. C. (2007). Rock hyraces: A cause of San rock art deterioration? *J Raman Spectrosc* 38:496–503.

Ferber, C. E. M., and H. E. Nursten (1977). The aroma of beeswax. *J Sci Food Agric* 28:511–18.

Nongmaithem, B. D., P. Mouatt, et al. (2017). Volatile and bioactive compounds in opercula from Muricidae molluscs. . . . *Sci Rep* 7:17404.

Kaiser, R. (2006). *Meaningful Scents around the World: Olfactory, Chemical, Biological, and Cultural Considerations*, 143–51 [pine resin]. Wiley.

New aromatics from chemists
David, O. R. P. (2017). Artificial nitromusks, stories of chemists and businessmen. *Eur J Org Chem* 2017(1):4–13.

De Nicolaï, P. (2008). A smelling trip into the past: The influence of synthetic materials on the history of perfumery. *Chem & Biodiversity* 5(6):1137–46.

Gupta, C. (2015). A biotechnological approach to microbial based perfumes and flavours. *J Microbiol Exp* 2(1):11–18.

Kraft, P., J. A. Bajgrowicz, et al. (2000). Odds and trends: Recent developments in the chemistry of odorants. *Angewandte Chem Int Ed* 31.

Kraft, P., and W. Eichenberger (2003). Conception, characterization and correlation of new marine odorants. *Eur J Org Chem* 19:3735–43.

Arctander, S. (1969). *Perfume and Flavor Chemicals*, 2 vols. Self-published.

Rimkus, G. G., ed. (2004). *Synthetic Musk Fragrances in the Environment.* Springer.

Perfumes

Aftel, M. (2001). *Essence and Alchemy: A Book of Perfume.* North Point Press.

Turin, L., and T. Sanchez (2008). *Perfumes: The Guide.* Viking.

Calkin, R. R., and J. S. Jellinek. (1994). *Perfumery: Practice and Principles.* Wiley.

Sell, C. (2008). *Understanding Fragrance Chemistry.* Allured.

Sell, C., ed. (2015). *The Chemistry of Fragrances: From Perfumer to Consumer.* Royal Society of Chemistry.

Ohloff, G., P. Kraft, et al. (2012). *Scent and Chemistry: The Molecular World of Odors.* Wiley.

Kapoor, B. (1991). Attars of India—a unique aroma. *Perfumer & Flavorist* 16(1):21–24.

Listening to smells

Aftel, M. (2014). *Fragrant: The Secret Life of Scent.* Riverhead.

Kaiser, R. (2006). *Meaningful Scents around the World: Olfactory, Chemical, Biological, and Cultural Considerations*, 59–64 [agarwood]. Wiley.

Dalby, L. (2009). *East Wind Melts the Ice: A Memoir through the Seasons.* Univ. California Press.

Dalby, L. (n.d.). Incense, http://www.lizadalby.com /LD/TofM_incense.html.

Morita, K. (2006). *The Book of Incense: Enjoying the Traditional Art of Japanese Scents.* Kodansha.

Fujii, N., D. Abla, et al. (2007). Prefrontal activity during koh-do incense discrimination. *Neuroscience Res* 59:257–64.

CHAPTER 18. COOKED FOODS

Bachelard, G. (1938). *La Psychanalyse du Feu.* Gallimard. Trans. A. C. M. Ross (1964), *The Psychoanalysis of Fire.* Beacon.

Wrangham, R. (2009). *Catching Fire: How Cooking Made Us Human.* Basic Books.

Kitchen alchemy

Paravisini, L., K. Gourrat-Pernin, et al. (2012). Identi-

fication of compounds responsible for the odorant properties of aromatic caramel. *Flavour Fragr J* 27:424–32.

Paravisini, L., C. Septier, et al. (2014). Caramel odor: Contribution of volatile compounds according to their odor qualities to caramel typicality. *Food Res Int* 57:79–88.

Dunkel, A., M. Steinhaus, et al. (2014). Nature's chemical signatures in human olfaction: A foodborne perspective for future biotechnology. *Angewandte Chem Int Ed* 53:7124–7143 [key food odorants].

Cooked bouquets, cooking methods

Belitz, H., W. Grosch, et al. (2009). *Food Chemistry.* Springer.

Bordiga, M., and L. M. L. Nollet, eds. (2019). *Food Aroma Evolution: During Food Processing, Cooking, and Aging.* CRC Press.

Fickert, B., and P. Schieberle (1998). Identification of the key odorants in barley malt (caramalt) . . . *Food/ Nahrung* 42(6):371–75.

Rahman, M. M., and K. H. Kim (2012). Release of offensive odorants from the combustion of barbecue charcoals. *J Hazardous Materials* 215:233–42.

Sung, W.-C. (2013). Volatile constituents detected in smoke condensates from the combination of the smoking ingredients sucrose, black tea leaves, and bread flour. *J Food and Drug Analysis* 21:292–300.

Watcharananun, W., K. R. Cadwallader, et al. (2009). Identification of predominant odorants in Thai desserts flavored by smoking with "tian op" . . . *J Agric Food Chem* 57:996–1005.

Choe, E., and D. B. Min (2007). Chemistry of deep-fat frying oils. *J Food Sci* 72:R77–86.

Thürer, A., and M. Granvogl (2016). Generation of desired aroma-active as well as undesired toxicologically relevant compounds during deep-frying of potatoes. . . . *J Agric Food Chem* 64:9107–15.

Piyachaiseth, T., W. Jirapakkul, et al. (2011). Aroma compounds of flash-fried rice. *Kasetsart J Nat Sci* 45:717–29.

Liu, T., Z. Liu, et al. (2017). Emission of volatile organic compounds . . . from stir-frying spices. *Sci Total Environ* 599–600:1614–21.

Oils and fats

Watanabe, K., and Y. Sato (1968). Aliphatic γ- and δ-lactones in meat fats. *Agric Biological Chem* 32: 1318–24.

Campestre, C., G. Angelini , et al. (2017). The compounds responsible for the sensory profile in monovarietal virgin olive oils. *Molecules* 22(1833):1–28.

Sambanthamurthi, R. (2000). Chemistry and biochemistry of palm oil. *Prog Lipid Res* 39:507–58.

Santos, J. E. R., B. J. Villarino, et al. (2011). Analysis of volatile organic compounds in virgin coconut oil. . . . *Philippine J Sci* 140(2):161–71.

Delort, E., A. Velluz, et al. (2011). . . . new volatile molecules found in extracts obtained from distinct parts of cooked chicken. *J Agric Food Chem* 59:11752–63.

Hwang, L. S., and C.-W. Chen (1994). Volatile compounds of lards from different treatments. In *Lipids in Food Flavors*, ed. C.-T. Ho and T. G. Hartman, 244–55. American Chemical Society.

Watanabe, A., Y. Ueda, et al. (2008). Analysis of volatile compounds in beef fat. . . . *J Food Sci* 73:C420–25.

Mallia, S., F. Escher, et al. (2008). Aroma-active compounds of butter: A review. *Eur Food Res Technol* 226:315–25.

Sarrazin, E., E. Frerot, et al. (2011). Discovery of new lactones in sweet cream butter oil. *J Agric Food Chem* 59:6657–66.

Wadodkar, U. R., J. S. Punjrath, et al. (2002). Evaluation of volatile compounds in different types of ghee . . . *J Dairy Res* 69:163–71.

Milk, cream, eggs

Cadwallader, K. R., and T. K. Singh (2009). Flavours and off-flavours in milk and dairy products. In *Advanced Dairy Chemistry*, ed. P. McSweeney and P. F. Fox, 631–90. Springer.

Schütt, J., and P. Schieberle (2017). Quantitation of nine lactones in dairy cream. . . . *J Agric Food Chem* 65:10534–41.

Schwendel, B. H., T. J. Wester, et al. (2017). Pasture feeding conventional cows removes differences between organic and conventionally produced milk. *Food Chem* 229:805–13.

Kaffarnik, S., Y. Kayademir, et al. (2014). Concentrations of volatile 4-alkyl-branched fatty acids in sheep and goat milk and dairy products. *J Food Sci* 79:C2209–14.

Jo, S.-H., K.-H. Kim, et al. (2013). Study of odor from boiled eggs over time using gas chromatography. *Microchemical J* 110:517–29.

Cerny, C., and R. Guntz (2004). Evaluation of potent odorants in heated egg yolk. . . . *Eur Food Res Technol* 219:452–54.

Meats and meatiness

Resconi, V., A. Escudero, et al. (2013). The development of aromas in ruminant meat. *Molecules* 18:6748–81.

Nollet, L. M. L., and F. Toldrá, eds. (2010). *Sensory Analysis of Foods of Animal Origin*. CRC Press.

Takakura, Y., T. Sakamoto, et al. (2014). Characterization of the key aroma compounds in beef extract. . . . *Meat Sci* 97:27–31.

Mahadevan, K., and L. Farmer (2006). Key odor impact compounds in three yeast extract pastes. *J Agric Food Chem* 54:7242–50.

Fraser, R., P. O. Brown, et al. (2014). Methods and compositions for affecting the flavor and aroma profile of consumables. WIPO Patent Appl WO2014110532 A2 [heme].

Rochat, S., and A. Chaintreau (2005). Carbonyl odorants contributing to the in-oven roast beef top note. *J Agric Food Chem* 53:9578–85.

Tang, W., D. Jiang, et al. (2013). Flavor chemistry of 2-methyl-3-furanthiol, an intense meaty aroma compound. *J Sulfur Chemistry* 34:38–47.

Snitkjær, P., M. B. Frøst, et al. (2010). Flavour development during beef stock reduction. *Food Chem* 122:645–55.

Zhao, J., M. Wang, et al. (2017). Volatile flavor constituents in the pork broth of black-pig. *Food Chem* 226:51–60.

Frank, D., P. Watkins, et al. (2016). Impact of brassica and lucerne finishing feeds and intramuscular fat on lamb eating quality and flavor. *J Agric Food Chem* 64:6856–68.

Takakura, Y., M. Mizushima, et al. (2014). Characterization of the key aroma compounds in chicken soup stock . . . *Food Sci Technol Res* 20:109–13.

Wu, C.-M., and S.-E. Liou (1992). Volatile components of water-boiled duck meat and Cantonese style roasted duck. *J Agric Food Chem* 40:838–41.

Straßer S., and P. Schieberle (2014). Characterization of the key aroma compounds in roasted duck liver. . . . *Eur Food Res Technol* 238:307–13.

Neethling, J., L. C. Hoffman, et al. (2016). Factors influencing the flavor of game meat: A review. *Meat Sci* 113:139–53.

Tansawat, R., C. A. J. Maughan, et al. (2013). Chemical characterization of pasture- and grain-fed beef related to meat quality and flavour attributes. *Int J Food Sci Technol* 48:484–95.

Fruet, A. P. B., F. Trombetta, et al. (2018). Effects of feeding legume-grass pasture and different concentrate levels on fatty acid profile, volatile compounds, and off-flavor. . . . *Meat Sci* 140:112–18.

Arshamian, A., M. Laska, et al. (2017). A mammalian blood odor component serves as an approach-avoidance cue across phylum border—from flies to humans. *Sci Rep* 7:13635.

Fish and shellfish

Shahidi, F., and K. R. Cadwallader, eds. (1997). *Flavor and Lipid Chemistry of Seafoods*. American Chemical Society.

Nollet, L. M. L., and F. Toldrá, eds. (2009). *Handbook of Seafood and Seafood Products Analysis*. CRC Press.

Tamura, T., K. Taniguchi, et al. (2009). Iron is an essential cause of fishy aftertaste formation in wine and seafood pairing. *J Agric Food Chem* 57:8550–56.

Carrascon, V., A. Escudero, et al. (2014). Characterisation of the key odorants in a squid broth (*Illex argentinus*). *LWT—Food Sci Technol* 57:656–62.

Selli, S., C. Rannou, et al. (2006). Characterization of aroma-active compounds in rainbow trout (*Oncorhynchus mykiss*) eliciting an off-odor. *J Agric Food Chem* 54:9496–9502.

Methven, L., M. Tsoukka, et al. (2007). Influence of

sulfur amino acids on the volatile and nonvolatile components of cooked salmon. . . . *J Agric Food Chem* 55:1427–36.

Vegetables

Hui, Y. H., ed. (2010). *Handbook of Fruit and Vegetable Flavors.* CRC Press.

Zhu, Y., H. J. Klee, et al. (2018). Development and characterization of a high quality plum tomato essence. *Food Chem* 267:337–43.

Koutidou, M., T. Grauwet, et al. (2017). Impact of processing on odour-active compounds of a mixed tomato-onion puree. *Food Chem* 228:14–25.

Buttery, R. G., and G. R. Takeoka (2013). Cooked carrot volatiles . . . linden ether as an important aroma component. *J Agric Food Chem* 61:9063–66.

Blank, I., and P. Schieberle (1993). Analysis of the seasoning-like flavour substances of a commercial lovage extract (*Levisticum officinale* Koch.). *Flavour Fragr J* 8(4):191–95.

Granvogl, M., M. Christlbauer, et al. (2004). Quantitation of the intense aroma compound 3-mercapto-2-methylpentan-1-ol in raw and processed onions . . . *J Agric Food Chem* 52:2797–2802.

Villière, A., S. Le Roy, et al. (2015). Evaluation of aroma profile differences between sué, sautéed, and pan-fried onions . . . *Flavour* 4:24.

Yang, P., H. Song, et al. (2019). Characterization of key aroma-active compounds in black garlic . . . *J Agric Food Chem* 67:7926–34.

Christlbauer, M., and P. Schieberle (2011). Evaluation of the key aroma compounds in beef and pork vegetable gravies à la chef. . . . *J Agric Food Chem* 59:13122–30.

Noe, F., J. Polster, et al. (2017). OR2M3: A highly specific and narrowly tuned human odorant receptor for the sensitive detection of onion key food odorant 3-mercapto-2-methylpentan-1-ol. *Chemical Senses* 42:195–210.

Naef, R., and A. Velluz (2000). The volatile constituents of extracts of cooked spinach leaves. . . . *Flavour Fragr J* 15(5):329–34.

Simian, H., F. Robert, et al. (2004). Identification and synthesis of 2-heptanethiol, a new flavor compound found in bell peppers. *J Agric Food Chem* 52:306–10.

Ulrich, D., E. Hoberg, et al. (2001). Contribution of volatile compounds to the flavor of cooked asparagus. *Eur Food Res Technol* 213:200–204.

Starkenmann, C., Y. Niclass, et al. (2019). Occurrence of 2-acetyl-1-pyrroline and its nonvolatile precursors in celtuce. . . . *J Agric Food Chem* 67:11710–17.

MacLeod, A. J., N. M. Pieris, et al. (1982). Aroma volatiles of *Cynara scolymus* and *Helianthus tuberosus.* *Phytochemistry* 21(7):1647–51.

Singh, J., and L. Kaur, eds. (2016). *Advances in Potato Chemistry and Technology.* Elsevier.

Oruna-Concha, M. J., S. C. Duckham, et al. (2001). Comparison of volatile compounds isolated from the skin and flesh of four potato cultivars after baking. *J Agric Food Chem* 49:2414–21.

Wang, Y., and S. J. Kays (2001). Effect of cooking method on the aroma constituents of sweet potatoes. . . . *J Food Quality* 24:67–78.

Chung, M.-J., S.-S. Cheng, et al. (2012). Profiling of volatile compounds of *Phyllostachys pubescens* shoots in Taiwan. *Food Chem* 134:1732–37 [bamboo].

Grosshauser, S., and P. Schieberle (2013). Characterization of the key odorants in pan-fried white mushrooms . . . *J Agric Food Chem* 61:3804–13.

Li, B., C. Liu, et al. (2019). Effect of boiling time on the contents of flavor and taste in *Lentinus edodes.* *Flavour Fragr J* 34:506–13.

MacLeod, A. J., and N. G. de Troconis (1983). Aroma volatiles of aubergine. . . . *Phytochemistry* 22:2077–79.

Shu, N., and H. Shen (2012). Identification of odour-active compounds in dried and roasted nori. . . . *Flavour Fragr J* 27:157–64.

Nuts, grains, beans

Siegmund, B., and M. Murkovic (2004). Changes in chemical composition of pumpkin seeds during the roasting process. *Food Chem* 84:367–74.

Zhou, M., K. Robards, et al. (1999). Analysis of volatile compounds and their contribution to flavor in cereals. *J Agric Food Chem* 47:3941–53.

Cho, S., and S. J. Kays (2013). Aroma-active compounds of wild rice. . . . *Food Res Int* 54:1463–70.

Buttery, R. G., L. C. Ling, et al. (1997). Studies on popcorn aroma and flavor volatiles. *J Agric Food Chem* 45:837–43.

Karahadian, C., and K. A. Johnson (1993). Analysis of headspace volatiles and sensory characteristics of fresh corn tortillas made from fresh masa dough. . . . *J Agric Food Chem* 41:791–99.

Buttery, R. G., and L. C. Ling (1998). Additional studies on flavor components of corn tortilla chips. *J Agric Food Chem* 46:2764–69.

Pozo-Bayón, M. A., E. Guichard, et al. (2006). Flavor control in baked cereal products. *Food Reviews Int* 22:335–79.

Kirchhoff, E., and P. Schieberle (2002). Quantitation of odor-active compounds in rye flour and rye sourdough. . . . *J Agric Food Chem* 50:5378–85.

Schoenauer, S., and P. Schieberle (2019). Characterization of the key aroma compounds in the crust of soft pretzels. . . . *J Agric Food Chem* 67:7110–19.

Kato, Y. (2005). Influence of butter and/or vegetable oil on flavors of roux prepared from wheat flour and fat/oil. *Food Sci Technol Res* 11:278–87.

Mishra, P. K., J. Tripathi, et al. (2019). GC-MS olfactometric characterization of odor active compounds in cooked red kidney beans. . . . *Heliyon* 5:e02459.

Matsui, K., H. Takemoto, et al. (2018). 1-Octen-3-ol is formed from its glycoside during processing of soybean seeds. *J Agric Food Chem* 66:7409–16.

Morisaki, A., N. Yamada, et al. (2014). Dimethyl sulfide as a source of the seaweed-like aroma in cooked soybeans . . . *J Agric Food Chem* 62:8289–94.

Murekatete, N., C. Zhang, et al. (2015). Soft tofu-type gels: Relationship between volatile compounds. . . . *Int J Food Engineering* 11:307–21.

Fruits, syrups, honeys

Sabarez, H. T., W. E. Price, et al. (2000). Volatile changes during dehydration of d'Agen prunes. *J Agric Food Chem* 48:1838–42.

Wang, D., C.-Q. Duan, et al. (2017). Free and glycosidically bound volatile compounds in sun-dried raisins made from different fragrance intensities grape varieties . . . *Food Chem* 228:125–35.

Nursten, H. E., and M. L. Woolfe (1972). An examination of the volatile compounds present in cooked Bramley's seedling apples. . . . *J Sci Food Agric* 23:803–22.

Lesschaeve, I., D. Langlois, et al. (1991). Volatile compounds in strawberry jam: Influence of cooking on volatiles. *J Food Sci* 56:1393–98.

Takei, Y. (1985). Comparison of the aroma of fresh peels or marmalades between several species of citrus fruits. *J Home Econ Japan* 36(10):754–62.

Urbanus, B. L., G. O. Cox, et al. (2014). Sensory differences between beet and cane sugar sources . . . *J Food Sci* 79:S1763–68.

Kabir, M. A., and Y. Lorjaroenphon (2014). Identification of aroma compounds in coconut sugar. In *Agric Sci. Proceedings of the 52nd Kasetsart Univ Annual Conference* 6:239–46.

Franitza, L., M. Granvogl, et al. (2016). Influence of the production process on the key aroma compounds of rum. . . . *J Agric Food Chem* 64:9041–53.

Asikin, Y., K. Wada, et al. (2018). Compositions, taste characteristics, volatile profiles, and antioxidant activities of sweet sorghum . . . and sugarcane . . . syrups. *Food Measure* 12:884–91.

Perkins, T. D., and A. K. van den Berg (2009). Maple syrup—production, composition, chemistry, and sensory characteristics. *Advances in Food and Nutrition Res* 56:101–43.

Zhou, Q., C. L. Wintersteen, et al. (2002). Identification and quantification of aroma-active components that contribute to the distinct malty flavor of buckwheat honey. *J Agric Food Chem* 50:2016–21.

Fearnley, L., D. R. Greenwood, et al. (2012). Compositional analysis of manuka honeys. . . . *Food Chem* 132:948–53.

Naef, R., A. Jaquier, et al. (2004). From the linden flower to linden honey . . . *Chem & Biodiversity* 1:1870–79.

Pita-Calvo, C., and M. Vázquez (2017). Differences between honeydew and blossom honeys. *Trends Food Sci Technol* 59:79–87.

Pastries, cakes

Gassenmeier, K., and P. Schieberle (1994). Comparison of important odorants in puff-pastries prepared with butter or margarine. *LWT—Food Sci Technol* 27:282–88, https://doi.org/10.1006/fstl.1994.1056.

Cepeda-Vázquez, M., B. Rega, et al. (2018). How ingredients influence furan and aroma generation in sponge cake. *Food Chem* 245:1025–33.

Pozo-Bayón, M. A., A. Ruíz-Rodríguez, et al. (2007). Influence of eggs on the aroma composition of a sponge cake and on the aroma release in model studies. . . . *J Agric Food Chem* 55:1418–26.

Rega, B., A. Guerard, et al. (2009). . . . endogenous aroma compounds released during the baking of a model cake. *Food Chem* 112:9–17.

Ait Ameur, L., B. Rega, et al. (2008). The fate of furfurals and other volatile markers during the baking process of a model cookie. *Food Chem* 111:758–63.

Mohsen, S. M., H. H. M. Fadel, et al. (2009). Effect of substitution of soy protein isolate on aroma volatiles, chemical composition and sensory quality of wheat cookies. *Int J Food Sci Technol* 44:1705–12.

Coffee, chocolate

Buffo, R. A., and C. Cardelli-Freire (2004). Coffee flavour: An overview. *Flavour Fragr J* 19:99–104.

Sunarharum, W. B., D. J. Williams, et al. (2014). Complexity of coffee flavor. *Food Res Int* 62:315–25.

Aprotosoaie, A. C., S. V. Luca, et al. (2016). Flavor chemistry of cocoa and cocoa products—an overview. . . . *Comp Revs Food Sci Food Safety* 15:73–91.

Beckett, S. T., M. Fowler, et al., eds. (2017). *Beckett's Industrial Chocolate Manufacture and Use*. Wiley.

Liu, J., M. Liu, et al. (2015). A comparative study of aroma-active compounds between dark and milk chocolate. . . . *J Sci Food Agric* 95:1362–72.

Stewart, A., A. S. Grandison, et al. (2017). Impact of the skim milk powder manufacturing process on the flavor of model white chocolate. *J Agric Food Chem* 65:1186–95.

CHAPTER 19. CURED AND FERMENTED FOODS

Feynman, R. P., R. B. Leighton, et al. (1970). *The Feynman Lectures on Physics*, vol. 1, ch. 3. Addison-Wesley.

Davidson, A. (1979). *North Atlantic Seafood*, 368. Macmillan.

Speth, J. D. (2017). Putrid meat and fish in the Eurasian middle and upper Paleolithic: Are we missing a key part of Neanderthal and modern human diet? *PaleoAnthropology* 2017:44–72.

Staleness and rancidity

Forss, D. A., E. G. Pont, et al. (1955). . . . volatile compounds associated with oxidized flavour in skim milk. *J Dairy Res* 22:345–48.

Czerny, M., and A. Büttner (2009). Odor-active compounds in cardboard. *J Agric Food Chem* 57:9979–84.

Warner, K., and T. Nelsen (1996). AOCS collaborative study on sensory and volatile compound analyses of vegetable oils. *J Am Oil Chem Soc* 73:157–66.

Shahidi, F., and J. A. John (2013). Oxidative rancidity in nuts. In *Improving the Safety and Quality of Nuts*, ed. L. J. Harris, 198–229. Elsevier.

Barrett, D. M., E. L. Garcia, et al. (2000). Blanch time and cultivar effects on quality of frozen and stored corn and broccoli. *J Food Sci* 65:534–40.

Asaduzzaman, M., M. Scampicchio, et al. (2020). Methanethiol formation during the photochemical oxidation of methionine-riboflavin system. *Flavour Fragr J* 35:34–41 [milk].

Baert, J. J., J. De Clippeleer, et al. (2012). On the origin of free and bound staling aldehydes in beer. *J Agric Food Chem* 60:11449–72.

Kerler, J., and W. Grosch (1996). Odorants contributing to warmed-over flavor (WOF) of refrigerated cooked beef. *J Food Sci* 61:1271–75.

Cured meats, fish, eggs

García-González, D. L., N. Tena, et al. (2008). Relationship between sensory attributes and volatile compounds qualifying dry-cured hams. *Meat Sci* 80:315–25.

Song, H., and K. R. Cadwallader (2007). Aroma components of American country ham. *J Food Sci* 73:C29–35.

Song, H., K. R. Cadwallader, et al. (2008). Odour-active compounds of Jinhua ham. *Flavour Fragr J* 23:1–6.

del Pulgar, J. S., C. García, et al. (2013). Study of the volatile compounds and odor-active compounds of dry-cured Iberian ham extracted by SPME. *Food Sci Technol Int* 19:225–33.

Dehaut, A., C. Himber, et al. (2014). Evolution of volatile compounds and biogenic amines throughout the shelf life of marinated and salted anchovies. . . . *J Agric Food Chem* 62:8014–22.

Caprino, F., V. M. Moretti, et al. (2008). Fatty acid composition and volatile compounds of caviar from farmed white sturgeon. . . . *Analytica Chimica Acta* 617:139–47.

Ganasen, P., S. Benjakul, et al. (2013). Effect of different cations on pidan composition and flavor in comparison to the fresh duck egg. *Korean J for Food Sci of Animal Resources* 33:214–20.

Tea

Zeng, L., N. Watanabe, et al. (2019). Understanding the biosyntheses and stress response mechanisms of aroma compounds in tea. . . . *Crit Revs Food Sci and Nutrition* 59:2321–34.

Feng, Z., Y. Li, et al. (2019). Tea aroma formation from six model manufacturing processes. *Food Chem* 285:347–54.

Baba, R., Y. Amano, et al. (2017). Characterization of the potent odorants contributing to the characteristic aroma of matcha. . . . *J Agric Food Chem* 65:2984–89.

Yang, Z., E. Kobayashi, et al. (2012). Characterisation of volatile and non-volatile metabolites in etiolated leaves of tea (*Camellia sinensis*) plants in the dark. *Food Chem* 135:2268–76.

Kumazawa, K., K. Kubota, et al. (2005). Influence of manufacturing conditions and crop season on the formation of 4-mercapto-4-methyl-2-pentanone in Japanese green tea (Sen-cha). *J Agric Food Chem* 53:5390–96.

Schuh, C., and P. Schieberle (2006). Characterization of the key aroma compounds in the beverage prepared from Darjeeling black tea. . . . *J Agric Food Chem* 54:916–24.

Lu, H.-P., Q.-S. Zhong, et al. (2012). Aroma characterisation of pu-erh tea using headspace-solid phase microextraction combined with GC/MS and GC–olfactometry. *Food Chem* 130:1074–81.

Li, Z., C. Feng, et al. (2018). Revealing the influence of microbiota on the quality of pu-erh tea during fermentation process. . . . *Food Microbiol* 76:405–15.

Soiled and spoiled

Sano, K.-I., and A. Anraku (2018). Draft genome sequence of *Brevibacillus reuszeri* strain NIT02, isolated from a laundered rental cloth hot towel. *Genome Announc* 6:e01353-17.

Adams, R. I., D. S. Lymperopoulou, et al. (2017). Microbes and associated soluble and volatile chemicals on periodically wet household surfaces. *Microbiome* 5:128.

Takeuchi, K., M. Yabuki, et al. (2013). Review of odorants in human axillary odour and laundry malodour. *Flavour Fragr J* 28:223–30.

Sperber, W. H., and M. P. Doyle, eds. (2009). *Compendium of the Microbiological Spoilage of Foods and Beverages*. Springer.

Fermentation microbes

Tamang, J. P., K. Watanabe, et al. (2016). Review: Diversity of microorganisms in global fermented foods and beverages. *Front Microbiol* 7.

Boiocchi, F., D. Porcellato, et al. (2017). Insect frass in stored cereal products as a potential source of *Lactobacillus sanfranciscensis* for sourdough ecosystem. *J Appl Microbiol* 123:944–55.

Hittinger, C. T., J. L. Steele, et al. (2018). Diverse yeasts for diverse fermented beverages and foods. *Curr Opinion Biotechnol* 49:199–206.

Bodinaku, I., J. Shaffer, et al. (2019). Rapid phenotypic and metabolomic domestication of wild *Penicillium* molds on cheese. *mBio* 10:e02445-19.

Gibbons, J. G., L. Salichos, et al. (2012). The evolutionary imprint of domestication on genome variation and function of the filamentous fungus *Aspergillus oryzae*. *Curr Biol* 22:1403–9.

Fermented vegetables

Dunlop, F. (2011). Rotten vegetable stalks, stinking bean curd and other Shaoxing delicacies. In *Cured, Fermented and Smoked Foods: Proceedings of the Oxford Symposium on Food and Cookery 2010*, ed. H. Saberi, 84–96. Prospect Books.

Liu, Y., Z. Miao, et al. (2011). Analysis of volatile flavor components in a simultaneous distillation-extraction extract of fermented amaranthus stem. *J Chinese Inst Food Sci Technol* 11(1):226–32.

Cheung, H.-M. (2005). Identification of odorous compounds in commercial chaw tofu and evaluation of the quality of model broths during fermentation. MS thesis, Chinese Univ. Hong Kong.

Paramithiotis, S., ed. (2017). *Lactic Acid Fermentation of Fruits and Vegetables*. CRC Press.

Sonmezdag, A. S., H. Kelebek, et al. (2019). Characterization of aroma-active compounds, phenolics, and antioxidant properties in fresh and fermented capers. . . . *J Food Sci* 84:2449–57.

Fermented seed foods: breads and beans

Pozo-Bayón, M. A., I. Andújar-Ortiz, et al. (2010). Application of supercritical CO_2 extraction for the elimination of odorant volatile compounds from winemaking inactive dry yeast preparation. *J Agric Food Chem* 58:3772–78.

Kirchhoff, E., and P. Schieberle (2002). Quantitation of odor-active compounds in rye flour and rye sourdough. . . . *J Agric Food Chem* 50:5378–85.

Hatzikamari, M., M. Yiangou, et al. (2007). Changes in numbers and kinds of bacteria during a chickpea submerged fermentation used as a leavening agent for bread production. *Int J Food Microbiol* 116:37–43.

McGee, H. (2014). The disquieting delights of salt-rising bread. *Lucky Peach* 11:16–19.

Huang, J., Y. Liu, et al. (2018). Characterization of the potent odorants contributing to the characteristic aroma of Beijing douzhi. . . . *J Agric Food Chem* 66:689–94.

Xie, C., H. Zeng, et al. (2018). Volatile flavour components, microbiota and their correlations in different sufu, a Chinese fermented soybean food. *J Appl Microbiol* 125:1761–73.

Liu, Y., H. Su, and H.-L. Song (2018). Comparison of four extraction methods . . . for the analysis of flavor compounds in nattō. *Food Analysis Methods* 11:343–54.

McGrath, C. (2018). Up in the air: The emerging science of dust and sandstorm microbes. *Genome Biol Evol* 10(8):2008–9.

Jeleń, H., M. Majcher, et al. (2013). Determination of compounds responsible for tempeh aroma. *Food Chem* 141:459–65.

*Fermented seed condiments:
soy sauce, miso, jiang*

Perry, C. (1988). Medieval Near Eastern rotted condiments. In *Taste: Proceedings of the Oxford Symposium on Food and Cookery 1987*, ed. T. Jaine, 169–77. Prospect Books.

Huang, H. T. (2000). *Fermentations and Food Science. Science and Civilisation in China*, vol. 6, part 5. Cambridge Univ. Press.

Feng, Y., G. Su, et al. (2015). Characterisation of aroma profiles of commercial soy sauce. . . . *Food Chem* 167:220–28.

Lee, S. M., B. C. Seo, et al. (2006). Volatile compounds in fermented and acid-hydrolyzed soy sauces. *J Food Sci* 71:C146–56.

Meng, Q., R. Kitagawa, et al. (2017). Contribution of 2-methyl-3-furanthiol to the cooked-meat-like aroma of fermented soy sauce. *Biosci Biotech Biochem* 81:168–72.

Zhao, J., X. Dai, et al. (2011). Comparison of aroma compounds in naturally fermented and inoculated Chinese soybean pastes. . . . *Food Control* 22:1008–13.

Kumazawa, K., S. Kaneko, et al. (2013). Identification and characterization of volatile components causing the characteristic flavor in miso (Japanese fermented soybean paste) and heat-processed miso products. *J Agric Food Chem* 61:11968–73.

Ohata, M., T. Tominaga, et al. (2009). Quantification and odor contribution of 2-furanmethanethiol in different types of fermented soybean paste miso. *J Agric Food Chem* 57:2481–85.

Chen, Q.-C., Y.-X. Xu, et al. (2011). Aroma impact compounds in Liuyang douchi, a Chinese traditional fermented soya bean product. *Int J Food Sci Technol* 46:1823–29.

Li, Z., L. Dong, et al. (2016). Bacterial communities and volatile compounds in doubanjiang. . . . *J Appl Microbiol* 120:1585–94.

Huang, M., Y. Li, et al. (2017). Correlation of volatile compounds and sensory attributes of Chinese traditional sweet fermented flour pastes. . . . *J Chemistry* 2017:1–8.

Kang, K.-M., and H.-H. Baek (2014). Aroma quality assessment of Korean fermented red pepper paste (gochujang). . . . *Food Chem* 145:488–95.

Meats and fish

Lapsongphon, N., J. Yongsawatdigul, et al. (2015). Identification and characterization of the aroma-impact components of Thai fish sauce. *J Agric Food Chem* 63:2628–38.

Giri, A., K. Osako, et al. (2010). Olfactometric characterization of aroma active compounds in fermented fish paste in comparison with fish sauce, fermented soy paste, and sauce products. *Food Res Int* 43:1027–40.

Kleekayai, T., S. Pinitklang, et al. (2016). Volatile components and sensory characteristics of Thai traditional fermented shrimp pastes. . . . *J Food Sci Technol* 53:1399–1410.

Amitsuka, T., M. Okamura, et al. (2017). The contribution of aromatic components in katsuobushi to preference formation and reinforcement effect. *Biosci Biotech Biochem* 81:1561–68.

Ishiguro, K., H. Wakabayashi, et al. (2001). Changes in volatile compounds . . . and evaluation of major aroma constituents of dried bonito (katsuo-bushi). *J Jap Soc for Food Sci Technol* 48:570–77.

Skåra, T., L. Axelsson, et al. (2015). Fermented and ripened fish products in the northern European countries. *J Ethnic Foods* 2:18–24.

Belleggia, L., L. Aquilanti, et al. (2020). Discovering microbiota and volatile compounds of surströmming, the traditional Swedish sour herring. *Food Microbiol* 91:103503.

Blank, I., S. Devaud, et al. (2001). Odor-active compounds of dry-cured meat: Italian-type salami and Parma ham. In *Aroma Active Compounds in Foods*, ed. G. R. Takeoka, M. Güntert, et al., 9–20. American Chemical Society.

Milk and cream

Cadwallader, K. R., and T. K. Singh (2009). Flavours and off-flavours in milk and dairy products. In *Advanced Dairy Chemistry*, ed. P. McSweeney and P. F. Fox, 631–90. Springer.

Routray, W., and H. N. Mishra (2011). Scientific and technical aspects of yogurt aroma and taste. *Comp Revs in Food Sci and Food Safety* 10:208–20.

Shepard, L., R. E. Miracle, et al. (2013). Relating sensory and chemical properties of sour cream to consumer acceptance. *J Dairy Sci* 96:5435–54.

Mallia, S., F. Escher, et al. (2008). Aroma-active compounds of butter: A review. *Eur Food Res Technol* 226:315–25.

Triqui, R., and H. Guth (2001). Potent odorants in "smen," a traditional fermented butter product. *Eur Food Res Technol* 212:292–95.

Cheese

McSweeney, P. L. H., P. F. Fox, et al., eds. (2017). *Cheese: Chemistry, Physics and Microbiology*. Elsevier.

Tunick, M. H. (2014). *The Science of Cheese*. Oxford Univ. Press.

Donnelly, C., ed. (2016). *The Oxford Companion to Cheese*. Oxford Univ. Press.

Bertuzzi, A. S., P. L. H. McSweeney, et al. (2018). Detection of volatile compounds of cheese and their contribution to the flavor profile. . . . *Comp Revs Food Sci Food Safety* 17:371–90.

Hannon, J. A., K. N. Kilcawley, et al. (2007). Flavour precursor development in Cheddar cheese due to lactococcal starters and the presence and lysis of *Lactobacillus helveticus*. *Int Dairy J* 17:316–27.

Brückner, A., and M. Heethoff (2016). Scent of a mite: Origin and chemical characterization of the lemon-like flavor of mite-ripened cheeses. *Exp Appl Acarol* 69:249–61.

Leal, W. S., Y. Kuwahara, et al. (1989). 2(*E*) . . . β-Acaridial: A new type of monoterpene from the mold mite *Tyrophagus putrescentiae*. . . . *Agric Biol Chem* 53(3):875–78.

Wines

Robinson, J., and J. Harding, eds. (2015). *The Oxford Companion to Wine*. Oxford Univ. Press.

Waterhouse, A. L., D. W. Jeffery, et al. (2016). *Understanding Wine Chemistry*. Wiley.

Bakker, J., and R. J. Clarke (2011). *Wine: Flavour Chemistry*. Wiley.

Peynaud, E. (1987). *The Taste of Wine: The Art and Science of Wine Appreciation*, trans. M. Schuster. Macdonald Orbis.

Ferreira, V., and R. Lopez (2019). The actual and potential aroma of winemaking grapes. *Biomolecules* 9:818.

Ferreira, V., A. Escudero, et al. (2007). The chemical foundations of wine aroma—a role game aiming at wine quality, personality, and varietal expression. *Proc 13th Austral Wine Ind Tech Conf*, 1–9.

Ferreira, V., M. P. Sáenz-Navajas, et al. (2016). Sensory interactions between six common aroma vectors. . . . *Food Chem* 199:447–56.

Magyar, I., and J. Soós (2016). Botrytized wines—current perspectives. *Int J Wine Res* 8:29–39.

Parker, M., D. L. Capone, et al. (2018). Aroma precursors in grapes and wine: Flavor release during wine production and consumption. *J Agric Food Chem* 66:2281–86.

Becher, P. G., S. Lebreton, et al. (2018). The scent of the fly. *J Chem Ecology* 44:431–35.

Beers

Oliver, G., ed. (2012). *The Oxford Companion to Beer*. Oxford Univ. Press.

Bamforth, C. (2009). *Beer: Tap into the Art and Science of Brewing*. Oxford Univ. Press.

Gallone, B., S. Mertens, et al. (2018). Origins, evolution, domestication, and diversity of *Saccharomyces* beer yeasts. *Curr Opinion Biotechnol* 49:148–55.

Gonzalez Viejo, C., S. Fuentes, et al. (2019). Chemical characterization of aromas in beer and their effect on consumers liking. *Food Chem* 293:479–85.

Kishimoto, T., A. Wanikawa, et al. (2006). Comparison of the odor-active compounds in unhopped beer and beers hopped with different hop varieties. *J Agric Food Chem* 54:8855–61.

Olaniran, A. O., L. Hiralal, et al. (2017). Flavour-active volatile compounds in beer: Production, regulation and control. *J Inst Brewing* 123:13–23.

Spitaels, F., A. D. Wieme, et al. (2015). The microbial diversity . . . reveals a core microbiota for lambic beer fermentation. *Food Microbiol* 49:23–32.

Asian grain alcohols

Liu, Y.-C., M.-J. Chen, et al. (2002). Studies on lao-chao culture filtrate. . . . *Asian Australas J Anim Sci* 15:602–9.

Yu, H., T. Xie, et al. (2019). Characterization of key

aroma compounds in Chinese rice wine. . . . *Food Chem* 293:8–14.

Chen, S., C. Wang, et al. (2019). Characterization of the key aroma compounds in aged Chinese rice wine. . . . *J Agric Food Chem* 67:4876–84.

Japan Sake and Shochu Makers Association (2011). *A Comprehensive Guide to Japanese Sake.*

Mimura, N., A. Isogai, et al. (2014). Gas chromatography/mass spectrometry–based component profiling and quality prediction for Japanese sake. *J Biosci Bioeng* 118:406–14.

Isogai, A., H. Utsunomiya, et al. (2005). Changes in the aroma compounds of sake during aging. *J Agric Food Chem* 53:4118–23.

Vinegars

Solieri, L., and P. Giudici, eds. (2009). *Vinegars of the World.* Springer.

Giudici, P., F. Lemmetti, et al. (2015). *Balsamic Vinegars: Tradition, Technology, Trade.* Springer.

Aurand, L. W., J. A. Singleton, et al. (1966). Volatile components in the vapors of natural and distilled vinegars. *J Food Sci* 31:172–77.

Corsini, L., R. Castro, et al. (2019). Characterization by gas chromatography-olfactometry of the most odour-active compounds in Italian balsamic vinegars. . . . *Food Chem* 272:702–8.

Liang, J., J. Xie, et al. (2016). Aroma constituents in Shanxi aged vinegar before and after aging. *J Agric Food Chem* 64:7597–7605.

Zhou, Z., S. Liu, et al. (2017). Elucidation of the aroma compositions of Zhenjiang aromatic vinegar. . . . *J Chromatogr A* 1487:218–26.

Villarreal-Soto, S. A., S. Beaufort, et al. (2018). Understanding kombucha tea fermentation: A review. *J Food Sci* 83:580–88.

Distillates

Forbes, R. J. (1948). *Short History of the Art of Distillation.* . . . Brill.

Allchin, F. R. (1979). India: The ancient home of distillation? *Man* 4(1):55–63.

Watts, V. A., and C. E. Butzke (2003). Analysis of microvolatiles in brandy: Relationship between methylketone concentration and Cognac age. *J Sci Food Agric* 83:1143–49.

Russell, I., ed. (2003). *Whisky: Technology, Production and Marketing.* Elsevier.

Poisson, L., and P. Schieberle (2008). Characterization of the most odor-active compounds in an American bourbon whisky. . . . *J Agric Food Chem* 56:5813–19.

Sakaida, H., N. Nakahara, et al. (2003). Characteristic flavor of buckwheat shochu and comparison of volatile compounds from variety cereal shochu. *J Jap Soc for Food Sci Technol* 50(12):555–62.

Fang, C., H. Du, et al. (2018). Compositional differences and similarities between typical Chinese baijiu and Western liquor. . . . *Metabolites* 9(2):1–18.

Zou, W., G. Ye, et al. (2018). Diversity, function, and application of *Clostridium* in Chinese strong flavor baijiu ecosystem. . . . *J Food Sci* 83:1193–99.

González-Robles, I. W., and D. J. Cook (2016). The impact of maturation on concentrations of key odour active compounds which determine the aroma of tequila. *J Inst Brew* 122:369–80.

Vera-Guzmán, A., R. Guzmán-Gerónimo, et al. (2018). Volatile compound profiles in mezcal spirits as influenced by agave species. . . . *Beverages* 4:9.

Cardoso, D. R., L. G. Andrade Sobrinho, et al. (2004). A rapid and sensitive method for dimethylsulphide analysis in Brazilian sugar cane sugar spirits. . . . *J Braz Chem Soc* 15:277–81.

de Souza, M. D. C. A., P. Vásquez, et al. (2006). Characterization of cachaça and rum aroma. *J Agric Food Chem* 54:485–88.

Franitza, L., M. Granvogl, et al. (2016). Influence of the production process on the key aroma compounds of rum: From molasses to the spirit. *J Agric Food Chem* 64:9041–53.

Heady footnote

Saerens, S. M. G., F. R. Delvaux, et al. (2010). Production and biological function of volatile esters in *Saccharomyces cerevisiae. Microbial Biotechnol* 3:165–77.

Becher, P. G., A. Hagman, et al. (2018). Chemical signaling and insect attraction is a conserved trait in yeasts. *Ecol Evol* 8:2962–74.

CONCLUSION: MY SECOND GROUSE

Barwich, A.-S. (2018). Measuring the world: Olfaction as a process model of perception. In *Everything Flows: Towards a Processual Philosophy of Biology,* ed. D. Nicholson and J. Dupré. Oxford Univ. Press.

Dunlop, F. (2011). Rotten vegetable stalks, stinking bean curd and other Shaoxing delicacies. In *Cured, Fermented and Smoked Foods: Proceedings of the Oxford Symposium on Food and Cookery 2010,* ed. H. Saberi, 92. Prospect Books.

Lorjaroenphon, Y., and K. R. Cadwallader (2015). Identification of character-impact odorants in a cola-flavored carbonated beverage. . . . *J Agric Food Chem* 63:776–86.

Romagny, S., G. Coureaud, et al. (2018). Key odorants or key associations? Insights into elemental and configural odour processing. *Flavour Fragr J* 33:97–105.

SELECTED REFERENCES

Clark, A. J., J. L. Calvillo, et al. (2011). Degradation product emission from historic and modern books. . . . *Analytical and Bioanalytical Chem* 399:3589–3600.

Bembibre, C., and M. Strlič (2017). Smell of heritage: A framework for the identification, analysis, and archival of historic odours. *Herit Sci* 5:2, 1–11.

Index

....

Page numbers in italics refer to tables and larger diagrams.